Aloes

The genus *Aloe*

Edited by Tom Reynolds

Medicinal and Aromatic Plants — Industrial Profiles

CRC PRESS

Boca Raton   London   New York   Washington, D.C.

Library of Congress Cataloging-in-Publication Data

Aloes:the genus Aloe / edited by Tom Reynolds.
    p. cm.—(Medicinal and aromatic plants—industrial profiles; v.35)
    Includes bibliographical references and index.
    ISBN 0-415-30672-8
    1. Aloe. 2. Aloe—Therapeutic use.
    [DNLM: 1. Aloe. 2. Phytotherapy. 3. Plants, Medicinal.] I. Reynolds,
Tom, 1933- II. Series.

SB295 .A45 A58 2003
584′.352—dc21
2002155804

British Library Cataloguing in Publication Data

A catalogue record for this book is available from the British Library

Visit the CRC Press Web site at www.crcpress.com

Aloes

The genus *Aloe*

Medicinal and Aromatic Plants — Industrial Profiles

Individual volumes in this series provide both industry and academia with in-depth coverage of one major genus of industrial importance.

Edited by Dr Roland Hardman

Volume 1
Valerian, edited by Peter J. Houghton
Volume 2
Perilla, edited by He-ci Yu, Kenichi Kosuna and Megumi Haga
Volume 3
Poppy, edited by Jenö Bernáth
Volume 4
Cannabis, edited by David T. Brown
Volume 5
Neem, edited by H.S. Puri
Volume 6
Ergot, edited by Vladimír Křen and Ladislav Cvak
Volume 7
Caraway, edited by Éva Németh
Volume 8
Saffron, edited by Moshe Negbi
Volume 9
Tea Tree, edited by Ian Southwell and Robert Lowe
Volume 10
Basil, edited by Raimo Hiltunen and Yvonne Holm
Volume 11
Fenugreek, edited by Georgios Petropoulos
Volume 12
Gingko biloba, edited by Teris A. Van Beek
Volume 13
Black Pepper, edited by P.N. Ravindran
Volume 14
Sage, edited by Spiridon E. Kintzios
Volume 15
Ginseng, edited by W.E. Court
Volume 16
Mistletoe, edited by Arndt Büssing
Volume 17
Tea, edited by Yong-su Zhen
Volume 18
Artemisia, edited by Colin W. Wright
Volume 19
Stevia, edited by A. Douglas Kinghorn
Volume 20
Vetiveria, edited by Massimo Maffei

Volume 21
Narcissus and Daffodil, edited by Gordon R. Hanks
Volume 22
Eucalyptus, edited by John J.W. Coppen
Volume 23
Pueraria, edited by Wing Ming Keung
Volume 24
Thyme, edited by E. Stahl-Biskup and F. Sáez
Volume 25
Oregano, edited by Spiridon E. Kintzios
Volume 26
Citrus, edited by Giovanni Dugo and Angelo Di Giacomo
Volume 27
Geranium and Pelargonium, edited by Maria Lis-Balchin
Volume 28
Magnolia, edited by Satyajit D. Sarker and Yuji Maruyama
Volume 29
Lavender, edited by Maria Lis-Balchin
Volume 30
Cardamom, edited by P.N. Ravindran and K.J. Madhusoodanan
Volume 31
Hypericum, edited by Edzard Ernst
Volume 32
Taxus, edited by H. Itokawa and K.H. Lee
Volume 33
Capsicum, edited by Amit Krish De
Volume 34
Flax, edited by Alister Muir and Niel Westcott
Volume 35
Urtica, edited by Gulsel Kavalali
Volume 36
Cinnamon and Cassia, edited by P.N. Ravindran, K. Nirmal Babu and M. Shylaja
Volume 37
Kava, edited by Yadhu N. Singh
Volume 38
Aloes, edited by Tom Reynolds

To my wife

Si Monumentum Requiris Circumspice

Attr. Son of Sir Christopher Wren

Contents

List of contributors   ix
Preface to the series   xii
Preface   xiv

PART 1
The plants   1

1   **Aloes in habitat**   3
    LEONARD E. NEWTON

2   **Taxonomy of Aloaceae**   15
    GIDEON F. SMITH AND ELSIE M.A. STEYN

PART 2
Aloe constituents   37

3   **Aloe chemistry**   39
    TOM REYNOLDS

4   **Aloe polysaccharides**   75
    YAWEI NI, KENNETH M. YATES AND IAN R. TIZARD

5   **Aloe lectins and their activities**   88
    HIROSHI KUZUYA, KAN SHIMPO AND HIDEHIKO BEPPU

6   **Analytical methodology: the gel-analysis of aloe pulp and its derivatives**   111
    YAWEI NI AND IAN R. TIZARD

7   **Analytical methodology: the exudate**   127
    TOM REYNOLDS

8   **Industrial processing and quality control of *Aloe barbadensis* (*Aloe vera*) gel**   139
    TODD A. WALLER, RONALD P. PELLEY AND FAITH M. STRICKLAND

PART 3

Therapeutic activity of aloes   207

  9   Healing powers of aloes   209
      NICOLA MASCOLO, ANGELO A. IZZO, FRANCESCA BORRELLI, RAFFAELE CAPASSO,
      GIULIA DI CARLO, LIDIA SAUTEBIN AND FRANCESCO CAPASSO

10   *Aloe vera* in wound healing   239
      GARY D. MOTYKIE, MICHAEL K. OBENG AND JOHN P. HEGGERS

11   *Aloe vera* in thermal and frostbite injuries   251
      MICHAEL K. OBENG, GARY D. MOTYKIE, AMER DASTGIR, ROBERT L. McCAULEY
      AND JOHN P. HEGGERS

12   Plant saccharides and the prevention of sun-induced skin cancer   265
      FAITH M. STRICKLAND AND RONALD P. PELLEY

13   Aloes and the immune system   311
      IAN R. TIZARD AND LALITHA RAMAMOORTHY

14   Bioactivity of *Aloe arborescens* preparations   333
      AKIRA YAGI

PART 4

Aloe biology   353

15   The chromosomes of *Aloe* – variation on a theme   355
      PETER BRANDHAM

16   Aloe leaf anatomy   361
      DAVID F. CUTLER

17   Pests of aloes   367
      MONIQUE S.J. SIMMONDS

      Index   381

Contributors

Hidehiko Beppu
Fujita Memorial Institute
of Pharmacognosy
Fujita Health University
Hisai, Mie 514-1296
Japan

Francesca Borrelli
Department of Experimental
Pharmacology
Via D. Montesano 49
80131 Naples
Italy

Peter Brandham
Jodrell Laboratory
Royal Botanic Gardens
Kew, Richmond, Surrey
TW9 3DS, UK

Francesco Capasso
Department of Experimental
Pharmacology
Via D. Montesano 49
80131 Naples
Italy

Raffaele Capasso
Department of Experimental
Pharmacology
University of Naples Federico II
Via D. Montesano 49
80131 Naples
Italy

Giulia Di Carlo
Department of Experimental
Pharmacology
Via D. Montesano 49
80131 Naples
Italy

David F. Cutler
Jodrell laboratory
Royal Botanic Gardens
Kew, Richmond, Surrey
UK

Amer Dastgir
University of Texas Medical Branch
Galveston, TX 77550
USA

John P. Heggers
Division of Plastic Surgery
Department of Surgery
Department of Microbiology
and Immunology
University of Texas Medical Branch
Galveston, TX 77550
USA

Angelo A. Izzo
Department of Experimental
Pharmacology
Via D. Montesano 49
80131 Naples
Italy

Hiroshi Kuzuya
Fujita Memorial Institute
  of Pharmacognosy
Fujita Health University
Hisai, Mie 514-1296
Japan

Robert L. McCauley
Division of Plastic Surgery
Department of Surgery
University of Texas Medical Branch
Galveston, TX 77550
USA

Nicola Mascolo
Department of Experimental
  Pharmacology
Via D. Montesano 49
80131 Naples
Italy

Gary D. Motykie
Division of Plastic Surgery
Department of Surgery
University of Texas Medical Branch
Galveston, TX 77550
USA

Leonard E. Newton
Department of Botany
Kenyatta University
P.O. Box 43844, Nairobi
Kenya

Yawei Ni
Department of Veterinary
  Pathobiology
Texas A&M University, College Station
TX 77840
USA

Michael K. Obeng
Division of Plastic Surgery
Department of Surgery
University of Texas Medical Branch
Galveston, TX 77550
USA

Ronald P. Pelley
Pangea Phytoceuticals
Suite 119, 306 E. Jackson
Harlingen, TX 78550
USA

Lalitha Ramamoorthy
Department of Veterinary Pathobiology
Texas A&M University, College
  Station
TX 77840
USA

Tom Reynolds
Jodrell Laboratory
Royal Botanic Gardens
Kew, Richmond, Surrey
UK

Lidia Sautebin
Department of Experimental
  Pharmacology
Via D. Montesano 49
80131 Naples
Italy

Kan Shimpo
Fujita Memorial Institute
  of Pharmacognosy
Fujita Health University
Hisai, Mie 514-1296
Japan

Monique S.J. Simmonds
Jodrell Laboratory
Royal Botanic Gardens
Kew, Richmond, Surrey
TW9 3AB, UK

Gideon F. Smith
Office of the Director
Research and Scientific Services
National Botanical Institute
Private Bag X101
Pretoria 0001
South Africa

Elsie M.A. Steyn
Office of the Director
Research and Scientific Services
National Botanical Institute
Private Bag X101
Pretoria 0001
South Africa

Faith M. Strickland
Department of Immunology
Box 178, M.D. Anderson Cancer Center
1515 Holcombe Boulevard
Houston, TX 77030
USA

Ian R. Tizard
Department of Veterinary Pathobiology
Texas A&M University, College Station
TX 77840
USA

Todd A. Waller
Pangea Phytoceuticals
Suite 119, 306 E. Jackson
Harlingen, TX 78550
USA

Akira Yagi
Faculty of Pharmacy and Pharmaceutical
    Sciences
Fukuyama University
Fukuyama, Hiroshima, 729-0292
Japan

Kenneth M. Yates
Carrington Laboratories Inc
1300 East Rochelle
Irving, Texas
USA

Preface to the series

There is increasing interest in industry, academia and the health sciences in medicinal and aromatic plants. In passing from plant production to the eventual product used by the public, many sciences are involved. This series brings together information which is currently scattered through an ever increasing number of journals. Each volume gives an in-depth look at one plant genus, about which an area specialist has assembled information ranging from the production of the plant to market trends and quality control.

Many industries are involved such as forestry, agriculture, chemical, food, flavour, beverage, pharmaceutical, cosmetic and fragrance. The plant raw materials are roots, rhizomes, bulbs, leaves, stems, barks, wood, flowers, fruits and seeds. These yield gums, resins, essential (volatile) oils, fixed oils, waxes, juices, extracts and spices for medicinal and aromatic purposes. All these commodities are traded worldwide. A dealer's market report for an item may say 'Drought in the country of origin has forced up prices'.

Natural products do not mean safe products and account of this has to be taken by the above industries, which are subject to regulation. For example, a number of plants which are approved for use in medicine must not be used in cosmetic products.

The assessment of safe to use starts with the harvested plant material which has to comply with an official monograph. This may require absence of, or prescribed limits of, radioactive material, heavy metals, aflatoxin, pesticide residue, as well as the required level of active principle. This analytical control is costly and tends to exclude small batches of plant material. Large scale contracted mechanized cultivation with designated seed or plantlets is now preferable.

Today, plant selection is not only for the yield of active principle, but for the plant's ability to overcome disease, climatic stress and the hazards caused by mankind. Such methods as *in vitro* fertilization, meristem cultures and somatic embryogenesis are used. The transfer of sections of DNA is giving rise to controversy in the case of some end-uses of the plant material.

Some suppliers of plant raw material are now able to certify that they are supplying organically-farmed medicinal plants, herbs and spices. The Economic Union directive (CVO/EU No. 2092/91) details the specifications for the *obligatory* quality controls to be carried out at all stages of production and processing of organic products.

Fascinating plant folklore and ethnopharmacology leads to medicinal potential. Examples are the muscle relaxants based on the arrow poison, curare, from species of *Chondrodendron*, and the anti-malarials derived from species of *Cinchona* and *Artemisia*.

The methods of detection of pharmacological activity have become increasingly reliable and specific, frequently involving enzymes in bioassays and avoiding the use of laboratory animals. By using bioassay linked fractionation of crude plant juices or extracts, compounds can be specifically targeted which, for example, inhibit blood platelet aggregation, or have anti-tumour, or anti-viral, or any other required activity. With the assistance of robotic devices, all the members of a genus may be readily screened. However, the plant material must be *fully* authenticated by a specialist.

The medicinal traditions of ancient civilizations such as those of China and India have a large armamentaria of plants in their pharmacopoeias which are used throughout South-East Asia. A similar situation exists in Africa and South America. Thus, a very high percentage of the World's population relies on medicinal and aromatic plants for their medicine. Western medicine is also responding. Already in Germany all medical practitioners have to pass an examination in phytotherapy before being allowed to practise. It is noticeable that throughout Europe and the USA, medical, pharmacy and health related schools are increasingly offering training in phytotherapy.

Multinational pharmaceutical companies have become less enamoured of the single compound magic bullet cure. The high costs of such ventures and the endless competition from 'me too' compounds from rival companies often discourage the attempt. Independent phytomedicine companies have been very strong in Germany. However, by the end of 1995, eleven (almost all) had been acquired by the multinational pharmaceutical firms, acknowledging the lay public's growing demand for phytomedicines in the Western World.

The business of dietary supplements in the Western World has expanded from the health store to the pharmacy. Alternative medicine includes plant-based, products. Appropriate measures to ensure the quality, safety and efficacy of these either already exist or are being answered by greater legislative control by such bodies as the Food and Drug Administration of the 'USA and the recently created European Agency for the Evaluation of Medicinal Products, based in London.

In the USA, the Dietary Supplement and Health Education Act of 1994 recognized the class of phytotherapeutic agents derived from medicinal and aromatic plants. Furthermore, under public pressure, the US Congress set up an Office of Alternative Medicine and this office in 1994 assisted the filing of several Investigational New Drug (IND) applications, required for clinical trials of some Chinese herbal preparations. The significance of these applications was that each Chinese preparation involved several plants and yet was handled as a *single* IND. A demonstration of the contribution to efficacy, of *each* ingredient of *each* plant, was not required. This was a major step forward towards more sensible regulations in regard to phytomedicines.

My thanks are due to the staffs of Harwood Academic Publishers and Taylor & Francis who have made this series possible and especially to the volume editors and their chapter contributors for the authoritative information.

Roland Hardman, 1997

Preface

Aloes provide a fascinating subject for research from a chemical, biochemical, pharmaceutical, taxonomic, horticultural and economic point of view. Their use as medicinal plants is mentioned in many ancient texts, including the Bible, although here as elsewhere there may be doubts as to the botanical identity of the material.

This multitude of medicinal uses described and discussed over the centuries is sometimes difficult to evaluate. Authors such as Crosswhite and Crosswhite (1984) have given detailed accounts of the drug in classical antiquity, concentrating on the species *Aloe vera* which seems to be the main one in use, with *Aloe perryi* from Socotra mentioned more rarely. While the acquisition of Socotra by Alexandra the Great to ensure supplies of *A. perryi* to treat his troops is well known and presumed true, the origins of *A. vera* are obscure. The plant is recorded from lands around the Mediterranean back to Mesopotamian times and was subsequently carried to the Atlantic islands and the West Indies to the west and India and China to the east. It is almost impossible to distinguish stands of plants to be of either introduced or native origin, although Hepper (personal communication) claims to have seen specimens growing in Yemen in regions so remote as to preclude the possibility of introduction, while Newton (personal communication) also places the origin somewhere in the Arabian peninsula. An interesting commentary on the value accorded to *A. vera* is its use as a 'door plant' where a detached branch is suspended over the threshold where it remains alive and even flowers, with the connotation of immortality. This practice has been observed, albeit with *A. arborescens*, in Rhodes at the present day (Reynolds, unpublished observation) and is no doubt still widespread. In other parts of the world aloes have almost as long a history. Bruce (1975) and Morton (1961) mention varied uses in India, China and the East Indies while its cultivation and use in South America and the Caribbean has been described by Hodge (1953). A note of caution is needed for New World records because sometimes *Agave* and its products have been mentioned as 'American aloes', although they are not related. Hodge also describes the Cape aloes industry in South Africa which provides the purgative drug aloes. Aloe products in one form or another from a multitude of species are used throughout Africa for a variety of folk medicinal purposes. (Watt and Breyer-Brandwyk, 1962). Although the efficacy of these folk remedies has not been substantiated, there must surely be some validity in some of the many claims, enough at least to stimulate further research.

The sticky aloe gel from the interior of the leaves of many species has been valued as a sovereign cure for skin ailments and is still treasured as such, either directly from the plant or over-the-counter in a host of preparations. Other therapeutic effects are claimed, some convincingly, some less so. Strangely, the active ingredients responsible for the healing have not been identified, nor has it been clear precisely what the biochemical targets are. Both these mysteries are gradually being resolved and are proving to have many aspects. Chapters in this book by researchers from Texas describe work which is starting to unravel this amazing story. At the same time precise descriptions of procedure are given on how and how not to prepare the commercial product from its best known source, *Aloe vera*. One wonders how many reports of therapeutic failure in the past have been due to the use of badly processed material.

The Japanese have long used *Aloe arborescens* medicinally and contributions from that country give much information about the plant and its products.

Quite apart from the gel, the outer layers of the leaf, when cut, produces an exudate which provides the drug bitter aloes, well known and perhaps dreaded as a purgative with a particular griping action. The source of this substance is usually either *Aloe ferox* from South Africa (Cape aloes) or *Aloe perryi* from Socotra (Socotran aloes) and sometimes *Aloe vera* (Curaçao aloes). Not all species produce bitter exudates nor are all purgative and in fact some species produce little or no exudate. Cells adjacent to the leaf vascular bundles contain the exudate which is released on cutting but it has not been established that these cells actually produce the exudate constituents. In some species fibres occupy the position of these secretary cells. The purgative principle is well known as an anthrone *C*-glucoside but many other compounds, mostly of a phenolic nature, have been recognized as zones on thin-layer chromatograms in one or other of the many species examined. From the 300 or so species so far investigated chemically, about 80 major chromatographic zones have been recognized. Of these entities, however, only about a third have chemical structures ascribed. The identification and distribution among the species of these compounds affords much stimulating research for the chemist and the taxonomist. Together with the wealth of folk medicine lore, apart from use of the gel, there is much scope for investigating the biological activity of the many aloe compounds.

We must not forget the plants themselves in this flurry of sophisticated research. About 400 species have been described so far, ranging in distribution from the Cape up to the Sahara and into Arabia and two well-known African botanists have contributed chapters describing them. Aloes occur in a wide range of sizes from a few inches to many tens of feet and have many habits. They are more or less succulent and although typical of semi-arid savannah, species are found in dry deserts, cliff faces and even under the spray of a waterfall. Some are very widespread in their distribution, while some occupy perhaps just a single hillside. The flowers are basically typical of the species but vary in form as well as in colour, being red, orange, yellow and even white. These attractive plants drew the attention of early explorers and several species were brought back to Europe for cultivation. In particular *Aloe arborescens* and *Aloe saponaria* are frequently seen as ornamentals, while *Aloe vera*, for its medicinal properties, was carried from an unknown provenance across the world from the West Indies to China.

So, this book is in a sense a milestone or perhaps a turning point. Together with other recent reviews it lays down knowledge of aloes acquired over the last half-century and also lays a foundation for an immense amount of further research into this fascinating genus, knowledge which will bring immense value, intellectually and practically.

This book was produced with the encouragement of Professor Monique Simmonds at Royal Botanic Gardens, Kew. I am grateful to all the authors for their painstaking efforts to prepare their chapters. I would like to thank Dr Roland Hardman, the series editor, for constant advice and reassurance during the book's production.

REFERENCES

Bruce, W.G.G. (1975) Medicinal properties in the Aloe. *Excelsa*, **5**, 57–68.

Crosswhite, F.S. and Crosswhite, C.D. (1984) Aloe vera, plant symbolism and the threshing floor: light, life and good in our heritage. *Desert plants*, **6**, 46–50.

Hodge, W.H. (1953) The drug aloes of commerce, with special reference to the cape species. *Economic Botany*, **7**, 99–129.

Morton, J.F. (1961) Folk uses and commercial exploitation of *Aloe* leaf pulp. *Economic Botany*, **15**, 311–319.

Watt, J.M. and Breyer-Brandwijk, M.G. (1962) *The Medicinal and Poisonous Plants of South and Eastern Africa*, pp. 679–687. Edinburgh and London: E. & S. Livingstone Ltd.

Aloe vera plant growing in Barranco de Infierno, Tenerife, apparently native but in reality an escape from cultivation. (Photo, L.A. Reynolds) (*see Colour Plate 1*).

Part 1

The plants

1 Aloes in habitat

Leonard E. Newton

ABSTRACT

Aloes are xerophytes with structural and physiological adaptations for survival in arid regions. They are widespread in sub-Saharan Africa, the Arabian Peninsula and a number of Indian Ocean islands. They occupy many different kinds of natural habitat, from forest to exposed rock surfaces, but they are absent from the moist lowland forests of mainland Africa. People use aloes in many ways, mostly at a non-commercial local level. As a result of over-exploitation and habitat destruction many species are endangered and in need of conservation. Aloes are protected by CITES.

LIFE FORMS

All *Aloe* species (Newton, 2001) are perennial, leaf-succulent xerophytes. Xerophytes are plants adapted to survive in areas of low or erratic precipitation, and the adaptations may include structural and physiological features. Succulence is one kind of xerophytic adaptation. It is the development of water storage tissue, or aqueous tissue, consisting of large thin-walled cells in which water is held in mucilage. The accommodation of the aqueous tissue requires enlargement of either leaves, as in aloes, or stems, as in cacti (Willert *et al.*, 1992).

As leaf-succulents, most aloes have thick and fleshy leaves, enlarged to accommodate the aqueous tissue. The leaf cuticle is thick and covered with a layer of wax. In most species examined, the surface wax has distinctive patterns of ridges and/or micro-papillae, which can be of taxonomic value (e.g. Newton, 1972; Cutler, 1982). Stomata are sunken, i.e. situated at the base of a suprastomatal cavity. Most species have groups of cells associated with the vascular bundles, variously called aloin or aloitic cells, that store and perhaps secrete a mixture of compounds with medicinal value, with the composition varying in different species. This appears as an exudate, usually yellow, when leaves are broken or cut. Leaves of a few species, such as *A. fibrosa* Lavranos and L.E. Newton, have fibres in place of the aloitic cells. In their physiology, aloes display crassulacean acid metabolism (CAM), involving the conversion of carbon dioxide to malic acid in darkness, and its release for use in photosynthesis in light. This process enables the plant to have its stomata open at night and closed in daytime, a reversal of the usual occurrence in plants, thus conserving water during the heat of the day (Kluge and Ting, 1978).

Aloes may be acaulescent or caulescent, the 'acaulescent' species having very short stems that are completely hidden by the leaves. In different species the leaves may be arranged in rosettes or spaced out along the stems. Acaulescent species have rosulate leaves, often occurring as solitary rosettes that can reach a metre or more in diameter (*A. rivae* Baker). Some species sucker from the base to form clumps of rosettes, in some cases with creeping stems so that the clumps can reach a couple of metre or more in diameter (e.g. *A. globuligemma* Pole-Evans). Some miniature plants, such as *A. myriacantha* Schult.f. and *A. saundersiae* Reynolds, have thin leaves that are scarcely succulent. Such plants are difficult to see in the grassland where they grow unless they are in flower or fruit. A few acaulescent species have underground bulbs (e.g. *A. buettneri* A.Berg.) whose level in the substrate is maintained by contractile roots. Caulescent species may be arborescent, shrubby, sprawling, climbing or pendulous. Arborescent species can reach 15 metres in height (e.g. *A. eminens* Reynolds and P.R.O.Bally) and may be branched or unbranched. Pendulous plants, such as *A. kulalensis* L.E.Newton and Beentje and *A. veseyi* Reynolds, hang down rock faces, sometimes rooted only into rock crevices with little soil.

Flowers are produced on racemose inflorescences, which may be either simple or paniculate. Usually the racemes are erect, but in some cases they are oblique or more or less horizontal, in which case the flowers are secund (e.g. *A. mawii* Christian). In a few species the peduncle grows downwards and the branches bend upwards to present the erect racemes (e.g. *A. penduliflora* Baker). Racemes may be dense or lax in different species. Individual flowers are usually nutant at anthesis, except in some of the species with secund flowers. In most species the flowers are brightly coloured and very conspicuous. Several species in northeast Africa and southwest Arabia, such as *A. tomentosa* Deflers and *A. trichosantha* A.Berg., are unusual in having hairy flowers.

GEOGRAPHICAL DISTRIBUTION

The majority of *Aloe* species occur naturally on mainland Africa, in tropical and sub-tropical latitudes. The genus is found almost throughout the African continent south of the Sahara Desert, except for the moist lowland forest zones and the western end of West Africa.

As will be seen in Figure 1.1 the majority of species occur in southern Africa and on the eastern side of the continent. Many other species are found on the Arabian Peninsula and on Madagascar and a few, mostly formerly in the genus *Lomatophyllum*, are known from some of the smaller Indian Ocean islands. The Arabian species have clear relationships with the species of northeast Africa. Madagascan species appear not to be closely related to those of mainland Africa and so active speciation seems to have occurred since the separation of these two land masses. Likewise, the former *Lomatophyllum* species form a group not represented on the African mainland.

Some species are very widespread in distribution. Reynolds (1966) cites *A. buettneri* as the most widespread species, with a range of at least 5,600 km, from Mali to Zambia and the species has since been recorded in Namibia. However, Carter (1994) regards this as a West African species only, with two related species in the rest of the range reported by Reynolds. Another very widespread species is

Figure 1.1 Geographical distribution of the genus *Aloe*, showing the number of species in each country.

A. myriacantha, with a range of about 4,800 km from Kenya and Uganda to the Republic of South Africa. Most other widespread species have more modest distribution ranges, amounting only to some hundreds of kilometres. At the other end of the scale there are species with very restricted distributions. They may be known only from one limited area (e.g. *A. reynoldsii* Letty) or only from the type locality (e.g. *A. murina* L.E.Newton). Some are known only from single isolated mountains (e.g. *A. kulalensis*).

Many countries have some endemic species, as shown in Table 1.1. The highest rate of endemism is in Madagascar and isolated Indian Ocean islands. Naturally, it is to be expected that very large countries, such as South Africa, and isolated islands, such as Madagascar, would have many endemic species. Conversely, it is not surprising that very small countries, such as Burundi and Rwanda have no endemic species. Of the small countries on the African mainland, only Lesotho and Swaziland have endemic species, with one each.

Table 1.1 Geographical distribution of the genus *Aloe*.

	Total species	Endemic species	Endemism (%)
Aldabra	1	–	0
Angola	24	13	54.2
Benin	3	–	0
Botswana	8	–	0
Burkina Faso	1	–	0
Burundi	1	–	0
Cameroon	1	–	0
Comoros	1	1	100
Democratic Republic of Congo	13	3	23
Eritrea	8	2	25
Ethiopia	34	16	47
Ghana	3	–	0
Kenya	55	24	43.6
Lesotho	8	1	12.5
Madagascar	77	77	100
Malawi	17	2	11.8
Mali	3	–	0
Mauritius	2	2	100
Mozambique	25	3	12
Namibia	26	8	30.8
Nigeria	3	–	0
Oman	5	3	60
Pemba	2	1	50
Réunion	1	1	100
Rodrigues	1	1	100
Rwanda	4	–	0
Saudi Arabia	22	10	45.5
Seychelles	1	–	0
Socotra	3	3	100
Somalia	33	25	75.8
South Africa	119	71	59.7
Sudan	12	3	25
Swaziland	18	1	5.6
Tanzania	40	13	32.5
Togo	1	–	0
Uganda	16	2	12.5
Yemen	26	13	50
Zambia	19	2	10.5
Zanzibar	2	–	0
Zimbabwe	27	5	18.5

Notes
[Pemba and Zanzibar are part of Tanzania but treated separately here because they are in the Indian Ocean, and one species on Pemba (*A. pembana*) belongs to the Lomatophyllum group, so far not known on the African mainland. Socotra is also listed separately, though it is part of Yemen.]

The long-cultivated *Aloe vera* (L.) Burm.f. has become naturalised in many tropical and sub-tropical countries, including some in the New World. The exact origin of *A. vera* is uncertain, but it seems likely that it is the Arabian Peninsula, home of the closely related, and possibly conspecific, *A. officinalis* Forssk.

HABITATS

Aloes occupy a wide range of habitats, varying from forests to exposed rock surfaces and cliff faces. The genus does not occur in moist lowland forest, but the dry coastal forests of eastern Africa include arborescent species such as *A. eminens* and *A. volkensii* Engl. subsp. *volkensii*. Not all arborescent species are forest trees, the huge *A. dichotoma* L. being a prominent feature of the arid Namaqualand landscape. Many shrubby species are found in *Acacia* scrub and other thickets, sometimes depending on the surrounding vegetation for support (e.g. *A. morijensis* S.Carter and Brandham). Grasslands offer another habitat, especially for many acaulescent species, such as *A. lateritia* Engl. and *A. secundiflora* Engl. Several species, including *A. chrysostachys* Lavranos and L.E.Newton and *A. classenii* Reynolds, occur on expanses of rocks, rooted into soil pockets or crevices. Most of the species formerly included in the genus *Lomatophyllum* occur in moist coastal forests of Madagascar and several Indian Ocean islands.

The genus occupies a considerable altitudinal range. Aloes may be found from sea level (e.g. *A. boscawenii* Christian, *A. kilifiensis* Christian) up to about 3,500 meters above sea level (e.g. *A. ankoberensis* M.G.Gilbert and Sebsebe, *A. steudneri* Schweinf. ex Penz.). On the east side of the continent tropical conditions occur at sea level, whilst at the higher altitudes the plants are subjected to cold conditions and may even be covered by snow for a while, especially in southern latitudes (e.g. *A. polyphylla* Schönl. ex Pillans). In South Africa *Aloe haemanthifolia* Marloth and A.Berg. grows on cliffs near waterfalls and for at least part of the year has water running down through the clumps of rosettes.

In the wild, aloes occur on a wide range of soil types and substrates. Some seem to be restricted to certain substrates, including dolomite (e.g. *A. alooides* (Bolus) van Druten), granite (e.g. *A. torrei* Verd. and Christian), gypsum (e.g. *A. breviscapa* Reynolds and Bally) and limestone (e.g. *A. calcairophila* Reynolds). However, the success of most species in cultivation suggests that they are tolerant of the many soil types to be found in gardens and potting mixtures.

Some *Aloe* species are found as dense populations (e.g. *A. falcata* Baker), whilst others occur as scattered individuals (e.g. *A. variegata* L.). They are often prominent, but rarely dominant in the ecological sense, except where the vegetation is sparse and they are the only large plants around (e.g. *A. pillansii* Guthrie).

POLLINATION AND DISPERSAL

Flowers of almost all *Aloe* species are diurnal, tubular, brightly coloured red or yellow, unscented and produce abundant nectar. These features point to ornithophily as the pollination syndrome, and sunbirds (Nectariniidae) are frequent visitors to aloe flowers in the field and in African gardens (Figure 1.2).

Although they are not typical melittophilous flowers, aloes are also visited by bees. In some areas, especially in South Africa, the flowering of aloes is important in apiculture, though it is reported that the nectar and pollen of some species can affect the behaviour of bees, making them vicious (Watt and Breyer-Brandwijk, 1962). The abundance of nectar produced by the flowers of some species is such that even baboons have been seen collecting flowers in order to suck the nectar (Reynolds, 1950). In West Africa some wasps were captured on flowers of *A. buettneri*, and dissection of the gut

Figure 1.2 Sunbird on *Aloe cheranganiensis* flower (*see Colour Plate 2*).

revealed that they had been eating pollen. Pollination of flowers of different shapes, such as the white campanulate flowers of *A. albiflora* Guillaumin, has yet to be investigated. The Madagascan *Aloe suzannae* Decary is exceptional in the genus in having nocturnal fragrant flowers, presumably pollinated by nocturnal animals such as bats and small lemurs.

Almost all aloes are self-incompatible, though oddly enough the flowers are protandrous (anthers ripen and pollen is dispersed before the stigma is receptive) and so self-pollination would not occur anyway. The species formerly in the genus *Lomatophyllum* are reported as exceptions in being self-compatible (Lavranos, 1998). In an area where two or more species of *Aloe* flower at the same time, the main pollen vectors, sunbirds, fly indiscriminately from one species to another. Consequently hybridisation is frequent. Reynolds (1950, 1966) reported many natural hybrids, especially in South Africa, and Newton (1998) reported a number in East Africa. In gardens and greenhouse collections, where there might be only one specimen of each species, almost all seeds produced are of hybrid origin. In a tropical garden, hybrid seedlings can become established quickly, and numerous garden aloes that are difficult to identify are undoubtedly of this origin.

Most aloes produce capsules, dry dehiscent fruits that split open at maturity to release the seeds. As the inflorescences sway in the wind the seeds, which are winged, are thrown out and blown away. Species that were formerly included in the small genus *Lomatophyllum* have berries, indehiscent fruits with fleshy walls. The seeds of these species are without wings. The berries have been observed to fall to the ground from the parent plant and decay on the ground, thereby releasing the seeds.

In the natural habitat young seedlings grow quickly in order to build up sufficient aqueous tissue to survive their first dry season. They are usually found amongst rocks or below shrubs (nurse plants), which give protection from the hot sun and from browsing animals.

Bulbil formation is reported as a regular occurrence in *A. bulbillifera* H.Perrier (Madagascar) and *A. patersonii* B.Mathew (Democratic Republic of Congo). The bulbils are formed on the peduncle and inflorescence branches. This phenomenon was also observed once on a cultivated plant of the Kenyan *A. lateritia*.

USES

Aloes have been used as medicinal plants for centuries. *Aloe vera* was mentioned in the herbal of Dioscorides, produced in the first century CE and an illustration appeared in the Codex Aniciae Julianae, produced in the year CE 512. It was formally named *Aloe perfoliata* var. *vera* by Linnaeus in 1753, which is the starting point for flowering plant nomenclature. Reynolds (1966) argued that it was first recognised as a distinct species in 1768 by Miller, who called it *Aloe barbadensis*. However, Reynolds had overlooked a publication by Burman slightly earlier, perhaps only a few days, in 1768, in which the taxon was called *Aloe vera*. Therefore the correct name is now *Aloe vera* (L.) Burm.f. (Newton, 1979). Several other species have also been used to treat both humans and domestic animals. Medicinal uses are covered in other chapters of the present book, but there are also other uses.

Leaves of aloes, especially *A. vera*, are used in the production of many cosmetic products. This is part of the 'back to nature' movement, whose adherents believe that using natural products derived from plants such as the well-known 'health plant' *Aloe vera* is a healthy way of life. The many kinds of product on the market include after-shaving gel, a mouthwash, hair tonic and shampoo, skin-moistening gel, and even a 'health drink.' In recent years a brand of washing powder and a brand of toilet paper with 'Aloe Vera' have appeared on the market, apparently with the aim of improving our well being. In some southern states of the U.S.A. vast plantations of *A. vera* can be seen, supporting this large and very successful business and within the last decade or so plantations of the same species have appeared in South Africa.

In Mali, leaf sap of *A. buettneri* is said to be used as an ingredient of arrow poison, though this is not a species known as being poisonous. Several species are poisonous, because of the presence in leaves of the hemlock alkaloid γ-coniceine (Nash *et al.*, 1992). These species have a characteristic smell usually described as that of mice or rats, for which reason the East African *Aloe ballyi* is known as 'the rat aloe.' In Kenya and Somalia it is reported that *Aloe ruspoliana* Baker is used to kill hyaenas by smearing meat with the leaf extract. There are published reports of human deaths resulting from the use of aloe leaves (Verdcourt and Trump, 1969; Drummond *et al.*, 1975; Newton, 2001). In some cases overdoses of the medicinally useful compounds are responsible (Neuwinger, 1960).

Some other uses are also based on the chemical content. An insect repellent can be made by drying and burning aloe leaves and similar preparations are used to protect animals against ticks and stored food against weevils (Reynolds, 1950; Watt and Breyer-Brandwijk, 1962; Newton and Vaughan, 1996). *Aloe maculata* Medik. was known as *A. saponaria* (Ait.) Haw. for many years, the specific epithet alluding to the use of the roots to make soap. In South Africa, leaf sap of *A. maculata* was used in the tanning of garments made from skins (Reynolds, 1950). The leaf exudate of aloes is usually yellow but in some species it rapidly turns purple on exposure to air. Some such species, such as *A. megalacantha* Baker and *A. confusa* Engl., are used to dye cloth and for making ink. The ash of dried *A. ferox* Miller and *A. marlothii* A.Berg. leaves is an ingredient in snuff prepared in some parts of South Africa.

With the bitter compounds in the leaves, aloes are not regarded as edible, but Reynolds (1950) reported that in South Africa the leaves of *A. ferox* were used to make a jam. However, flowers of various species are eaten in different parts of Africa. Young flowering shoots of *A. kraussii* Baker and *A. minima* Baker are eaten as raw vegetables by Zulus (Gerstner in Reynolds, 1950), who cook the flowers of *A. boylei* Baker, *A. cooperi* Baker, and other species as a vegetable (Watt and Breyer-Brandwijk, 1962). It is reported that the flowers of *A. zebrina* Baker have been used to make cakes (Reynolds, 1950). In West Africa, flowers of *A. macrocarpa* Tod. are eaten by various tribes, and used as a seasoning herb in cooking (Chevalier in Reynolds, 1966). Newton and Vaughan (1996) report that dried leaf material may be mixed with tea leaves in South Africa.

In many African countries aloes are used in gardens as decorative plants and this may be seen on a smaller scale in other tropical or sub-tropical countries. In Nairobi, the capital of Kenya, the indigenous *Aloe ballyi* Reynolds and the exotic *A. barberae* Dyer are used as street trees. Smaller-growing species are popular pot plants for enthusiasts who grow succulent plants as a hobby in many temperate region countries. There has also been some degree of artificial hybridisation, to produce new cultivars. Hybrids of smaller-growing species, suitable for pot cultivation, have been produced in Australia, the U.K. and the U.S.A. The miniature Madagascan species are especially popular for this purpose. Larger-growing species have been hybridised in South Africa to produce some spectacular garden plants. Many cultivar names have been published, and many cultivars have been registered with the South African Aloe Breeders' Association, which is the recognised international registration agency.

Where the climate is suitable, some shrubby species are grown as hedge plants. The use of aloes for hedging in South Africa was reported by Carl Peter Thunberg in 1795, and today *A. arborescens* Miller and *A. ferox* are seen planted around cattle pens. In East Africa *Aloe dawei* A.Berg. and *A. kedongensis* Reynolds are used as hedges (Figure 1.3) and closely planted *A. rivae* plants have also been used as boundaries.

In Kenya, *A. chrysostachys* is planted in rows on eroded slopes in an attempt to protect the soil (Figure 1.4).

Arborescent aloes are not regarded as timber plants, but in the absence of more suitable plants in the area, stems of the Madagascan *A. vaotsanda* Decary are used locally for building huts. Branches of the South African *A. dichotoma* are used to make quivers for arrows, giving this species the vernacular name 'quiver tree' (first reported by Simon van der Stel in 1685). In more modern times the hollow dead stems are cut into pieces to make various decorative items, such as ash trays and other small containers. The spiny leaves of *A. marlothii* are used for scraping and thinning animal hides to prepare them for making garments (Reynolds, 1950).

As with many other succulent plants noted for their unusual or even bizarre appearance, aloes have been used in various activities relating to superstition (Reynolds, 1966). Aloe products, such as dried leaves, have been found amongst the items used for fetish purposes by traditional priests and witch-doctors. As uprooted aloes can survive for years, and even flower in this condition, they are often hung over doors of houses as charms intended to ensure long life for the occupants. An uprooted *A. aristata* Haw. plant

Figure 1.3 Hedge of *Aloe kedongensis* near Naivasha, Kenya (*see Colour Plate 3*).

Figure 1.4 Erosion control using *Aloe chrysostachys* near Mwingi, Kenya (*see Colour Plate 4*).

in the home of a childless woman in Botswana is supposed to indicate whether or not the woman will bear a child, according to whether the plant flowers or dries up. Several species, such as *A. rivae* and *A. vera*, are planted on graves. In southern Africa a Sotho man maintained a plant of *A. arborescens* as the home of the spirits of his male ancestors (Jackson, 1964). *Aloe maculata* is used to prepare a charm against lightning. Another supposed protective function is that *A. ecklonis* Salm-Dyck is used as a charm to turn enemy bullets into water drops (Watt and Breyer-Brandwijk, 1962). Watt and Breyer-Brandwijk (1962) report that young female initiates in South Africa bathe with a lotion prepared from *A. kraussii*, though perhaps this overlaps with a possible medicinal value for the skin.

CONSERVATION

Many aloes are regarded as endangered species. Various threats exist, and they can be placed into three main categories: over-collection of plants for cultivation, destruction of plants in harvesting leaf exudates and destruction of natural habitats.

Some species, especially the miniature species of Madagascar and smaller-growing plants of South Africa, are possibly over-collected by people supplying the nursery trade (Newton and Chan, 1998; Oldfield, 1997). The rising popularity of 'field trip holidays' by amateur growers as a new kind of tourism, also results in some collection of wild plants. An individual might collect only one or two plants of each species and may feel that little or no harm is being done to the population. However, as the numbers of such travellers grow the result can be a major depletion of well-known populations to which access is easy.

There is also a lucrative trade in leaf exudates, required mainly for medicinal and cosmetic purposes, and these are frequently harvested from wild plants. Much of this activity is well organised, but there is also a large unofficial exploitation of wild plants (Newton, 1994; Newton and Vaughan, 1996). In South Africa 'aloe tapping' is a well-established industry, going back for over 200 years. The main species used is *A. ferox*, with export records dating back to 1761. The species is widespread in Western Cape and Eastern Cape Provinces. Newton and Vaughan (1996) estimated that a total of about 700 tons of crystalline bitters is harvested each year from about 17 million plants, 95% of which are in the wild. Much of this harvesting and export is illegal or undocumented. However, they concluded that the harvesting is carried out on a sustainable basis, thanks to traditions established in the communities of aloe tappers. In contrast, the harvesting of leaf exudates is more recent in Kenya, with local people being paid by 'outsiders' with no attempt to ensure sustainability. With no traditional or other controls in place, various species may be harvested without regard to chemical composition. In some areas this harvesting might be done on a sustainable basis (Newton, 1994) but cases are known where whole populations are destroyed in the process.

A third threat is the destruction of habitats. One problem is overgrazing. Many people in arid areas have herds of domestic animals in numbers far greater than the carrying capacity of the land, and the land becomes increasingly denuded of vegetation. In many countries where aloes are native, the rise in human population levels results in an increased demand for land to use for agriculture, building, etc. This has led to wholesale clearing of natural vegetation. In some areas, the continued expansion of human populations is forcing people to move into arid areas, where many aloes occur.

Attempts to protect aloes as endangered species have been made at two levels — national and international. Many countries have signed various international agreements on the conservation of biodiversity, though the will to act, which is implied by the signature, is not always translated into action. Most countries have national legislation aimed at protecting endangered species, animals and plants and at preserving habitats in selected localities, such as in national parks. Unfortunately, enforcement of the legislation is poor or wanting in most African countries, because of a lack of enforcement personnel and a need to concentrate national effort on solving enormous economic and social problems. Only South Africa has strong enforcement activity, and even there much illegal activity is known to occur (Newton and Chan, 1998).

The most effective protection attempt at the international level is provided by the Convention on International Trade in Endangered Species of Wild Fauna and Flora (CITES), started in 1976. This convention aims at controlling the movement of endangered species and derivatives between countries, prohibiting trade in some species (listed in CITES Appendix I) and requiring official documentation for numerous other species (listed in CITES Appendix II). Currently, 22 species of *Aloe* (mostly Madagascan) are listed in CITES Appendix I and the rest are in Appendix II, with the sole exception of *Aloe vera*, which is free of restrictions. The documentation requirement provides an opportunity for monitoring the international trade in the listed plants, or at least the legal trade. Data for the period 1983–1989 suggest that the most heavily traded aloes are not the species regarded as threatened in the wild, and much artificially propagated material is involved (Oldfield, 1997). However, there is concern about the demand for rarer species of Madagascar and South Africa. To meet the demand in the horticultural trade many nurserymen and individual growers propagate their stock plants by seeds or by vegetative means. This is positive action to reduce the pressure on wild populations. There are also attempts to re-introduce some rare species of succulent plants to their natural habitats, using material propagated in cultivation. One such scheme involves plans to plant out seedlings of Madagascar's largest aloe, *A. suzannae*, raised in South Africa from field-collected seeds, though the bureaucratic formalities have proved to be formidable (Smith and Swartz, 1997–1999). The most promising development is the use of tissue culture techniques, by means of which thousands of plants can be produced within a few months. Relatively few aloes have been propagated by tissue culture, but there are reports of success (Barringer *et al.*, 1996; Fay and Gratton, 1992; Fay *et al.*, 1995). It is to be hoped that future work in this field will ensure the survival of these fascinating plants.

REFERENCES

Barringer, S.A., Mohamed-Yasseen, Y. and Splittstoesser, W.E. (1996) Micropropagation of endangered *Aloe juvenna* and *A. volkensii* (Aloaceae). *Haseltonia*, 4, 43–45.

Carter, S. (1994) Aloaceae. In *Flora of Tropical East Africa*, edited by R.M. Polhill, 60 pp. Rotterdam: Balkema.

Cutler, D.F. (1982) Cuticular sculpturing and habitat in certain *Aloe* species (Liliaceae) from Southern Africa. In *The Plant Cuticle*, edited by D.F. Cutler, K.L. Alvin and C.E. Price, pp. 425–444. London: Academic Press.

Drummond, R.B., Gelfand, M. and Mavi, S. (1975) Medicinal and other uses of succulents by the Rhodesian African. *Excelsa*, 5, 51–56.

Fay, M.F. and Gratton, J. (1992) Tissue culture of cacti and other succulents: a literature review and a report on micropropagation at Kew. *Bradleya*, **10**, 33–48.

Fay, M.F., Gratton, J. and Atkinson, P.J. (1995) Tissue culture of succulent plants – an annotated bibliography. *Bradleya*, **13**, 38–42.

Jackson, A.O. (1964) An unusual Bantu household altar. *Aloe*, **2**, 14–15.

Kluge, M. and Ting, I.P. (1978) *Crassulacean Acid Metabolism*. 209 pp. Berlin, Heidelberg and New York: Springer-Verlag.

Lavranos, J.J. (1998) Neues aus der Gattung *Aloe* in Madagaskar. *Kakteen und andere Sukkulenten*, **49**, 157–164.

Nash, R.J., Beaumont, J., Veitch, N.C., Reynolds, T., Benner, J., Hughes, C.N.G., Dring, J.V., Bennett, R.N. and Dellar, J.E. (1992) Phenylethylamine and piperidine alkaloids in *Aloe* species. *Planta Medica*, **58**, 84–87.

Neuwinger, H.D. (1996) *African Ethnobotany. Poisons and Drugs*. xviii, 941 pp. London: Chapman & Hall.

Newton, D.J. and Chan, J. (1998) *South Africa's Trade in Southern African Succulent Plants*. 162 pp. Johannesburg: TRAFFIC East/Southern Africa.

Newton, D.J. and Vaughan, H. (1996) *South Africa's Aloe ferox Plant, Parts and Derivatives Industry*. 61 pp. Johannesburg: TRAFFIC East/Southern Africa.

Newton, L.E. (1972) Taxonomic use of the cuticular surface features in the genus *Aloe* (Liliaceae). *Botanical Journal of the Linnean Society*, **65**, 335–339.

Newton, L.E. (1979). In defence of the name *Aloe vera. Cactus and Succulent Journal of Great Britain* 41: 29–30.

Newton, L.E. (1994) Exploitation and conservation of aloes in Kenya. *Proceedings of the XIIIth Plenary Meeting of A.E.T.F.A.T.*, Malawi, **1**, 219–222.

Newton, L.E. ("1995", published 1998) Natural hybrids in the genus *Aloe* (Aloaceae) in East Africa. *Journal of East African Natural History*, **84**, cover + 141–145.

Newton, L.E. (2001) *Aloe*. In *Illustrated Handbook of Succulent Plants: Monocotyledons*, edited by U. Eggli, pp. 103–186. Berlin, Heidelberg and New York: Springer Verlag.

Newton, L.E. (2001) Poisonous aloes. *East Africa Natural History Society Bulletin*, **31**, 8–9.

Oldfield, S. (ed.). (1997) *Status Survey and Conservation Action Plan. Cactus and Succulent Plants*. x, 214 pp. Gland: IUCN.

Reynolds, G.W. (1950) *The Aloes of South Africa*. xxiv, 520 pp. Johannesburg: The Aloes of South Africa Book Fund.

Reynolds, G.W. (1966) *The Aloes of Tropical Africa and Madagascar*. xxii, 537 pp. Mbabane: Aloes Book Fund.

Smith, G.F. and Swartz, P. (1997–1999) Re-establishing *Aloe suzannae* in Madagascar. *British Cactus and Succulent Journal*, **15**, 88–93, 149–155; **17**, 45–49.

Verdcourt, B. and Trump, E.C. (1969) *Common Poisonous Plants of East Africa*. London: Collins.

Watt, J.M. and Breyer-Brandwijk, M.G. (1962) *The Medicinal and Poisonous Plants of Southern and Eastern Africa*, 2nd edn. 1457 pp. Edinburgh and London: Livingstone.

Willert, D.J., von, Eller, B.M., Werger, M.J.A., Brinckmann, E. and Ihlenfeldt, H.-D. (1992) *Life Strategies of Succulents in Deserts with special reference to the Namib Desert*. xix, 340 pp. Cambridge: Cambridge University Press.

2 Taxonomy of Aloaceae

Gideon F. Smith and Elsie M.A. Steyn

ABSTRACT

The taxonomy of Aloaceae is discussed at the family, genus and species levels. A key is provided to the seven genera accepted in the present chapter.

INTRODUCTION

There is no doubt that ancient peoples classified plants. However, the folk-taxonomic groupings that they used were much simpler and more robust than those used today (Stace, 1980). Typical mental abstractions (categories) into which plants were placed were whether they were useful (e.g. as sources of food or building materials), poisonous, or could be worshipped. The species of *Aloe* L., particularly the widely known and used *A. vera* (L.) Burm.f., would no doubt have been included in the first-mentioned of these categories. The remarkable healing powers of the leaf juices of various species of *Aloe* have been known for very long and were used before and in biblical times; one of their uses was as an ingredient in embalming ointments (Bible, New Testament: John 19: 39; Admiraal, 1984; Smith, 1993). However, as with all plants, the formal taxonomy of species of *Aloe* and their relatives dates back only to 1753 when Linnaeus proposed his sexual system of plant classification, based primarily on the number of stamens and pistils in a flower. Since species of *Aloe* have six stamens and a single pistil, they fitted comfortably into the Hexandria Monogynia of Linnaeus (1753). The artificiality of the Linnean System was widely realised and between 1753 and the early 1800s various attempts at a Natural Classification of plants, thus classifying them to reflect overall similarities, were made. The aloes and their relatives were also subjected to these proposals and they saw a number of familial reshufflements over the past 250-odd years. In addition, alooid genus delimitation and the circumscription of species have also been the subject of taxonomic argument. This chapter provides a synoptic overview of the taxonomy of, especially, higher-category alooid taxa. The currently accepted genera are also briefly discussed and a key to facilitate their identification is provided.

WHAT ARE THE ALOACEAE?

A wide range of different classifications for the Aloaceae have been proposed during the past few decades, and seven genera are widely accepted (Table 2.1). These genera

Table 2.1 Synonymy of genera of the Aloaceae.

Genera recognized in present chapter	Recent taxonomic treatment	Synonyms
Aloe Linnaeus: 319 (1753)	Reynolds (1966, 1982)	*Catevala* Medikus: 67 (1786) *pro parte*; *Kumara* Medikus: 69 (1786); *Rhipidodendrum* Willdenow: 164 (1811); *Pachidendron* Haworth: 35 (1821); *Bowiea* Haworth: 299 (1824) *non* J.D. Hooker: t. 5619 (1867); *Agriodendron* Endlicher: 144 (1836); *Succosaria* Rafinesque: 137 (1840); *Busipho* Salisbury: 76 (1866); *Ptyas* Salisbury: 76 (1866); *Chamaealoe* A.Berger: 43 (1905); *Leptaloe* Stapf: t. 9300 (1933) *Aloinella* Lemée: 27 (1939) *non* Cardot: 76 (1909); *Guillauminia* Bertrand: 41 (1956); P.V.Heath: 153 (1993)
Gasteria Duval: 6 (1809)	Van Jaarsveld (1994)	*Atevala* Rafinesque: 136 (1840); *Papilista* Rafinesque: 137 (1840)
Haworthia Duval: 7 (1809) nom. cons.	Bayer (1999)	*Catevala* Medikus: 67 (1786) *pro parte*; *Apicra* Willdenow: 167 (1811) *non* Haworth: 61 (1819); *Kumaria* Rafinesque: 137 (1840); *Tulista* Rafinesque: 137 (1840)
Lomatophyllum Willdenow: 166 (1811)	Jacobsen (1986)	*Phylloma* Ker: t. 1585 (1813)
Chortolirion A.Berger: 72 (1908)	Smith (1995a)	–
Poellnitzia Uitewaal: 61 (1940)	Smith (1995b)	–
Astroloba Uitewaal: 53 (1947)	Roberts Reinecke (1965)	*Apicra* Haworth: 61 (1819) *non* Willdenow: 167 (1811)

constitute a natural assemblage that differs fairly consistently from those included in the closely related Asphodelaceae Juss. in their conspicuous succulent leaf consistency, crescentiform or cymbiform leaf outline in cross-section and the markedly bimodal karyotype consisting of $2n=14$ chromosomes. Further characteristics that unify the Aloaceae are the presence of 1-methyl-8-hydroxyanthraquinones in the roots and anthrone-C-glucosides in the leaves, and leaf vascular bundles containing a parenchymatous inner bundle sheath (Smith and Van Wyk, 1998). In circumscribing the Aloaceae, the genera *Bulbine* Wolf and *Kniphofia* Moench of the Asphodelaceae appear to be superficially problematical, particularly since some of the species of *Bulbine* have karyotypes and morphologies similar to that of certain taxa of Aloaceae [*cf.* Spies and Hardy (1983) on *B. latifolia* (L.f.) Schult. and J.H. Schult. and Rowley (1954a) on *Bulbine* in general]. However, *Bulbine* can be clearly distinguished from the genera of the Aloaceae on the basis of its open, yellow or only very rarely white or orange flowers, free perianth segments, bearded filaments, lack of nectar production and the annual nature of some of its species (e.g. *B. alata* Baijnath). Furthermore, *Bulbine* has an African-Australian distribution whereas the Aloaceae are absent from this austral continent. Mainly for these reasons *Bulbine* was not considered to be a constituent of the Aloaceae. In contrast to representatives of *Bulbine* and the Aloaceae, leaf succulence is virtually absent in *Kniphofia*, *K. typhoides* Codd being a notable exception (Retief, 1997). Also, in *Kniphofia* the leaf outline in cross-section is V-shaped and representatives of the genus have a chromosome base-number of six. On the other hand, synapomorphous characteristics of the tubular flowers and fusion of the perianth segments suggest an apparently strong relationship between *Kniphofia* and the Aloaceae. These characteristics probably provide sufficient evidence to justify the choice of *Kniphofia* as an outgroup (sister group) to the Aloaceae (Smith and Van Wyk, 1991).

The Aloaceae is fundamentally an Old World group with most genera occurring in subsaharan Africa. The genus *Aloe* is also found on the Arabian Peninsula, Madagascar and Socotra, while the fleshy-fruited *Lomatophyllum* Willd. has been reported from the Aldabra Islands, Madagascar and Mauritius. However, the greatest concentration of genera apart from *Aloe*, *Astroloba* Uitewaal, *Chortolirion* A.Berger, *Gasteria* Duval, *Haworthia* Duval and *Poellnitzia* Uitewaal and species is in southern Africa, that is, roughly, the region south of the Kunene, Okavango and Zambezi Rivers.

FAMILY CONCEPTS IN THE ALOACEAE

The German botanist, August J.G.C. Batsch, is accredited as the first person to afford the alooid plants segregate suprageneric status at the familial rank. However, Batsch's (1802: 138) Aloaceae (published as Alooideae) were not widely accepted and fell into disuse for approximately 180 years, most taxonomists referring the aloes and related genera to the Linnaean Hexandria Monogynia (e.g. Schultes and Schultes, 1829: 631) or, later, to the Liliaceae (e.g. Baker, 1880: 148). 'Aloeaceae' were again popularized by Cronquist (1981: 1215) and the taxon is now widely accepted, although there is some controversy over the spelling of the name, Aloeaceae (e.g. Cronquist, 1981: 1215; Forster and Clifford, 1986: 66; Mabberley, 1987: 21) versus Aloaceae (e.g. Cronquist, 1988: 485, 516; Brummit, 1992; Smith, 1993; Glen *et al.*, 1997). Following Smith (1993), we here accept Aloaceae as the correct spelling of the name.

However, by far the most popular and traditionally widely accepted family classification used for aloes and their generic relatives during the late nineteenth and greater part of the twentieth centuries, has been their inclusion as Aloineae, one of 28 tribes recognized by Hutchinson (1959) in the all-embracing Liliaceae *sensu lato*. Since the 1950s, the Liliaceae has been subjected to significant taxonomic reassessment in terms of its constituent infrafamilial taxa. The different interpretations of the circumscription of the Liliaceae by various taxonomists have resulted in, amongst others, the tribe Aloineae *sensu* Hutchinson (1959) being segregated from the family.

Despite the wide variation in morphology and anatomy within many species (Schelpe, 1958; Reynolds, 1966, 1982; Newton, 1972; Cutler *et al.*, 1980), remarkable cytogenetic uniformity exists within the Aloaceae as a whole (Sharma and Mallick, 1966; Brandham, 1971, 1974; Riley and Majumdar, 1979). Moreover, the chromosomes are few and large enough to enable any deviation from the rule to be detected easily (Brandham, 1976, 1977). These similarities, along with the clear monophyletic origin of this natural assemblage of genera, led Cronquist (1981) to remove them from the Liliaceae *sensu lato*. He reclassified the genera *Aloe* (including *Aloinella* Lem. *non* Cardot, *Chamaealoe* A.Berger, *Guillauminia* Bertrand and *Leptaloe* Stapf), *Gasteria, Haworthia* (including *Astroloba, Chortolirion* and *Poellnitzia*), *Kniphofia* (including *Notosceptrum* Bentham) and *Lomatophyllum* in the family 'Aloeaceae' Batsch, the latter being regarded as a derivative of the Liliaceae *sensu stricto* (Table 2.2). The relative ease with which species of most of the alooid genera interbreed indicates that they are closely related. The absence of barriers to hybridization remains one of the most important points of evidence for phylogeny, especially since barriers may

Table 2.2 Summary of the taxonomic treatments of alooid genera of Cronquist (1981), Dahlgren and Clifford (1982) and Dahlgren *et al.* (1985), compared with the proposals of the present chapter. Only the genera *Bulbine, Kniphofia* and those traditionally included in the tribe Aloineae, family Liliaceae *sensu* Hutchinson (1959) are listed.

	Cronquist (1981)	Dahlgren & Clifford (1982)	Dahlgren et al. (1985)	Proposals of this chapter
Family	Aloeaceae	Asphodelaceae	Asphodelaceae	Aloaceae
Infrafamilial classification	–	Subfamily Asphodeloideae	Subfamily Alooideae	–
	Aloe	Aloe	Aloe	Aloe
	–	Astroloba	Astroloba	Astroloba
		Chamaealoe	Chamaealoe	–[1]
		Chortolirion	–	Chortolirion
Genera upheld	Gasteria	Gasteria	Gasteria	Gasteria
	Haworthia	Haworthia	Haworthia	Haworthia
	Lomatophyllum	Lomatophyllum	Lomatophyllum	Lomatophyllum
	–	–	Poellnitzia	Poellnitzia
	–	Bulbine	–	–[2]
	–	Kniphofia	–	–[2]

Notes
1 Included in the synonymy of *Aloe* as *A. bowiea* Schult. & J.H. Schult. (Smith and Van Wyk, 1993).
2 Retained in the family Asphodelaceae.

evolve rapidly even in closely related groups (see for example Riley and Majumdar, 1979).

The proposal of Cronquist (1981) was primarily based on the fact that the closely related, but New World family, Agavaceae is currently recognized as a separate family owing to its specialized growth habit, among other characteristics. Since the Aloaceae differ from the Liliaceae on essentially the same grounds, recognition of the Agavaceae as a family warrants the acceptance of the Aloaceae as a lilioid segregate. Although the inclusion of at least *Yucca* Dill. *ex* L. (superior ovary) and *Agave* L. (inferior ovary) in the Agavaceae is supported by the common occurrence of a very distinctive bimodal karyotype consisting of five large chromosomes and 25 small ones, the karyology of other genera referred to the Agavaceae on grounds of their growth habit differ significantly from that of *Yucca* and *Agave*. Cronquist (1981) concluded that in the Agavaceae a karyological characteristic '... does not correlate well enough with other features to be of critical taxonomic importance.' In contrast, the Aloaceae is characterised by a distinctive karyotype ($2n=14$) which consists of four pairs of long chromosomes and three pairs of short chromosomes.

It is noteworthy that Cronquist (1981) retains *Bulbine* in the Liliaceae. Several species of *Bulbine* have morphological and cytological characters in common with species that he included in the 'Aloeaceae' (Rowley, 1954a, 1967). Furthermore, Reynolds (1950) recorded a plant 10 km north of Alice in the eastern Cape Province, which gave every indication of being a hybrid between *Aloe tenuior* Haworth var. *decidua* Reynolds and *Bulbine alooides* (L.) Willd. Whilst retaining *Bulbine* in the Liliaceae, Cronquist (1981) included *Kniphofia* along with the alooid genera in the 'Aloeaceae.' This genus is generally regarded as transitional between the Liliaceae and 'Aloeaceae' (Cronquist, 1981).

In a somewhat different reclassification of the inclusive family Liliaceae, Dahlgren and Clifford (1982) included the genera traditionally classified in the tribe Aloineae, along with many others, in the type subfamily of the reinstated Asphodelaceae. In their restructuring of the Liliaceae *sensu lato*, they made use of characteristics derived from a wide range of disciplines and attempted an arrangement of all monocotyledonous families according to overall similarities obtained from comparative studies. Although Dahlgren and Clifford (1982) did not elaborate on the infrasubfamilial classification of the Asphodeloideae, they did mention that, within the latter taxon, the so-called *Aloe*-group forms a coherent unit. Thus, recognition of the alooid genera as a discreet entity within the Asphodelaceae would accord with the generally accepted requirement that supraspecific taxa should ideally be monophyletic.

In a comprehensive book on the structure, evolution and taxonomy of the monocotyledons, Dahlgren *et al.* (1985) made further refinements to their 1982 asphodeloid subfamily concept. They removed seven alooid genera from the Asphodeloideae and placed these in a separate subfamily, Alooideae. Dahlgren *et al.* (1985) therefore recognised two asphodeloid subfamilies, Asphodeloideae and Alooideae. The alooid genera show distinct phenetic similarities and these authors interpreted it as a monophyletic group derived from a common ancestor, also shared with other Asphodelaceae. The eleven genera included by Dahlgren and Clifford (1982) in the Asphodeloideae do not display the same marked African concentration as in the case of those transferred to the Alooideae, and probably represent a paraphyletic group. Of the genera originally upheld in the *Aloe*-group of the Asphodeloideae (Dahlgren and Clifford, 1982), only *Chortolirion* was not transferred to the Alooideae, probably as an oversight (Table 2.2). However, the genus *Poellnitzia*, which was omitted from their initial comparative study (Dahlgren and Clifford, 1982), was subsequently included in the Alooideae (Dahlgren *et al.*, 1985). In both publications the monotypic genus *Chamaealoe* A.Berger is upheld (Table 2.2). In accordance with more recent proposals by Smith (1990) and Smith and Van Wyk (1993), *Chamaealoe* should be included in the synonymy of *Aloe*. Dahlgren and Clifford (1982) included the alooid genera in the Asphodeloideae along with, amongst others, the apparently related *Bulbine* and *Kniphofia*. In their subsequent rearrangement of the Asphodelaceae (Dahlgren *et al.*, 1985) both these last-mentioned genera were retained in the subfamily Asphodeloideae. The treatment of Dahlgren *et al.* (1985) of the subfamily Alooideae (in Asphodelaceae) therefore approximates Cronquist's (1981) concept of the 'Aloeaceae' with the exception of the placement of *Kniphofia*: Cronquist (1981) included this non-succulent genus in the 'Aloeaceae.' We basically support Cronquist's (1981) treatment, but regard the asphodeloid genus *Kniphofia* as a probable sister group of the Aloaceae (Smith and Van Wyk, 1991) and recognize seven genera in the family (Tables 2.1 and 2.2). The circumscription of the lilioid tribe Aloineae of Hutchinson (1959) has therefore undergone comparatively little change in terms of generic content, the major controversies surrounding its taxonomy having centred on family and species concepts.

GENERIC CONCEPTS AND TAXONOMIC RESEARCH IN THE ALOACEAE

Although 29 genus names are available for taxa included in the Aloaceae, only seven are widely recognized (Table 2.1).

The taxonomic study of a group of plants, such as the Aloaceae, must necessarily include an analysis of the botanical way of thinking at the time of publication of its constituent generic names and their subsequent histories. Such an investigation gives insight into the circumstances prevalent when the names were proposed, as well as their current relevance. The chronology of genus name proliferation and the history of the genus concept in the Aloaceae are summarized in Table 2.3. The chronology (Table 2.3) is compared to the historical phases of plant classification and to the major periods of systematic biology. Although the latter phases and periods cannot be sharply delimited (Turner, 1967), it is clear that early nineteenth century attempts by some European botanists to reflect natural affinities amongst plants in general resulted in the publication of 17 of the 29 genus names available for alooid taxa. However, the circumscription of only three of these genera, namely *Aloe, Gasteria* and *Haworthia* are generally accepted.

Table 2.3 Chronology of genus name proliferation in the Aloaceae.

Major periods of systematic biology (Alston and Turner, 1963; Merxmüller, 1972)	Historical phases of plant classification (Lawrence, 1951)	Anno Domini	Reference	Genus names generally upheld[1]	Genus names not upheld
	Artificial systems based on numerical classifications	1753	– Linnaeus (1753)	*Aloe*	
			– Medikus (1786)		*Catevala; Kumara*
		1800			
Micromorphic			– Duval (1809)	*Gasteria; Haworthia*	
			– Willdenow (1811)	*Lomatophyllum*	*Apicra[2] Rhipidodendrum*
			– Ker (1813)		*Phylloma*
			– Haworth (1819)		*Apicra[3]*
			– Haworth (1821)		*Pachidendron*
	Natural systems based on form relationships		– Haworth (1824)		*Bowiea*
			– Endlicher (1836)		*Agriodendron[4]*
			– Rafinesque (1840)		*Atevala; Kumaria; Papilista; Succosaria; Tulista*
		1860			
			– Salisbury (1866)		*Busipho; Ptyas*

Evolutionary				
↓		1880		
↑		1900		
				Chamaealoe[5]
		– Berger (1905)		
		– Berger (1908)	*Chortolirion*	
Cytogenetical				
	Systems based on phylogeny			
		– Stapf (1933)		*Leptaloe*
		– Lemée (1939)		*Aloinella*
		– Uitewaal (1940)	*Poellnitzia*	
		– Uitewaal (1947)		*Guillauminia*
		– Bertrand (1956)		
↓		1993 – Heath (1993)		*Lemeea*

Notes

1 These seven genera are also recognized in the present chapter.

2 *Apicra* Willd. is a superfluous name for *Haworthia* Duval.

3 *Apicra* Haw. is a later homonym of *Apicra* Willd., applied to a group of haworthioid species, currently upheld as a separate genus. Although an attempt was made to conserve *Apicra* Haw. against *Apicra* Willd. (Stearn, 1939a, b), the genus was later renamed *Astroloba* by Uitewaal (1947).

4 Endlicher (1936) attributed *Agriodendron* to Haworth. However, this name could not be located in any of Haworth's publications. Jackson (1895) justifiably cites Endlicher as author of the name *Agriodendron*.

5 *Chamaealoe*, although published as a substitute name for *Bowiea* Haw., was superfluous at the time of publication. If Berger's (1905) generic concept is revived, a new name would therefore have to be given to what has been called *Bowiea* Haw. (*nom. rej.*) and *Chamaealoe* (*nom. illeg.*).

The opening up of distant lands, such as the southern African interior, by, amongst others Masson, Burchell and Bowie (Smith and Van Wyk, 1989), resulted in a steady stream of novel material reaching European botanical gardens. Botanists who had to deal with the wealth of new and undescribed specimens evidently did not know how to fit them into existing classifications and reverted to basing new genera on non-diagnostic, floral and/or vegetative structures or combinations of structures. However, in some circles the recognition of only a single genus, *Aloe*, for all the succulent-leaved, rosulate, alooid taxa persisted until at least the 1880s (Table 2.3). Clearly, a genus lies somewhere between these two extremes and, as Rowley (1976a) so aptly put it, '. . . the best we can hope to do is to avoid gross inconsistencies in our chosen unit,' – in this case the Aloaceae.

The taxonomic history of the Aloaceae started out conservatively in 1753. Of the genera currently classified in this subfamily, Linnaeus (1753) recognized only one, namely *Aloe*, which he included in his Class Hexandria Order Monogynia. Up to the time that Linnaeus proposed his sexual classification system, the few known alooid taxa, mainly comprised *A. vera* (L.) Burm.f., possibly of Arabia (Forster and Clifford, 1986), *A. perryi* Baker of Socotra (Lavranos, 1969; Horwood, 1971) and the South African *A .maculata* All. (Dandy, 1970), *A. arborescens* Miller, *A. brevifolia* Miller, *A. commixta* A.Berger, *A. ferox* Miller, *A. glauca* Miller, *A. humilis* (L.) Miller, *A. plicatilis* (L.) Miller, *A. succotrina* Lam. and *A. variegata* L. (Wijnands, 1983; Reynolds, 1950). These aloes were grouped mainly on the basis of their characteristic succulent leaves (Bauhin, 1651; Uitewaal, 1947; Rowley, 1960; 1976b). Although Linnaeus did not intentionally utilize vegetative characters in his classification system, the latter line of thought did, to some degree, precipitate in his treatment of *Aloe*. This is clearly illustrated by the inclusion in *Aloe* of taxa currently classified in *Aloe, Astroloba, Gasteria, Haworthia* (Reynolds, 1950; Bayer, 1976), *Kniphofia* (Codd, 1968) and *Sansevieria* Thunb. (Brenan, 1963; Wijnands, 1973). With the knowledge of hindsight, Linnaeus (1753) afforded *Aloe* and the New World *Agave* separate generic status, in contrast to other early taxonomists, such as Bradley (1716–1727) who, somewhat understandably, confused the two genera. Later researchers in the Aloaceae, for example Reynolds (1950, 1966) and Holland (1978) on *Aloe sensu stricto*, Van Jaarsveld (1994) on *Gasteria* and Bayer (1976, 1982, 1999) on *Haworthia*, have attempted natural classifications for these genera. Scott (1985), on the other hand, proposed an entirely artificial classification system for *Haworthia*, based mainly on vegetative characteristics.

Since the publication of the genus name *Aloe* (Linnaeus, 1753), this taxon has been plagued by taxonomic confusion. Linnaeus preferred a broad circumscription of this genus and his use of a limited number of reproductive characteristics as criteria for classification could not possibly provide conclusive evidence for generic circumscription in the Aloaceae. Furthermore, Linnaeus made extensive use of infraspecific categories for classification, one of his species, *A. perfoliata*, being burdened by 16 varieties. Although Reynolds (1950) established the identity of ten of these varieties, some remain obscure and cannot with certainty be linked to field populations.

Attempts to subdivide *Aloe sensu* Linnaeus (1753) started some 30 years after this heterogeneous entity was proposed (Table 2.3). Although this initial attempt (Medikus, 1786) to split *Aloe* into smaller, more homogeneous units was unsuccessful, the present-day circumscription of the four comparatively large genera, *Aloe, Gasteria, Haworthia* and *Lomatophyllum* dates from the early nineteenth century (Table 2.3). However, the Aloaceae does not consist of large genera only. Especially in the first half of the twentieth century genus names were proposed for several smaller units segregated from *Aloe, Haworthia* and *Astroloba*, the latter then being known as *Apicra* Haw. non Willd. (see footnote 3 of Table 2.3). This period coincided with the publication of the first plant classification systems based on phylogeny and probably represented attempts to display patterns of evolution within the alooid taxa. Of the genera proposed during this period, only *Chortolirion* and *Poellnitzia* are generally upheld, *Astroloba* having been recognized as a segregate of *Aloe* by the late nineteenth century (Baker, 1880).

TAXONOMIC NOTES ON THE GENERA OF THE ALOACEAE

Brief notes are given here on the genera of the Aloaceae upheld in this study, with emphasis on recent taxonomic contributions. For additional information on the genera, Smith and Van Wyk (1991, 1998) should be consulted.

Core genera (*Aloe, Lomatophyllum*)

In the taxonomy of the Aloaceae, *Aloe* is of central importance. Not only is it the oldest genus in the family, but of all the genera of the Aloaceae it has been studied most extensively, both in terms of taxonomy and systematics. [For the taxonomy versus systematics controversy the argumentative contributions of Gilmartin (1986, 1987), La Duke (1987), Donoghue (1987) and Small (1989) should be consulted]. Furthermore, since it was the only genus recognized by Linnaeus (1753), all seven genera generally upheld in this monophyletic group are in effect segregates of *Aloe*. Although *Aloe* was monographed on a number of occasions in the nineteenth century, the botanical community and students of the genus owe most credit to Dr. G.W. Reynolds who travelled widely and studied *Aloe* intensively for more than thirty years on a continental scale. Thus, not only did he survey the southern African species of *Aloe* (Reynolds, 1950), but he expanded his research to include the tropical African and Madagascan species as well (Reynolds, 1958, 1966). However, at the time of his death in 1967, many taxonomic problems remained unsolved and numerous newly discovered species awaited formal description. Many of the latter taxa were eventually published by L.C. Leach and D.S. Hardy. Both had a working knowledge of Reynolds' methodology and were in a good position to add further to our knowledge of the genus. In a recent revision of *Aloe*, Glen and Hardy (2000) reduced the number of species recognized from southern Africa from about 150 to 119. One of these authors (H.F. Glen, personal communication) is of the opinion that an account of *Aloe* sect. *Pictae*, paying attention to the appearance of the plants rather than the requirements of collectors, would result in a further reduction in the number of species recognized.

The origin of the currently accepted infrageneric classification of *Aloe* goes back to researchers such as Baker (1880) working in England, and Berger (1905, 1908) operating from Germany and Italy. Berger's classification, in particular, was very popular and gained wide acceptance, mostly because at least one of his contributions on *Aloe* (Berger, 1908) was published as part of Engler's *Das Pflanzenreich* which was an 'update' of Linnaeus's (1753) *Species Plantarum*. This basically meant that Berger's work had a broad circulation. It is therefore not surprising that the infrageneric classifications of *Aloe* of both Groenewald (1941) and Reynolds (1950, 1966) were fundamentally based on Berger's proposals. As a result of the comprehensiveness of the two works of Reynolds, these became the standard works on the genus for the next 50 years. In this way the groups of Berger (1905, 1908) are essentially still the most widely used infrageneric treatment to which all *Aloe* taxonomists revert.

Not surprisingly, Berger's (1905, 1908) classification of *Aloe* was based primarily on a combination of vegetative and reproductive morphological characters. For many families and genera, classifications based on such characters still hold true, even today. The reason is simple: these characters are the most easily observable and readily available when a classification is attempted. Berger therefore established in *Aloe* such groups as sect. *Maculatae* (nomenclaturally more accurately the *Pictae*) for the spotted-leaved aloes and sect. *Graminialoe* for the grass-leaved aloes.

More recently, other characters, many of which are cryptic and not easily observable have been employed by taxonomists to study relationships in the genus. Some of these, such as pollen morphological characters, are predictably rather uninformative at the species rank (Steyn *et al.*, 1998), while others, such as chemical characteristics are in many instances more useful (Viljoen and Van Wyk, 1996; Viljoen *et al.*, 2001) for proposing new or confirming existing relationships. However, as with many other plant groups the most accurate classifications can be derived from using a multifaceted matrix of numerous characters. For some time to come, in *Aloe* at least, gross morphology will remain the most important source of characters for identifying preserved and living material. As laboratory technology and methodology become more sophisticated and accessible, infrageneric relationships may be clarified by employing more obscure characters.

The second 'typically' alooid genus, *Lomatophyllum*, is inadequately known from a taxonomic perspective. It was recently proposed that it should be included in the synonymy of *Aloe* as a section (Rowley, 1996), but this may be an oversimplification of a rather intricate taxonomy. Pending comprehensive revision, the genus is here maintained separately from *Aloe*, due mainly to its baccate and not capsular, fruit.

Gasteria

For many years species concepts in *Gasteria* were poorly defined (Schelpe, 1958). The genus was a taxonomist's nightmare with more than 100 names available for some 20 discreet taxa. The revision of *Gasteria* by Van Jaarsveld (1994) provides a clear picture of relationships amongst species and the phytogeography of the genus.

Haworthioid genera (*Haworthia, Astroloba, Poellnitzia, Chortolirion*)

This informal infrasubfamilial group is characterised by a combination of the following three characteristics: rather insignificant, greenish-white flowers, except in one species, *Poellnitzia rubriflora* (Bolus) Uitewaal, which has red flowers; flowers slanted upwards and, a generally small stature. All the species are small, rosulate succulents. Although these characteristics are not unique to the haworthioid genera, they do occur in almost all the species of this group and are generally rare in the other alooid genera.

Since the recognition of *Haworthia* as a genus separate from *Aloe* (Duval, 1809), it has been burdened by a profusion of species names often based on specimens of unknown or garden origin. Furthermore, most of the species names available in the genus have been subdivided into numerous varieties and forms. Probably few of these warrant recognition at any rank in a hierarchical classification system. Although *Haworthia*, like *Aloe*, has been monographed on a number of occasions during the past 180 years, most of these 'revisions' can be discarded as of limited value due to the fact that they reflect little more than horticultural fashion and have little or no connection with habitat collected material (Smith, 1989). Indeed, the first comprehensive overviews of *Haworthia* based on detailed field work (Bayer, 1976, 1982, 1999; Scott, 1985; Breuer, 1998) were initiated as recently as 30 years ago.

The uncertain state of the taxonomy of *Haworthia* is amply illustrated by the fact that Bayer (1982) recognised 68 species in this genus and Scott (1985) 88, while Gibbs Russell *et al.* (1985) listed 166 names for *Haworthia*, the last-mentioned work including *Chortolirion, Astroloba* and *Poellnitzia*. Bayer (1986) has expressed the view that a truly botanical dispensation could reduce to 33 the number of species of *Haworthia* worthy of recognition. However, this suggestion should be seen in the correct perspective, some of the reasons for this confusing situation having been set out by Bayer (1978, 1986). In his taxonomic update of *Haworthia* published in 1999, he recognised many more species and infraspecific taxa.

The second haworthioid genus, *Astroloba*, was monographed by Roberts Reinecke (1965) and Groen (1986, 1987a, b, c, d, e, f). *Astroloba* is a small genus consisting of seven (Roberts Reinecke) or four (Groen) species, depending on which revision is preferred (see Table 1 in Smith and Marx, 1990). Characteristically, species of *Astroloba* have five-tiered, pungent-leaved, caulescent rosettes. At least in comparison to *Haworthia*, the flowers of *Astroloba* appear regular rather than bilabiate. Two species of *Astroloba*, *A. herrei* Uitewaal and *A. spiralis* (L.) Uitewaal, have flowers with marked inflation of the perianth tube (Bayer, 1975). This characteristic, a unique apomorphy in cladistic terminology, is absent from all other genera of the Aloaceae.

In the past *Poellnitzia*, the third haworthioid genus, was either inadvertently excluded from treatments of the Aloaceae or its only species, *P. rubriflora*, was simply shifted from genus to genus without adequate supporting evidence being offered (Smith, 1994). Manning and Smith (2000) recently argued that there is little cladistic support for the retention of the generic status of *Poellnitzia* and proposed its inclusion in the entomophilous *Astroloba* as an ornithophilous entity. However, it is one of the few alooid genera with at least one distinctive and unique apomorphy, namely the connivence of the tips of the perianth segments. For this reason it is here retained tentatively as a monotypic genus.

The fourth and final haworthioid genus, *Chortolirion*, is also monotypic. It is a perennial, deciduous and herbaceous form, widely distributed in the summer rainfall region of southern Africa. The genus is morphologically quite distinct from *Haworthia*, especially with regard to the presence of an underground bulbous rootstock. Furthermore, *Chortolirion* is unique amongst haworthioid species in that it is the only grassland taxon of which the leaves are deciduous and die back to ground level after fires or frost.

SPECIES LEVEL TAXONOMY OF THE ALOACEAE

From its modest beginning as a single, heterogeneous genus, *Aloe*, the number of both genera and species in the Aloaceae has increased substantially. Apart from the 29 genus names which have been proposed for taxa of the family at one time or another (Rowley, 1976c, d), *Aloe* alone currently boasts about 400 species (Smith and Van Wyk, 1998; Newton, 2001) and numerous infraspecific taxa. However, the taxonomy of the Aloaceae remains in flux as the existence of some genera is questioned and boundaries of others debated. This can primarily be attributed to the fact that, although being a genetically homogeneous group, the Aloaceae display unusual patterns of variation among populations and species and inconsistent intergradations among genera. However, the generic concepts as proposed in this study best reflect the current state of knowledge. Pending detailed taxonomic revisions of, particularly *Aloe*, *Haworthia* and *Lomatophyllum*, any reclassification of 'groups of species' (*fide* Funk, 1985) would be premature, and not in the best interest of nomenclatural stability.

As has been shown by Osborne *et al.* (1988) for the South African cycads, the aloes and aloe-like plants have deservedly been admired and studied by botanists, horticulturists and laymen, especially in view of their aesthetic appeal and medicinal value. Increasing attention should now be paid to their taxonomy, micro- and macromorphology, vegetative and reproductive, phytogeography, chemistry and molecular systematics. In many cases this work is in its infancy and some of the avenues of research may yet cast additional light on the alooid genus concept.

Numerous difficulties are encountered when dealing with taxonomic literature and herbarium specimens pertaining to succulent plants in general, the Aloaceae being no exception. Especially in the eighteenth and nineteenth centuries species were described from single plants grown under artificial conditions in private collections, especially in Europe, thousands of kilometers from their natural habitats. In cultivation succulent plants often become etiolated and chlorotic and may bear only a slight resemblance to plants in the field. Furthermore, authors of new species had no idea of the variability of these taxa in their natural habitats (Smith, 1948; Schelpe, 1958). Succulent plants easily survived long sea journeys to botanical gardens in Europe and the relative ease with which specimens can be rooted made them popular horticultural objects. Many of the early botanical explorers sent to the Cape of Good Hope to collect plants for, amongst others, Kew Gardens, had a special interest in succulent plants, resulting in a steady stream of such specimens reaching Europe (Gunn and Codd, 1981; Smith and Van Wyk, 1989).

However, succulent plants make poor herbarium specimens (Chudovska, 1979a, b; Fuller and Barbe, 1981; Leuenberger, 1982; Baker *et al.*, 1985; Logan, 1986) and type specimens were usually not prepared. Bradley (1716–1727), in the first ever publication devoted entirely to succulents, used this difficulty with which herbarium specimens are made as a criterion for distinguishing them from other plants (see also Higgins, 1940; Rowley, 1952, 1954b, 1977, 1983; Thomas, 1952; Butterfield, 1968; Stafleu and Cowan, 1976). Succulent plants also tend to grow in a press, Marloth (1925) mentioning *Crassula barbata* Thunb. which sprouted and flowered after having been pressed for nine months! Unfortunately, new species such as *Haworthia venteri* Poelln. and *H. paradoxa* Poelln. were described from specimens which elongated during the sea voyage from the Cape to Europe (Von Poellnitz, 1939; Smith, 1948).

Scientific communication amongst succulent plant enthusiasts working on the same taxa was often unsatisfactory, resulting in multiple names being given to the same taxon or one name being published for different taxa, *Apicra* Haw. *non* Willd. being a good example. Species descriptions usually consisted of brief Latin diagnoses only, which, if unaccompanied by accurate illustrations, could be applied to a suite of species. Furthermore, much of the early work on alooid taxa was done by amateur collectors with little or no botanical background and no knowledge of the growing conditions of the plants in habitat. However, it is noteworthy that botanical training by no means guarantees success as a taxonomist specialising in the family Aloaceae, the often rather peculiar works of Professor Dr Flavio Resende (e.g. Resende, 1943) being good examples. In contrast, the published work on *Aloe* of amateurs such as Dr G.W. Reynolds (Anonymous, 1967; D'Ewes, 1967) and Mr L.C. Leach (Smith, 1990 and references therein) are of excellent quality. Clearly, the difference between Reynolds and Leach and other, earlier amateurs interested in alooid taxonomy, is that they had field experience of the plants and based their studies on populations and not individuals. As Bayer (1972) justifiably surmises '…many the ship of a prospective taxonomist has been ship-wrecked on the rocks of the Liliaceae.'

No matter how carefully a private collection of succulent plants is kept, living material can never replace herbarium specimens as a primary source of taxonomic information. Although Rowley (1951) claims that at least some of Adrian Hardy Haworth's (1768–1833) original alooid material has survived to the present day, these plants can hardly be authenticated in terms of either original ownership or their status as having been the material originally described. All the *Haworthia* material, including some original Haworth clonotypes, which John Thomas Bates (1884–1966), an avid English succulent plant collector, possessed was lost during the First World War (Roan, 1948; Rowley, 1985). Although this collection, which Bates subsequently reconstructed, is currently administered scientifically, it no doubt differs considerably in content from Bates' original collection (Roberts, 1983). It is also likely that during the previous century only small quantities of what eventually proved to be new species were available and authors of plant names were reluctant to press the only specimen of a new element available to them. With no idea of variability along a habitat range, authors of names of succulent plants must also have thought it easy to eventually link published names to natural populations. This is of course often impossible. Furthermore, not even the deposition of a herbarium specimen is always sufficient to assure the correct application of a published name, the confused interpretation of the name *H. pearsonii* C.H. Wright, which was published as recently as 1907, being a good example (Scott, 1980; Bayer, 1982). As Bayer (1982) and Heath (1989, 1990) have clearly shown with regard to the application of the name '*Haworthia pumila*' to the largest known species of *Haworthia*, confusion usually reigns where Linnaeus and other early taxonomists working on succulent plants did not give a clear lead (Scott, 1978; Tjaden, 1985; Wijnands, 1985). In many of the genera included in the Aloaceae, nomenclatural and taxonomic confusion still exist, Bayer (1982), for example, recognizing 68 species in *Haworthia* while Scott (1985) upholds 88 species in the same genus. A comparison of two major monographs on *Aloe* which appeared shortly apart, namely that of Groenewald (1941) and Reynolds (1950) also illustrates how different researchers could draw vastly different conclusions with regard to species concepts and the correct application of previously published names in the same alooid genus.

The value of early taxonomic work on the Aloaceae is often difficult to assess since it might be valuable in one group but valueless in another. Some of the publications of these early workers were also not as valuable as other publications by the same author. Remarking on Salm-Dyck's *Catalogue raisonne' des Especes de' Aloes* that appeared in 1817, Reynolds (1950) concluded that 'The taxonomist and *Aloe* student would have been spared many puzzles had this work never been published.' In contrast, Schelpe (1958) claims that Salm-Dyck's *Monographia Generum Aloes et Mesembryanthemi* which appeared periodically between 1836 and 1863 '...is undoubtedly the most valuable single publication to the modern student of *Gasteria*.'

KEY TO THE GENERA

The following key and illustrations (Figures 2.1–2.7) should facilitate identification of the seven genera here included in the Aloaceae. From a phytogeographical perspective it should be noted that, although some of the genera included in the family are found elsewhere—mainly *Aloe* which ranges from the southern tip of Africa to the Arabian Peninsula and some of the islands, such as Socotra and Madagascar, off the east African coast and *Lomatophyllum* which occurs on the Mascarene Islands and Madagascar—the present-day centre of generic and species diversity of the Aloaceae fundamentally is in southern Africa.

1a.	Fruit a berry	1. *Lomatophyllum*
1b.	Fruit a capsule	2
2a.	Perianth segments apically connivent	2. *Poellnitzia*
2b.	Perianth segments apically spreading or recurved	3
3a.	Flowers pendulous at anthesis, perianth tube curved upwards	3. *Gasteria*
3b.	Flowers erect, suberect or spreading at anthesis, perianth tube straight or curved downwards	4
4a.	Capsule apically acuminate, underground parts distinctly bulbous	4. *Chortolirion*
4b.	Capsule apically rounded or obtuse, underground parts rhizomatous	5
5a.	Perianth bilabiate, <15 mm long, mouth not upturned	5. *Haworthia*

Figure 2.1 Lomatophyllum occidentale: A, habit (reduced); B, leaf, ×0.4: C, terminal part of inflorescence, ×0.4. Artist: G. Condy.

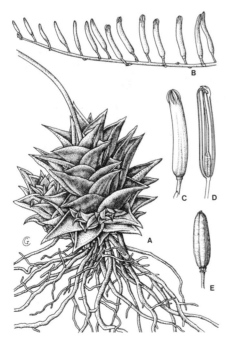

Figure 2.2 Poellnitzia rubriflora: A, habit, ×0.4; B, terminal part of inflorescence, ×0.4; C & D, flower, ×0.8: E, fruit, ×0.6. Artist: G. Condy.

Figure 2.3 Gasteria baylissiana: A, habit, ×0.5; B, terminal part of infloresence, ×0.5. Artist: G. Condy.

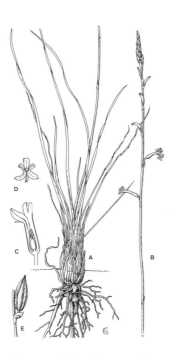

Figure 2.4 Chortolirion angolense: A, habit, ×0.24; B, inflorescence, ×0.24; C, longitudinal section of flower, one stamen removed, ×1; D, flower face showing reflexed segments, ×0.6; E, fruit, ×0.5. Artist: G. Condy.

Figure 2.5 Haworthia geraldii: A, habit, ×0.6; B, terminal part of inflorescence, ×0.6. Artist: G. Condy.

Figure 2.6 Aloe arborescens: A, habit (reduced); B, leaf, ×0.1; C, inflorescence, ×0.1. Artist: R. Holcroft.

Figure 2.7 Astroloba bullulata: A, habit, ×0.5; B, terminal part of raceme, ×0.5. Artist: G. Condy.

5b. Perianth regular (if rarely weakly bilabiate then flowers
   >15 mm long, mouth upturned)   6
6a. Flowers usually brightly coloured, fleshy, stamens as long as
   or longer than the perianth   6. *Aloe*
6b. Flowers dull-coloured, flimsy, stamens included   7. *Astroloba*

REFERENCES

Admiraal, J. (1984) *Die psalmodie van die plante. Plante in en om die Bybel*. pp. 11–12. Mediese
   Universiteit van Suid-Afrika, Medunsa.
Alston, R.E. and Turner, B.L. (1963) Biochemical systematics. In *Plant taxonomy. A brief history
   of major developments in the field*. Englewood Cliffs, New Jersey: Prentice-Hall, Inc.
Anonymous. (1967) Editorial: Gilbert Reynolds. *The Cape Times*, 10 April 1967.
Baker, J.G. (1880) A synopsis of Aloineae and Yuccoideae. *Journal of the Linnean Society of London
   (Botany)*, 18, 148–241.
Baker, M.A., Mohlenbrock, M.W. and Pinkawa, D.J. (1985) A comparison of two new methods of
   preparing cacti and other stem succulents for standard herbarium mounting. *Taxon*, 34, 118–121.
Batsch, A.J.G.C. (1802) *Tabula affinitatum regni vegetabilis*, p.138. Weimar: Landes-Industrie
   Comptair.
Bauhin, J. (1651) Historiae universalis plantarum. In *Herbae crassifoliae et succulentae*, Vol. 3,
   Book 35, pp. 677–712. Ebroduni.
Bayer, M.B. (1972) Anguish among the haworthias. *Cactus and Succulent Journal of Great Brittain*,
   34, 35–37.
Bayer, M.B. (1975) Notes on *Astroloba herrei* Uitewaal. *Aloe*, 13, 99–100.

Bayer, M.B. (1976) Haworthia *handbook*. Kirstenbosch: National Botanic Gardens of South Africa.

Bayer, M.B. (1978) Letter to the editor. *Cactus and Succulent Journal (U.S.)*, **50**, 72.

Bayer, M.B. (1982) *The new* Haworthia *handbook*. Kirstenbosch: National Botanic Gardens of South Africa.

Bayer, M.B. (1986) *Haworthia* and nomenclatural confusion. *British Cactus and Succulent Journal*, **4**, 45–47.

Bayer, M.B. (1999) Haworthia *revisited*. Pretoria: Umdaus Press.

Berger, A. (1905) Über die systematische Gliederung der Gattung *Aloë*. *Botanische Jahrbücher für Systematik, Planzengeschichte und Pflanzengeographie*, **36**, 42–68.

Berger, A. (1908) Liliaceae–Asphodeloideae–Aloineae. Vegetationorgane. In *Das Pflanzenreich*, **33**, edited by A. Engler and K. Prantl, pp. 5–6. Leipzig: Engelmann.

Bertrand, A. (1956) *Guillauminia albiflora*. *Cactus (Paris)*, **49**, 41–42.

Bradley, R. (1716–1727) *The history of succulent plants*. Vol. 1–5; Vol. 1–3 published by the author, Vol. 4 and 5 published by William Mears, London, Vol. 2, p.1, t. 11.

Brandham, P.E. (1971) The chromosomes of the Liliaceae: II. Polyploidy and karyotype variation in the Aloineae. *Kew Bulletin*, **25**, 381–399.

Brandham, P.E. (1974) The chromosomes of the Liliaceae: III. New cases of interchange hybridity in the Aloineae. *Kew Bulletin*, **28**, 341–348.

Brandham, P.E. (1976) The frequency of spontaneous structural change. In *Kew chromosome conference I, Current chromosome research* edited by K. Jones and P.E. Brandham, pp. 77–87. Amsterdam: Elsevier.

Brandham, P.E. (1977) The inheritance of leaf pigmentation in *Gasteria* (Liliaceae). *Kew Bulletin*, **32**, 13–17.

Brenan, J.P.M. (1963) Agavaceae. *Sansevieria guineensis* (L.) Willd. *Kew Bulletin*, **17**, 174–177.

Breuer, I. (1998) *The world of haworthias. Vol 1. Bibliography and annotated text*. Niederzier: Ingo Breuer and Abkreis für Mammillarienfreunde e.V.

Brummit, R.K. (1992) *Vascular plant families and genera*, pp. 696–697. Kew: Royal Botanic Gardens.

Butterfield, H.M. (1968) Richard Bradley's names for succulent plants and some names for flower parts (1716–1726). *Cactus and Succulent Journal*, **40**, 53–56.

Cardot, J. (1909) *Aloinella* (Musci–Pottiaceae). *Revue Bryologique*, **36**, 76.

Chudovska, O. (1979a) Ein neues Konservierungsverfahren für Sukkulenten. I. Die konservierung von Blattsukkulenten. *Kakteen und andere Sukkulenten*, **30**, 94–96.

Chudovska, O. (1979b) Ein neues Konservierungsverfahren für Sukkulenten. II. Die konservierung von Stammsukkulenten. *Kakteen und andere Sukkulenten*, **30**, 101–107.

Codd, L.E. (1968) The South African species of *Kniphofia*. *Bothalia*, **9**, 363–513.

Cronquist, A. (1981) *An integrated system of classification of flowering plants*, p. 1215. New York: Columbia University Press.

Cronquist, A. (1988) *The evolution and classification of flowering plants*, 2nd edn, pp. 485, 516. New York: New York Botanical Gardens.

Cutler, D.F., Brandham, P.E., Carter, S. and Harris, S.J. (1980) Morphological, anatomical, cytological and biochemical aspects of evolution in East African shrubby species of *Aloe* L. (Liliaceae). *Journal of the Linnean Society (Botany)*, **80**, 293–317.

Dahlgren, R.M.T. and Clifford, H.T. (1982) *The monocotyledons: a comparative study*, p. 28. London London: Academic Press.

Dahlgren, R.M.T., Clifford, H.T. and Yeo, P.F. (1985) *The families of the monocotyledons: structure, evolution and taxonomy*, pp. 179–182. Berlin: Springer-Verlag.

Dandy, J.E. (1970) Annotated list of the names published in Allioni's *Auctarium ad Sydnopsim Stirpium Horti Reg. Taurinensis. Taxon*, **19**, 626–627.

D'Ewes, D. (1967) Still a place for an amateur. *Journal of the Botanical Society of South Africa*, **53**, 29–30.

Donoghue, M.J. (1987) Experiments and hypotheses in systematics. *Taxon*, **36**, 584–587.

Duval, H.A. (1809) *Plantae succulentae, in Horto Alenconio*, p. 6. Paris: Gabon et Socios.

Endlicher, S. (1836) *Genera plantarum secundum ordines naturales disposa*, p. 144. Wien: Beck.

Forster, P.I. and Clifford, H.T. (1986) *Aloeaceae*. In *Flora of Australia*, Vol. 46, edited by A.S. George, pp. 66–77. Canberra: Australian Government Publishing Service.

Fuller, T.C. and Barbe, G.D. (1981) A microwave-oven method for drying succulent plant specimens. *Taxon*, 30, 867.

Funk, V.A. (1985) Cladistics and generic concepts in the Compositae. *Taxon*, 34, 72–80.

Gibbs-Russell, G.E., Reid, C., Van Rooy, J. and Smook, L. (1985) List of species of southern African plants (2nd edn). Recent literature and synonyms. Part 1. Cryptograms, gymnosperms and monocotyledons. In *Memoirs of the botanical survey of South Africa*, No. 51, pp. 1–152. Pretoria: Department of Agriculture and Water Supply.

Gilmartin, A.J. (1986) Experimental systematics today. *Taxon*, 35, 118–119.

Gilmartin, A.J. (1987) Experimental systematics–La Duke's refreshing counterpoint of view. *Taxon*, 36, 65.

Glen, H.F. and Hardy, D.S. (2000) Aloaceae. In *Flora of southern Africa*, edited by G. Germishuizen, Vol. 5, Fasc. 1, Part 1, pp. 1–167. Pretoria: National Botanical Institute.

Glen, H.F., Meyer, N.L., Van Jaarsveld, E.J. and Smith, G.F. (1997) Aloaceae. In *List of southern African succulent plants*, edited by G.F. Smith, E.J. Van Jaarsveld, T.H. Arnold, F.E. Steffens, R.D. Dixon and J.A. Retief, pp. 6–11. Pretoria: Umdaus Press.

Groen, L.E. (1986) *Astroloba* Uitewaal. *Succulenta*, 65, 19–23.

Groen, L.E. (1987a) *Astroloba* Uitewaal. (II). *Succulenta*, 66, 51–55.

Groen, L.E. (1987b) *Astroloba* Uitewaal. (III). *Succulenta*, 66, 82–87.

Groen, L.E. (1987c) *Astroloba* Uitewaal. (IV). *Succulenta*, 66, 110–113.

Groen, L.E. (1987d) *Astroloba* Uitewaal. (V). *Succulenta*, 66, 162–167.

Groen, L.E. (1987e) *Astroloba* Uitewaal. (VI). *Succulenta*, 66, 171–174.

Groen, L.E. (1987f) *Astroloba* Uitewaal. (conclusion). *Succulenta*, 66, 261–263.

Groenewald, B.H. (1941) *Die aalwyne van Suid-Afrika, Suidwes-Afrika, Portugees Oos-Afrika, Swaziland, Basoetoeland en 'n spesiale ondersoek van die klassifikasie, chromosome en areale van die Aloe maculatae*. Bloemfontein: Nasionale Pers.

Gunn, M.D. and Codd, L.E. (1981) *Botanical exploration of southern Africa.* Cape Town: Balkema.

Haworth, A.H. (1819) *Supplementum plantarum succulentarum.* London: Harding.

Haworth, A.H. (1821) *Revisiones plantarum succulentarum.* London: Taylor.

Haworth, A.H. (1824) Decas secunda novarum plantarum succulentarum. *Philosophical Magazine*, 64, 298–302.

Heath, P.V. (1989) Yet a further example of the autonym rules – the case of *Haworthia margaritifera* (Asphodelaceae). *Taxon*, 38, 481–483.

Heath, P.V. (1990) Proposals to emend the Code. (2–6) Five proposals to clarify the Code. Proposal (2). Amend Article 6.6. *Taxon*, 39, 138–139.

Heath, P.V. (1993) New generic names in the Asphodelaceae. *Calyx*, 3, 153.

Higgins, V. (1940) In the Lindley Library. II. An early book on succulent plants. *Journal of the Royal Horticultural Society*, 65, 115–117.

Holland, P.G. (1978) An evolutionary biography of the genus *Aloe*. *Journal of Biogeography*, 5, 213–226.

Hooker, J.D. (1867) *Bowiea volubilis* W.H. Harvey ex J.D. Hooker. *Curtis's Botanical Magazine*, 93, t. 5619.

Horwood, F.K. (1971) Exotic xerophytes. 7. Some notes on the succulents of Socotra, Part 2. *National Cactus and Succulent Journal*, 26, 106–110.

Hutchinson, J. (1959) The families of flowering plants, Vol. 2, Monocotyledons, 2nd edn, pp. 601–602. Oxford: Clarendon Press.

Jackson, B.D. (1895) *Index Kewensis, an enumeration of the genera and species of flowering plants*, Vol. 1. Oxford: Clarendon Press.

Jacobsen, H. (1986) *A handbook of succulent plants*, 2, Ficus *to* Zygophyllum, 4th edn, pp. 674–676. Poole, Dorset: Blandford Press.

Ker, J.B. (1813) *Phylloma aloiflorum*. The Bourbon Aloe. Class and Order Hexandria Monogynia. *Curtis's Botanical Magazine*, **37**, t. 1585.

La Duke, J.C. (1987) The existence of hypotheses in plant systematics or biting the hand that feeds you. *Taxon*, **36**, 60–64.

Lawrence, G.H.M. (1951) *Taxonomy of vascular plants*. pp. 13–42. New York: MacMillan Publishing Co.

Lavranos, J.J. (1969) The genus *Aloe* in the Socotra Archipelago, Indian Ocean; a revision. *Cactus and Succulent Journal (U.S.)*, **41**, 202–207.

Lemeé, A.M.V. (1939) *Dictionnaire descriptif et synonymique des genres de plantes phanérogames*, 7, Suppl. Paris: Chevalier.

Leuenberger, B.E. (1982) Microwaves: a modern aid in preparing herbarium specimens of succulents. *Cactus and Succulent Journal of Great Britain*, **44**, 42–44.

Linnaeus, C. (1753) *Species plantarum*, Vol. 1. Stockholm: Impensis Laurentii Salvii.

Logan, J. (1986) A pre-pressing treatment for *Begonia* species and succulents. *Taxon*, **35**, 671–684.

Mabberley, D.J. (1987) *The plant book. A portable dictionary of the higher plants*, pp. 21. Cambridge: Cambridge University Press.

Manning, J.C. and Smith, G.F. (2000) The genus *Poellnitzia* included in *Astroloba*. *Bothalia*, **30**, 53.

Marloth, R. (1925) *The flora of South Africa with synoptical tables of the genera of the higher plants*, Vol. 2, Section 1, p. 17, Podostemonaceae–Umbelliferae. Cape Town: Darter Bros. and Co.

Medikus, F.K. (1786) *Theodora speciosa, ein neues Pflanzen Geschlecht*, pp. 67, 69. Mannheim: Neue Hof- und Akademische Buchhandlung.

Merxmüller, H. (1972) Systematic botany – an achieved synthesis. *Biological Journal of the Linnean Society*, **4**, 311–321.

Newton, L.E. (1972) Taxonomic use of the cuticular surface features in the genus *Aloe* (Liliaceae). *Journal of the Linnean Society (Botany)*, **65**, 335–339.

Newton, L.E. (2001) *Aloe*. In *Illustrated handbook of succulent plants: Monocotyledons*, edited by U. Eggli, pp. 103–186. Berlin: Springer-Verlag.

Osborne, R., Grobbelaar, N. and Vorster, P. (1988) South African cycad research: progress and prospects. *South African Journal of Science*, **84**, 891–896.

Rafinesque, C.S. (1840) *Autikon botanikon*, 2nd century, vi–x. Philadelphia: Privately published.

Resende, F. (1943) Succulentas Africanas III. Contribucao para o estudo da morfologia, da fisiologia da floracao e da geno-sistemática das Aloineae. *Memórias da Sociedade Broteriana*, **3**, 1–119.

Retief, E. (1997) Asphodelaceae Juss. In *List of southern African succulent plants*, pp. 34–36, edited by G.F. Smith, E.J. Van Jaarsveld, T.H. Arnold, F.E. Steffens R.D. Dixon and J.A. Retief Pretoria: Umdaus Press.

Reynolds, G.W. (1950) *The aloes of South Africa*. Johannesburg: The aloes of South Africa book fund.

Reynolds, G.W. (1958) *Les aloes de Madagascar. Revision*. Tananarive: Institut de Recherche Scientifique de Madagascar.

Reynolds, G.W. (1966) *The aloes of tropical Africa and Madagascar*. Mbabane: The trustees of the aloes book fund.

Reynolds, G.W. (1982) *The aloes of South Africa*, 4th edn. Cape Town: Balkema.

Riley, H.P. and Majumdar, S.K. (1979) *The Aloineae: a biosystematic survey*. Lexington: University Press of Kentucky.

Roan, H.M. (1948) An appreciation of John Thomas Bates. *National Cactus and Succulent Journal*, **3**, 99–100.

Roberts, M. (1983) Experience with a reference collection. *Cactus and Succulent Journal*, **55**, 212–215.

Roberts Reynecke, P. (1965) The genus *Astroloba* Uitewaal (Liliaceae). M.Sc. Thesis, University of Cape Town, Cape Town, South Africa.

Rowley, G.D. (1951) Studies in the Ficoidaceae. II. Adrian Hardy Haworth and his observations. *National Cactus and Succulent Journal*, **6**, 48–49.

Rowley, G.D. (1952) News from the herbaria. A history of succulent plants. *Taxon*, 1, 134.

Rowley, G.D. (1954a) Cytology and the succulent. *National Cactus and Succulent Journal*, 9, 15–19.

Rowley, G.D. (1954b) Richard Bradley and his 'History of succulent plants' (1716–1727). *Cactus and Succulent Journal of Great Britain*, April–July, 1954, 1–7. (Reprint with corrections).

Rowley, G.D. (1960) A short history of succulent plants. In *A Handbook of Succulent Plants*, H. Jacobsen, Vol. 1, Chapter. 1, pp. 1–66. London: Blandford Press.

Rowley, G.D. (1967) A numerical survey of the genera of Aloineae. *National Cactus and Succulent Journal*, 22, 71–77.

Rowley, G.D. (1976a) In quest of the genus. *Cactus and Succulent Journal of Great Britain*, 38, 53–56.

Rowley, G.D. (1976b) The rise and fall of the 'Succulentae'. *Cactus and Succulent Journal (U.S.)*, 48, 184–189.

Rowley, G.D. (1976c) Generic concepts in the Aloineae. Part 1. The genus – to split or to lump? *National Cactus and Succulent Journal*, 31, 26–31.

Rowley, G.D. (1976d) Generic concepts in the Aloineae. Part 2. Generic names in the Aloineae. *National Cactus and Succulent Journal*, 31, 54–56.

Rowley, G.D. (1977) Was Richard Bradley an evolutionist? *Cactus and Succulent Journal of Great Britain*, 39, 95–96.

Rowley, G.D. (1983) Dedication to Richard Bradley, F.R.S. *Bradleya*, 1, 1–2.

Rowley, G.D. (1985) *The* Haworthia *drawings of John Thomas Bates.* Essex: The Succulent Plant Trust.

Rowley, G.D. (1996) The berried aloes: *Aloe* Section *Lomatophyllum*. *Excelsa*, 17, 59–62.

Salisbury, R.A. (1866) *The genera of plants. A fragment containing part of Liriogamae. Order 11. Aloeae.* London: Van Voorst.

Schelpe, E.A.C.L.E. (1958) *Gasteria* – a problem genus of South African succulent plants. *Journal of the Botanical Society of South Africa*, 44, 17–22.

Schultes, J.A. and Schultes, J.H. (1829) Classis VI, Hexandria Monogynia Genera, 1417. *Aloe*. In *Systema vegetabilium*, edited by J.J. Roemer and J.A. Schultes, Vol. 7, 1, pp. 631–715. Stuttgardtiae: Sumptibus J.G. Cottae.

Scott, C.L. (1978) The correct application of the name *Haworthia pumila* (L.) Duval. *Taxon*, 16, 44–46.

Scott, C.L. (1980) *Haworthia pearsonii* C.W. Wright. *Aloe*, 18, 7–8.

Scott, C.L. (1985) *The genus* Haworthia *(Liliaceae): a taxonomic revision.* Johannesburg: Aloe Books.

Sharma, A.K. and Mallick, R. (1966) Interrelationships and evolution of the tribe Aloineae as reflected in its cytology. *Journal of Genetics*, 59, 20–47.

Small, E. (1989) Systematics of biological systematics (or, taxonomy of taxonomy). *Taxon*, 38, 335–356.

Smith, G.F. (1989) Notes on *Haworthia glabrata* (Salm-Dyck) Baker. *Aloe*, 26, 18–22.

Smith, G.F. (1990) Nomenclatural notes on the subsection *Bowieae* in *Aloe* (Asphodelaceae: Alooideae). *South African Journal of Botany*, 56, 303–308.

Smith, G.F. (1993) Familial orthography: Aloeaceae vs. Aloaceae. *Taxon*, 42, 87–90.

Smith, G.F. (1994) Taxonomic history of *Poellnitzia* Uitewaal, a unispecific genus of Alooideae (Asphodelaceae). *Haseltonia*, 2, 74–78.

Smith, G.F. (1995a) *FSA contributions* 2: Asphodelaceae/Aloaceae, 102910 *Chortolirion*. *Bothalia*, 25, 1, 31–33.

Smith, G.F. (1995a) *FSA contributions* 2: Asphodelaceae/Aloaceae, 1028010 *Poellnitzia*. *Bothalia*, 25, 1, 35–36.

Smith, G.F. and Marx, G. (1990) Notes on the vegetation and succulent flora of the eastern Cape Province, South Africa. *Aloe*, 27, 56–66.

Smith, G.F. and Van Wyk, A.E. (1989) Biographical notes on James Bowie and the discovery of *Aloe bowiea* Schult and J.H. Schult. (Alooideae: Asphodelaceae). *Taxon*, 38, 557–568.

Smith, G.F. and Van Wyk, A.E. (1993) Notes on the pollen morphology and taxonomy of *Aloe bowiea* (Asphodelaceae: Alooideae). *Madoqua*, 18, 93–99.

Smith, G.F. and Van Wyk, B.-E. (1991) Generic relationships in the Alooideae (Asphodelaceae). *Taxon*, 40, 557–581.

Smith, G.F. and Van Wyk, B.-E. (1998) Asphodelaceae. In *Vascular plant genera of the world*, edited by K. Kubitzki, Volume 3, Lilianae, pp. 130–140. Berlin: Springer-Verlag.

Smith, G.G. (1948) Views on the naming of haworthias. *Journal of South African Botany*, 14, 55–62.

Spies, J.J. and Hardy, D.S. (1983) A karyotype and anatomical study of an unidentified liliaceous plant. *Bothalia*, 14, 215–217.

Stace, C.A. 1980. *Plant taxonomy and biosystematics*, p. 22. London: Edward Arnold.

Stafleu, F.A. and Cowan, R.S. (1976) *Taxonomic literature*, Vol. 1: A–G, 2nd edn. London: Bohn, Scheltema and Holkema.

Stapf, O. (1933) *Leptaloë albida. Curtis's Botanical Magazine*, 156, t. 9300.

Stearn, W.T. (1939a) Generic name proposed for conservation. 1028 (Liliac.) *Apicra* Haworth, Suppl.Pl. Succ. (1819). *Bulletin of Miscellaneous Information, Royal Botanic Gardens, Kew*, 7, 329–330.

Stearn, W.T. (1939b) Conservation of the name *Apicra. The Cactus Journal (Cactus & Succulent Society of Great Britain)*, 8, 27–28.

Steyn, E.M.A., Smith, G.F., Nilsson, S. and Grafstrom, E. (1998) Pollen morphology in *Aloe* (Aloaceae). *Grana*, 37, 23–27.

Thomas, H.H. (1952) Richard Bradley, an early eighteenth century biologist. *Bulletin. British Society for the History of Science*, 1, 176–178.

Tjaden, W. (1985) A choice of name: *Haworthia margaritifera* or *Haworthia pumila?* British *Cactus and Succulent Journal*, 3, 88.

Turner, B.L. (1967) Plant chemosystematics and phylogeny. *Pure and Applied Chemistry*, 14, 198–213.

Uitewaal, A.J.A. (1940) Een nieuw geslacht der Aloineae. *Succulenta*, 22, 61–64.

Uitewaal, A.J.A (1947) Revisie van de nomenclatuur der genera *Haworthia* en *Apicra. Succulenta (Amsterdam) 1947*, (5), 51–54.

Van Jaarsveld, E.J. 1994. *Gasterias of South Africa. A new revision of a major succulent group.* Cape Town: Fernwood Press in association with the National Botanical Institute.

Viljoen, A.M. and Van Wyk, B.-E. (1996) The evolution of aloes: new clues from their leaf chemistry. *Aloe*, 33, 30–33.

Viljoen, A.M., Van Wyk, B.-E. and Newton, L.E. (2001) The occurrence and taxonomic distribution of the anthrones aloin, aloinoside and microdontin. *Biochemical Systematics and Ecology*, 29, 53–67.

Von Poellnitz, J.K.L.A. (1939) A new species of *Haworthia. Cactus Journal*, 8, 19.

Wijnands, D.O. (1973) Typification and nomenclature of two species of *Sansevieria* (Agavaceae). *Taxon*, 22, 109–114.

Wijnands, D.O. (1983) *The botany of the Commelins.* Rotterdam: Balkema.

Wijnands, D.O. (1985) Nomenclatural aspects of the plants pictured by Jan and Caspar Commelin with three proposals to conserve or reject. *Taxon*, 34, 307–315.

Willdenow, C.L. (1811) Bemerkungen über die Gattung *Aloë. Der Gesellschaft naturforschender Freunde zu Berlin Magazin für die neuesten Entdeckungen in der gesammten Naturkunde*, 5 (**Neue Schreibe**), 163–168; 269–283.

Part 2

Aloe constituents

3   Aloe chemistry

Tom Reynolds

ABSTRACT

When the leaves of most species of *Aloe* are cut a more or less copious exudate appears, yellow at first but rapidly darkening to brown or in a few species dark red. This exudate contains phenolic compounds which can be distinguished chromatographically as over 80 major zones staining characteristic colors with fast blue B, a dye reacting with phenols and coupling amines. Some of the compounds in these zones have been characterized. Most of the exudate compounds identified so far are chromone, anthraquinone or anthrone derivatives. Some are widespread in the genus, and some are confined to a few species and therefore of potential chemotaxonomic value.

These phenolics do not occur in the parenchyma cells within the leaf, where polysaccharides and glycoproteins are characteristic. These substances are dealt with in other chapters, so are only summarised here.

INTRODUCTION

The chemistry of the aloe plant has been studied for many years from a number of viewpoints. The leaf, the most frequently studied organ, can be divided into the outer green mesophyll, including the vascular bundles and the inner colourless parenchyma containing, to various degrees, the well known aloe gel. When a typical aloe leaf is cut there appears on the cut surface an exudate arising from cells adjacent to the vascular bundles (Beaumont *et al.*, 1985), which is usually yellow-brown, which in a few species, eg. *A. confusa* Engl. can change to a deep blood red. This exudate from certain species, when dried, is the bitter aloes of commerce, used as a bittering agent or as a somewhat violent purgative. Early work identified the bitter, purgative factor as an anthrone-C-glucoside, barbaloin (Birch and Donovan, 1965; Hay and Haynes, 1956). Then, other compounds in the product were recognized and eventually characterized (reviewed Reynolds, 1985a). Subsequently, when a wider range of species was examined using thin-layer chromatography, a larger number more than 80-chromatographic zones were revealed, representing mostly unidentified compounds (reviewed Reynolds, 1985b). These zones were given code numbers to distinguish them and to aid further discussion. Since then many of these constituents have been isolated and characterized (reviewed Dagne, 1996). The distribution of these entities, either as chromatographic zones or as characterized compounds has been used for chemotaxonomic discussions. Their biological activities remain largely unknown but might be expected to yield interesting results.

Interest in the parenchyma gel has centered on its well-known therapeutic properties, although there is some evidence of chemotaxonomic variation in the polysaccharides and no doubt in the lectins. A certain amount of research has established the characteristics of some of these gel components but they are largely unknown in species away from the few well-known, widely distributed ones.

LEAF EXUDATE COMPOUNDS

Chromones

Most of the chromones so far described from aloe leaf exudates are derivatives of 8-C-glucosyl-7-hydroxy-5-methyl-2-propyl-4-chromone. Variation arises from the degree of oxidation in the propyl side-chain, methylation of the hydroxyl group on C7 and esterification of the glucose moiety.

Of the aglycones (Figure 3.1), aloesone was reported as a minor component from 11 *Aloe* species (Holdsworth, 1972), while the reduction product, aloesol occurs in *Rheum* (Kashiwada *et al.*, 1984). Methylation of the 7-hydroxyl group has been achieved synthetically (Gramatica *et al.*, 1986) but has not been observed among aglycones from aloe exudates. A derivative in which the 7-hydroxyl group is cyclised into a furan ring at C8 of the chromone ring has been observed in Cape Aloes (*A. ferox* Mill.) and named furoaloesone (Figure 3.2) (Speranza *et al.*, 1993b). A simpler 7-hydroxy-5-methyl-chromone with a methyl group on C2 has been described from Polygonaceae, an ascomycete and then more recently from Cape Aloe (Speranza *et al.*, 1993).

Two 5,7-dihydroxy-4-chromones have been found in aloes. One from *A. vera* was the 8-C-glucoside of the 2-methyl derivative, 8-C-glucosylnoreugenin (Okamura *et al.*, 1998). The other, an aglycone from *A. cremnophila* Reynolds and P.R.O Bally ('cremno-chromone'), had an acrylic acid residue on C2 (Conner *et al.*, 1990a).

Aloesin (Figure 3.3), which could be regarded as the parent compound of the aloe chromones was described first (Haynes *et al.*, 1970) and is widespread through the genus, occurring in 35% (Reynolds, 1985b) or 46% (Rauwald *et al.*, 1991) of species

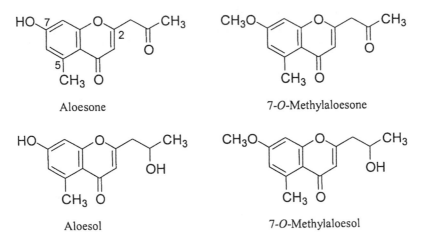

Aloesone

7-*O*-Methylaloesone

Aloesol

7-*O*-Methylaloesol

Figure 3.1 Chromone aglycones of aloe compounds.

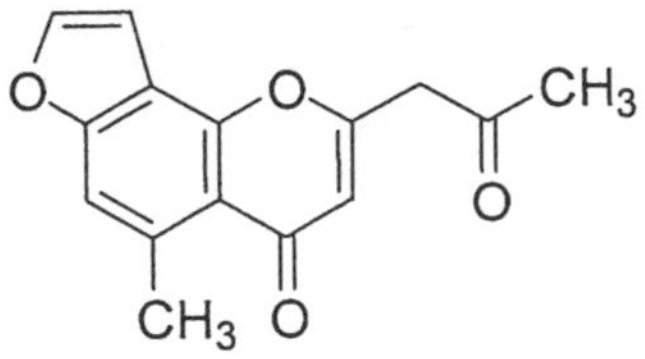

Furoaloesone

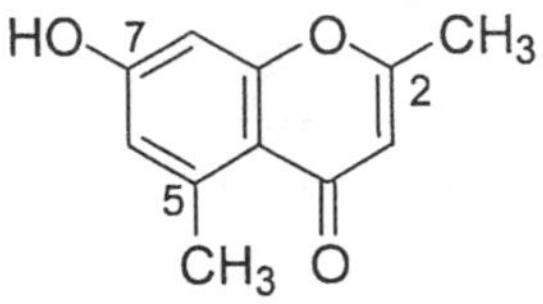

7-Hydroxy-2,5-dimethylchromone

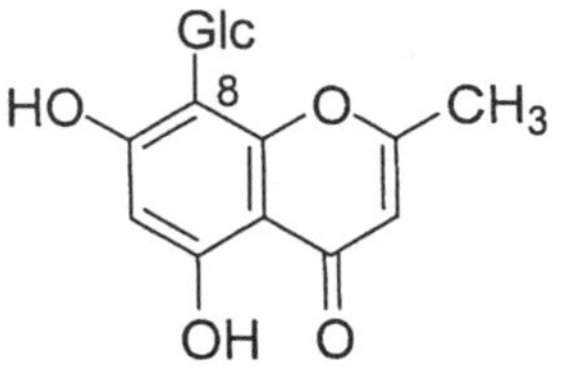

8-*C*-Glucosylnoreugenin

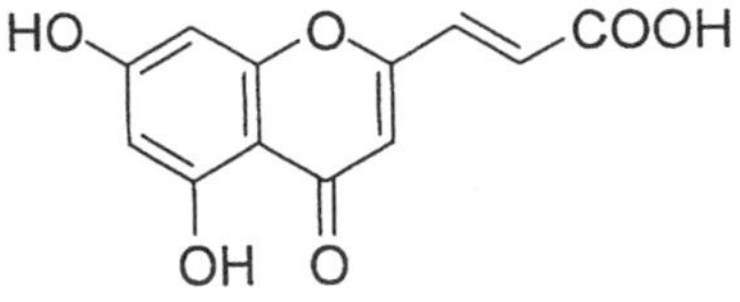

Cremnochromone

Figure 3.2 Other chromone aglycones from aloes.

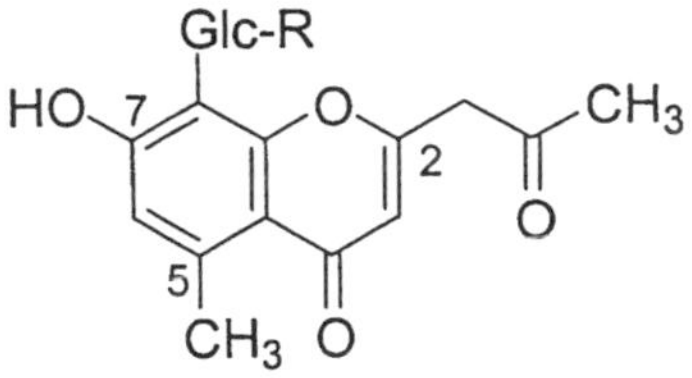

Aloesin

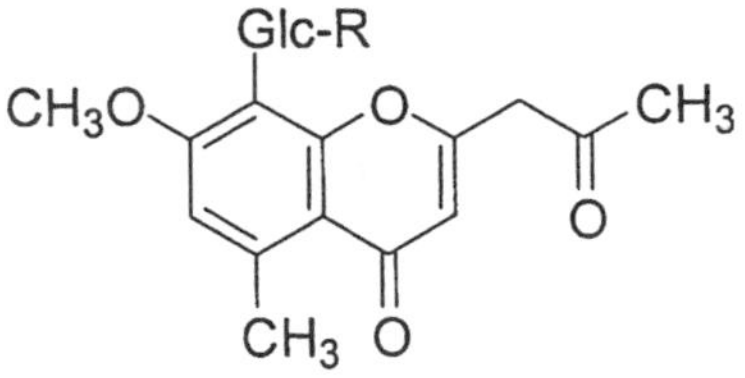

7-*O*-Methylaloesin

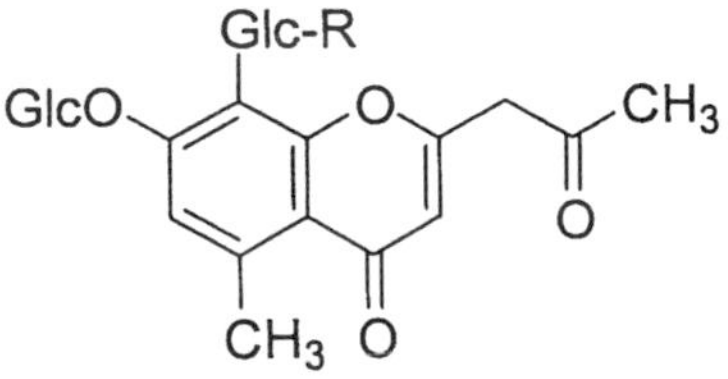

7-*O*-Glucosylaloesin

Figure 3.3 Aloesin and its derivatives.

examined and often in some quantity. A variant with the *C*-glucosyl residue in the furanose form has been described from *A. vera* (L.)Burm.f. as neoaloesin A (Park *et al.*, 1996). An isomer with the *C*-glucosyl group on the 6-position was described from *A. vera* var.*chinensis* (Haw.) A.Berger and named *iso*-aloesin (Axing, 1993).

The 7-*O*-methyl derivative has only recently been found (Bisrat *et al.*, 2000) in *A. rupestris* Baker. Esterification of the glucose moiety of both these forms and of the 7-*O*-glucosyl molecule has been observed with cinnamic, *p*-coumaric, *O*-methyl-*p*-coumaric, caffeic, ferulic and tiglic acids, although not every combination occurs (Table 3.1), the 2'-*p*-coumaric acid ester of aloesin being the one most commonly

Table 3.1 Acylated derivatives of aloesin.

Acyl group	7-Hydroxy-	7-Methoxy-	7-Glucosyl-
Unsubstituted	Aloesin (Aloeresin B) 35–46% of *Aloe* species Haynes *et al.*, 1970	7-*O*-Methoxyaloesin *A. rupestris* Bisrat *et al.*, 2000 *A. rubroviolacea* *A. perryi* Schmidt *et al.*, 2001	
	neoAloesin A *A. vera* Park *et al.*, 1996		
2'-*p*-Coumaroyl	Aloeresin A Makino *et al.*, 1974	7-*O*-Methylaloeresin A *A. marlothii* Bisrat *et al.*, 2000 *A. rubroviolacea* *A. perryi* Schmidt *et al.*, 2001	Aloeresin C *A. ferox, A. vera* Speranza *et al.*, 1985 Rauwald *et al.*, 1997
	isoAloeresin A *A. ferox* Speranza *et al.*, 1988		
3'-*p*-Coumaroyl	*A. lutescens* Van Heerden *et al.*, 2002		
6'-*p*-Coumaroyl	*A. castanea* Van Heerden *et al.*, 2000		
2', 6-Di-*p*-coumaroyl	*A. speciosa* Holzapfel *et al.*, 1997		
3', 6'-Di-*p*-coumaroyl	*A. lutescens* Van Heerden *et al.*, 2002		
2'-*O*-Methyl-*p*-coumaroyl	*A. excelsa* Mebe, 1987		
2'-Cinnamoyl	Aloeresin F *A. peglerae* van Heerden *et al.*, 1996 *A. rubroviolacea* *A. perryi* Schmidt *et al.*, 2001	*A. broomii* Holzapfel *et al.*, 1997 *A. rubroviolacea* *A. perryi* Schmidt *et al.*, 2001	Aloeresin E(sic) *A. peglerae* van Heerden *et al.*, 1996
2'-(*E*)-Cinnamoyl	*A. vera* Rauwald *et al.*, 1997		
2'-Caffeoyl		*A. broomii* Holzapfel *et al.*, 1997	
2'-Feruloyl	*A. arborescens* Makino *et al.*, 1974	*A. africana* Holzapfel *et al.*, 1997 *A. rubroviolacea* *A. perryi* Schmidt *et al.*, 2001	
2'-Tiglyl	*A. cremnophila, A. jacksonii* Conner *et al.*, 1990a		
2'-[3'', 4''-Dimethylcaffeoyl]		*A. rubroviolacea* *A. perryi* Schmidt *et al.*, 2001	

found. The coumaric acid residue occurs most usually in the *E* form, although the *Z* form has been observed as a minor component of Cape Aloes (*A. ferox*) and termed isoaloeresin A (Speranza *et al.*, 1988). The conformations of many of these structures have been determined in some detail (Manito *et al.*, 1990a). Some of the chromones have been given trivial names based on 'aloeresin x.' Thus aloesin was designated aloeresin B and its 2'-*O*-(*E*)-*p*-coumaroyl ester named aloeresin A. Some confusion has arisen with aloeresin E which was described in September 1996 as the 2'-*O*-cinnamoyl ester of 7-*O*-methylaloesol (Okamura *et al.*, 1996) and then in November 1996 as the 2'-*O*-cinnamoyl ester of 7-*O*-glucosylaloesin (van Heerden *et al.*, 1996).

Reduction of the keto group on carbon 10 gives rise conceptually to another series of compounds, derivatives of aloesol, some of which have been recognized in aloes but not to the extent of the aloesin derivatives (Table 3.2) (Okamura *et al.*, 1997). Further oxidation of the propyl side-chain to give a propanediol structure has also been observed in aloe compounds (Table 3.3) (Okamura *et al.*, 1997, 1998). Methylation of the hydroxyl group on carbon 7 occurs and gives rise to yet another series of aloe compounds (Table 3.1). *O*-glucosylation has been recorded in two instances (Table 3.1), although one imagines that some more will come to light.

A simpler structure where the propyl side-chain is replaced by a methyl group has been reported from *Polygonum* (Kimura *et al.*, 1983) and *Rheum* (Kashiwada *et al.*, 1984) and more recently from Cape Aloes drug (Speranza *et al.*, 1993).

Variants in which the methyl group on carbon 5 is replaced by an hydroxyl group is represented in aloes by cremnochromone and its 8-*C*-glucoside (Conner *et al.*, 1990a) and by the 8-*C*-glucoside of noreugenin (Okamura *et al.*, 1998) (Figure 3.2).

Anthraquinones and anthrones

Free anthraquinones and anthrones have been observed in some *Aloe* species but are not a major component of leaf exudates as they are found in greater variety in the roots and subterranean stems (Yagi *et al.*, 1974; Dagne, 1994; Van Wyk *et al.*, 1995).

The compounds are derivatives of either 1, 8-dihydroxy-3-methyl-anthraquinone (chrysophanol) (Figure 3.4) or 3, 8-dihydroxy, 1-methyl-anthraquinone (aloesaponarin II) (Figure 3.5). Although it was early noted in drug aloes (Hörhammer *et al.*, 1965), chrysophanol had rarely been reported from aloe leaves (e.g. *A. vera*, Chopra and Ghosh, 1938; *A. saponaria*, Rheede van Oudtshoorn, 1963, 1964) but was found in the underground stems of *A. saponaria* Haw. (Yagi *et al.*, 1977), while a summary claimed its presence in unspecified organs of 8 *Aloe* species (Hammouda *et al.*, 1977). It was then found in the leaves of *A. berhana* Reynolds (=*A. debrana* Christian) *A. rivae* Baker, *A. megalacantha* Baker and *A. pulcherrima* M.G.Gilbert and Sebsebe (Dagne and Alemu, 1991) and then in the roots of *A. berhana* (Dagne *et al.*, 1992) and *A. graminicola* Reynolds (Yenesaw *et al.*, 1993), and then a larger survey showed it in 32 species (Dagne *et al.*, 1994). This was followed by an even larger survey, where it was demonstrated in 162 species (Van Wyk *et al.*, 1995a). Chrysophanol was subsequently found in the roots of the related genera, *Lomatophyllum*, *Asphodelus*, *Asphodeline*, *Bulbine*, *Bulbinella* and *Knifophia* (Van Wyk *et al.*, 1995b, c). Chrysophanol-8-*O*-methyl ether was found recently in the roots of *A. berhena* (Dagne *et al.*, 1992) and then in 20 other species (Dagne *et al.*, 1994).

Aloe emodin itself and its anthrone have infrequently been described from *Aloe* species but are known in quite unrelated plants (Reynolds, 1985a), although the glycosides are well known and widely distributed (Groom and Reynolds, 1989). This

Table 3.2 Derivatives of 8-*C*-glucosylaloesol.

Acyl group	7- Hydroxy-	7- Methoxy-
Unsubstituted	Aloesinol *A. vera* Okamura *et al.*, 1997	7-*O*-methylaloesinol *A. vera* Okamura *et al.*, 1996 *A. rubroviolacea* *A. perryi* Schmidt *et al.*, 2001
2'-*p*-Coumaroyl		Aloeresin D *A. vera* Speranza *et al.*, 1986b Lee *et al.*, 2000 *A. rabaiensis* Conner *et al.*, 1989 *A. rubroviolacea* *A. perryi* Schmidt *et al.*, 2001 isoAloeresin D *A. vera* Okamura *et al.*, 1996
4'-Glucosyl-2'-*cis*-*p*-coumaroyl		4'-*O*-glucosyl-isoaloeresin DI
4'-Glucosyl-2'-*trans*-*p*-coumaroyl		4'-*O*-glucosyl-isoaloeresin DII *A. vera* Okamura *et al.*, 1998
2'-Cinnamoyl	*A. rubroviolacea* *A. perryi* Schmidt *et al.*, 2001	Aloeresin E *A. vera* Hutter *et al.*, 1996, (E) form Okamura *et al.*, 1996, (S) form
2'-Methoxycinnamoyl	*A. rubroviolacea* *A. perryi* Schmidt *et al.*, 2001	
2'-Caffeoyl		Rabaichromone *A. rabaiensis* Conner *et al.*, 1989
2'[3'',4''-Dimethyl-caffeoyl]		*A. rubroviolacea* *A. perryi* Schmidt *et al.*, 2001
2'-Tiglyl		*A. rubroviolacea* *A. perryi* Schmidt *et al.*, 2001
2'-Feruloyl		*A. rubroviolacea* *A. perryi* Schmidt *et al.*, 2001
2',4'-*p*-Methoxybenzoyl ('Dragonyl')		*A. rubroviolacea* *A. perryi* Schmidt *et al.*, 2001

Table 3.3 Derivatives of 8-*C*-glucosylaloediol (aloesindiol).

Acyl group	7- Hydroxy-	7- Methoxy-	7-Glucosyl
Unsubstituted		7-*O*-Methylaloesindiol *A. vera* Okamura *et al.*, 1997	
2'-Cinnamoyl		2'-*O*-Cinnamoyl-7-*O*- methyl-aloesindiol A 2'-*O*-Cinnamoyl-7-*O*- methyl-aloesindiol B *A. vera* Okamura *et al.*, 1998 *A. rubroviolacea* *A. perryi* Schmidt *et al.*, 2001	

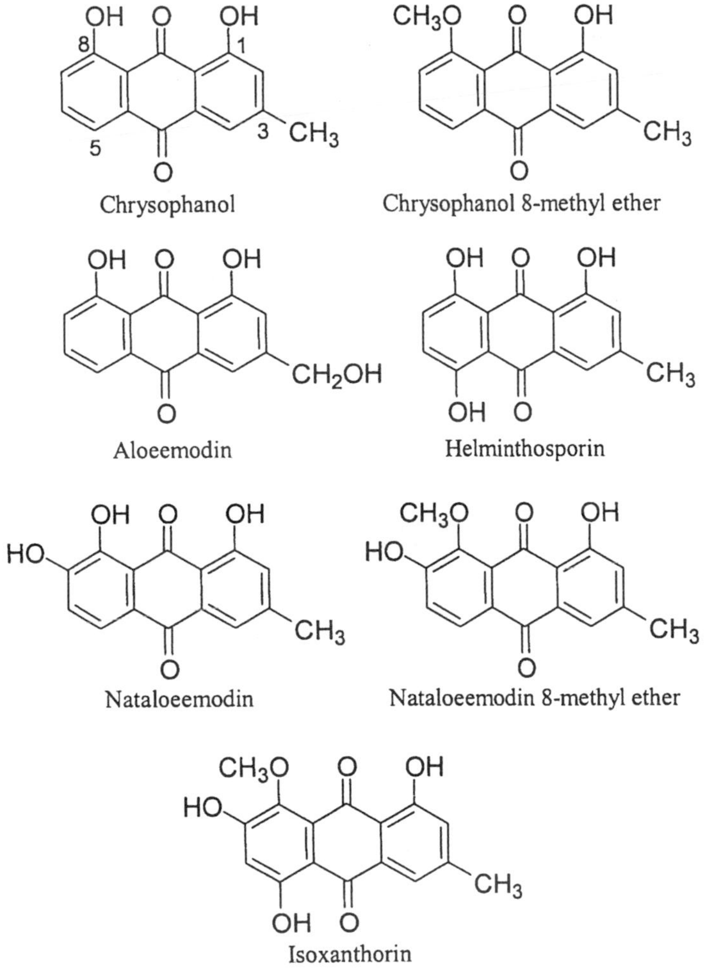

Figure 3.4 Chrysophanol (3-methyl,1,8-dihydroxy-anthraquinone) and its derivatives.

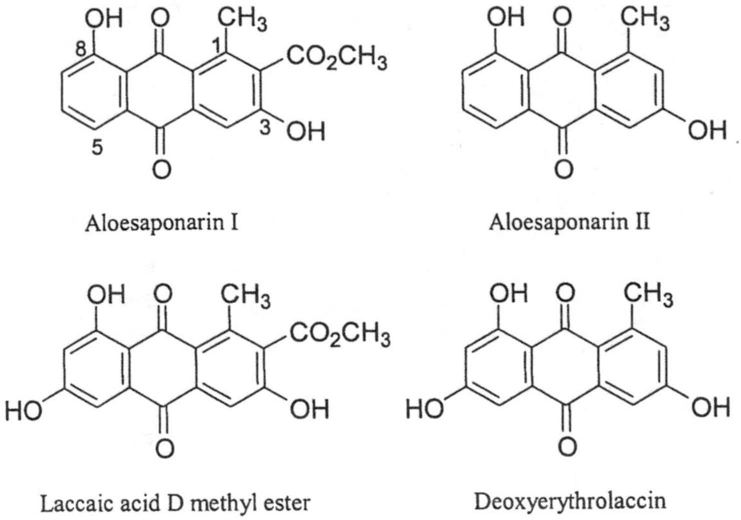

Aloesaponarin I

Aloesaponarin II

Laccaic acid D methyl ester

Deoxyerythrolaccin

Figure 3.5 Derivatives of 1-methyl-3,8-dihydroxy-anthraquinone.

may be because anthraquinones are a minor component of leaf exudates and tend to be overlooked in analyses. Early observations, usually by paper chromatography, showed it in Cape aloes drug (*A. ferox*) (Awe *et al.*, 1958; Hörhammer *et al.*, 1965) and it was then isolated from this product (Koyama *et al.*, 1994). It was demonstrated in the leaves of *A. africana* Mill., *A. marlothii* Berger and *A. pretoriensis* Pole Evans (Rheede van Oudtshoorn, 1964) and also in *A. elgonica* Bullock (Conner *et al.*, 1990b) and in *A. arborescens* Mill. (Constantinescu *et al.*, 1969; Hirata and Suga, 1977; Kodym, 1991; Yamamoto *et al.*, 1991). There are also records of it being found in *A. vera* leaves (Choi *et al.*, 1996; Saleem *et al.*, 1997b; Strickland *et al.*, 2000; Pecere *et al.*, 2000), but not in the roots of any species.

The first record of helminthosporin and isoxanthorin in aloes was in underground stems of *A. saponaria* (Yagi *et al.*, 1977a). Helminthosporin was then found in the roots of 27 species (Dagne *et al.*, 1994). The anthraquinone aglycone of the other common anthrone-*C*-glucoside, homonataloin, is the 8-*O*-methyl ether of nataloe-emodin and has been reported from *A. lateritia* Engl. (Rauwald and Niyonzima, 1991a), while nataloe-emodin itself was isolated from *A. nyeriensis* var. *kedongensis* (Reynolds) S. Carter leaves (Conner *et al.*, 1987).

Aloesaponarin II, an isomer of chrysophanol, where the position of the groups on carbon atoms 1 and 3 is reversed, was reported from *A. saponaria* (Yagi *et al.*, 1974) and then in the roots of 97 *Aloe* species (Van Wyk *et al.*, 1995a).

It was also found in five species of *Lomatophyllum* (Van Wyk *et al.*, 1995c). Its 6-hydroxy derivative, deoxyerythrolaccin was also found in *A. saponaria* and then in *A. ferox* (as Cape aloes) (Koyama *et al.*, 1994). Aloesaponarin I and laccaic acic D methyl ester, the methyl esters of derivatives of these compounds containing a carboxyl group on carbon atom 2, were first found in *A. saponaria* (Yagi *et al.*, 1974), and were subsequently observed in the roots of over 100 species (Van Wyk *et al.*, 1995a) including some in *Lomatophyllum*.

Both chrysophanolanthrone and aloe-emodinanthrone occur in aloe flowers (Rauwald and Beil, 1993; Sigler and Rauwald, 1994b). The first occurs in *A. bakeri* Scott-Elliot, *A. vaombe* Decorse and Poiss., *A. khamiesensis* Pillans and *A. dawei* A.Berger and the second in *A. bakeri*, *A. vaombe*, *A. ballyi* Reynolds and *A. dawei*. It is noteworthy that no trace was found in the leaf exudates.

Tetrahydroanthracenones

The compounds in which the C-ring is reduced are typical of the subterranean stems and roots of aloes and mirror in some of their substitution patterns the leaf anthraquinones (Figure 3.6). In fact it was postulated that the conversion of aloesaponol I to aloesapon-

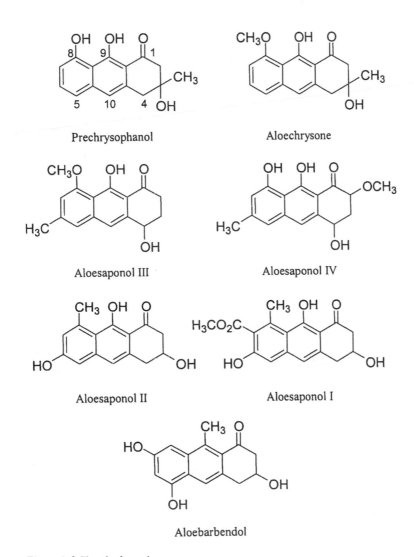

Prechrysophanol

Aloechrysone

Aloesaponol III

Aloesaponol IV

Aloesaponol II

Aloesaponol I

Aloebarbendol

Figure 3.6 Tetrahydroanthracenes.

arin I was a step in the biosynthetic process (Yagi *et al.*, 1978a). Subsequently it was shown that tetrahydroanthracene glycosides were converted to anthraquinone glycosides in the light (Yagi *et al.*, 1983). In contrast, however, aloe emodin appears to be synthesized directly from acetate units (Simpson, 1980; Grün and Franz, 1982).

Aloechrysone was isolated from roots of *A. berhana* Reynolds (Dagne *et al.*, 1992) and subsequently found in 66 species (Van Wyk *et al.*, 1995a) and then in *A. vera* (Saleem *et al.*, 1997b). This and prechrysophanol from subterranean stems of *A. graminicola* Reynolds (Yenesew *et al.*, 1993) are seen as having the chrysophanol pattern subjected to reduction, a double keto-enol transformation and migration of a hydroxyl group and it is postulated that prechrysophanol may be a biosynthetic precursor. Similarly aloesaponol I is related to aloesaponarin I and aloesaponol II to aloesaponarin II. These two compounds were originally isolated from *A. saponaria* (Yagi *et al.*, 1974) and then found in the roots of 32 species (Dagne *et al.*, 1994). In a further survey, aloesaponol I was found in 70 species and aloesaponol II in 113 species (Van Wyk *et al.*, 1995a). Aloesaponol III and aloesaponol IV (Yagi *et al.*, 1977) relate structurally to helminthosporin and isoxanthorin, respectively. Again, these were first isolated from *A. saponaria,* while aloesaponol III was then found in 31 other species (Dagne *et al.*, 1994). Recently three tetrahydroanthracenones, gasteriacenones A, B and C, with related substitution patterns have been isolated from the neighbouring genus *Gasteria* (Figure 3.6) (Dagne *et al.*, 1996b). A more remote structure, aloebarbendol (Figure 3.6), was recently described from roots of *A. vera* (Saleem *et al.*, 1997b).

From the investigations described above it appears that tetrahydroanthracenes occur in the roots accompanied by several anthraquinones (cf. Sigler and Rauwald, 1994a) and that the leaves contain some anthraquinones, notably aloe-emodin, while the flowers contain free anthrones. Much more prominent in the leaves are the *O*- and *C*-glycosides. The next research step might be a quantitative survey of the compounds in the various organs of a few species to complement the chemotaxonomic surveys.

Anthraquinone and anthrone O-glycosides

The *O*-glycosides are surprisingly not often reported as aloe constituents although they occur in several other plants. Aloe-emodin-*O*-galactoside and aloe-emodin-11-*O*-rhamnoside were observed on leaf exudate chromatograms (reviewed Reynolds, 1985). More recently, 7-*O*-glucosylnataloe-emodin was isolated from *A. nyerienis* Christian ex I.Verd. (Conner *et al.*, 1987) and 11-*O*-rhamnosyl aloe-emodin from *A. rabaiensis* Rendle (Conner *et al.*, 1989). The 1-*O*-glucoside of homonataloin B has been characterized from *A. lutescens* Groenew.A (Van Heerden *et al.*, 2002) and identified in a further 13 species (Viljoen *et al.*, 2002), while a diglycoside of *O*-methoxy-nataloe-emodin-8-methyl ether has been previously described from *A. vera*.

Tetrahydroanthracenone glycosides

The 6-*O*-glucosides of aloesaponol I and II were described from *A. saponaria* (Yagi *et al.*, 1977) as well as the 8-*O*-glucoside aloesaponol III (Yagi *et al.*, 1977). Subsequently the 4-*O*-glucosides of aloesaponol III and IV were isolated from *A. barbadensis* callus tissue (Yagi *et al.*, 1998).

Anthrone-C-glycosides

These compounds are considered typical of aloe leaf exudate constituents although they do not, in fact, occur in all species (Reynolds, 1985b). Where they do occur they are usually the main component of the exudate and are mostly represented by either barbaloin or homonataloin (reviewed Reynolds, 1985a). These two compounds appear to be mutually exclusive in the leaf exudates but have been observed together in *A. mutabilis* Pillans (Reynolds, 1990; Chauser-Volfson and Gutterman, 1998). Barbaloin (≡aloin) is the bitter principle in drug aloes and was characterized as the C-glycoside of aloe-emodin anthrone (Figure 3.7).

Barbaloin has been noted in at least 68 *Aloe* species at levels from 0.1 to 6.6% of leaf dry weight, (making between 3% and 35% of the total exudate) (Groom and

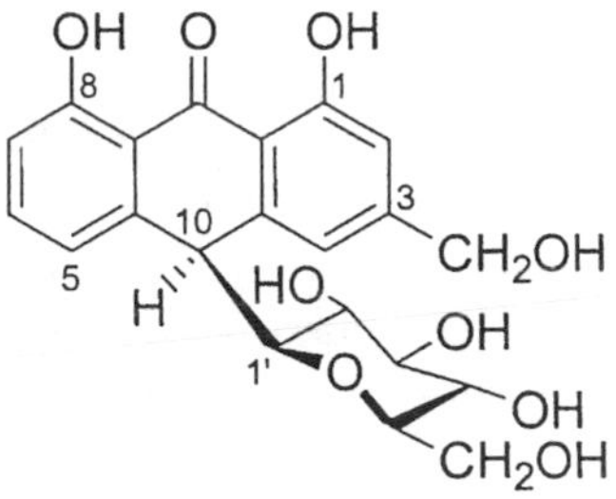

Aloin A (barbaloin A) (10*S*, 1'*S*)

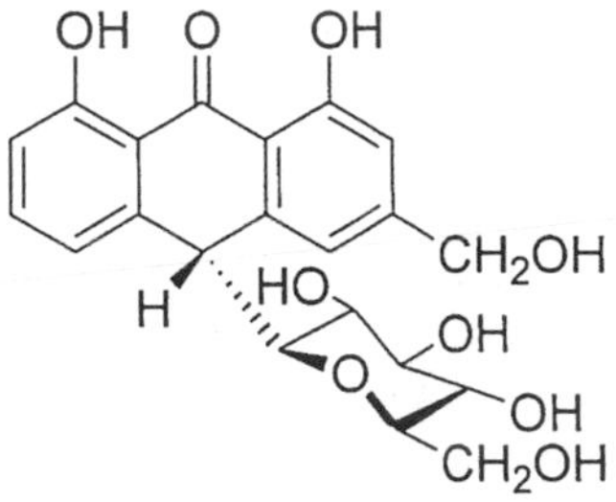

Aloin B (barbaloin B) (10*R*, 1'*S*)

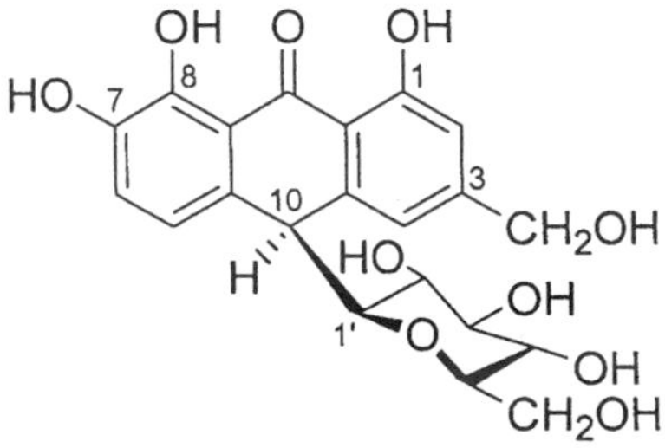

7-Hydroxyaloin A (isobarbaloin)

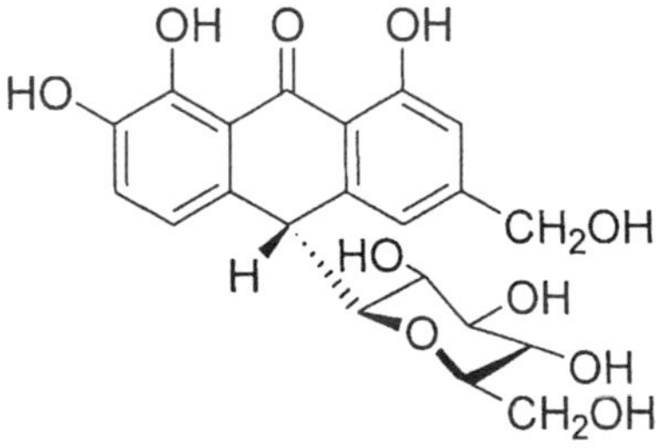

7-Hydroxyaloin B

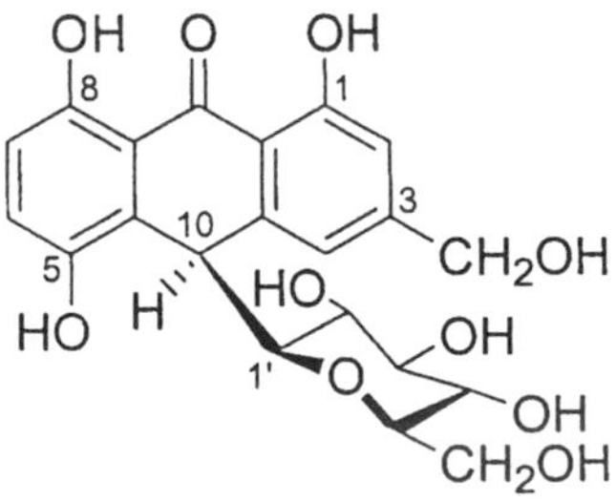

5-Hydroxyaloin A ('periodate-positive substance')

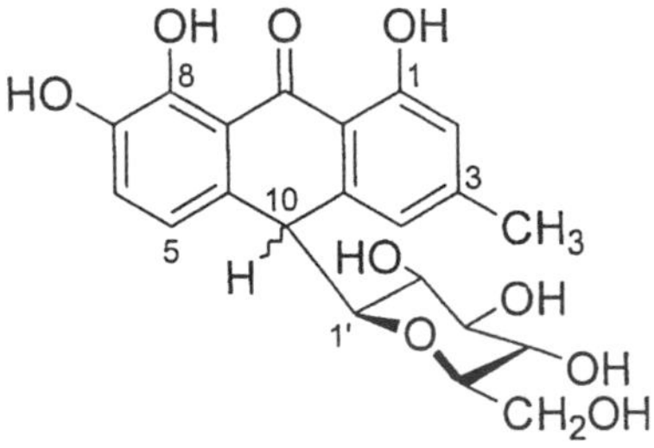

Nataloin

Figure 3.7 C-Glycosides of aloe-emodin anthrone.

Reynolds, 1987) and in another 17 species at indeterminate levels (Reynolds, 1995b). In one species, *A. pubescens* Reynolds the barbaloin content was higher in the younger leaves. In another study using *A. arborescens*, levels of 0.4% in the younger leaves and 0.2% in the older leaves were observed (Chauser-Volfson and Gutterman, 1996). Barbaloin is easily separated by HPLC or DCCC into two stereoisomers (Auterhoff *et al.*, 1980; Rauwald, 1982). The chromatographically faster component, barbaloin B was shown to have the 10*R*, 1'*S* configuration at C10, while the slower barbaloin A was the 10*S*, 1'*S* diastereomer (Rauwald *et al.*, 1989; Manitto *et al.*, 1990b), confirmed by molecular modelling (Höltje *et al.*, 1991). Barbaloin B slowly changes to barbaloin A *in vitro*.

The *C*-glycoside of nataloe-emodin anthrone, nataloin (Figure 3.8) was claimed to have been found along with homonataloin in obsolete Natal aloes (Leger, 1917) and then isolated again (Rosenthaler, 1931), although the analysis figures given do not tally with subsequent work. Much later a structure was assigned to the anthraquinone and its 8-*O*-methyl ether (Haynes and Henderson, 1960). It was not noted in plants until a partially characterized anthrone-*C*-glycoside was isolated from *A. nyeriensis* spp. *nyeriensis* Christian (Reynolds, 1986) and subsequently characterized by NMR spectroscopy (Conner *et al.*, 1987). The compound has since been reported from *A. pulcherima* M.G. Gilbert and Sebsebe (Dagne and Alemu, 1991) and from some plants of *A. mutabilis*, where it replaces homonataloin (Chauser-Volfson and Gutterman, 1998). Much more common among *Aloe* spp is its 8-*O*-methyl ether homonataloin.

Homonataloin is the *C*-glycoside of the 8-*O*-methyl ether of nataloe-emodin anthrone and is the principle of the obsolete Natal aloes (Figure 3.8) (Haynes *et al.*, 1960). It was observed in 34 *Aloe* species (Reynolds, 1985b) making up between 14% and 47% of the dried exudate (Beaumont *et al.*, 1984). As found for barbaloin, levels were higher in the younger leaves. Elsewhere, exudate from leaves of *A. hereroensis* Engler was found to contain homonataloin at levels from 8% in older leaves to 39% in young leaves (Chauser-Volfson and Gutterman, 1997). This compound also proved to exist as two stereoisomers, described from *A. cremnophila* as (−) and (+) isomers (Conner *et al.*, 1990a) and also from *A. lateritia* Engler (Rauwald, 1990). The configuration of the two diastereomers was subsequently assigned as 10*R*, 1'*S* (homonataloin B) and 10*S*, 1'*S* (homonataloin A) using material isolated from *A. lateritia* (Rauwald and Niyonzima, 1991a). A hydroxylated derivative of homonataloin isolated from *A. vera* was characterized as the 8-*O*-methyl ether of 7-hydroxybarbaloin, also existing in the two diastereomeric

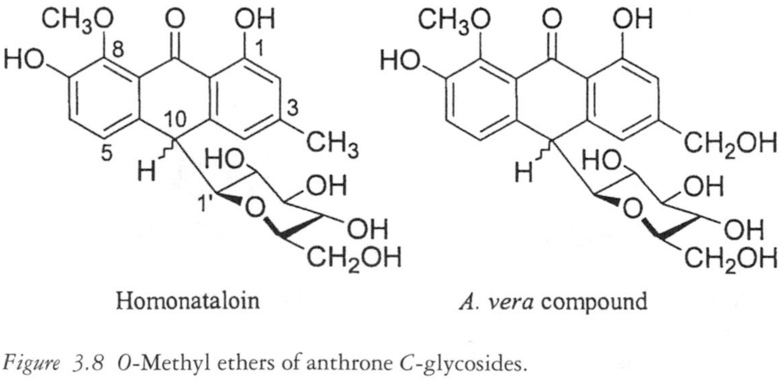

Figure 3.8 *O*-Methyl ethers of anthrone *C*-glycosides.

forms mentioned above (Rauwald and Niyonzima, 1991b). They were accompanied by their 6'-cinnamoyl esters.

The barbaloin molecule is subject to hydroxylation at several different carbon atoms and some of these entities have been observed in aloe exudates (Figure 3.7). A chromatographic zone from *A. vera* was named 'isobarbaloin' (Barnes and Holfeld, 1956) and later characterized as the 7-hydroxy derivative (Rauwald and Voetig, 1982), or perhaps its 8-O-methyl ether (Rauwald, 1990). It was also shown to occur as two diastereomers, 7-hydroxybarbaloin A (10S, 1'S) and 7-hydroxybarbaloin B (10R, 1'S) (Rauwald, 1990) and was found in 13% of 183 species examined (Rauwald *et al.*, 1991c). Another previously unidentified chromatographic zone, 'periodate-positive substance' (Böhme and Kreutzig, 1963) was characterized as 5-hydroxybarbaloin which occurred only as the B-configuration, 10R, 1'S (Rauwald, 1987, 1990; Rauwald and Beil, 1993a), confirmed subsequently by NMR data (Dagne *et al.*, 1997). It was found in 4% of 183 species examined (Rauwald *et al.*, 1991c) and was later isolated from *A. broomii* Schonl. (Holzapfel *et al.*, 1997). HPLC analysis revealed levels of 0.1% to 6.6% of leaf dry weight in seven *Aloe* species (Rauwald and Beil, 1993a, b). The levels from different plants vary considerably, however, as was shown by analysis of a range of *A. ferox* individuals (Van Wyk *et al.*, 1995).

The *C*-glycoside of aloe-emodin anthranol, 10-hydroxybarbaloin, was found in *Rhamnus purshiana* DC (Rauwald *et al.*, 1991) and made by treating barbaloin with ammonia (Rauwald and Lohse, 1992). It occurs also as diastereomers, 10R, 1' R 10-hydroxybarbaloin A and 10S, 1' R 10-hydroxybarbaloin B (Rauwald, 1990) (Figure 3.9). The latter was then isolated from *A. littoralis* Baker (Dagne *et al.*, 1996a) and recognized by HPLC in eight other species (Viljoen *et al.*, 1996). Both were then isolated from *A. vera* (Okamura *et al.*, 1997). This transformation had been reported previously but the product was then identified as 4-hydroxybarbaloin (Graf and Alexa, 1980).

Esterification of the sugar residue has been reported in several species (Figure 3.10). Acetate esters at the 6' and 4', 6' positions of 7-hydroxyaloin from *A. succotrina* Lam. were observed in both leaves and flowers (Rauwald and Diemer, 1986; Rauwald, 1990; Rauwald and Beil, 1993a; Sigler and Rauwald, 1994b). The 6'-O-acetate of 10-hydroxybarbaloin B was isolated and characterized from *A. claviflora* Burch. (Dagne *et al.*, 1998b). The *p*-coumaroyl ester at C6' of barbaloin A and B was reported from *A. vera* (Rauwald, 1990) and then characterized as microdontin A and B from *A. microdonta* Chiov. (Farah and Andersson, 1992), being then determined in 35 other *Aloe* species or 10% of the aloes surveyed (Viljoen *et al.*, 2001). The same esters of 7-hydroxybarbaloin were reported from *A. vera* (Rauwald, 1987, 1990). The 6'-O-caffeoyl ester of 5-hydroxybarbaloin A was characterized from *A. microstigma* Salm-Dyck (Dagne *et al.*, 1997) as microstigmin A and then found in four other species (Viljoen and Van Wyk, 2001). The 6'-O-cinnamoyl ester of 5-hydroxybarbaloin A was isolated from *A. broomii* (Holzapfel *et al.*, 1997) and the 6'-acetyl ester from *A. marlothii* (Bisrat *et al.*, 2000).

Esterification at C15 of 10-hydroxy barbaloin has been observed with the unusual substance nilic acid (3-hydroxy-2-methylbutanoic acid) in littoraloin, from *A. littoralis*, which also has an acetate group on C6' (Dagne *et al.*, 1996a). It was accompanied by the deacetylated compound (deacetyllittoraloin). Subsequently the 3'-O-glucoside of littoraloin was characterized as littoraloside (Dagne *et al.*, 1998a) (Figure 3.11). Two stereoisomeric 15-O-rhamnosides of barbaloin have long been known from *A. ferox* as aloinosides A and B (Hörhammer *et al.*, 1964; Rauwald, 1990) (Figure 3.11) and were

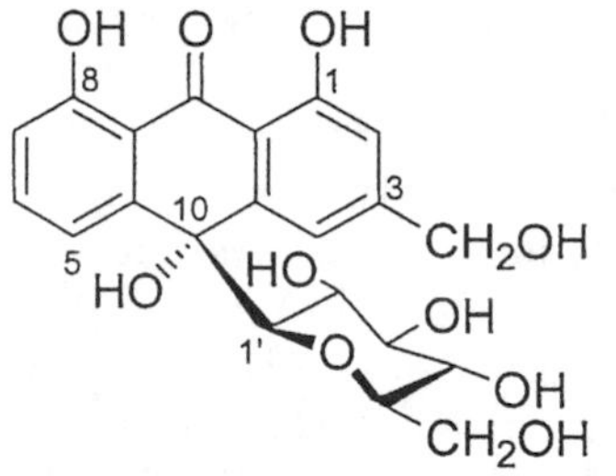

10-Hydroxyaloin A

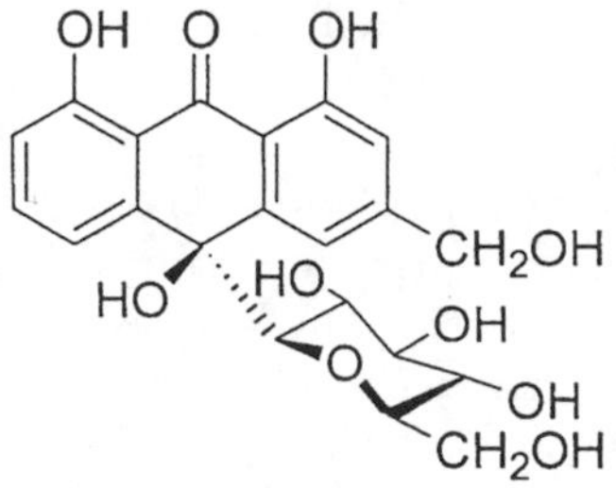

10-Hydroxyaloin B

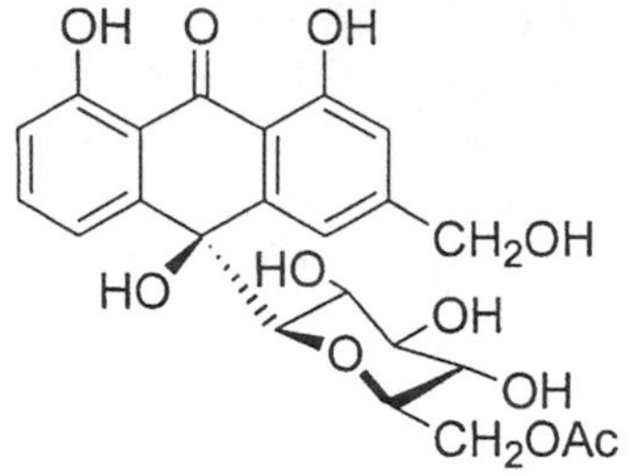

A. claviflora compound

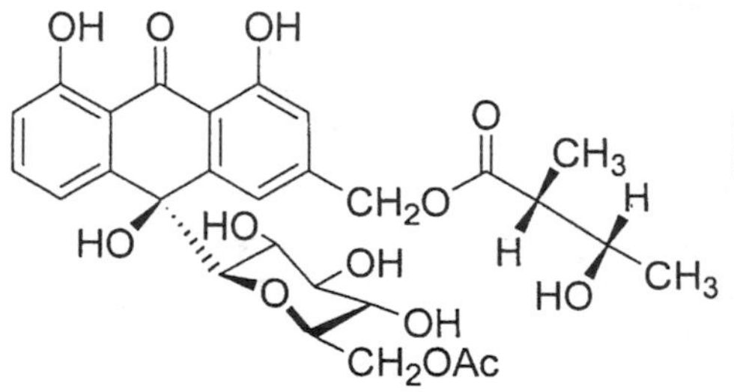

Littoraloin

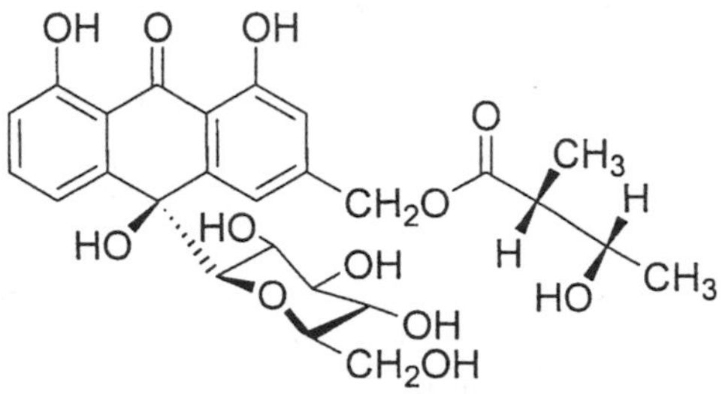

Deacetyllittoraloin

Figure 3.9 Derivatives of aloe-emodin anthranol.

subsequently reported from 16 *Aloe* species (Rauwald *et al.*, 1991c) and then from 33 species (Viljoen *et al.*, 2001).

An artifact found in some drug aloe samples could be prepared from barbaloin and acetone in acidic solution and was shown to be barbaloin-dimethylketal (Graf *et al.*, 1980). An artifact of aloenin, produced by hot-air drying of an *A. arborescens* sample, was characterized as 4',6'-O-ethylidene-aloenin (Woo *et al.*, 1994)

Dimeric compounds

A series of compounds have been described from aloes and other plants (reviewed Reynolds, 1985) where instead of a sugar, a second substituted anthracene unit is attached by a C-C bond, although not always at the same position on the ring as with the C-glycosides. These compounds have been found first in two *Aloe* species, four

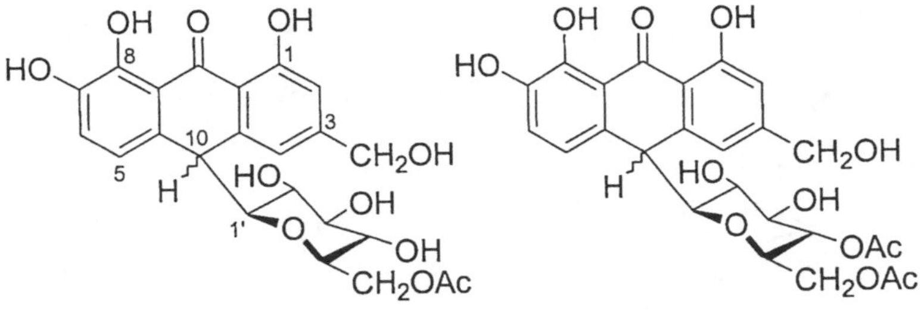

A. succotrina compounds

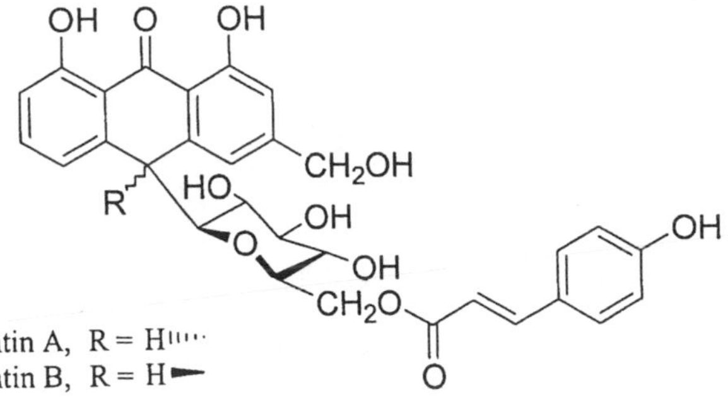

Microdontin A, R = H︙
Microdontin B, R = H►

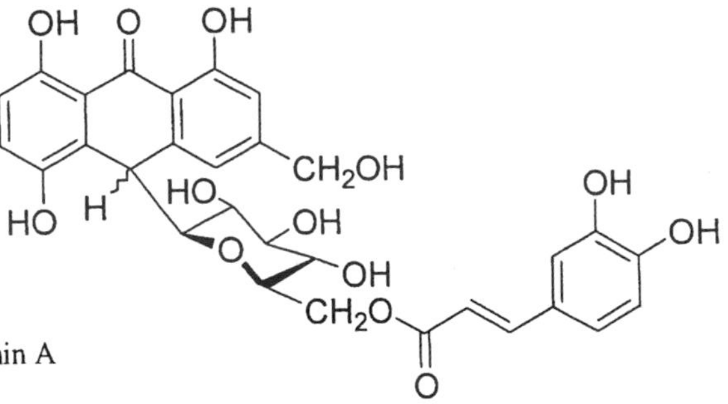

Microstigmin A

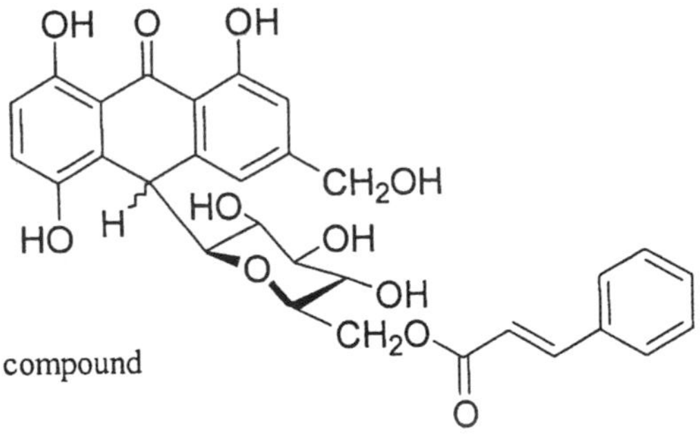

A. broomi compound

Figure 3.10 Esters of barbaloin derivatives.

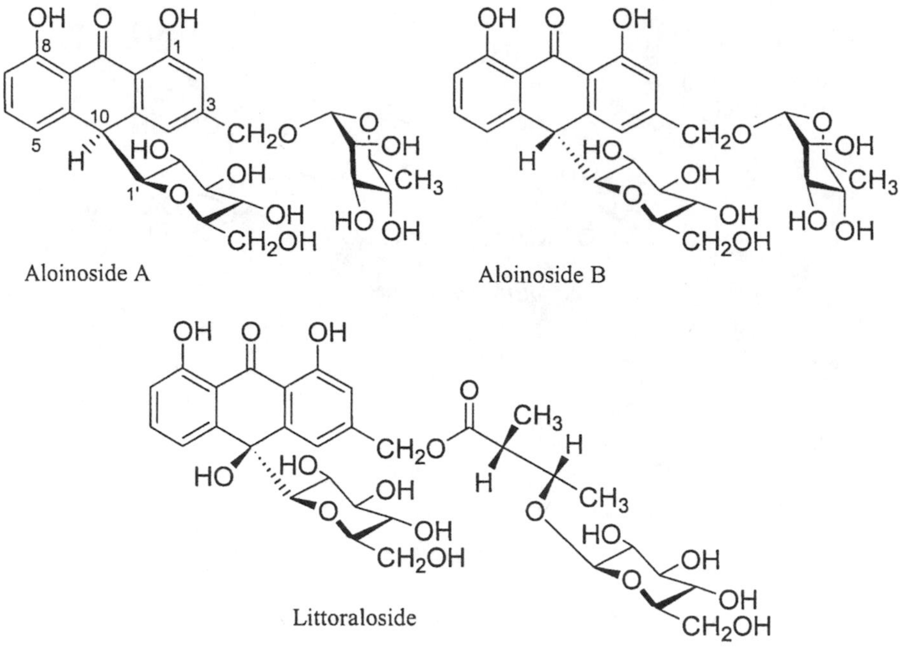

Aloinoside A

Aloinoside B

Littoraloside

Figure 3.11 C,O,-Diglucosylated compounds.

chrysophanol dimers, A, B, C and D from *A. saponaria* (Figure 3.12) (Yagi *et al.*, 1978b) and a pair of diastereomers based on aloe-emodin from *A. elgonica*, present as gluco-sides (Conner *et al.*, 1990b) and then found in *A. vera* (Choi *et al.*, 1996).

Dimer A was found to be identical to (+) asphodelin (4,7'-bichrysophanol) (Gonzalez *et al.*, 1973) and was identified in the roots of 32 *Aloe* species (Dagne *et al.*, 1994). Another type of dimeric molecule has been found in two distantly related genera, *Kniphofia* and *Bulbine*. Here 2',4'-dihydroxy-6'-methoxyacetophenone is attached to either chrysophanol (knipholone) (Dagne and Steglich, 1984) or chrysophanol anthrone (knipholone anthrone) (Dagne and Yenesew, 1993).

Phenolic compounds

Other phenolic compounds have been isolated from *Aloe* species and it is likely that many more will follow. Flavonoids are very widely distributed in plants but have not featured in aloe exudates until recently, when the flavones isovitexin and apigenin, the dihydroflavonol dihydroisorhamnetin and the flavanone narigenin (Figure 3.13) were reported from 31 *Aloe* species accompanied by a number of unidentified compounds (Viljoen *et al.*, 1998). Both aglycones and glycosides were present. Isovitexin was the major phenolic in Sections *Graminaloe* and *Leptoaloe* while naringenin and dihydro-isorhamnetin occured in Series *Superpositae*, *Rhodacanthae* and *Echinatae*.

Aloenin, first isolated from *A. arborescens* and then found in 12% of over 200 *Aloe* species surveyed (Reynolds, 1985b), is the *O*-glucoside of a phenol-pyran-2-one dimer (Figure 3.14) (Suga *et al.*, 1974; Hirata *et al.*, 1976; Hirata and Suga, 1978).

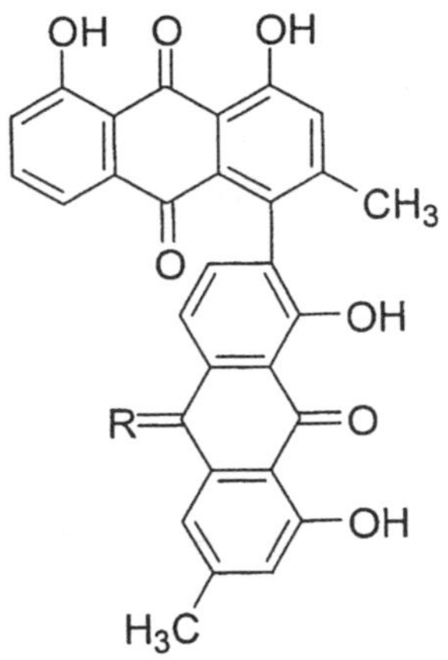

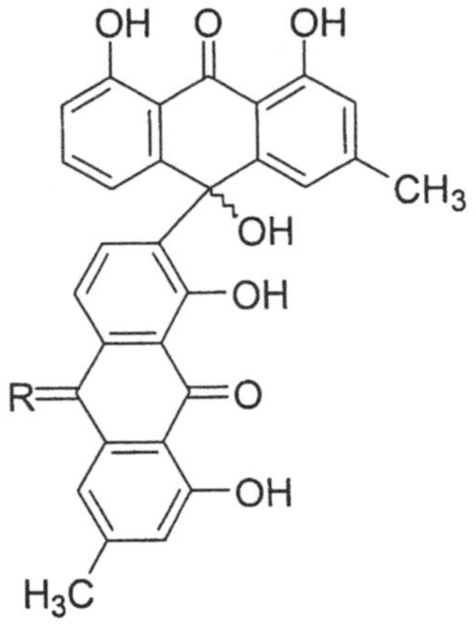

A. saponaria Pigment A, R = O,
Pigment C, R = CH$_2$

Pigment B, R = O
Pigment D, R = CH$_2$

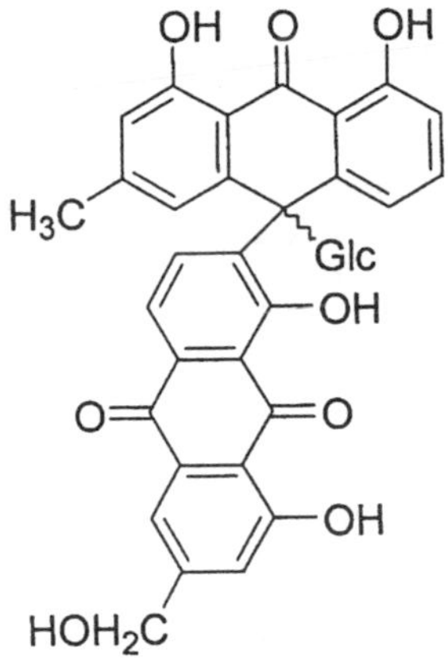

A. elgonica dimer A

Figure 3.12 Dimeric compounds.

The coumaroyl ester of an *O-O*-diglucoside was subsequently found in a Kenya aloe drug sample and named aloenin B (Speranza *et al.*, 1986a). Later the coumaroyl ester of aloenin itself was described from *A. nyeriensis*, together with the aglycone (Conner *et al.*, 1987). A breakdown product ('process product') isolated from Cape aloes was shown to be orcinol linked by a methylated methylene bridge to a phenyl residue (Figure 3.14) reflecting part of the aloenin structure (Speranza *et al.*, 1994). An even simpler compound, methyl-*p*-coumarate, was isolated from Cape aloes (Graf and Alexa, 1982). Another dimeric compound isolated from the roots of *A. vera* consisted of two 4-hydroxy-6-methoxybenzopyran moieties (Figure 3.15) joined by a C-C bond (Saleem *et al.*, 1997a) which relates distantly to the chromones. Feralolide (Figure 3.15), isolated as a minor component of Cape aloes, was shown to be a dimer with a methylene bridge of 2, 4-

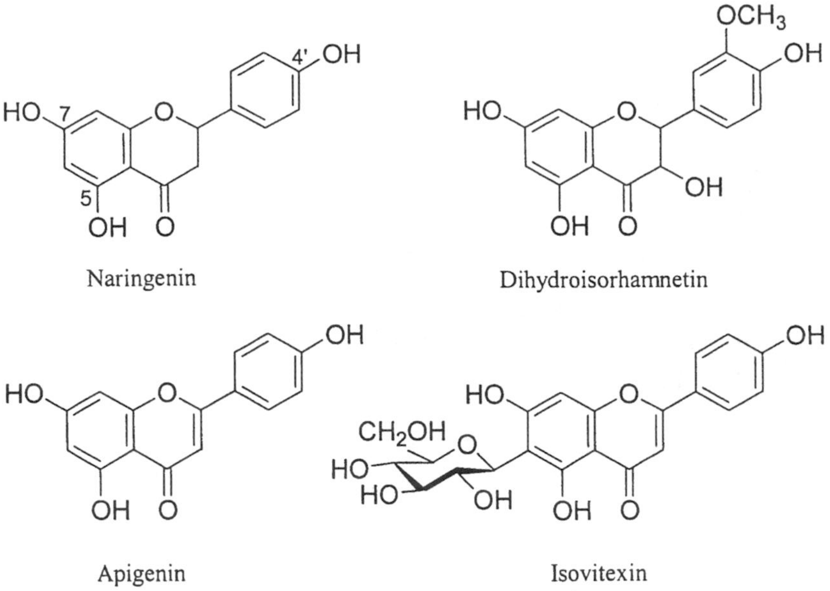

Figure 3.13 Flavonoids from aloes.

dihydroxyacetophenone and 6, 8-dihydroxyisocoumarin (Speranza *et al.*, 1993a). It was subsequently found in *A. vera* (Choi *et al.*, 1996) while an *O*-glucoside was isolated from *A. hildebrandtii* Baker (Veitch *et al.*, 1994).

A number of structures based on the naphthalene and tetralin nuclei have been assigned to aloe components from time to time. Tetrahydroanthracenes could be regarded as naphthalene derivatives with a fused cyclohexane ring but because their hydroxylation patterns relate to those of the anthraquinones they are more conveniently described separately. A substituted naphthalene diglucoside has been isolated from *A. plicatilis* (L.)Miller and named plicatiloside (Wessels *et al.*, 1996) (Figure 3.16). It was then found in 19 other *Aloe* species (Viljoen *et al.*, 1999). Underground stems of *A. saponaria* contain isoeleutherol-5-*O*-glucoside (Figure 3.16), a derivative of a substituted naphthoic acid (Yagi *et al.*, 1977b), while the aglycone was found more recently in eight *Aloe* species (Dagne *et al.*, 1994) and then in 18 species (Van Wyk *et al.*, 1995a). Three 1-methyltetralins (derivatives of 5,6,7,8-tetrahydronaphthalene) were isolated from commercial Cape aloes (Figure 3.16). First the aglycone, feroxidin, was characterized as 3,6,8-trihydroxy-1-methyltetralin (1,3,6-trihydroxy-8-methyl-5,6,7,8-tetrahydronaphthalene) (Speranza *et al.*, 1990, 1991) with 6*S*, 8*S* configuration. Then the 3-*O*-glucoside (Feroxin A) and its *p*-coumaric acid ester (Feroxin B) were described (Speranza *et al.*, 1992). Three naphtho [2,3-C] furans bearing some structural resemblance to a reduced isoeleutherol were characterized from Cape aloes (Figure 3.16) (Koyama *et al.*, 1994).

Alkaloids

Aloes are not known as poisonous plants even though the bitter anthrone glycosides make some of them unpleasant to taste. However a few reports of toxic effects may be found in the literature. Consumption of infusions of *A. chabaudii* Schönl., *A. globuligemma*

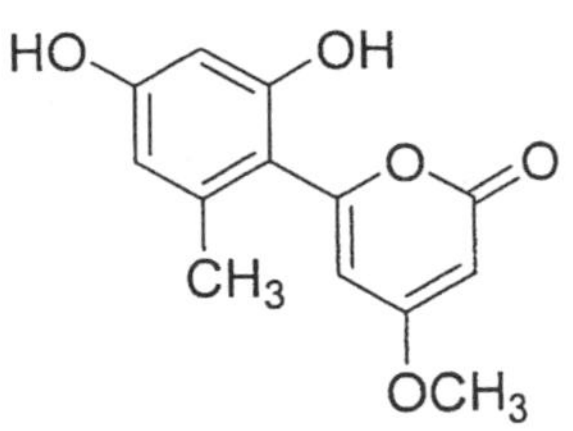

Aloenin aglycone

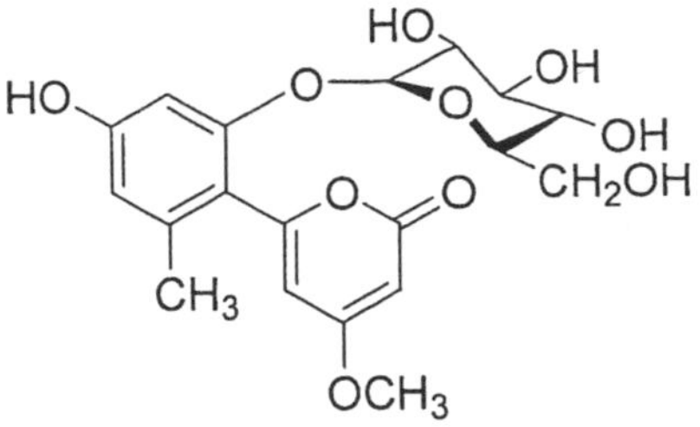

Aloenin

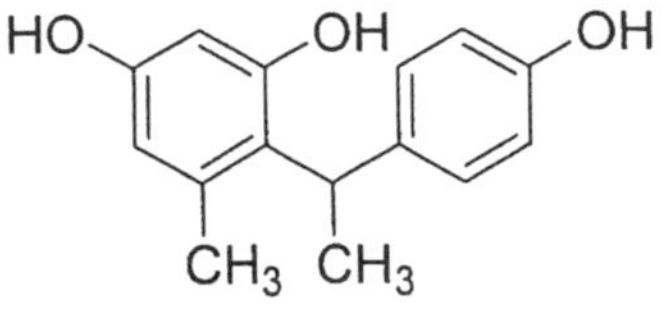

'Process product' from Cape aloe

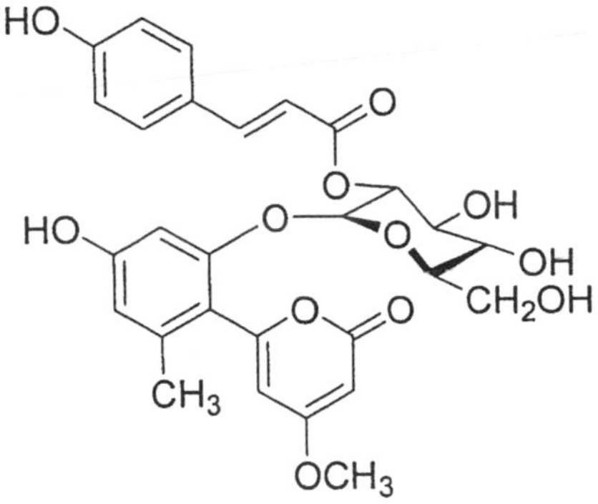

A. nyeriensis compound

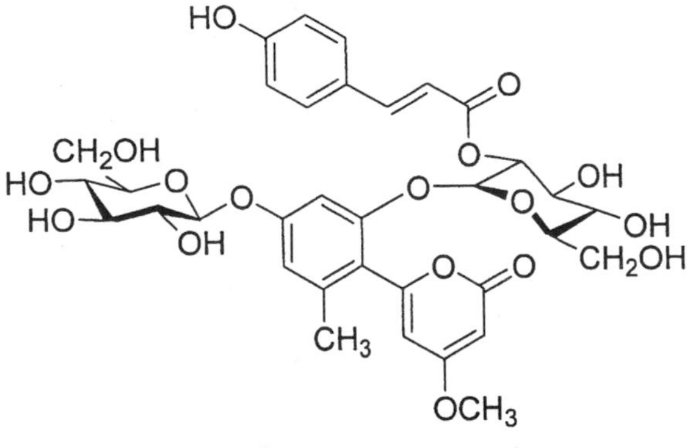

Aloenin B

Figure 3.14 Phenyl-pyrone derivatives from aloes.

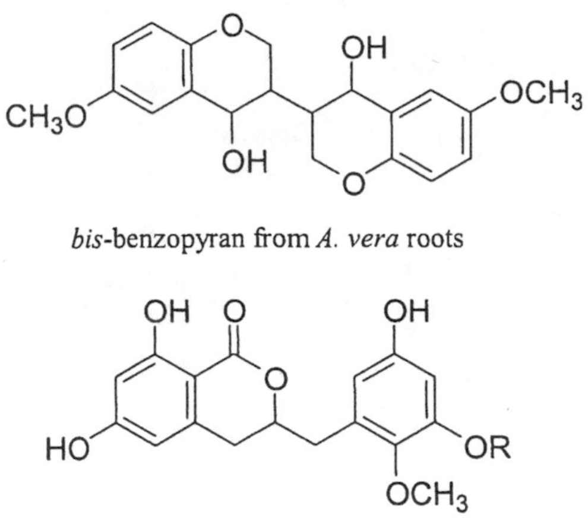

bis-benzopyran from *A. vera* roots

R = H   Feralolide
R = Glc *A. hildebrandtii* leaf glucoside

Figure 3.15 Phenolic dimers from aloes.

Pole Evans and *A. ortholopha* Christian et Milne-Redh. was said to be responsible for deaths in Africa (Drummond *et al.*, 1975). The first two, together with *A. christianii* Reynolds, are listed as poisonous in Zimbawe (Nyazema, 1984) and the same two again, listed as highly toxic (Parry and Matambo, 1992; Parry and Wenyika, 1994). A smell of mice or rats was associated with some of these plants and elsewhere, *A. ballyi* Reynolds is referred to for this reason as the 'Rat Aloe' (Reynolds, 1966). This clue led to the identification of a hemlock (*Conium maculatum* L.; Umbeliferae) alkaloid, γ-coniceine, in seven *Aloe* species (Dring *et al.*, 1984) and then in a further two (Nash *et al.*, 1992), out of over 200 examined. The related coniine was found in four species, including *A. globuligemma* and *A. ortholopha*, mentioned above. In addition, N-methyltyramine was found in 12 species and N-methyl-2-(4-methoxyphenyl) ethylamine in ten species. (The odiferous principle in rodents is said to be alloxan, not a poisonous alkaloid!) The presence of hemlock alkaloids was confirmed in *A. sabaea* Schweinf.(=*A. gillilandii* Reynolds) and was accompanied by N-4'-chlorobutylbutyramide (Blitzke *et al.*, 2000).

GEL COMPONENTS

Turning now to the central parenchymatous tissues of the aloe leaf, these are tasteless and colourless and of a more or less glutinous nature. These tissues in *A. vera* yield the aloe gel now so widespread in commercial therapeutic and cosmetic preparations (reviewed elsewhere in this volume). Although not accepted medically because of the difficulty in establishing the dosage, there are very many reports of healing properties, especially for skin complaints. It seems that activity may take place through the immunological system (reviewed Reynolds and Dweck, 1999; also Tizard and Ramamoorthy, chapter 13). Various components have been described from the gel, the principle ones being polysaccharides which give the substance its glutinous nature. Also present are

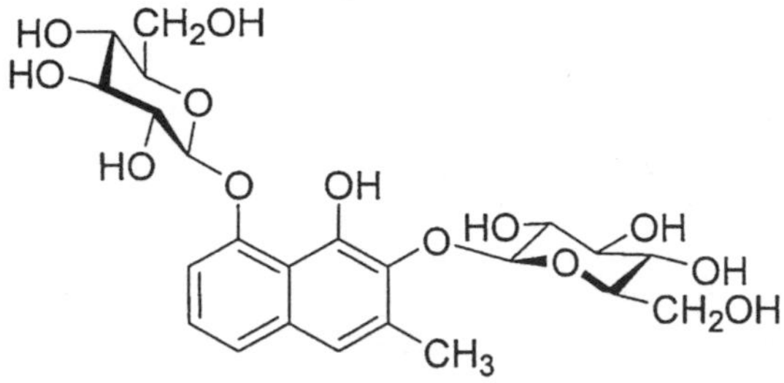

Plicatiloside

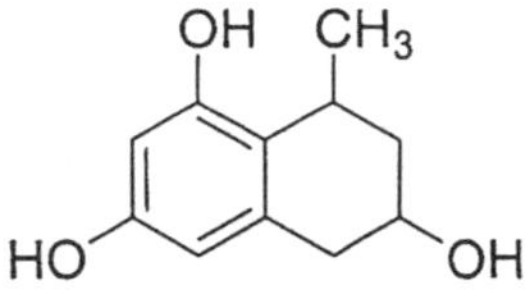

Feroxidin

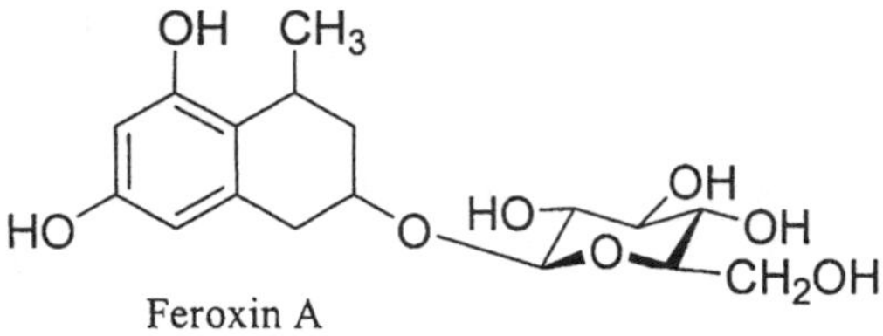

Feroxin A

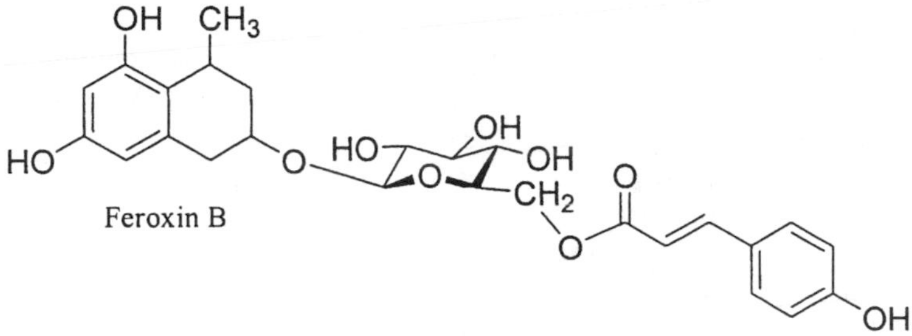

Feroxin B

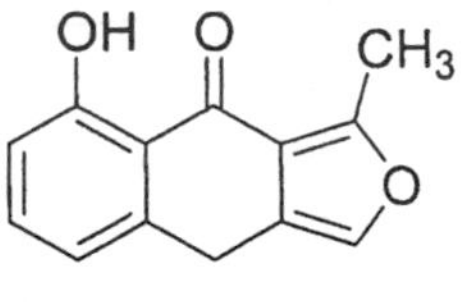

Cape aloes compound 1

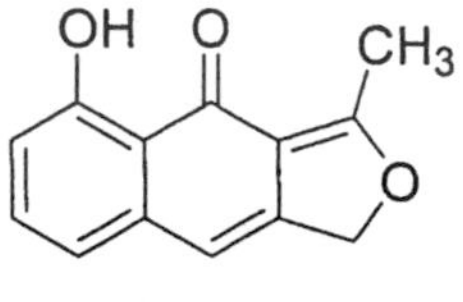

Cape aloes compound 3

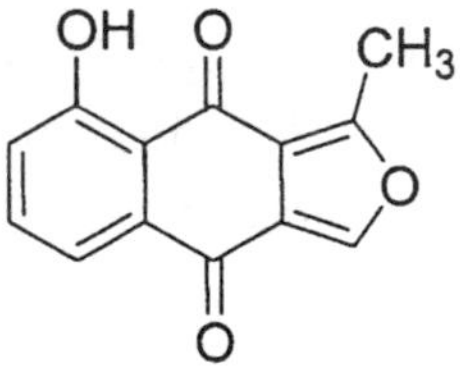

Cape aloes compound 2

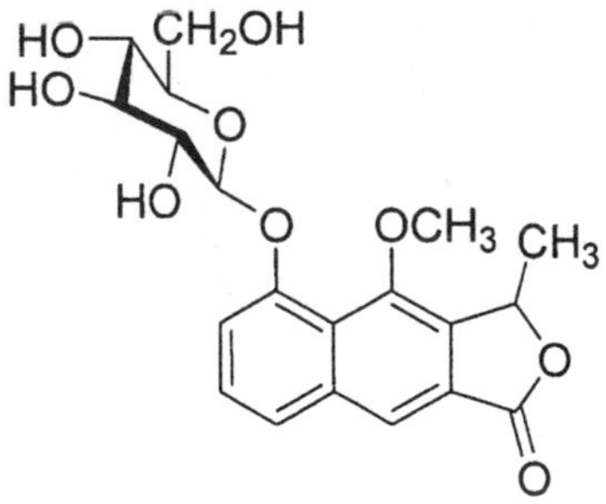

Isoeleutherol 5-*O*-glucoside

Figure 3.16 Naphthalene and tetralin derivatives.

glycoproteins for which biological activity has been reported. From time to time mention of the presence of various small molecules is made.

Polysaccharides

As might be expected much work has been carried out on *A. vera* although a few other species have been used. In the first modern preparation of polysaccharides from aloe gel mucilage, a water soluble material was produced which was precipitated by ethanol and dialysed to remove the high ash content (13%). It represented 30% of the dry weight of the leaf parenchyma (Robez and Haagen-Smith, 1948). On acid hydrolysis equal quantities of glucose and mannose were formed with a small amount of uronic acid (2%). A similar material was prepared in the course of a patented procedure (Farkas, 1967), which contained about 15% chemically bound calcium and had a molecular weight of around 450,000 daltons, which varied depending on variations in the process. A further short report confirmed the presence of mannose and glucose but in the molar ratio of about 10:1 with added arabinose, galactose and xylose in trace amounts (Segal *et al.*, 1968). A later study reported a mannose glucose ratio of 6:1 for a similar product and fractionated the substance by graded ethanol precipitation into three fractions A_1, A_2 and B and then A_1 into A_{1a} and A_{1b} (Gowda *et al.*, 1979). They were shown to be partially acetylated glucomannans of which B had the highest *O*-acetyl content and formed the most mucilagenous mixture with water. Further study showed them to be $1 \rightarrow 4$ linked and to have molecular weights in excess of 2×10^5 daltons.

In contrast analysis of a gel from a different *A. vera* individual demonstrated a high content of pectic acid (70–85%) separated by ethanol precipitation. The remaining polysaccharides separated by DEAE–cellulose chromatography were shown to be a galactan, a glucomannan and an arabinan (Mandal and Das, 1980a). The D-galactan contained $1 \rightarrow 4$ and $1 \rightarrow 6$ linkages. The glucomannan contained glucose and mannose in the molar ratio of 1:22 joined by $1 \rightarrow 4$ linkages with some side-chains joined by $1 \rightarrow 6$ linkages (Mandal and Das, 1980b). The pectic acid fraction contained mainly galacturonic acid together with galactose and traces of glucose and arabinose. The galactose units were joined by $1 \rightarrow 3$ linkages to a linear $1 \rightarrow 4$ linked galacturonic acid chain (Mandal *et al.*, 1983). A cruder preparation using just fractional precipitation by ethanol gave a single product characterized as a linear glucogalactomannan with glucose, mannose and galactose in a molar ratio of 2:2:1 (Haq and Hannan, 1981).

An activity-guided purification of gel polysaccharides using both anion-exchange and gel filtration chromatography identified two fractions, B-I and B-II showing inhibition of classical pathway complement activity ('t Hart *et al.*, 1989). They contained mainly mannose together with galactose, glucose and arabinose in ratios 89:4:3:1 for B-I and 22:2:1:1 for B-II, with a molecular weight between 1.5×10^5 and 4.8×10^5 daltons. From the data it appears that there is also considerable activity in non-polysaccharide fractions.

In 1987 an acetylated mannan was reported from *A. vera* gel as a commercial product, acemannan or carrisyn™ (McDaniel *et al.*, 1987) to which several types of biological activity were attributed (reviewed Reynolds and Dweck, 1999). Subsequent work established the position of the *O*-acetyl group as equally on either C_2 or C_3 and on C_6 of the mannose moiety (Manna and McAnalley, 1993). A very detailed account of the preparation and purification of this substance and its chemical and physical properties is the subject of two U.S. patents (McAnalley, 1988, 1990), which also mention therapeutic activity. Extensive details of the whole process are given elsewhere in this volume (Chapter 8).

The evidence above indicates that the presence of certain specific polysaccharides is critical for healing. These can vary markedly from product to product, so much so that accurate analysis is necessary to predict activity (Ross *et al.*, 1997).

Aloe arborescens is another species very widespread in cultivation and used medicinally, especially in the Far East. A partially acetylated mannan was isolated from the gel of this plant as a single substance precipitated by acetone and having a molecular weight of 15,000 daltons (Yagi *et al.*, 1977). Separation of the polysaccharides by gel filtration yielded three fractions A, B and C (Yagi *et al.*, 1986). Fraction A was a $1 \rightarrow 6$ linked glucan, with a molecular weight of 15,000 daltons. Fraction B contained arabinose and glucose in the molar ratio 3:2 with $2 \rightarrow 6$ linkages, with a molecular weight of 30,000 daltons. Fraction C was a $1 \rightarrow 4$ linked acetylated mannan containing 10% acetyl groups, with a molecular weight of 40,000 daltons. Acetyl groups were on the C_2 or C_3 and C_6 portions, recalling the structure of acemannan. Meanwhile elsewhere, an acidic polysaccharide isolated by fractional precipitation of copper complexes was shown to consist of $1 \rightarrow 3$ and $1 \rightarrow 4$ linked glucose and glucuronic acid in the ratio 9:1 and to have a molecular weight of around 3.6×10^3 daltons (Hranisavlyevič-Jakovlyevic and Miljkovic-Stojunovic, 1981). It was accompanied by an unspecified polyglucan.

In another study, ion exchange chromatography separated two fractions named arboran A and arboran B, said to diminish plasma glucose level in mice (Hikino *et al.*, 1986). Both fractions contained protein, 2.5% and 10.4% respectively, although it is not certain if this was joined to the polysaccharide. Arboran A contained mainly galactose, glucose and rhamnose in the molar ratios 10:3:3 and traces of fucose, arabinose, xylose and mannose with 16.7% O-acetyl groups and a molecular weight of 1.2×10^4 daltons. Arboran B contained only glucose and mannose in the molar ratio 10:3 with 5.3% O-acetyl groups and a molecular weight of 5.7×10^4 daltons. Another preparation using separation on an ion exchange gel resulted in three polysaccharide fractions. The acidic fraction with a molecular weight of 5×10^4 daltons contained arabinose and galactose in a molar ratio of 1:1 with traces of rhamnose and glucose and 6% glucuronic acid. The two neutral fractions had molecular weights of 1.2×10^4 and 1×10^6 daltons and a mannose – glucose molar ratio of 95:5 with $1 \rightarrow 4$ linking. Both contained O-acetyl groups located at C_6 and $C_{2,3}$ (Wozniewski *et al.*, 1990).

Aloe saponaria is another species with reputed therapeutic properties and its gel yielded three polysaccharides separated by fractional precipitation. The main component (77%) was an acetylated mannan. One of the minor components was a polyglucan and the other contained mannose, glucose and galactose in the ratio 5:4:1 (Gowda, 1980). Elsewhere a polysaccharide was isolated as the main component (c70%) of the gel. Here, the material harvested earlier yielded a $1 \rightarrow 4$ linked mannan with acetyl groups (18%) on C6 and a molecular weight of 1.5×10^4 daltons, while material harvested later was a $1 \rightarrow 4$ linked acetylated mannan with 5% glucose units and a molecular weight of 6.6×10^4 daltons (Yagi *et al.*, 1984). Interestingly, immunoadjuvant properties had been reported for the gel of a Madagascan endemic, *A. vaombe*, Decorse et Poiss. (incorrectly cited as *A. vahombe* in text) and a polysaccharide isolated by ethanol precipitation. Further purification by gel filtration yielded a fraction with a glucose to mannose ratio of 1:3 and one acetyl unit on each glucose unit (Radjabi *et al.*, 1983; Radjabi-Nassab *et al.*, 1984). Furthermore, critical separation by these methods separated this entity, with a molecular weight of 1×10^5 daltons, from three other acetylated glucomannans with weights of 2.5×10^3, 2×10^4 and above 10^5 daltons. This latter large molecule also contained protein (Vilkas and Radjobi-Nassab, 1986). A very ornamental aloe with strap-shaped leaves, *A. plicatilis* Mill., contained an acetylated glucomannan with a molecular weight of 1.2×10^4 daltons, and a glucose to mannose ratio of 1:2.8 (Paulsen *et al.*, 1978). On the other hand a pure acetylated mannan was the major component of *A. vanbalenii* gel (Gowda, 1980). The minor polysaccharides were a glucan and a galactoglucomannan. A different array of compounds was reported from the gel of *A. ferox* Mill. where 14 distinct polysaccharide entities were distinguished, most of which were arabinogalactans or rhamnogalacturonans (Mabsuela *et al.*, 1990).

The picture emerges from the six *Aloe* species investigated so far of a number of polysaccharides with mannose as the predominant monomer and molecular weights ranging of several orders of magnitude from 10^3 to 10^6 daltons. The patented compound with the greatest number of reported therapeutic activities is acemannan which is an acetylated mannan about 8×10^4 daltons in size (see Chapter 4).

Nitrogenous compounds

Nitrogen analysis of leaf extracts and of crude or partially purified aloe gel preparations has always yielded positive results. Thus Rowe and Parks (1941) reported 1.39% dry weight of nitrogen in the outer leaf tissues (rind) of *A.vera* but gave no figure for the gel. Another early report of *A. vera* gel reported 2.9% protein of the leaf parenchyma dry weight (Roboz and Haagen-Smit, 1948). A later assay gave a protein content of 0.01% for aloe 'juice' (sic) (Gjerstad, 1971) and recognized hydroxyproline, histidine and cystine as free acids. Another analysis recognized 17 common amino acids in the free state, of which arginine was the most abundant (Waller *et al.*, 1978). Similar results were obtained elsewhere (Khan, 1983), with arginine again being especially of note, together with glutamic acid. In *A. ferox*, on the other hand, asparagine was reported as the most abundant, followed by glutamine, alanine and histidine (Ishikawa *et al.*, 1987). Alanine, proline, lysine and glutamic acid were prominent in *A. arborescens* leaves, accompanied by other free protein amino acids (Yagi *et al.*, 1987). Elsewhere, aspartic acid, glutamic acid and serine were noted in *A. vera* gel (Baudo *et al.*, 1992).

Lectin activity had been reported in aloe preparations (e.g. Fujita *et al.*, 1978) and is the subject of Chapter 5, so the presence of glycoproteins could be presumed. Two such substances were separated by Sephadex chromatography from the whole leaves of *A. arborescens* and given codes P-2 and S-1. Both were confirmed as glycoproteins, with 18% and 50% neutral carbohydrate and molecular weights of 1.8×10^4 and 2.4×10^4 daltons (Suzuki *et al.*, 1979). They were designated as Aloctin A and Aloctin B. A quite separate entity was isolated about the same time and named ATF 1011, later shown to activate T cells (Yoshimoto *et al.*, 1987). Another preparation was subsequently made and examined by polyacrylamide gel electrophoresis which confirmed the molecular weights and extended the range of biological activities (Saito, 1993). Separation of a protein fraction from *A. arborescens* 'fresh leaf juice' (sic) by DEAE cellulose and Sepharose 6B chromatography was also described (Yagi *et al.*, 1986). In addition to the expected amino acids the hydrolysed protein fraction contained glucose, mannose, galactose, glucosamine, galactosamine and N-acetylglucosamine in the ratio 2:2:1:1:4:1. Two glycoproteins were separated by ammonium sulphate precipitation, both of which contained mannose, arabinose, glucose, galactose and glucosamine (Kodym, 1991). The outer green tissues of the leaves yielded several proteins separated on DEAE cellulose, of which one had lectin activity and a molecular weight of 3.5×10^4 daltons (Koike *et al.*, 1995).

Lectins were also reported from *A. vera* gel (Winters, 1993). Polyacrylamide gel electrophoresis revealed at least 23 polypeptides from the mature leaf, of which 13 occurred in the gel (Winters and Bouthet, 1995). One of these which had a molecular weight of 1.8×10^4 daltons resembled the Aloctin A, mentioned above (Bouthet *et al.*, 1996), while six others appeared common to this species and to *A. arborescens* which itself had a total of nine. In the same investigation 11 polypeptides were noted in *A. saponaria* reflecting those observed in *A. vera*. Elsewhere a glycoprotein fraction promoting cell proliferation *in vitro* was shown to contain a single entity of molecular weight 2.9×10^4 daltons and to contain 11% carbohydrate (Yagi *et al.*, 1997).

A different type of protein has been isolated from *A. arborescens* leaves. Ultrafiltration of extracts yielded a fraction with a molecular weight in excess of 1.0×10^5 which had bradykininase activity (Fujita *et al.*, 1976) and it was characterized as a carboxypeptidase (Fujita *et al.*, 1979), specifically a serine carboxypeptidase (Ito *et al.*, 1993). In a separate study, two glycoprotein fractions were separated by Sepharose chromatography, one of which was shown to have bradykininase activity (Yagi *et al.*, 1987). The molecular weight was 4.0×10^4 daltons and the other properties appeared to be identical with an entity previously described as glycoprotein A (Yagi *et al.*, 1986). These authors had previously obtained a bradykininase from *A. saponaria* containing mainly mannose as the carbohydrate moiety (Yagi *et al.*, 1982).

A number of enzymes were extracted and separated by starch gel electrophoresis for use as genetic markers to identify hybrids between *A. arborescens* and *A. ferox* (van der Bank and van Wyk, 1996).

OTHER COMPONENTS

Many analyses have been carried out on aloes in search of constituents which might be responsible for beneficial properties and various organic compounds additional to those described above reported from time to time. In a detailed study the composition of various fractions of an *A. arborescens* plant was investigated as shown in Table 3.4 (Hirata and Suga, 1977). Another very detailed analysis, of *A. vera* this time, revealed a lipid content of c.5% of the dry gel (Femenia *et al.*, 1999). The common plant sterol, β-sitosterol was found in whole *A. vera* leaves, accompanied by lesser amounts of cholesterol, campestrol and lupeol (Waller *et al.*, 1978). Also observed were a number of unidentified volatiles, recognized by thin-layer chromatography. A further determination of β-sitosterol was made in *A. arborescens* leaf (Yamamoto *et al.*, 1990; Yamamoto *et al.*, 1991) and again in *A. vera* leaf (Ando and Yamaguchi, 1990). Then, sitosterol glucoside and its palmitic acid ester were found in whole leaves of *A. vera* together with, again, lupeol (Kinoshita *et al.*, 1996). A later study also showed β-sitosterol and a variety of n-alkanes in the gel of *A. vera* with n-octadecane predominating, as well as fatty acids and their methyl esters (Yamaguchi *et al.*, 1993). The main compnent of a steam distillate of a methanol extract of *A. arborescens* was 3-hydroxymethyl furan (Kameoka *et al.*, 1981). Later, a milder steam distillation extracted a number of volatiles of which (Z)-3-hexanol and (Z)-3-hexanal were the major components, while headspace extraction yielded (Z)-3-hexanal

Table 3.4 Analysis of an *A. arborescens* plant (Hirata and Suga, 1977).

Leaves		Roots	
'Juice'		*Methanol Extract*	
Succinic acid	0.05%	*Ether soluble Fraction*	0.02%
Magnesium lactate	0.01%	n-alkanes	0.0002%
Glucose	0.014%	fatty acid methyl esters	0.007%
		sitosterol	0.002%
		1-linoleyl monoglyceride	0.002%
		fatty acids	0.0004%
		Ether insoluble Fraction	0.08%
Methanol Extract of Residue		sitosterol glucoside	0.002%
Acidic Fraction		glucose	0.02%
Fatty acids	0.017%	magnesium lactate	0.03%
Neutral Fraction	0.16%	n-triacontanol (C_{30})	0.002%
n-alkanes	0.0008%	n-dotriacontanol (C_{32})	0.0007%
fatty acid methyl esters	0.06%		
sitosterol	0.001%		
Principal Fatty Acids		*Principal Fatty Acids*	
palmitic acid	18% of acid fraction	palmitic acid	49% of acid fraction
linoleic acid	28%	linoleic acid	10%
linolenic acid	35%	stearic acid	21%
Principal Fatty Acid Methyl Esters		*Principal Fatty Acid Methyl Esters*	
linoleic	35% of methyl ester fraction	linoleic	10% of methyl ester fraction
linolenic	5%	palmitic	46%
		stearic	24%
Principal n-Alkanes		*Principal n-Alkanes*	
C_{29}	23% of n-alkane fraction	C_{29}	12% of n-alkane fraction
C_{31}	20%	C_{19}	11%
		C_{15}	11%

and (E)-2-hexanal (Umano *et al.*, 1999). Elsewhere, citric, malic and formic acids were found in various aloes (Ishikawa *et al.*, 1987) and again, malic acid was determined as 1.3% of unspecified 'raw material' (Bereshvili *et al.*, 1989). Malic acid has been used as a marker to validate gels offered for sale as aloe-derived (Chapter 6). Previously, an extensive survey of leaf and perianth waxes of 63 *Aloe* species showed that the major component in most samples was hentriacontane (C_{31}). Occurrence of other hydrocarbons had a taxonomic correlation (Herbin and Robins, 1968). An extract of whole *A. vera* plants yielded 0.7% non-polar lipids, of which the major components were stigmasterol (18.4% of sample) and its stearate (21.3%), with lesser amounts of cholesterol (12.5%), methyl oleate (7.1%), triolein (2%) and oleic acid (1.3%). The polar lipids (0.9%) contained principally phosphatidic acid (47.3% of sample), with some sulfoquinovosyl diglyceride (16.8%), phosphatidylcholine (12.1%) and phosphatidylethanolamine (12%) (Afzal *et al.*, 1991). Hydrolysis of the lipid fraction yielded mainly γ-linolenic acid (42% of fraction), with a lesser amount of arachidonic acid (3.1%) which together were postulated as precursors of prostaglandins. Another analysis of *A. vera* gel by thin-layer chromatography claims the presence of sterols, saponins, triterpenoids and naphthoquinones (Vazquez *et al.*, 1996). The simple sugars, glucose, fructose and sucrose are present in large amounts in the nectar of aloes (Van Wyk *et al.*, 1993) and a survey of 82 species revealed two groups in which sucrose was found at very high or very low levels.

As well as the organic substances described above there is of course an inorganic component in any plant material. An early report showed 3.3% ash content of *A. vera* leaf 'rind' and 0.2% (wet weight) ash in leaf 'pulp', which contained calcium oxalate (Rowe and Parks, 1941). Another report at that time gave the ash content of a purified gel as 12.9% (dry weight) (Roboz and Haagen-Smith, 1948). Twenty years later an analysis of *A. vera freeze-dried* 'juice' revealed a high level of chlorine (12.2%), accompanied by potassium (6.6%) and calcium (4.7%) (Bouchey and Gjersted, 1969). Analysis of an *A. arborescens, incinerated,* whole-leaf 'juice' also showed potassium (57%) as the main metallic component, accompanied by sodium (32%), manganese (9%), magnesium (2%) and calcium (1%) (Hirata and Suga, 1977). Calcium and magnesium lactates were found in fresh leaves of *A. arborescens*, together with 11 other metallic cations (Kodym, 1988). Calcium (0.3%) was a major component of a sample of *A. vera* gel (Baudo, 1992). A later analysis of *A. vera, freeze-dried,* gel gave prominence to calcium (3.5%), followed by magnesium (0.7%) and sodium (0.2%) (Yamaguchi *et al.*, 1993). The detailed analysis of Femenia *et al.* (1999) also gave potassium (4.1%) as a major component, followed by sodium (3.7%) and calcium (3.6%) based on the *freeze-dried* gel. It is noteworthy that they also found a relatively high calcium content (3.3%) of the alcohol insoluble polysaccharides.

ACKNOWLEDGEMENT

The author is grateful to Dr N.C. Veitch for the meticulous drawing of the structural formulae and for critically reading the text.

REFERENCES

Afzal, M., Ali, M., Hassan, R.A.H., Sweedan, N. and Dhami, M.S.I. (1991) Identification of some prostanoids in Aloe vera extracts. *Planta Medica*, 57, 38–40.

Ando, N. and Yamaguchi, I. (1990) ß-Sitosterol from Aloe vera (Aloe vera(L.) Burm. f.) gel. *Kenkyu Kiyo-Tokyo Kasei Daigaku*, 30, 15–20.

Auterhoff, H., Graf, E., Eurisch, G. and Alexa, M. (1980) Trennung des Aloins in Diastereomere und deren Charakterisierung. *Archiv der Pharmazie*, 313, 113–120.

Awe, W., Auterhoff, H. and Wachsmuth-Melm, C.L. (1958) Beitrage zur papierchromatographischen untersuchung von *Aloe*-drogen, *Arzneimittel-Forschung*, 8, 243–245.

Axing, Y. (1993) The molecular structure of iso-aloesin isolated from the leaves of *Aloe vera* var.*chinensis* (Haw.) Berge. *Journal of Chinese Medicine*, 18, 609–611.

Barnes, R.A. and Holfeld, W. (1956) The structure of barbaloin and *iso*barbaloin. *Chemistry and Industry*, 873–874.

Baudo, G. (1992) Aloe vera. *Erboristeria Domani*, 2, 29–33.

Beaumont, J., Reynolds, T. and Vaughan, J.G. (1984) Homonataloin in Alöe species. *Planta Medica*, 50, 505–508.

Beaumont, J., Cutler, D.F., Reynolds, T. and Vaughan, J.G. (1985) The secretory tissue of aloes and their allies. *Israel Journal of Botany*, 34, 265–282.

Bereshvili, D.T., Shemeryankina, M.I., Kosova, N.G. and Komarova, E.I. (1989) Determination of carboxylic acids in aloe. *Farmatsiya*, 28, 38–40.

Birch, A.J. and Donovan, F.W. (1955) Barbaloin 1. Some observations on its structure. *Australian Journal of Chemistry*, 8, 523–528.

Bisrat, D., Dagne, E., Van Wyk, B.-E. and Viljoen, A. (2000) Chromones and anthrones from *Aloe marlothii* and *Aloe rupestris*. *Phytochemistry*, 55, 949–952.

Blitzke, T., Porzel, A., Masaoud, M. and Schmidt, J. (2000) A chlorinated amide and piperidine alkaloids from *Aloe sabaea*. *Phytochemistry*, 55, 979–982.

Böhme, H. and Kreutzig, L. (1963) Zur Papier- und Dünnschichtchromatographie von Aloe-Drogen. *Deutsche Apotheker-Zeitung*, 103, 505–508.

Bouchey, D.G. and Gjerstad, G. (1969) Chemical studies of *Aloe vera* juice II: Inorganic constituents. *Quarterly Journal of Crude Drug Research*, 9, 1445–1453.

Bouthet, C.F., Schirf, V.R. and Winters, W.D. (1996) Semi-purification and characterization of haemagglutinin substance from *Aloe barbadensis* Miller. *Phytotherapy Research*, 10, 54–57.

Chauser-Volfson, E. and Gutterman, Y. (1996) The barbaloin content and distribution in *Aloe arborescens* leaves according to the leaf part, age, position and season. *Israel Journal of Plant Sciences*, 44, 289–296.

Chauser-Volfson, E. and Gutterman, Y. (1997) Content and distribution of the secondary phenolic compound homonataloin in *Aloe hereroensis* leaves according to leaf part, position and monthly changes. *Journal of Arid Environments*, 37, 115–122.

Chauser-Volfson, E. and Gutterman, Y. (1998) Content and distribution of anthrone C-glyco-sides in the South African arid plant species *Aloe mutabilis* growing in direct sunlight and in shade in the Negev Desert of Israel. *Journal of Arid Environments*, 40, 441–451.

Choi, J.-S., Lee, S.-K., Sung, C.-K. and Jung, J.-H. (1996) Phytochemical study on *Aloe vera*. *Archives of Pharmaceutical Research*, 19, 163–167.

Chopra, R.N. and Ghosh, N.N. (1938) Chemische Untersuchung der indischen aloearten Aloe vera, Aloe indica, Boyle. *Archiv der Pharmazie*, 276, 348–350.

Conner, J.M., Gray, A.I., Reynolds, T. and Waterman, P.G. (1987) Anthraquinone, anthrone and phenylpyrone components of *Aloe nyeriensis* var. *kedongensis* leaf exudate. *Phytochemistry*, 26, 2995–2997.

Conner, J.M., Gray, A.I., Reynolds, T. and Waterman, P.G. (1989) Anthracene and chromone derivatives in the exudate of *Aloe rabaiensis*. *Phytochemistry*, 28, 3551–3553.

Conner, J.M., Gray, A.I., Reynolds, T. and Waterman, P.G. (1990a) Anthracene and chromone components of *Aloe cremnophila* and *A. jacksonii* leaf exudates. *Phytochemistry*, 29, 941–944.

Conner, J.M., Gray, A.I., Reynolds, T. and Waterman, P.G. (1990b) Novel anthrone-anthraquinone dimers from Aloe elgonica. *Journal of Natural Products*, 53, 1362–1364.

Constantinescu, E., Palade, M., Grasu, A. and Rotaru, E. (1969) Contributii la studiul chimic al plantei Aloe arborescens Mill. *Farmacia*, 17, 591–600.

Dagne, E. (1996) Overview of the chemistry of aloes of Africa. *Proceedings of the 1st International IOCD Symposium, Victoria Falls*, 143–157.

Dagne, E. and Alemu, M. (1991) Constituents of the leaves of four *Aloe* species from Ethiopia. *Bulletin of the Chemical Society of Ethiopia*, 5, 87–91.

Dagne, E., Bisrat, D., Codina, C. and Bastida, J. (1998a) A C,0-Diglucosylated oxanthrone from *Aloe littoralis*. *Phytochemistry*, 48, 903–905.

Dagne, E., Bisrat, D., Van Wyk, B.-E. and Viljoen, A. (1998b) 10-Hydroxyaloin B 6'-O-acetate from *Aloe claviflora*. *Journal of Natural Products*, 61, 256–257.

Dagne, E., Bisrat, D., Van Wyk, B.-E., Viljoen, A., Hellwig, V. and Steglich, W. (1997) Anthrones from *Aloe microstigma*. *Phytochemistry*, 44, 1271–1274.

Dagne, E., Casser, I. and Steglich, W. (1992) Aloechrysone, a dihydroanthracenone from *Aloe berhana*. *Phytochemistry*, 31, 1791–1793.

Dagne, E. and Steglich, W. (1984) Knipholone: a unique anthraquinone derivative from *Kniphofia foliosa*. *Phytochemistry*, 23, 1729–1731.

Dagne, E., Van Wyk, B.-E., Mueller, M. and Steglich, W. (1996b) Three dihydroanthracenones from *Gasteria bicolor*. *Phytochemistry*, 41, 795–799.

Dagne, E., Van Wyk, B.-E., Stephenson, D. and Steglich, W. (1996a) Three oxanthrones from *Aloe littoralis*. *Phytochemistry*, 42, 1683–1687.

Dagne, E. and Yenesew, A. (1993) Knipholone anthrone from *Kniphofia foliosa*. *Phytochemistry*, 34, 1440–1441.

Dagne, E., Yenesew, A., Asmellash, S., Demissew, S. and Mavi, S. (1994) Anthraquinones, pre-anthraquinones and isoeleutherol in the roots of *Aloe* species. *Phytochemistry*, 35, 401–406.

Dring, J.V., Nash, R.J., Roberts, M.F. and Reynolds, T. (1984) Hemlock alkaloids in aloes. Occurrence and distribution of γ-coniceine. *Planta Medica*, 50, 442–443.

Drummond, R.B., Gelfand, M. and Mavi, S.9. (1975) Medicinal and other uses of succulents by the Rhodesian African. *Excelsa*, No.5, 51–56.

Farah, M.H., Andersson, R. and Samuelsson, G. (1992) Microdontin A and B: two new aloin derivatives from *Aloe microdonta*. *Planta Medica*, 58, 88–93.

Farkas, A. (1967) Aloe polysaccharide compositions. *U.S. 3, 360, 511 (Cl. 260–209)*, 0.

Femenia, A., Sánchez, E.S., Simal, S. and Rosselló, C. (1999) Compositional features of poly-saccharides from Aloe vera (*Aloe barbadensis* Miller) plant tissues. *Carbohydrate Polymers*, 39, 109–117.

Fujita, K., Ito, S., Teradaira, R. and Beppu, H. (1979) Properties of a carboxypeptidase from aloe. *Biochemical Pharmacology*, 28, 1261–1262.

Fujita, K., Suzuki, I., Ochiai, J., Shinpo, K., Inoue, S. and Saito, H. (1978) Specific reaction of aloe extract with serum proteins of various animals. *Experientia*, 34, 523–524.

Fujita, K., Teradaira, R. and Nagatsu, T. (1976) Bradykinase activity of aloe extract. *Biochemical Pharmacology*, 25, 205.

Gjerstad, G. (1969) An appraisal of the Aloe vera juice. *American Perfumer and Cosmetics*, 84, 43–46.

Gonzalez, A.G., Freire, R., Hernandez, R., Salazar, J.A. and Suarez, E. (1973) Asphodelin and microcarpin, two new bianthraquinones from *Asphodelus microcarpus*, *Chemistry and Industry*, 851–852.

Gowda, D.C. (1980) Structural studies of polysaccharides from *Aloe saponaria* and *Aloe vanbalenii*. *Carbohydrate Research*, 83, 402–405.

Gowda, D.C., Neelisiddaiah, B. and Anjaneyalu, Y.V. (1979) Structural studies of polysaccharides from *Aloe vera*. *Carbohydrate Research*, 72, 201–205.

Graf, E. and Alexa, M. (1980) Die Bestimmung der Aloine und Aloeresine durch HPLC. *Archiv der Pharmazie*, 313, 285–286.

Graf, E. and Alexa, M. (1982) p-Cumarsäure-methylester in Kap-Aloe. *Archiv der Pharmazie*, 315, 969–970.

Graf, E., Breitmaier, E. and Alexa, M. (1980) Aloin-dimethylketal in aloin. *Planta Medica*, 40, 197.

Gramatica, P., Gianotti, M.P., Speranza, G. and Manitto, P. (1986) Synthesis of naturally occurring 2,5-dialkylchromones. Part 1. Synthesis of aloesone and aloesol. *Heterocycles*, 24, 743–750.

Groom, O.J. and Reynolds, T. (1987) Barbaloin in *Aloe* species. *Planta Medica*, 53, 345–348.

Grün, M. and Franz, G. (1982) Untersuchungen zur Biosynthese der Aloine in *Aloe arborescens* Mill. *Archiv der Pharmazie*, 315, 231–241.

Hammouda, F.M., Rizk, A.M. and Seif El-Nasr, M.M. (1977) Naturally occurring anthra-quinones with special emphasize to those of family Liliaceae. *Herba Hungarica*, 16, 79–102.

Haq, Q.N. and Hannan, A.(1981) Studies on glucogalactomannan from the leaves of *Aloe vera*, Tourn.(ex Linn.). *Bangladesh Journal of Scientific and Industrial Research*, 16, 68–72.

't Hart, L.A., van den Berg, A.J.J., Kuis, L. van Dijk, H. and Labadie, R.P. (1989) An anti-complementary polysaccharide with immunological adjuvant activity from the leaf paren-chyma gel of Aloe vera. *Planta medica*, 55, 509–512.

Hay, J.E. and Haynes, L.J. (1956) The aloins. Part I. The structure of barbaloin. *Journal of the Chemical Society*, 3141–3147.

Haynes, L.J. and Henderson, J.I. (1960) Structure of homonataloin. *Chemistry and Industry*, 50.

Haynes, L.J., Henderson, J.I. and Russell, R. (1970) *C*-Glycosyl compounds. Part VI. Aloesin, a *C*-glucosylchromone from *Aloe* sp. *Journal of the Chemical Society*, 2581–2586.

Haynes, L.J., Henderson, J.I. and Tyler, J.M. (1960) *C*-glycosyl compounds. Part IV. The structure of homonataloin and the synthesis of nataloe-emodin. *Journal of the Chemical Society*, 4879–4885.

Herbin, G.A. and Robins, P.A. (1968) Studies on plant cuticular waxes I. The chemotaxonomy of alkanes and alkenes of the genus *Aloe* (Liliaceae). *Phytochemistry*, 7, 239–255.

Hikino, H., Takahashi, M., Murakami, M., Konno, C., Mirin, Y., Karikura, M. and Hayashi, T. (1986) Isolation and hypoglycemic activity of arborans A and B, glycans of *Aloe arborescens* var. *natalensis* leaves. *International Journal of Crude Drug Research*, 24, 183–186.

Hirata, T. and Suga, T. (1977) Biologically active constituents of leaves and roots of *Aloe arborescens* var. *natalensis*. *Zeitschrift für Naturforschung*, 32c, 731–734.

Holdsworth, D.K. (1972) Chromones in *Aloë* species. part 2. Aloesone. *Planta Medica*, 22, 54–58.

Höltje, H.-D., Stahl, K., Lohse, K. and Rauwald, H.W. (1990) Molecular-Modelling-Untersuchungen zur Konformation und Konfiguration natürlich vorkommender, diastereomerer 10-*C*-glucosylanthron- und -oxanthronverbindungen (Aloin- und 10-hydroxyaloin-Typ). *Archiv der Pharmazie*, 324, 859–861.

Holzapfel, C.W., Wessels, P.L., Van Wyk, B.-E., Marais, W. and Portwig, M. (1997) Chromone and aloin derivatives from *Aloe broomii, A. africana* and *A. speciosa*. *Phytochemistry*, 45, 97–102.

Hörhammer, L., Wagner, H. and Bittner, G. (1964) Aloinosid B, ein neues Glycosid aus Aloe. *Zeitschrift fur Naturforschung*, 196, 222–226.

Hörhammer, L., Wagner, H., Bittner, G. and Graf, E. (1965) Neue methoden im pharmakognostischen unterricht 10. Mitteilung:Unterscheidung handelsblicher aloesorten mittels Dünnschichtchromatographie. *Deutsche Apotheker-Zeitung*, 105, 827–830.

Hranisavljevic-Jakovljevic, M. and Miljkovic-Stojanovic, J. (1981) Structural study of an acidic polysaccharide isolated from *Aloe arborescens* Mill. I. Periodate oxidation and partial acid hydrolysis. *Bulletin de la Societe Chimique Beograd*, 46, 269–273.

Hutter, J.A., Salman, M., Stavinoha, W.B., Satsangi, N., Williams, R.F., Streeper, R.T. and Weintraub, S.T. (1996) Antiinflammatory *C*-glucosyl chromone from *Aloe barbadensis*. *Journal of Natural Products*, 59, 541–543.

Ishikawa, M., Yamamoto, M. and Masui, T. (1987) Studies on analysis of organic acids and amino acids in various aloe species. *Shizuoka-ken Eisei Kankyo Senta Hokoku*, 30, 25–30.

Ito, S., Teradaira, R., Beppu, H., Obata, M., Nagatsu, T. and Fujita, K. (1993) Properties and pharmacological activity of carboxypeptidase in *Aloe arborescens* Mill. var. *natalensis* Berger. *Phytotherapy Research*, 7, S26-S29.

Kameoka, H., Maruyama, H. and Miyazawa, M. (1981) The constituents of the steam volatile oil from *Aloe arborescens* Mill. var. *natalensis* Berger. *Nippon Nogeikagaku Kaishi*, 55, 997–999.

Kashiwada, Y., Nonaka, G.-I. and Nishioka, I. (1984) Studies on rhubarb (Rhei Rhizoma). V. Isolation and characterization of chromone and chromanone derivatives. *Chemical and Pharmaceutical Bulletin*, 32, 3493–3500.

Khan, R.H. (1983) Investigating the amino acid content of the exudate from the leaves of Aloe barbadensis (Aloe vera). *Erde International*, 1, 19–25.

Kimura, Y., Kozawa, M., Baba, K. and Hata, K. (1983) New constituents of roots of Polygonum cuspidatum. *Planta Medica*, 48, 164–168.

Kinoshita, K., Koyama, K., Takahashi, K., Noguchi, Y. and Amano, M. (1996) Steroid glucosides from *Aloe barbadensis*. *Journal of Japanese Botany*, 71, 83–86.

Kodym, A. (1988) Mineral analysis of the dry extract obtained from fresh leaves of Aloe arborescens. *Farmacja Polska*, 44, 71–74.

Kodym, A. (1991) The main chemical components contained in fresh leaves and in a dry extract from three years old *Aloe arborescens* Mill.grown in hothouses. *Pharmazie*, 46, 217–219.

Koike, T., Beppu, H., Kuzuya, H., Maruta, K., Shimpo, K., Suzuki, M., Titani, K. and Fujita, K. (1995) A 35 kDa mannose-binding lectin with hemagglutinating and mitogenic activities from 'Kidachi Aloe' (*Aloe arborescens* Miller var. *natalensis* Berger). *Journal of Biochemistry*, **118**, 1205–1210.

Koyama, J., Ogura, T. and Tagahara, K. (1994) Naphtho [2,3-c] furan-4,9-dione and its derivatives from *Aloe ferox*. *Phytochemistry*, **37**, 1147–1148.

Lee, K.J., Weintraub, S.T. and Yu, B.P. (2000) Isolation and identification of a phenolic anti-oxidant from *Aloe barbadensis*. *Free Radical Biology and Medicine*, **28**, 261–265.

Léger, M.E. (1917) Les aloïnes. Deuxième partie. *Annales de Chimie*, **8**, 265–302.

Mabusela, W.T., Stephen, A.M. and Botha, M.C. (1990) Carbohydrate polymers from *Aloe ferox* leaves. *Phytochemistry*, **29**, 3555–3558.

Makino, K., Yagi, A. and Nishioka, I. (1974) Studies on the constituents of *Aloe arborescens* Mill. var. *natalensis* Berger.II. The structures of two new aloesin esters. *Chemical and Pharmaceutical Bulletin*, **22**, 1565–1570.

Mandal, G. and Das, A. (1980a) Structure of the D-galactan isolated from *Aloe barbadensis* Miller. *Carbohydrate Research*, **86**, 247–257.

Mandal, G. and Das, A. (1980b) Structure of the glucomannan isolated from the leaves of *Aloe barbadensis* Miller. *Carbohydrate Research*, **87**, 249–256.

Mandal, G., Ghosh, R. and Das, A. (1983) Characterisation of polysaccharides of *Aloe barbadensis* Miller:Part III-Structure of an acidic oligosaccharide. *Indian Journal of Chemistry*, **22B**, 890–893.

Manitto, P., Monti, D. and Speranza, G. (1990a) Conformational studies of natural products. III. Conformation of natural 8-glucosyl-7-hydroxy-5-methylchromones and their derivatives. *Gazzetta Chimica Italiana*, **120**, 641–646.

Manitto, P., Monti, D. and Speranza, G. (1990b) Studies on Aloe. Part 6. Conformation and absolute configuration of aloins A and B and related 10-*C*-glucosyl-9-anthrones. *Journal of the Chemical Society, Perkin Transactions*, **1**, 1297–1300.

Manna, S. and McAnalley, B.H. (1993) Determination of the position of the *O*-acetyl group in a beta (1 → 4) mannan (acemannan) from *Aloe barbardensis* Miller. *Carbohydrate Research*, **241**, 317–319.

McAnalley, B.H. (1988) Process for preparation of aloe products products, produced thereby and composition thereof. *U.S. Patent 4, 735, 935.*

McAnalley, B.H. (1990) Processes for preparation of aloe products products, produced thereby and composition thereof. *U.S. Patent 4, 917 890.*

McDaniel, H.R., Perkins, S. and McAnalley, B.H. (1987) A clinical pilot study using Carrsyn in the treatment of aquired immunodeficiency syndrome (AIDS). *American Journal of Clinical Pathology*, **88**, 534.

Mebe, P.P. (1987) 2'-*p*-Methoxycoumaroylaloeresin, a *C*-glucoside from *Aloe excelsa*. *Phytochemistry*, **26**, 2646–2647.

Nash, R.J., Beaumont, J., Veitch, N.C., Reynolds, T., Benner, J., Hughes, C.N.G., Dring, J.V., Bennett, R.N., Dellar, J.E. (1992) Phenyethylamine and piperidine alkaloids in *Aloe* species. *Planta Medica*, **58**, 84–86.

Nyazema, N.Z. (1984) Poisoning due to traditional remedies. *The Central African Journal of Medicine*, **30**, 80–83.

Okamura, N., Hine, N., Harada, S., Fujioka, T., Mihashi, K. and Yagi, A. (1996) Three chromone components from *Aloe vera* leaves. *Phytochemistry*, **43**, 495–498.

Okamura, N., Hine, N., Tateyama, Y., Nakazawa, M., Fujioka, T., Mihashi, K. and Yagi, A. (1997) Three chromones of *Aloe vera* leaves. *Phytochemistry*, **45**, 1511–1513.

Okamura, N., Hine, N., Tateyama, Y., Nakazawa, M., Mihashi, K. and Yagi, A. (1998) Five chromones from *Aloe vera* leaves. *Phytochemistry*, **49**, 219–223.

Park, M.K., Park, J.H., Shin, Y.G., Kim, W.Y., Lee, J.H. and Kim, J.H. (1996) Neoalosin A: a new *C*-glucofuranosyl chromone from *Aloe barbadensis*. *Planta medica*, **62**, 363–365.

Parry, O. and Matambo, C. (1992) Some pharmacological actions of aloe extracts and *Cassia abbreviata* on rats and mice. *Central African Journal of Medicine*, 38, 409–414.

Parry, O. and Wenyika, J. (1994) The uterine effect of *Aloe chabaudii*. *Fitoterapia*, 65, 253–259.

Paulsen, B.S., Fagerheim, E. and Overbye, E. (1978) Structural studies of the polysaccharide from *Aloe plicatilis* Miller. *Carbohydrate research*, 60, 345–351.

Pecere, T., Gazzola, M.V., Mucignat, C., Parolin, C., Vecchia, F.D. and 6 others. (2000) Aloe-emodin is a new type of anticancer agent with selective activity against neuroectodermal tumors. *Cancer Research*, 60, 2800–2804.

Radjabi, F., Amar, C. and Vilkas, E. (1983) Structural studies of the glucomannan from *Aloe vahombe*. *Carbohydrate research*, 116, 166–170.

Radjabi-Nassab, F., Ramiliarison, C., Monneret, C. and Vilkas, E. (1984) Further studies of the glucomannan from *Aloe vahombe* (liliaceae). II. Partial hydrolyses and NMR 13C studies. *Biochimie*, 66, 563–567.

Rauwald, H.-W. (1982) Präparative Trennung der diastereomeren Aloine mittels Droplet-Counter-Current-Chromatography (DCCC). *Archiv der Pharmazie*, 315, 769–772.

Rauwald, H.-W. (1987) New hydroxyaloins: the 'peroidate-positive substance' from Cape Aloes and cinnamoyl esters from Curacao Aloes. *Pharmaceutisch Weekblad*, 9, 215.

Rauwald, H.-W. (1990) Naturally occurring quinones and their related reduction forms: analysis and analytical methods. *Pharmazeutische Zeitung Wissenschaft*, 3, 169–181.

Rauwald, H.-W. and Beil, A. (1993a) 5-Hydroxyaloin A in the genus *Aloe* thin layer chromatographic screening and high performance liquid chromatographic determination. *Zeitschrift für Naturforschung*, 48c, 1–4.

Rauwald, H.-W. and Beil, A. (1993b) High-performance liquid chromatographic separation and determination of diastereomeric anthrone-*C*-glucosyls in Cape aloes. *Journal of Chromatography*, 639, 359–362.

Rauwald, H.-W. Beil, A. and Prodöhl, C.P. (1991c) Occurrence, distribution and taxonomic significance of some *C*-glucosylanthrones of the aloin-type and *C*-glucosylchromones of the aloeresin-type in *Aloe* species. *Planta medica*, 57, A129.

Rauwald, H.-W. and Diemar, J. (1986) The first naturally occuring esters in aloin-type glycosyls. *Planta medica*, 52, 570.

Rauwald, H.-W. and Lohse, K. (1992) Strukturrevision des 4-hydroxyaloin:10-hydroxyaloin A und B als Haupt-*in vitro*-Oxidationsprodukte der diastereomeren aloine. *Planta medica*, 58, 259–262.

Rauwald, H.-W., Lohse, K. and Bats, J.W. (1989) Establishment of configurations for the two diastereomeric *C*-glucosylanthrones aloin A and aloin B. *Angewandte Chemie-International Edition in English*, 28, 1528–1529.

Rauwald, H.-W., Maucher, R. and Niyonzima, D.-D. (1997) Three 8-*C*-glucosyl-5-methyl-chromones from *Aloe barbadensis* (Ph.Eur.1997). *Pharmazie*, 52, 962–964.

Rauwald, H.-W. and Niyonzima, D.-D. (1991a) A new investigation on constituents of *Aloe* and *Rhamnus* species.XV Homonataloins A and b from *Aloe lateritia*: isolation, structure and configurational determination of the diastereomers. *Zeitschrift für Naturforschung*, 46c, 177–182.

Rauwald, H.-W. and Niyonzima, D.-D. (1991b) Free and cinnamoylated 8-*O*-methyl-7-hydroxyaloins from *Aloe barbadensis*:isolation, structure and configurational determination of the diastereomers. *Planta medica*, 57, A129.

Rauwald, H.-W. and Voetig, R. (1982) 7-Hydroxy-Aloin:die Leitsubstanz aus *Aloë barbadensis* in der Ph.Eur.III. *Archiv der Pharmazie*, 315, 477–478.

Reynolds, G.W. (1966) *The Aloes of Tropical Africa and Madagascar*, p. 326. Mbabane, Swaziland: The Aloes Book Fund.

Reynolds, T. (1985a) The compounds in *Aloë* leaf exudates: a review. *Botanical Journal of the Linnean Society*, 90, 157–177.

Reynolds, T. (1985b) Observations on the phytochemistry of the *Aloë* leaf-exudate compounds. *Botanical Journal of the Linnean Society*, 90, 179–199.

Reynolds, T. (1986) A contribution to the phytochemistry of the East African tetraploid shrubby aloes and their diploid allies. *Botanical Journal of the Linnean Society*, **92**, 383–392.

Reynolds, T. (1990) Comparative chromatographic patterns of leaf exudate compounds from shrubby aloes. *Botanical Journal of the Linnean Society*, **102**, 273–285.

Reynolds, T. and Dweck, A.C. (1999) Aloe vera leaf gel: a review update. *Journal of Ethnopharmacology*, **68**, 3–37.

Roboz, E. and Haagen-Smit, A.J. (1948) A mucilage from Aloe vera. *Journal of the American Chemical Society*, **70**, 3248–3249.

Rosenthaler, L. (1931) Untersuchungen über Bestandteile von Abführdrogen III. *Pharmaceutica Acta Helvetiae*, **6**, 115–117.

Ross, S.A., ElSohly, M.A. and Wilkins, S.P. (1997) Quantitative analysis of Aloe vera mucilaginous polysaccharides in commercial Aloe vera products. *Journal of AOAC International*, **80**, 455–457.

Rowe, T.D. and Parks, L.M. (1941) Phytochemical study of *Aloe vera* leaf. *Journal of the American Pharmaceutical Association*, **30**, 262–266.

Saito, H. (1993) Purification of active substances of *Aloe arborescens* Miller. and their biological and pharmacological activity. *Phytotherapy Research*, 7, S14–S19.

Saleem, R., Faizi, S., Deeba, F., Siddiqui, B.S. and Qazi, M.H. (1997a) A new bisbenzopyran from *Aloe barbadensis* roots. *Planta medica*, **63**, 454–456.

Saleem, R., Faizi, S., Deeba, F., Siddiqui, B.S. and Qazi, M.H. (1997b) Anthrones from *Aloe barbadensis*. *Phytochemistry*, **45**, 1279–1282.

Schmidt, J., Blitzke, T. and Masaoud, M. (2001) Structural investigations of 5-methylchromone glycosides from *Aloe* species by liquid chromatography/electrospray tandem mass spectrometry. *European Journal of Mass Spectrometry*, 7, 481–490.

Segal, A., Taylor, J.A. and Eoff, J.C. (1968) A re-investigation of the polysaccharide material from *Aloe vera* mucilage. *Lloydia*, **31**, 423.

Sigler, A. and Rauwald, H.W. (1994a) Tetrahydroanthracenes as markers for subterranean anthranoid metabolism in *Aloe* species. *Journal of Plant Physiology*, **143**, 596–600.

Sigler, A. and Rauwald, H.W. (1994b) First proof of anthrone aglycones and diastereomeric anthrone-C-glycosides in flowers and bracts of *Aloe* species. *Biochemical Systematics and Ecology*, **22**, 287–290.

Simpson, T.J. (1980) Biosynthesis of polyketides. *Biosynthesis*, **6**, 1–39.

Speranza, G., Corti, S. and Manitto, P. (1994) Isolation and chemical characterization of a new constituent of Cape Aloe having the 1,1-diphenylethane skeleton. *Journal of Agricultural and Food Chemistry*, **42**, 2002–2006.

Speranza, G., Dada, G., Lunazzi, L., Gramatica, P. and Manitto, P. (1986a) Aloenin B, a new diglucosylated 6-phenyl-2-pyrone from Kenya aloe. *Journal of Natural Products*, **49**, 800–805.

Speranza, G., Dada, G., Lunazzi, L., Gramatica, P. and Manitto, P. (1986b) A C-glucosylated 5-methylchromone from Kenya aloe. *Phytochemistry*, **25**, 2219–2222.

Speranza, G., Gramatica, P., Dada, G. and Manitto, P. (1985) Aloeresin C, a bitter C,O-diglucoside from Cape aloe. *Phytochemistry*, **24**, 1571–1573.

Speranza, G., Manitto, P., Cassara, P. and Monti, D. (1993a) Feralolide, a dihydroisocoumarin from Cape Aloe. *Phytochemistry*, **33**, 175–178.

Speranza, G., Manitto, P., Cassara, P. and Monti, D. (1993b) Studies on aloe, 12. furoaloesone, a new 5-methylchromone from cape aloe. *Journal of Natural Products*, **56**, 1089–1094.

Speranza, G., Manitto, P., Monti, D. and Lianza, F. (1990) Feroxidin, a novel 1-methyltetralin derivative isolated from cape aloes. *Tetrahedron letters*, **31**, 3077–3080.

Speranza, G., Manitto, P., Monti, D. and Pezzuto, D. (1992) Studies on aloe, Part 10. Feroxins A and B, two O-glucosylated 1-methyltetralins from cape aloe. *Journal of Natural Products*, **55**, 723–729.

Speranza, G., Manitto, P., Pezzuto, D. and Monti, D. (1991) Absolute configuration of feroxidin: an experimental support to Snatzke's helicity rules for tetralins. *Chirality*, 3, 263–267.

Speranza, G., Martignoni, A. and Manitto, P. (1988) Iso-aloeresin A, a minor constituent of cape aloe. *Journal of Natural Products*, 51, 588–590.

Strickland, F.M., Muller, H.K., Stephens, L.C., Bucana, C.D., Donawho, C.K., Sun, Y. and Pelley, R.P. (2000) Induction of primary cutaneous melanomas in C3H mice by combined treatment with ultraviolet radiation, ethanol and aloe emodin. *Photochemistry and Photobiology*, 72, 407–414.

Suga, T., Hirata, T. and Tori, K. (1974) Structure of aloenin, a bitter glucoside from *Aloe species. Chemistry Letters*, 715–718.

Suzuki, I., Saito, H., Inoue, S., Migita, S. and Takahashi, T. (1979) Purification and characterization of two lectins from *Aloe arborescens. Journal of Biochemistry*, 85, 163–171.

Umano, K., Nakahara, K., Shoji, A. and Shibamoto, T. (1999) Aroma chemicals isolated and identified from leaves of *Aloe arborescens* Mill. var. *natalensis* Berger. *Journal of Agricultural and Food Chemistry*, 47, 3702–3705.

Van der Bank, F.H. and Van Wyk, B.-E. (1996) Biochemical genetic markers to identify hybrids between *Aloe arborescens* and *A. ferox* (Aloaceae). *South African Journal of Botany*, 62, 328–331.

Van Heerden, F.R., Van Wyk, B.-E. and Viljoen, A. (1996) Aloeresins E and F, two chromone derivatives from *Aloe peglerae. Phytochemistry*, 43, 867–869.

Van Heerden, F.R., Viljoen, A. and Van Wyk, B.-E. (2000) 6'-*O*-Coumaroylaloesin from *Aloe castanea*-a taxonomic marker for *Aloe* section *Anguialoe, Phytochemistry*, 55, 117–120.

Van Heerden, F.R., Viljoen, A. and Van Wyk, B.-E. (2002) Homonataloside B, an anthrone diglucoside and two new chromones from *Aloe lutescens. Journal of Natural Products*, in the press.

Van Rheede van Oudtshoorn, M.C.B. (1963) Preliminary chemotaxonomical observations on *Aloe* juices and on *Bulbine* species. *Planta Medica*, 11, 332–337.

Van Rheede van Oudtshoorn, M.C.B. (1964) Chemotaxonomic investigations in Asphodeleae and Aloineae (Liliaceae). *Phytochemistry*, 3, 383–390.

Van Wyk, B.-E., Van Rheede van Oudtshoorn, M.C.B. and Smith, G.F. (1995) Geographical variation in the major compounds of *Aloe ferox* leaf exudate. *Planta Medica*, 61, 250–253.

Van Wyk, B.-E., Whitehead, C.S., Glen, H.F., Hardy, D.S., Van Jaarsveld, E.J. and Smith, G.F. (1993) Nectar sugar composition in the subfamily Alooideae (Asphodelaceae). *Biochemical Systematics and Ecology*, 21, 249–253.

Van Wyk, B.-E., Yenesew, A. and Dagne, E. (1995a) Chemotaxonomic survey of anthraquinones and pre-anthraquinones in roots of *Aloe* species. *Biochemical Systematics and Ecology*, 23, 267–275.

Van Wyk, B.-E., Yenesew, A. and Dagne, E. (1995b) Chemotaxonomic significance of anthraquinones in the roots of Asphodeloideae (Asphodelaceae). *Biochemical Systematics and Ecology*, 23, 277–281.

Vazquez, B., Avila, G., Segura, D. and Escalante, B. (1996) Anti inflammatory activity of extracts from *Aloe vera* gel. *Journal of Ethnopharmacology*, 55, 69–75.

Van Wyk, B.-E., Yenesew, A. and Dagne, E. (1995c) The chemotaxonomic significance of root anthraquinones and pre-anthraquinones in the genus *Lomatophyllum* (Aspodelaceae). *Biochemical Systematics and Ecology*, 23, 805–808.

Veitch, N.C., Simmonds, M.S.J., Blaney, W.M. and Reynolds, T. (1994) A dihydroisocoumarin glucoside from *Aloe hildebrandtii. Phytochemistry*, 35, 1163–1166.

Viljoen, A.M. and Van Wyk, B.-E. (2001) A chemotaxonomic and morphological appraisal of *Aloe* series Purpurascentes, *Aloe* section Anguialoe and their hybrid, *Aloe broomii. Biochemical Systematics and Ecology*, 29, 621–631.

Viljoen, A.M., Van Wyk, B.-E. and Dagne, E. (1996) The chemotaxonomic value of 10-hydroxyaloin B and its derivatives in *Aloe* species Asperifoliae Berger. *Kew Bulletin*, 51, 159–168.

Viljoen, A.M., Van Wyk, B.-E. and Newton, L.E. (1999) Plicataloside in *Aloe*—a chemotaxonomic appraisal. *Biochemical Systematics and Ecology*, 27, 507–517.

Viljoen, A.M., Van Wyk, B.-E. and Newton, L.E. (2001) The occurrence and taxonomic distribution of the anthrones aloin, aloinoside and microdontin in *Aloe*. *Biochemical Systematics and Ecology*, 29, 53–67.

Viljoen, A.M., Van Wyk, B.-E. and Van Heerden, F.R. (1998) Distribution and chemotaxonomic significance of flavanoids in *Aloe* (*Asphodelaceae*). *Plant Systematics and Evolution*, 211, 31–42.

Viljoen, A.M., Van Wyk, B.-E. and Van Heerden, F.R. (2002) The chemotaxonomic value of the diglucoside anthrone homonataloside B in the genus *Aloe*. *Biochemical Systematics and Ecology*, 30, 35–43.

Vilkas, E. and Radjabi Nassab, F. (1986) The glucomannan system from *Aloe vahombe (liliaceae)*. III. Comparative studies on the glucomannan components isolated from the leaves. *Biochimie*, 68, 1123–1127.

Waller, G.R., Mangiafico, S. and Ritchey, C.R. (1978) A chemical investigation of *Aloe barbadensis* Miller. *Proceedings of the Oklahoma Academy of Science*, 58, 69–76.

Wessels, P.L., Holzapfel, C.W., Van Wyk, B.-E. and Marais, W. (1996) Plicatiloside, an 0,0-diglycosylated naphthalene derivative from *Aloe plicatilis*. *Phytochemistry*, 41, 1547–1551.

Winters,W.D. (1993) Immunoreactive lectins in leaf gel from *Aloe barbadensis* Miller. *Phytotherapy Research*, 7, S23–S25.

Winters, W.D. and Bouthet, C.F. (1995) Polypeptides of *Aloe barbadensis* Miller. *Phytotherapy Research*, 9, 395–400.

Woo, W.S., Shin, K.H., Chung, H.S., Lee, J.M. and Shim, K.S. (1994) Isolation of an unusual aloenin-acetal from Aloe. *Korean Journal of Pharmacognosy*, 25, 307–310.

Wozniewski, T., Blaschek, W. and Franz, G. (1990) Isolation and structure analysis of a glucomannan from the leaves of *Aloe arborescens* var. Miller. *Carbohydrate Research*, 198, 387–391.

Yagi, A., Egusa, T., Arase, M., Tanabe, M. and Tsuji, H. (1997) Isolation and characterization of the glycoprotein fraction with a proliferation-promoting activity on human and hamster cells *in vitro* from *Aloe vera* gel. *Planta medica*, 63, 18–21.

Yagi, A., Hamada, K., Mihashi, K., Harada, N. and Nishioka, I. (1984) Structure determination of polysaccharides in *Aloe saponaria* (Hill.)Haw. (Liliaceae). *Journal of Pharmaceutical Sciences*, 73, 62–65.

Yagi, A., Harada, N., Shimomura, K. and Nishioka, I. (1987) Bradykinin-degrading glycoprotein in *Aloe arborescens* var. *natalensis*. *Planta medica*, 53, 19–21.

Yagi, A., Harada, N., Yamada, H., Iwadare, S. and Nishioka, I. (1982) Antibradykinin active material in *Aloe saponaria*. *Journal of Pharmaceutical Sciences*, 71, 1172–1174.

Yagi, A., Shida, T. and Nishimura, H. (1987) Effect of amino acids in Aloe extract on phagocytosis by peripheral neutrophil in adult bronchial asthma. *Arerugi*, 36, 1094–1101.

Yagi, A., Hine, N., Asai, M., Nakazawa, N., Tateyama, Y., Okamura, N., Fujioka, T., Mihashi, K. and Shimomura, K. (1998) Tetrahydroanthracene glucosides in callus tissue from *Aloe barbadensis* leaves. *Phytochemistry*, 47, 1267–1270.

Yagi, A., Makino, K. and Nishioka, I. (1974) Studies on the constituents of *Aloe saponaria* Haw. I. The structures of tetrahydroanthracene derivatives and the related anthraquinones. *Chemical and Pharmaceutical Bulletin*, 22, 1159–1166.

Yagi, A., Makino, K. and Nishioka, I. (1977a) Studies on the constituents of *Aloe saponaria* Haw. II. The structures of tetrahydroanthracene derivatives, aloesaponol III and IV. *Chemical and Pharmaceutical Bulletin*, 25, 1764–1770.

Yagi, A., Makino, K. and Nishioka, I. (1977b) Studies on the constituents of Aloe saponaria Haw. III. The structures of phenol glucosides. *Chemical and Pharmaceutical Bulletin*, 25, 1771–1776.

Yagi, A., Makino, K. and Nishioka, I. (1978b) Studies on the constituents of Aloe saponaria Haw. IV. The structures of bianthraquinoid pigments. *Chemical and Pharmaceutical Bulletin*, 26, 1111–1116.

Yagi, A., Makino, K., Nishioka, I. and Kuchino, Y. (1977) Aloe mannan, polysaccharide from Aloe *arborescens* var. *natalensis. Planta medica*, 31, 17–20.

Yagi, A., Nishimura, H., Shida, T. and Nishioka, I. (1986) Structure determination of polysaccharides in *Aloe arborescens* var. *natalensis. Planta Medica*, 52, 213–217.

Yagi, A., Shoyama, Y. and Nishioka, I. (1983) Formation of tetrahydroanthracene glucosides by callus tissue of *Aloe saponaria. Phytochemistry*, 22, 1483–1484.

Yagi, A., Yamanouchi, M. and Nishioka, I. (1978a) Biosynthetic relationship between tetrahydroanthracene and anthraquinone in *Aloe saponaria. Phytochemistry*, 17, 895–897.

Yamaguchi, I., Mega, N. and Sanada, H. (1993) Components of the gel of *Aloe vera* (L.) Burm.f. *Bioscience, Biotechology, Biochemistry*, 57, 1350–1352.

Yamamoto, M., Akimoto, N. and Masui, T. (1986) Studies on determination of β-sitosterol in aloe. *Shizuoka-ken Eisei Kankyo Senta Hokoku*, 29, 47–51.

Yamamoto, M., Masui, T., Sugiyama, K., Yokota, M., Nakagomi, K. and Nakazawa, H. (1991) Anti-inflammatory active constituents of *Aloe arborescens* Miller. *Agricultural and Biological Chemistry*, 55, 1627–1629.

Yenesew, A., Ogur, J.A. and Duddek, H. (1993) (R)-Prechrysophanol from *Aloe graminicola. Phytochemistry*, 34, 1442–1444.

Yoshimoto, R., Kondoh, N., Isawa, M. and Hamuro, J. (1987) Plant lectin, ATF1011, on the tumor cell surface augments tumor-specific immunity through activation of T cells specific for the lectin. *Cancer Immunology Immunotherapy*, 25, 25–30.

4   Aloe polysaccharides

Yawei Ni, Kenneth M. Yates and Ian R. Tizard

ABSTRACT

It has been widely believed for several years, that many of the beneficial effects of aloe leaf extracts lie in their carbohydrates. The thick fleshy leaves contain both cell wall carbohydrates such as celluloses and hemicelluloses, as well as storage carbohydrates such as acetylated mannans, arabinans and arabinogalactans. Like all cells, they also contain many diverse glycoproteins. Acetylated mannan is the primary polysaccharide in the pulp, the inner clear portion of the leaf and has been most widely studied. It has been claimed to possess many therapeutic properties, including immune stimulation. The general structure of the mannan ($\beta 1 \rightarrow 4$ linked mannose residues and acetylation) has been well defined, although many structure features such as degree of acetylation, glucose content, and molecular weight remain to be defined, especially in relation to functional properties, extraction conditions, and harvesting times. There may well be two types of mannans in the pulp, a pure mannan and a glucomannan. There is very little information available on the enzymes used to modify or synthesize the mannan or other aloe carbohydrates. Aloe mannan is indeed a unique polysaccharide. Future studies on the structure–function relationship will certainly yield more insight into its chemical and functional properties.

INTRODUCTION

Aloe leaves, especially their inner clear part or pulp, are widely used in various medical, cosmetic, and nutraceutical applications. Many of the beneficial effects of these plants have been attributed to the polysaccharides present in the leaf pulp (Grindlay and Reynolds, 1986; Reynolds and Dweck, 1999). Among the many constituents that have been identified in the pulp, the polysaccharides are most abundant and widely studied. The earliest description of polysaccharides from aloe leaves was probably made by Rowe and Parks (1941). They found that an alcohol-insoluble fraction of the pulp contains 4.7% uronic acid along with other sugars. This was followed by a study by Roboz and Haagen-Smit (1948) that identified a polysaccharide consisting of mannose and glucose at a ~1:1 ratio. Subsequently, a partially acetylated $\beta 1$-4 linked mannan was identified based on the comparison of its infra-red (IR) spectrum to that of a known plant $\beta 1$-4 linked mannan. Since then, various other polysaccharides have been detected or isolated from the pulp, including mannans (Segal *et al.*, 1968; Yagi *et al.*, 1977, 1984; Gowda *et al.*, 1979; Rajabi-Nassad *et al.*, 1983; t'Hart *et al.*, 1989), galactans (Mandal and Das, 1980a), arabinogalactans (Mabusela *et al.*, 1990), pectic substances (Ovodova *et al.*, 1975; Mandal and Das, 1980a, b), and others that may contain a mixture of sugars (Hranisavljevic-Jakovljevic and Miljkovic-Stojanovic, 1981; Hikino *et al.*, 1986). However, the most often isolated and the most widely studied is the mannan.

Significant variations exist in the major pulp polysaccharides. For example, several studies have identified the mannan as the major polysaccharide of the pulp. However, other studies found that a pectic substance was the primary polysaccharide (Ovodova *et al.*, 1975; Mandal and Das, 1980a, b). The reason for such discrepancy is not understood, but has been attributed to seasonal changes and/or different geographic locations (Mandal and Das, 1980a, b; Grindley and Reynolds, 1986).

Aloe polysaccharides have been well reviewed in previous publications (Grindley and Reynolds, 1986; Reynolds and Dweck, 1999). Here we will concentrate on examination and comparison of the detailed structures of the identified polysaccharides, especially the mannan, from different *Aloe* species. Efforts have been made to indicate the *Aloe* species wherever a polysaccharide or its structure feature is described. Several different types of mannans have been identified in aloes, including mannans, glucomannans, and glucogalactomannans. Here, the term mannan will be used to describe all of them unless otherwise indicated.

Since aloes are members of a plant family (Liliaceae), we will first describe the general features of plant polysaccharides before addressing those from *Aloe* so that the latter will not be treated in an isolated manner.

Plant polysaccharides

There exist an enormous variety of plant polysaccharides. Different polysaccharides can be found in different plants and may be associated with different plant structures, such as leaves, seeds, roots, and tubers, etc. Polysaccharides may be very complex molecules, more so than any protein, lipid, or nucleic acid. This is due to two unique structural features that are found only in polysaccharides: (1) any two monosaccharides can be liked together in many different ways ($1 \rightarrow 2$, $1 \rightarrow 3$, $1 \rightarrow 4$ and so on in an α or β configuration); and, (2) the presence of branched side-chains. Thus, it is not surprising that there are so many different plant polysaccharides (Roybt, 1998). However, looking from the cell structure and functional point of view, there are some common polysaccharides associated with each plant structure (Kennedy and White, 1983; Robyt, 1998), although their structural details may vary among different plants. Aloes would therefore be expected to have at least some of these common plant polysaccharides although their particular structural details may be unique.

Plant cell walls contain celluloses, hemicelluloses, and pectins. Cellulose is a $\beta 1 \rightarrow 4$ linked glucan and synthesized by enzymes located on plasma membranes. Hemicelluloses include the xylans, xyloglucans, mannans, and galactans, all of which have $\beta 1 \rightarrow 4$ linked main chains. Pectins are acidic polysaccharides consisting primarily of $\alpha 1 \rightarrow 4$ linked polygalacturonic acids with intra-chain rhamnose insertion, an attachment of neutral sugar side-chains, and methyl esterification.

In plant cells or seeds, there are storage polysaccharides. The most important is starch ($\alpha 1 \rightarrow 4$ linked glucan) which is synthesized by cellular organelles in the cytoplasm called plastids. Starch can be separated into two components, amylose (a linear

$\alpha 1 \rightarrow 4$ linked glucan) and amylopectin (an $\alpha 1 \rightarrow 4$ linked glucan with $\alpha 1 \rightarrow 6$ linked branches). There are other types of storage polysaccharides, e.g. mannans, galactomannans, and glucomannans found in the thickened cell walls of storage tissues in plant seeds (Bewley and Reid, 1985; Reid, 1985), all of which have the $\beta 1 \rightarrow 4$ linked main chain. In galactomannans, single-unit galactose residues are $\alpha 1 \rightarrow 6$ linked to the mannose residues in the main chain as branches. In glucomannans, both glucose and mannose residues are $\beta 1 \rightarrow 4$ linked in a linear chain. Examples of these mannans include ivory nut mannan, Konjac mannan (a glucomannan), and locust bean gum (a galactomannan).

Gums are viscous exudates produced by plants to seal wounds in their bark. They consist of highly branched polysaccharides. Their composition and structure vary widely. Their major components can be arabinans, arabinogalactans, galacturonans, or xylans. Typically, uronic acid is also present. A closely related group of polysaccharides is found in mucilage. Mucilage is also a viscous liquid which functions as a water reservoir and is found in plant mucilage tissues or secreted by the seed covers or root tips of various plants. Uronic acid, especially the galacturonic acid (Gal A), is also a common constituent of mucilage polysaccharides.

In plant cell membranes, there are many glycoproteins and proteoglycans. One major type of proteoglycan is the arabinogalactan protein. The glycan part of this proteoglycan or arabinogalactan constitutes as much as 90% of the mass and consists of arabinose and galactose along with uronic acid and other sugars as minor components (Knox, 1995). In addition, the arabinogalactans constitute as much as 70% of total polysaccharide in the plant Golgi apparatus.

Total carbohydrate composition of the aloe pulp

Before dealing with the individual polysaccharides in the pulp, it is useful to have an overview of the total carbohydrate composition of pulp. The predominant sugars found in the pulp are mannose and glucose, accounting for >70% of the total sugars. Waller *et al.* (1978) determined the composition of lyophilized pulp of *Aloe vera* (L.)Burm.f. and found six sugars (arabinose, galactose, glucose, mannose, rhamnose, and xylose), among which mannose and glucose accounted for 85% with a glucose/mannose ratio of 1:1.3. Femenia *et al.* (1999) determined the sugar composition of alcohol insoluble residues (AIRs) prepared from the pulp of *A. vera* and found eight sugars (arabinose, galactose, glucose, mannose, rhamnose, xylose, fucose and uronic acid), among which mannose and glucose accounted for 73% with a glucose/mannose ratio of 1:1.7. In this latter case, the free monosaccharides are most likely excluded because they are alcohol soluble, i.e., the result reflects more on the composition of polysaccharides.

The monosaccharides are a significant component of the pulp, accounting for as much as 20–30% of the total dry matter from the liquid gel prepared from the pulp of *A. vera* by centrifugation or extrusion (Yaron, 1993; Femenia *et al.*, 1999; Paez *et al.*, 2000). Glucose is the dominant monosaccharide, accounting for ~95% of the total monosaccharides (Femenia *et al.*, 1999). This is probably the reason that there is less glucose in AIRs than in whole gel.

Mannan

Mannans are the most widely studied polysaccharides from aloes. This is because they are the most often isolated and have been shown to have several biological effects

(Marshall *et al.*, 1993; Tizard *et al.*, 1989; Zhang and Tizard, 1996; Ramamoorthy and Tizard, 1998). The mannan from *A. vera* is a partially acetylated $\beta1\rightarrow4$ linked polymannose (Yagi *et al.*, 1977; Paulsen *et al.*, 1978; Gowda *et al.*, 1979). A mannan with these two primary characteristics (the $\beta1\rightarrow4$ linkage and acetylation) has been detected in all *Aloe* species analyzed. In some preparations, the mannan also contains a significant amount of glucose and is therefore a glucomannan. The aloe mannan appears to be structurally unique among those widely known plant $\beta1\rightarrow4$ mannans including galactomannan, Konjac mannan, and ivory mannan (Kennedy and White, 1983), thus making it a compound characteristic of *Aloe* species (and possibly other members of the Liliaceae family). These other plant mannans either have distinct side-chains or are unacetylated and insoluble.

Composition and linkage

Structures of the aloe $\beta1\rightarrow4$ linked mannan as described in the literature vary greatly with regard to their degree of acetylation (d.s.), presence of glucose and/or other sugars, and existence of branches (Table 4.1). Paulsen *et al.* (1978) described an acetylated $\beta1\rightarrow4$ linked glucomannan with a glucose:mannose ratio of 1:2.8 and a degree of substitution by acetyl groups (d.s.) of 0.67, from *A. plicatilis* (L.)Miller. The substitution by the acetyl groups occurred randomly among the available positions. Several $\beta1\rightarrow4$ linked acetylated glucomannans with a glucose:mannose ratio of 1:0.6–1:19 and a d.s. of up to 0.78 were isolated under different extraction conditions from *A. vera* by Gowda *et al.* (1979). In both cases, the glucose residues were found to be $\beta1\rightarrow4$ linked with mannose residues in the main chain with no branching. A similar acetylated $\beta1\rightarrow4$ glucomannan was described in *A. vahombe* (sic, =*A. vaombe* Decorse et Poiss.) (Radjabi *et al.*, 1983), which had a glucose:mannose ratio of 1:3 and a d.s. of 0.33. It was also suggested that acetylation primarily occurred on glucose residues. This group of researchers further isolated several glucomannans under different conditions from the same *Aloe* species which differed in the glucose:mannose ratio (1:2–1:4) and acetylation (d.s. of 0.17–0.57) (Vilkas and Radjabi-Nassab, 1986). Wozniewski *et al.* (1990) also isolated a glucomannan from *A. arborescens* Miller that had a glucose:mannose ratio of 1:19 and a d.s. of 1.3.

On the other hand, Mandal and Das (1980b) described a glucomannan from *A. vera*, which had a glucose:mannose ratio of 1:20, but was non-acetylated, insoluble, and branched. Gowda (1980), after repeated purification steps, isolated a pure $\beta1\rightarrow4$ linked mannan preparation containing no glucose with a d.s. of 0.81–0.87 from both *A. saponaria* (Ait.)Haw and *A. vanbalenii* Pillans. Another acetylated $\beta1$-4 linked mannan without glucose was identified in *A. saponaria* and *A. arborescens* by Yagi *et al.* (1977, 1984, 1986). Haq and Hannan (1981) isolated a glucogalactomannan with a glucose:mannose:galactose ratio of 2:2:1 from *A. vera*.

Several other reports described isolation of mannose-rich polysaccharides from aloes, but no linkage study was performed (Hikino *et al.*, 1986; t'Hart *et al.*, 1989; Mabusela *et al.*, 1990; Yaron, 1993). The two polysaccharides isolated by Hikino *et al.* (1986) from *A. arborescens* appeared to be very different from the others. One contained glucose and mannose at a ratio of 1:0.3 and the other contained rhamnose, fucose, arabinose, xylose, mannose, galactose, and glucose at a ratio of 0.3:0.2:0.1:0.1:0.2:1:0.3. Both contained a small amount of protein and were acetylated. t'Hart *et al.* (1989) isolated a mannose-rich polysaccharide from *A. vera* by ion exchange chromatography, which

Colour Plate 1 Aloe vera plant growing in Barranco de Infierno, Tenerife, apparently native but in reality an escape from cultivation. (Photo, L.A. Reynolds.)

Colour Plate 2 Sunbird on *Aloe cheranganiensis* flower (*see page 8*).

Colour Plate 3 Hedge of *Aloe kedongensis* near Naivasha, Kenya (*see page 11*).

Colour Plate 4 Erosion control using *Aloe chrysostachys* near Mwingi, Kenya (*see page 11*).

Colour Plate 5 Sectioned leaf of the yellow-flowered *A. barbadensis* Miller (cv. RGV). The rind (R) and attached mesophyll (M) are readily stripped away from the gel parenchyma (G). This photograph was taken of leaves at the original 'Hilltop' plantation in Lyford, Texas (*see page 154*).

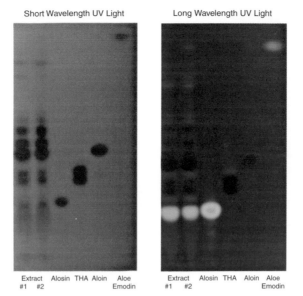

Short Wavelength UV Light   Long Wavelength UV Light

Extract Alosin THA Aloin   Aloe   Extract Alosin THA Aloin   Aloe
#1   #2   Emodin   #1   #2   Emodin

Colour Plate 6 Anthraquinones and chromones from *A. barbadensis* gel and representative purified and partially purified compounds. Compounds were placed on Merck silica gel G-60 plates and developed in toluene-ethyl acetate – methanol – water – formic acid = 10:40:12:6:3. Plates were photographed under short (280 nm) or long (350 nm) wavelength UV light without any specific staining. Identity of the standards, aloesin (2 μg), aloin A or barbaloin (10 μg) and aloe emodin (0.2 μg) were confirmed by mass spectroscopy. Other zones contained 100 μg of extract. Extracts were replicate acetone extracts of <5,000 molecular weight cut-off membrane (MWCO) dialysates of ARF'92B Standard Sample Gel (*see page 172*).

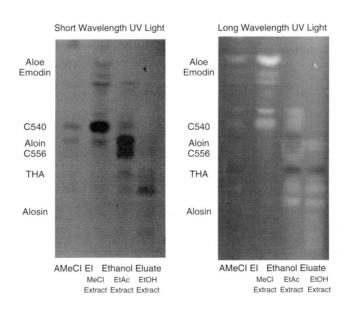

Short Wavelength UV Light   Long Wavelength UV Light

Aloe   Aloe
Emodin   Emodin

C540   C540
Aloin   Aloin
C556   C556

THA   THA

Alosin   Alosin

AMeCl EI Ethanol Eluate   AMeCl EI Ethanol Eluate
      MeCl   EtAc   EtOH   MeCl   EtAc   EtOH
    Extract Extract Extract   Extract Extract Extract

Colour Plate 7 TLC analysis of acetone/methylene chloride (AcMeCl) eluates (lane 1) and ethanol eluates (lanes 2–4) of aloe-adsorbed activated charcoal. The ethanol eluate has been fractionated by successive extractions with methylene chloride (MeCl), ethyl acetate (EtAc), and hot ethanol (EtOH). Along the right side of the plate are indicated the Rfs of reference compounds: C540, a chromone ester of mass 540 daltons; C556, a chromone ester of mass 556 d.; THA, tetrahydroxyanthraquinone) (*see page 176*).

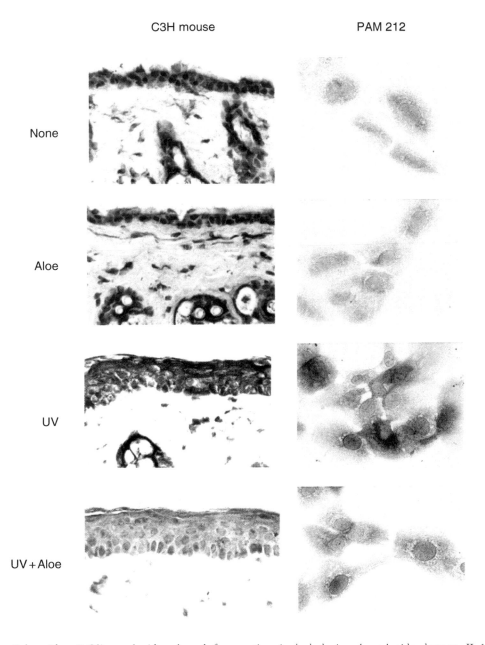

C3H mouse   PAM 212

None

Aloe

UV

UV + Aloe

Colour Plate 8 Oligosaccharides cleaved from native *A. barbadensis* polysaccharide decrease IL-10 production by UV irradiated murine keratinocytes. From Beyeon *et al.* (1998). Left Panel, mice were irradiated with 15,000 J/m^2 UV and treated topically with 500 mg per ml oligosaccharide. Four days later skin was removed and cryosections stained for IL-10 by immunohistochemistry. Right Panel, subconfluent cultures of PAM 212 cells were irradiated with 300 J/m^2 UV and were then treated for 1 hr with 10 mg per ml oligosaccharide. Cells were washed and IL-10 was detected 24 hrs later by immuno-histochemistry (*see page 300*).

15 kJ/m² UV (no 1° Ab)

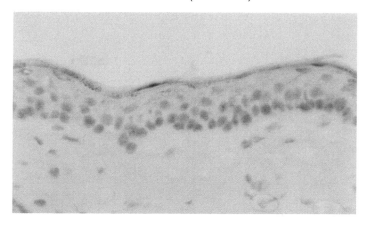

15 kJ/m² UV Only

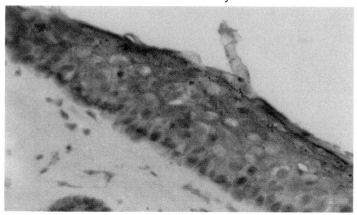

15 kJ/m² UV + Tamarind (1 μg)

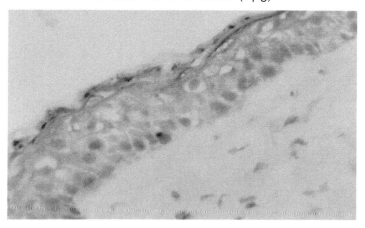

Colour Plate 9 Tamarind polysaccharides decrease IL-10 production by UV irradiated murine keratinocytes. From Strikland *et al.* (1999). Mice were irradiated with 15,000 J/m² UV and treated topically with Tamarind polysaccharide. Four days later skin was removed and cryosections stained for IL-10 by immunohistochemistry (*see page 301*).

Colour Plate 10 Leaf showing discoloration and erupting pustules of aloe rust (*Uromyces aloes*), a severe disease of aloes, particularly Series Saponariae. (Photo P. Brandham) (*see page 369*).

Colour Plate 11 Severe infestation of aloe plant with scale showing bluish-white overall appearance. Close-up shows more scale insects than leaf surface. (Photo P. Brandham) (*see page 375*).

Colour Plate 12 Cut leaves of *Aloe ferox* placed in a pile with cut surfaces inwards for the collection of exudates. (Photo, P. Brandham.)

Table 4.1 Mannans isolated from various *Aloe* species*.

Reference	Aloe species	Description or fraction	Linkage	Presence of glucose (glucose/mannose)	Branched	Acetylation (d.s.)
Roboz and Haagen-Smit (1948)	vera	None	ND	Yes (1:1)	ND	ND
Yagi et al. (1977)	arborescens	Mannan	β1-4	No	No	Yes (?)
Paulsen et al. (1978)	plicatilis	Glucomannan	1-4	Yes (1:2.8)	No	0.67
Gowda et al. (1979)	vera	Glucomannan	1-4	Yes (1.5:1–1:19)		≤0.78
Mandal and Das (1980b)	vera	Glucomannan	1-4	Yes (1:20)	Yes	No
Gowda et al. (1980)	saponaria	Mannan	ND	No	No	0.87
	vanbalenii	Mannan	ND	No	No	0.81
Haq and Hannan (1981)	vera	Glucogalacto-mannan	1-4	Yes (2:2:1; Glc:Gal:Man)	No	ND
Radjabi et al. (1983) and Radjabi-Nassab et al. (1984)	vahombe	Glucomannan	β1-4	Yes (1:3)	No	0.33
Yagi et al. (1984)	saponaria	Mannan	β1-4	No	No	0.83
Vilkas and Radjabi-Nassab (1986)	vahombe	Glucomannan	β1-4	Yes (1:2–1:4)	ND	0.17–0.57
Hikino et al. (1986)	arborescens	Arboran B	ND	Yes (1:0.3)	ND	0.26[a]
Yagi et al. (1986)	arborescens	Mannan (As mannan 1)	β1-4	No	ND	0.43
t'Hart et al. (1989)	vera	Mannan B-I	ND	Yes (1:28)	ND	ND
		Mannan B-II	ND	Yes (1:21)	ND	ND
Wozniewski et al. (1990)	arborescens	Glucomannan	β1-4	Yes (1:19)	No	1.3
Femenia et al. (1999)	vera	Mannan (H$_2$O; pulp)	1-4	Yes (1:4.6)	ND	ND

Notes

* The mannans listed are those whose linkage and/or degree of acetylation have been determined besides sugar composition.

a The d.s. value is converted from the % content (5.3) of acetyl groups along with a protein content of 10.4% (assuming rest of the preparation is carbohydrate).

was further divided into two fractions (B-I and B-II) by gel filtration chromatography. The polysaccharide was isolated as the fraction bound to DEAE-Sephacel beads. However, no uronic acid or others that would provide the basis for the ionic character was detected. A glucomannan consisting of mainly mannose and glucose at a glucose/mannose ratio of ~1:1 was isolated from *A. ferox* Miller by Mabusela *et al.* (1990).

Acetylation

All mannans or glucomannans described so far have been acetylated with a single exception (Mandal and Das, 1980b). Variations are found with respect to the degree of acetylation and the locations of acetyl groups. The reported d.s. values ranged from 0.17 to 1.3. Paulsen *et al.* (1978) found that acetylation occurred randomly among all available positions, whereas Manna and McAnalley (1993) showed that acetylation occurred at positions 2/3 and 6 at a ratio of 1:1 and Radjabi *et al.* (1983) suggested that acetylation primarily occurred on glucose residues.

Variations in the composition and degree of acetylation of aloe mannans

It is evident among all these studies that the most consistent feature of aloe mannans is the β1-4 linkage of the main chains and their acetylation. It is important to note that these two features are conserved among all *Aloe* species analyzed. However, significant variations in the composition and degree of acetylation exist as described above (Table 4.1). The reason for such variations could be many, including the differences in *Aloe* species, geographic locations (soil and climate), and extraction and analytical methods. As evidenced in Table 4.1, it seems clear that the *Aloe* species certainly plays an important role in these variations. No mannans from two different species showed the same characteristics. Furthermore, even within the same species, the mannans described by different investigators may also be very different. This clearly leads to another major factor that may contribute to these variations, i.e. the natural presence of different mannan species within the same *Aloe* species. Two major types of mannans have been identified in *Aloe*, the one with glucose (maybe along with other sugars) and the other without glucose (a pure mannan). It has often been questioned whether the glucose residues are simply from a contaminating polysaccharide or other components. However, the presence of glucose within the mannan chain or a glucomannan has been unequivocally shown for at least some *Aloe* species (Radjabi-Nassab *et al.*, 1984) and strongly suggested for several others (Table 4.1). Studies using sequential extraction under increasing ethanol concentrations, clearly suggest that different mannan species are present, although only one type appears to be dominant (Gowda *et al.*, 1979, 1980; Vilkas and Radjabi-Nassab, 1986). Thus, it is very interesting to note that following the sequential extraction with increasing ethanol concentration the glucose content in the isolated polysaccharide gradually decreased along with a gradual increase in the mannose content (Gowda *et al.*, 1979, 1980; Vilkas and Radjabi-Nassab, 1986) and ultimately only mannose or mannan was present in the case of Gowda *et al.* (1979). It appears that the mannan or glucomannan with a low glucose content (glucose:mannose <1:4) is only precipitated at a high ethanol concentration. The reason for this observation is not clear and may be related to the degree of acetylation (see below) and/or size. The mannan or glucomannan with a low glucose content is the dominant fraction in all the studies that dealt with the following four *Aloe* species: *A. vera*, *A. saponaria*, *A. vanbalenii*, and *A. vahombe* (Gowda *et al.*, 1979, 1980; Vilkas and Radjabi-Nassab, 1986). Thus, it is not surprising that when polysaccharides or mannans were isolated as whole or further purified by chromatography, only mannans having no or very few glucose residues were obtained (Yagi *et al.*, 1977, 1984, 1986; Gowda *et al.*, 1980; Mandal and Das, 1980b; t'Hart *et al.*, 1989; Wozniewski *et al.*, 1990).

Another trend that becomes clear from Table 4.1 is that the higher the glucose content, the lower the degree of acetylation (d.s.). Thus, the glucomannan with a high glucose content has a lower d.s. than the mannan or glucomannan with a low glucose content. This may suggest that acetylation occurs primarily at the mannose residues or that only the mannan or glucomannan with a low glucose content is preferentially acetylated. These are likely reasons for the observed variation in the d.s. of different mannan preparations.

Together, these analyses strongly suggests that: (1) both mannan and glucomannan are present within the same *Aloe* species and whether these two components are separated depends on the extraction method; and, (2) the mannan or glucomannan with a low glucose content (glucose:mannose <1:4) is the predominant polysaccharide and is more heavily acetylated. Evidence also suggests that the distinction in the composition between mannan and glucomannan could be gradual, i.e. the glucose content may increase gradually from zero (pure mannan) to as high as 50% or more.

Molecular weight

So far, the molecular weights of aloe mannans have been determined in most cases by size exclusion chromatography using neutral polysaccharides with known molecular weights as standards (Paulsen *et al.*, 1978; Gowda *et al.*, 1979; Yagi *et al.*, 1984; Wozniewski *et al.*, 1990; Ross *et al.*, 1997). However, other methods have also been used, including equilibrium ultracentrifugation (Yagi *et al.*, 1977) and osmometry (Mandal and Das, 1980b). Apparently, the mannan isolated from fresh pulp is a very large molecule with a size estimated to be $>1 \times 10^3$ kDa (Paulsen *et al.*, 1978; Wozniewski *et al.*, 1990; Ross *et al.*, 1997). In many reports, the sizes of isolated mannans were considerably smaller (Yagi *et al.*, 1984, 1986; t'Hart *et al.*, 1989). This may be related to the conditions of extraction and purification and the analytical methods used. In the case of t'Hart *et al.* (1989), the pulp was lyophilized and stored before use.

Stability

Aloe mannan is inherently unstable and can undergo degradation in a rapid manner (Yaron *et al.*, 1992; Yaron, 1993). The degradation can be caused by elevated temperature, pH change, bacterial contamination, or enzymes such as mannanases that may be present in the pulp. Bacteria are also a source of mannanase. In some commercial aloe products, the pulp preparation is treated with cellulase. The cellulase preparations, dependent on the source and purity, are often contaminated with mannanase. Thus, the mannanase present in cellulase preparations may cause the degradation of the mannan. The higher pH can cause deacetylation which renders the mannan insoluble, as the acetylation is a key factor in making the mannan soluble. This is evident with the fractions A1a and A3 described by Gowda *et al.* (1979) and by Mandal *et al.* (1980b), respectively, which were non-acetylated and insoluble. Gowda *et al.* (1979) further observed that alkaline treatment of the acetylated glucomannan causes the mannan to become insoluble and lose the ability to form a viscous solution. Similar observations were also made by Vilkas and Radjabi-Nassab (1986).

McAnalley (1988, 1990) described a substantially non-degradable $\beta 1 \rightarrow 4$ mannan, acemannan, from *A. vera* and the processes to obtain it. Acemannan (CarraVet Acemannan Immunostimulant) has recently been approved by the USDA as a biologic for treatment of fibrosarcoma in cats and dogs. Much effort has been made to study the chemistry and function of acemannan (Reynolds and Dweck, 1999). The term acemannan has been often used to describe in general the mannan isolated from *A. vera*.

When dissolved, the mannan isolated from fresh pulp produces a very viscous solution. This was observed very early on (Roboz and Haagen-Smit, 1948). Rheological studies showed that the mannan is the basis for the pseudoplastic flow behavior of the liquid gel from the pulp (Yaron, 1991, 1993). But once the mannan is degraded, the gel becomes less viscous and exhibits a Newtonian flow property.

Mannan from the rind

So far, the mannan has been mostly isolated from the pulp of all the *Aloe* species analyzed. It is interesting that Femenia (1999) reported a mannan in the rind that may be significantly different from that found in the pulp with respect to acetylation, branching, and molecular weight. The presence of a mannan in the rind was also suggested by Masubela *et al.* (1990). This raises an interesting question as to where the mannan is synthesized. It might be possible that the mannan is produced in the rind and then transported to the pulp. However, we need to be cautious about the interpretation of these results since they are heavily dependent on how well the rind is separated from the pulp. Ideally, there should be no trace of pulp remaining attached to the rind after cutting, scraping, and washing.

The role of mannan in evaluating pulp-based products

As described above, the mannan in the aloe pulp is unique among the known and widely available plant mannans. It is for this reason that aloe mannan has been used as an important criterion for evaluating the quality and authenticity of aloe pulp-based products (Ross *et al.*, 1997; Diehl and Teichmuller, 1998). One common pulp-based product is the so-called 'Aloe gel' or 'Aloe vera gel.' It is a whole pulp preparation. Its production process has been well outlined by Agarwala (1997). For quality control and authentication, the size, acetylation, and glycosidic linkage of the mannan in a product may be measured. The results shed light on the existence of an acetylated β1-4 linked mannan in a product and its integrity (size). If a product is not properly preserved or produced, the size of the mannan would be smaller or the typical mannan peak would be absent. If the product is falsified, the typical signal from acetyl groups would be absent. So the mannan is an indicator for the quality and also the source.

Pectic substance

The term pectic substance refers to a group of closely associated polysaccharides including pectin, pectic acid, and certain neutral polysaccharides such as arabinogalactan, often found in association with pectin (Aspinall, 1980). A pectin is an α1-4 linked polygalacturonic acid (Gal A) with intra-chain rhamnose insertion, neutral sugar side-chains, and methyl esterification. It is generally accepted that a pectin should have a Gal A content of >65%. A pectin free of methoxyl groups is a pectic acid.

The presence of uronic acid in various aloe leaf extracts has been described, usually at low levels (<10%), by several investigators (Rowe and Parks, 1941; Roboz and Haagen-Smit, 1948; Farkas, 1963; Gowda *et al.*, 1979; Hranisavljevic-Jakovljevic and Miljkovic-Stojanovic, 1981; Mabusela *et al.*, 1990). A pectic substance or a pectin with a high Gal A content was isolated from the whole leaf or pulp of *A. arborescens* and *A. vera* by Ovodova *et al.* (1975) and Mandal and Das (1980a), respectively. In both cases, extraction was performed with boiling water with or without ammonium oxalate. The pectic substance isolated by Mandal and Das (1980a) had a Gal A content of 85%. The one by Ovodova *et al.* (1975) had a Gal A content of 56% and was found to be partially esterified by methanol and cleaved by polygalacturonase (an enzyme specific for α1-4 linked polygalacturonic acid) and was thus identified as a pectin. Both studies, in light of the absence of the mannan, identified the pectic substance as the primary polysaccharide in the aloe pulp. However, the origin of these pectic substances was not clear. The presence of pectic substances as a dominant polysaccharide was explained by the seasonal change and difference in plant species and geographical location. This led Grindlay and Reynolds (1986) to summarize the understanding of pulp polysaccharides at that time by correctly stating 'the (mucilaginous) gel contains various carbohydrate polymers, notably glucomannans or pectic acid, along with....' in a comprehensive review on aloes and their uses.

Mandal *et al.* (1983) further hydrolyzed the isolated pectic substance and found that the resulting oligosaccharides consisted of a $1 \rightarrow 4$ linked polygalacturonic acid with galactose residues attached at the O3 position. Recently, Femenia *et al.* (1999) also isolated several pectic substances from *A. vera* with a chelating agent or under alkaline conditions. However, many of the pectic substance preparations still contained a significant amount of mannose or mannan.

A unique pectin has been isolated from the cell walls of *A. vera* leaves under special conditions following a detailed analysis of pulp structure (Ni *et al.*, 1999). This pectin is distinct from all known pectins or pectic substances in both chemical and functional properties.

Arabinan and arabinogalactan

Arabinan and arabinogalactan are commonly found in plant cells. Arabinogalactan is often associated with a small amount of protein and is therefore called arabinogalactan protein (Fincher *et al.*, 1983; Knox, 1995). Its carbohydrate composition is dominated by arabinose and galactose along with other sugars, including glucuronic acid and/or galacturonic acid. Certain arabinans and arabinogalactans are also often found as the neutral side-chains in pectins.

Mandal and Das (1980b) detected an arabinan in an *A. vera* pulp extract. Yagi *et al.* (1986) isolated a branched arabinogalactan from fresh leaf juice of *A. vera*. It had a galactose:arabinose ratio of 1:1.5 and the main chain consisted of $1 \rightarrow 6$ linked arabinose and $1 \rightarrow 2$ linked galactose. The branching occurred at O2 and O6 positions of the galactose residue. Mabusela *et al.* (1990) identified an arabinogalactan from *A. ferox* in which most of arabinose residues are terminal and linked at O4 or O5 positions, whereas the galactose was mostly 3,6-disubstituted. It also contained a small amount of other sugars including galacturonic acid.

The amount of arabinogalactan detected in the aloe pulp is low compared to the mannan or glucomannan. It is apparent that the mucilage-like nature of the pulp is not caused by the arabinan or arabinogalactan, but instead by the mannan. This notion is re-enforced by the fact that the mannan, when dissolved, produces a viscous solution that exhibits a distinct flow property. Thus, it appears unique that the mucilage nature of aloe pulp is constituted by a different kind of polysaccharide, a neutral acetylated mannan, instead of the common, acidic, uronic acid-containing polysaccharides including arabinogalactan (Kennedy and White, 1983).

Other Aloe polysaccharides

There are several other polysaccharides that have been identified or isolated from *Aloe*. However, results for most of them have not been independently reproduced. Hranisavljevic-Jakovljevic and Miljkovic-Stojanovic (1981) described an acidic, branched polysaccharide consisting of glucose and glucuronic acid from homogenized whole *A. arborescens* leaves. The main chain is a $\beta1 \rightarrow 3$ and $\beta1 \rightarrow 4$ linked glucose with glucuronic acid residues either located as an intra-chain insertion or a branch. Woznieski *et al.* (1990) also isolated an acidic glucuronic acid-containing polysaccharide, but with arabinose and galactose as dominant sugars. Mandal and Das (1980a) described a branched galactan from *A. vera* in which galactose residues were $\beta1 \rightarrow 4$ linked and the branch unit was linked at O1, O4, and O6 positions. The fact that this galactan had a $\beta1 \rightarrow 4$ linked main chain suggested that it might have been derived from the cell walls. Treatment with β-D-galactosidase released the D-galactose. In addition, a linear $1 \rightarrow 6$ linked glucose polymer from *A. arborescens* was reported by Yagi *et al.* (1986). Very interestingly, these same authors also isolated an $\alpha1 \rightarrow 4$ linked mannan with single branched glucose residues from *A. saponaria* (Yagi *et al.*, 1984). Hikino *et al.* (1986) described isolation of two unique polysaccharides (arborans A and B) from whole leaf of *A. arborescens* by methanol/water extraction. Arboran A contained several neutral sugars, with galactose being the most abundant, and arboran B contained glucose and mannose at a ratio of 1:0.3. In addition, Mabusela *et al.* (1990) isolated a xylan and a xyloglucan from *A. ferox*.

Glycoproteins

Glycoproteins have also been isolated from aloes (Fujita *et al.*, 1978; Yagi *et al.*, 1986, 1987, 1997; Vilkas and Radjabi-Nassab, 1986; Winters and Yang, 1996; Yoshimoto *et al.*, 1987). Many of these glycoproteins are found to be lectins (Fujita *et al.*, 1978; Suzuki *et al.*, 1979; Suzuki *et al.*, 1979; Yoshimoto *et al.*, 1987; Saito, 1993; Winters, 1993; Koike *et al.*, 1995; Akev and Can, 1999). Some of these were shown to have an enzymatic activity (Yagi *et al.*, 1986). However, the structure of the sugar side-chains in these glycoproteins have not been elucidated. Wozniewsky *et al.* (1990) reported a protein-containing acidic arabinogalactan with glucuronic acid, arabinose, galactose and glucose in *A. arborescens*. The glycoprotein from *A. arborescens* described by Yagi *et al.* (1986) contained glucosamine, galactosamine, and N-actylglucosamine along with other sugars. Another glycoprotein from *A. arborescens*, capable of degrading bradykinin, contained 28.5% carbohydrates consisting of mannose, galactose, glucose, glucosamine, galactosamine and N-acetylglucosamine in a ratio of 2:2:1:1:4:1 (Yagi *et al.*, 1987).

Interestingly, some mannans isolated from the pulp of *A. arborescens* and *A. vahombe* (*A. vaombe*) might also be linked to a protein (Hikino *et al.*, 1986; Vilkas and Radjabi-Nassab, 1986). In both studies, the mannan (glucomannan) was purified by chromatography. In the case of Hikino *et al.* (1986), the preparation was purified through three different columns, but still proteins were detected.

CONCLUSION

A significant amount of work has been done so far on the mannan from the aloe pulp. In comparison, research on other aloe pulp polysaccharides has been sporadic and even fewer efforts have been made to analyze the polysaccharides in the rind. It is clear that much effort is needed in this regard.

Two features of the mannan, the $\beta 1 \rightarrow 4$ linkage and acetylation, are highly conserved among all *Aloe* species. Significant variations exist regarding the presence of other sugars, primarily glucose and the degree of acetylation. The research so far strongly suggests that there are two types of mannans in the pulp, a pure mannan that is essentially free of glucose and another mannan (glucomannan) that contains various amounts of glucose, maybe along with other sugars. Much effort is still needed to determine proportions of these two types of mannans in any given *Aloe* species and the differences in chemical and biological properties between them. Evidence has suggested that the mannan or glucomannan with a low glucose content is more heavily acetylated than the glucomannan with a high glucose content.

Effort is also needed to identify and isolate the enzymes involved in the synthesis and modification of the mannan. Such information is crucial to understand the location of its synthesis and how the chemical and functional properties of the mannan is controlled.

REFERENCES

Agarwala, O.P. (1997) Whole leaf aloe gel vs. standard aloe gel. *Drug and Cosmetics Industry*, February, 22–28.

Akev, N. and Can, A. (1999) Separation and some properties of *Aloe vera* L. leaf pulp lectins. *Phytotherapy Research*, 13, 489–493.

Aspinall, G.O. (1980) Chemistry of cell wall polysaccharides. In *The Biochemistry of Plants*, edited by J. Preiss, p. 473. New York, Academic Press.

Bewley, J.D. and Reid, J.S.G. (1985) Mannana and glucomannans. In *Biochemistry of Storage Carbohydrates in Green plants*, pp. 289–304, edited by P.M. Dey and R.A. Dixon. London, Academic Press.

Diehl, B. and Teichmuller, E.E. (1998) Aloe vera, quality inspection and identification *Agro-food-industry Hi-Tech.*, 9, 14–16.

Farkas, A. (1963) Topical medicament including polyuronide derived from Aloe. *US Patent* 3, 103, 466.

Femenia, A., Sanchez, E.S., Simal, S. and Rossello, C. (1999) Compositional features of polysaccharides from Aloe vera (Aloe barbadensis Miller) plant tissues. *Carbohydrate Polymers*, 39, 109–117.

Fincher, G.B., Stone, B.A. and Clark, A.E. (1983) Arabinogalactan-proteins: structure, biosynthesis, and function. *Annual Review of Plant Physiology*, 34, 47–70.

Fujita, K., Suzuki, I., Ochiai, J., Shinpo, K., Inoue, S. and Saito, H. (1978) Specific reaction of aloe extract with serum proteins of various animals. *Experientia*, 34, 523–524.

Gowda, D.C., Neelisidaiah, B. and Anjaneyalu, Y.V. (1979) Structural studies of polysaccharides from *Aloe vera*. *Carbohydrate research*, 72, 201–205.

Gowda, D.C. (1980) Structural studies of polysaccharides from *Aloe saponaria* and *Aloe vanbalenii*. *Carbohydrate Research*, 83, 402–405.

Grindlay, D. and Reynolds, T. (1986) The aloe vera phenomenon: a review of the properties and modern uses of the leaf parenchyma gel. *Journal of Ethnopharmacology*, 16, 117–151.

Haq, Q.N. and Hannan, A. (1981) Studies on glucogalactomannan from the leaves of *Aloe vera*, TOURN. (EX LINN.). *Bangladesh Journal of Science and Industry Research*, XVI, 68–72.

Hikino, H., Takahashi, M., Murakami, M. and Konno, C. (1986) Isolation and hypoglycemic activity of arborans A and B, glycans of Aloe *arborescens* var. *natalensis* leaves. *International Journal of Crude Drug Research*, 24, 183–186.

Hranisavljevic-Jakovljevic, M. and Miljkovic-Stojanovic, J. (1981) Structural study of an acidic polysaccharide isolated from *Aloe arborescens* Mill. I. Periodate oxidation and partial acid hydrolysis. *Bulletin de la Societé Chimique Beograd*, 46, 269–273.

Kennedy, J.F. and White, C.A. (1983) *Bioactive carbohydrates: in Chemistry, Biochemistry and Biology.* pp. 142–181. New York: Ellis Horwood Ltd.

Knox, J.P. (1995) Developmentally regulated proteoglycans and glycoproteins of the plant cell surface. *FASEB Journal*, 9, 1004–1012.

Koike, T., Beppu, H., Kuzuya, H., Maruta, K., Shimpo, K., Suzuki, M., Titani, K. and Fujita, K. (1995) A 35 kDa mannose building lectin with hemagglutinating and mitogenic activities from "Kidachi Aloe" (Aloe *arborescens* Miller var. *natalensis* Berger). *Journal of Biochemistry*, 118, 1205–1210.

Mabusela, W.T., Stephen, A.M. and Botha, M.C. (1990) Carbohydrate polymers from *Aloe ferox* leaves. *Phytochemistry*, 29, 3555–3558.

Mandal, G. and Das, A. (1980a) Structure of the D-galactan isolated from *Aloe barbadensis* Miller. *Carbohydrate research*, 86, 247–257.

Mandal, G. and Das, A. (1980b) Structure of the glucomannan isolated from the leaves of *Aloe barbadensis* Miller. *Carbohydrate Research*, 87, 249–256.

Mandal, G., Ghosh, R. and Das, A. (1983) Characterization of polysaccharides of *Aloe barbadensis* Miller: Part III-structure of an acidic oligosaccharide. *Indian Journal of Chemistry*, 22B, 890–893.

Manna, S. and McAnalley, B.H. (1993) Determination of the position of the O-acetyl group in a β-(1 → 4)-mannan (acemannan) from *Aloe barbadensis* Miller. *Carbohydrate Research*, 241, 317–319.

Marshall, G.D., Gibbons, A.S. and Parnell, L.S. (1993) Human cytokines induced by acemannan. *Journal of Allergy and Clinical Immunology*, 91, 295.

McAnalley, B.H. (1988) Process for preparation of aloe products, produced thereby and composition thereof. *US patent*, 4,735,935.

McAnalley, B.H. (1990) Processes for preparation of aloe products, produced thereby and composition thereof. *US patent*, 4,917,890.

Ni, Y., Yates, K.M. and Zarzycki, R. (1999) Aloe pectins. *US patent* 5929051.

Ovodova, R.G., Lapchik, V.F. and Ovodova, Y.S. (1975) Polysaccharides in Aloe arborescens. *Khimija Prirodnykh Soedinenii*, 11, 3–5.

Paez, A., Gebre, G.M., Gonzalez, M.E. and Tschaplinski, T.J. (2000) Growth, soluble carbohydrates, and aloin concentration of Aloe vera plants exposed to three irradiance levels. *Environmental & Experimental Botany*, 44, 133–139.

Paulsen, B.S., Fagerheim, E. and Overbye, E. (1978) Structural studies of the polysaccharide from *Aloe plicatilis* Miller. *Carbohydrate Research*, 60, 345–351.

Radjabi, F., Amar, C. and Vilkas, E. (1983) Structural studies of the glucomannan from *Aloe vahombe*. *Carbohydrate Research*, 116, 166–170.

Radjabi-Nassab, F., Ramiliarison, C., Monneret, C. and Vilkas, E. (1984) Further studies of the glucomannan from *Aloe vahombe* (liliaceae). II. Partial hydrolyses and NMR ^{13}C studies. *Biochimie*, 66, 563–567.

Ramamoorthy, L. and Tizard, I.R. (1998) Induction of apoptosis in a macrophage cell line RAW 264.7 by acemannan, a β-(1 → 4)-acetylated mannan. *Molecular Pharmacology*, 53, 415–421.

Reid, J.S.D. (1985) Galactomannan. In *Biochemistry of Storage Carbohydrates in Green plants*, pp. 265–288, edited by P.M. Dey and R.A. Dixon. London: Academic Press.

Reynolds, T. and Dweck, A.C. (1999) Aloe vera leaf gel: a review update. *Journal of Ethnopharmacology*, 68, 3–37.

Roboz, E. and Haagen-Smit, A.J. (1948) A mucilage from Aloe vera. *Journal of American Chemical Society*, 70, 3248–3249.

Ross, S.A., ElSohly, M.A. and Wilkins, S.P. (1997) Quantitative analysis of Aloe vera mucilagenous polysaccharides in commercial Aloe vera products. *Journal of AOAC International*, 80, 455–457.

Robyt, J.F. (1998) *Essentials of carbohydrate chemistry*. New York: Springer-Verlag.

Rowe, T.D. and Parks, L.M. (1941) Phytochemical study of *Aloe vera* leaf. *Journal of the American Pharmaceutical Association*, 30, 262–266.

Saito, H. (1993) Purification of active substances of *Aloe arborescens* Miller and their biological and pharmacological activity. *Phytotherapy Research*, 7, S14–S19.

Segal, A., Taylor, J.A. and Eoff, J.C. (1968) A re-investigation of the polysaccharide material from *Aloe vera* mucilage. *Lloydia*, 31, 423.

Suzuki, I., Saito, H. and Inoue, S. (1979) A study of cell agglutination and cap formation on various cells with Aloctin A. *Cell Structure and Function*, 3, 379.

Suzuki, I., Saito, H., Inoue, S., Migita, S. and Takahashi, T. (1979) Purification and characterization of two lectins from *Aloe aborescens*. *Journal of Biochemistry*, 85, 163–171.

t'Hart, L.A., van den Berg, A.J.J., Kuis, L., van Dijk, H. and Labadie, R.P. An anti-complementary polysaccharide with immunological adjuvant activity from the leaf parenchyma gel of *Aloe vera*. *Planta Medica*, 55, 509–512.

Tizard, I., Carpenter, R.H., McAnalley, B.H. and Kemp, M. (1989) The biological activity of mannans and related complex carbohydrates. *Molecular Biotherapy*, 1, 290–296.

Vilkas, E. and Radjabi-Nassab, F. (1986) The glucomannan system from *Aloe vahombe* (*liliaceae*). III. Comparative studies on the glucomannan components isolated from the leaves. *Biochimie*, 68, 1123–1127.

Waller, G.R., Mangiafico, S. and Ritchey, C.R. (1978) A chemical investigation of *Aloe barbadensis* Miller. *Proceedings of Oklohoma Academy of Sciences*, 58, 69–76.

Winters, W.D. (1993) Immunoreactive lectins in leaf gel from *Aloe barbadensis* Miller. *Phytotherapy Research*, 7, S23–S25.

Winters, W.D. and Yang, P.B. (1996) Polypeptides of the three major medicinal aloes. *Phytotherapy Research*, 10, 5736–576.

Wozniewski, T., Blaschek, W. and Franz, G. (1990) Isolation and structure analysis of a glucomannan from the leaves of *Aloe arborescens* var. Miller. *Carbohydrate Research*, 198, 387–391.

Yagi, A., Makino, K., Nishioka, I. and Kuchino, Y. (1977) Aloe mannan, polysaccharide, from *Aloe arborescens* var. *natalensis*. *Plant Medica*, 31, 17–20.

Yagi, A., Hamada, K., Mihashi, K., Harada, N. and Nishioka, I. (1984) Structure determination of polysaccharides in *Aloe saponaria* (Hill.) Haw. (Liliaceae). *Journal of Pharmaceutical Science*, 73, 62–65.

Yagi, A., Harada, N., Shimomura, K. and Nishioka, I. (1987) Bradykinin-degrading glycoprotien in *Aloe arborescens* var. *natalensis*. *Planta Medica*, 53, 19–21.

Yagi, A., Nishimura, H., Shida, T. and Nishioka, I. (1986) Structure determination of polysaccharides in *Aloe arborescens* var. *natalensis*. *Planta Medica*, 3, 213–218.

Yagi, A., Egusa, T., Arase, M., Tanabe, M. and Tsuji, H. (1997) Isolation and characterization of the glycoprotein fraction with a proliferation-promoting activity on human and hamster cells *in vitro* from *Aloe vera* gel. *Planta Medica*, 63, 18–21.

Yaron, A. (1991) Aloe vera: chemical and physical properties and stabilization. *Israel Journal of Botany*, 40, 270.

Yaron, A. (1993) Characterization of *Aloe vera* gel before and after autodegredation, and stabilization of the natural fresh gel. *Phytotherapy Research*, 7, S11–S13.

Yaron, A., Cohen, E. and Arad, S.M. (1992) Stabilization of Aloe vera gel by interaction with sulfated polysaccharides from red microalgae and with xanthan gum. *Journal of Agricultural and Food Chemistry*, 40, 1316–1320.

Yoshimoto, R., Kondoh, N., Isawa, M. and Hamuro, J. (1987) Plant lectin, ATF1011, on the tumor cell surface augments tumor specific immunity through activation of T cells specific for lectin. *Cancer Immunology Immunotherapy*, 25, 25–30.

Zhang, L. and Tizard, I.R. (1996) Activation of a mouse macrophage line by acemannan: the major carbohydrate fraction from *Aloe vera* gel. *Immunopharmocology*, 35, 119–128.

5 Aloe lectins and their activities

Hiroshi Kuzuya, Kan Shimpo and Hidehiko Beppu

ABSTRACT

Since the presence of a lectin-like substance in *Aloe* was first reported in 1978, several kinds of aloe lectins have been characterized. Lectins are basically characterized by their cell agglutinating activities, particularly by their hemagglutinating activities. However, biological and pharmacological activities differ among the respective lectins. To date, characteristic activities and chemical properties of aloe lectins have not yet been sufficiently evaluated. Therefore, this paper reviews the characteristic features of the respective aloe lectins along the course of studies on these substances.

INTRODUCTION

What are lectins?

Stillmark (1888) first discovered that seed extracts of castor-oil plant agglutinate red blood cells from various animals. Since then, several different hemagglutinins were discovered in seeds of various plants. Lately, it was found that the respective hemagglutinins have binding specificities and their hemagglutinating activities are generally inhibited by monosaccharides or oligosaccharides. Boyd and Shapleigh (1954) in Boston University discovered ABO (blood group type A, B and O) blood type-specific hemagglutinins among these plant hemagglutinins and then proposed to name these hemagglutinins as lectins after the Latin, 'legere,' which means to choose. Thereafter, various kinds of lectins or lectin-like substances were isolated from bacteria and animals in addition to those isolated from plant seeds. Therefore, Goldstein *et al.* proposed the following definition of lectins in 1980, to clarify their characteristic features: 'Lectins are sugar-binding proteins or glycoproteins of non-immune origin which agglutinate cells and/or precipitate glyco-conjugates.' This definition of lectins is currently accepted in general. Simultaneously, they summarized the properties of lectins as follows: 'Lectins bear at least two sugar-binding sites, agglutinate animal and plant cells (most commonly erythrocytes, unmodified or enzyme-treated) and/or precipitate polysaccharides, glycoproteins and glycolipids. The specificity of a lectin is usually defined in terms of the monosaccharide(s) or simple oligosaccharides that inhibit lectin-induced agglutination, or precipitation, or aggregation reactions' (Goldstein *et al.*, 1980).

Based on their sugar-binding specificities, lectins are generally classified as follows: fucose-binding lectins, galactose-binding lectins, N-acetylglucosamine-binding lectins,

N-acetylgalactosamine-binding lectins, and N-acetylneuraminic acid (sialic acid)-binding lectins.

Concerning the natural roles of lectins, many studies have been conducted since the protective actions of wheat germ agglutinin (WGA) against pathogenic bacterial infection were reported by Mirelman (1986). In addition, it was demonstrated that the cell adhesion molecule family in vertebrate animals (selectin family; lymphocyte homing receptors, vascular endothelium-leukocyte adhesion molecules, etc.) have lectin activities and these cell adhesion molecules were involved in the recognition mechanism between cells and glycoconjugates or among cells. Moreover, the involvement of lectins in the following subjects have been studied to date: transportation of glycoproteins, control of lymphocyte migration, activation of the complement system, tumor metastasis, differentiation and development of organs and fertilization. Furthermore, the natural roles of lectins in invertebrate animals, myxomycetes, paracytes and viruses have also been studied (Sharon and Lis, 1989).

As with concanavalin A (Con A), phytohemagglutinin and pokeweed mitogen (PWM), lectins are used as substances that enhance the division of peripheral lymphocytes, as well as in the detection, analysis, or separation procedures in which sugar-binding specificities of lectins are required.

Animal lectins are roughly classified into C type lectins and S type lectins. C type lectins bind to sugars depending on the Ca^{2+} ion levels, while S type lectins bind to sugars without Ca^{2+} ions in the presence of reagents that protect the SH group.

General properties of common lectins are summarized in Table 5.1 and bibliographies and reviews for lectins are listed in the References (Etzler, 1985; Franz, 1988; Kocourek, 1986; Liener *et al.*, 1986; Lis and Sharon, 1986; Mirelman, 1986; Olden and Parent, 1987; Sharon and Lis, 1987).

Lectins in aloes

Since Fujita *et al.* (1978) reported their study of a 'lectin-like substance which reacts with serum proteins of various animals.' Studies of lectin in *Aloe arborescens* Miller var. *natalensis* ('Kidachi' aloe in Japanese) were regularly performed. *A. arborescens* was selected for their studies because it was called 'Isha-irazu' in Japanese, which means 'requiring no doctor' and has been widely used from ancient times as a folk medicine to treat mild burn, as well as a bitter stomachic. Because Fujita *et al.* (1958a, 1958b) were originally interested in cancers, they conducted their study on *A. arborescens* expecting some therapeutic values against cancers. As a result, they found a novel hemagglutinin that agglutinates ascites tumor cells and intact host erythrocytes in rats with transplanted Yoshida's sarcoma. This hemagglutinin did not agglutinate intact erythrocytes in healthy rats. They also demonstrated that this hemagglutinin non-specifically precipitated sera from various vertebrate animals (Fujita *et al.*, 1975). It was subsequently found that this hemagglutinin binds to $\alpha 2$-macroglubulin and $\alpha 1$-antitrypsin in the serum (Fujita *et al.*, 1978). This study was conducted based on the following observations: lectins are mainly useful for studying cell surface chemical structures and malignant changes in cells (Sharon and Lis, 1972); red kidney beans contain substances that do not agglutinate erythrocytes but agglutinate sarcoma-180 cells and leukocytes (Tunis, 1964; Weber, 1969; Allen *et al.*, 1969; Inbar and Sachs, 1969). Concanavalin A isolated from horse beans also agglutinates erythrocytes. Thus, the history of the study of aloe lectins is not very old. Therefore, this paper reviews the course of studies on aloe lectins performed by different groups.

Table 5.1 Properties of several lectins.

Lectin	Source	Sugar specificity	Mitogenic activity	Binding to serum glycoprotein	Chemical property
Lotus agglutinin	Lotus (*Lotus tetragonalobus*)	L-Fucose			
Peanut agglutinin	Peanut (*Arachis hypogaea*)	D-Galactose, N-Acetyl-D-galactosamine	+		
Soybean agglutinin (SBA)	Soybean (*Glycine max*)	D-Galactose, N-acetyl-D-galactosamine	+	+	4 subunits (two 5.8 kDa and two 17 kDa)
Ricin	Castor-oil plant (*Ricinus communis*)	D-Galactose, N-Acetyl-D-galactosamine	+	+	2 subunits (32 kDa and 34 kDa)
Bush bean agglutinin (PHA)	Bush bean (*Phaseolus vulgaris*)	D-Galactose, N-Acetyl-D-galactosamine	+		
Abrin	Indian licorice (*Abrus precatorius*)	D-Galactose, N-Acetyl-D-galactosamine			65 kDa consisting of 2 different subunits
Concanavalin A (Con A)	Jack bean (horse bean) (*Conavalia ensiformis*)	D-Mannose, D-Glucose	+	+	2 or 4 identical subunits (26 kDa)
Lentil lectin	Lentil (*Lens esculenta*)	D-Mannose, D-Glucose	+	+	46 kDa consisting of 5.7 and 17 kDa subunits
Pisum sativum agglutinin (PSA)	Garden pea (*Pisum satinum*)	D-Mannose, D-Glucose	+	+	
Wheat germ agglutinin (WGA)	Wheat germ (*Triticum vulgaris*)	N-Acetyl-D-glucosamine	+		2 subunits (21.6 kDa)
Pokeweed mitogen (PWM)	Pokeweed (*Phytolacca americana*)	N-Acetyl-D-glucosamine	+	+	5 isolectins (1.9 kDa–220 kDa)

Note

The agglutination toward human erythrocytes of the lectins listed in this table does not exhibit the specificity for ABO blood group except for lotus agglutinin which specifically reacts with O blood group.

LECTIN-LIKE PROTEINS WHICH REACT WITH
α2-MACROGLOBULIN AND α1-ANTITRYPSIN

As described in the Introduction, the presence of a novel hemagglutinin was demonstrated, which binds to ascites tumor cells and intact host erythrocytes in rats with transplanted Yoshida's sarcoma but does not agglutinate intact erythrocytes in healthy rats (Fujita *et al.*, 1975). It was also confirmed that this hemagglutinin binds to α2-macroglubulin and α1-antitrypsin in the serum (Fujita *et al.*, 1978).

A lyophilized powder containing components with molecular masses above 10 kDa was prepared from fresh leaves of *A. arborescens*. Two kg of the fresh aloe leaves were homogenized using a Polytron and filtered through Whatman GF/A paper. Then the filtrate was dialyzed and concentrated by an Amicon hollow fibre dialyzer concentrator DC-2 and the concentrate was lyophilized. The average yield was 691 mg. The lyophilized powder was used for immunodiffusion (Ouchterlony) and immunoelectrophoresis studies.

Sera from humans, rabbits, sheep, dogs, cats, horses, pigs, rats, bovines, mice, carps, snakes and frogs were examined together with eggs. As a result, precipitin lines were observed between aloe extracts and all sera and egg yolk. However, precipitin lines were not observed between aloe extracts and egg white. Subsequently, immunoelectrophoresis was skillfully performed after applying human sera between troughs containing aloe extracts and monospecific antisera (rabbit antihuman α2-macroglubulin and rabbit antihuman α1-antitrypsin). As a result, it was demonstrated that aloe extracts contain substances that react with human serum α2-macroglubulin and α1-antitrypsin.

Lectin proposed by Boyd and Shapleigh (1954) is the term for proteins that possess the ability to agglutinate erythrocytes, and the representative lectin, Con A, can agglutinate erythrocytes as well as precipitate serum protein. Based on these findings, it was speculated that substances in aloe extracts that reacted with α2-macroglobulin and α1-antitrypsin were lectin-like substances. The authors speculated that a lectin-like substance may be implicated in the possible anti-inflammatory action and the therapeutic effects for burns, as serum proteins reacting with aloe extract were α2-macroglobulin and α1-antitrypsin which are known to be the most representative protease inhibitors.

ALOCTIN A AND ALOCTIN B

Aloctin A and Aloctin B are lectins purified from *A. arborescens* and biochemically characterized by Suzuki *et al.* (1979) for the first time. Since then, biological and pharmacological activities of Aloctin A were aggressively studied. Therefore, Aloctin A is one of the most well studied lectins in *Aloe*. Suzuki *et al.* (1979) purified two different lectins from whole leaves of *Aloe* and initially called these lectins P2 and S1, respectively. Lately, these two lectins were named Aloctin A and Aloctin B. Aloctin A is contained in a fraction with hemagglutinating and mitogenic activities obtained by Sephadex G-200 chromatography of the acidic precipitate of aloe extracts. This fraction was rechromatographed by Sephadex G-200 to obtain Aloctin A. Aloctin B is contained in a fraction with hemagglutinating activity obtained by Sephadex G-100 chromatography of the acidic supernatant of aloe extracts.

Preparation

Aloe leaves were crushed in a commercial juicer. After removing coarse materials from the juice by centrifugation, solid ammonium sulphate was added to give 40% saturation. The precipitate was collected by centrifugation, dissolved in 0.05 M carbonate-bicarbonate buffer (pH 9.5) and then the solution was centrifuged. Acetic acid (1 M) was added to the supernatant to give a pH of 4.4. After centrifugation, the supernatant and the precipitate were separated. The precipitate was dissolved in 0.05 M phosphate buffer (pH 8.0) and was then chromatographed on a Sephadex G-200 column. A portion of each fraction was tested for detection of hemagglutinating and mitogenic activities and the active fractions were pooled, condensed with a Spectrapor membrane tube (Spectrum Medical Industries, Inc.) and rechromatographed on the same column. The active fraction was called Aloctin A. The acidic supernatant was lyophilized and dissolved in 0.05 M phosphate buffer (pH 8.0) at a suitable concentration and then chromatographed on a Sephadex G-100 column.

Chemical properties

Aloctin A has a molecular mass of about 18 kDa, consists of two subunits (α, β) linked by a disulfide bound and contains more than 18% neutral sugar by weight. The smaller subunit (α) has a molecular mass of about 7.5 kDa and the larger subunit (β) a molecular mass of about 10.5 kDa Aloctin B has a molecular mass of about 24 kDa, consists of two subunits ($\gamma2$), linked by a disulfide bond, with a molecular mass of about 12 kDa and contains more than 50% by weight of neutral sugar. An interesting feature of the amino acid compositions of these lectins is the high proportion of acidic amino acids, such as aspartic acid and glutamic acid and the low proportion of methionine and histidine. A sugar-binding specificity, an important property, has not yet been determined.

Biological and pharmacological activities

Many biological and pharmacological activities of Aloctin A were studied, including hemagglutinating activity, mitogenic activity, precipitate-forming reactivity with serum proteins, one of which is $\alpha2$-macroglobulin (Saito, 1993) and affects on the immune system (Imanishi, 1993; Imanishi *et al.*, 1987; Imanishi and Suzuki, 1984; Imanishi and Suzuki, 1986).

Hemagglutinating activity

Aloctin A (Aloctin B also) agglutinated erythrocytes of various species such as human, sheep and rabbit and did not show ABO blood group specificity in hemagglutination tests in the human system. Treatment of human erythrocytes with 0.1% trypsin increased the agglutination by Aloctins A and B five-fold compared with that of non-treated erythrocytes. Aloctin B had four-fold stronger hemagglutinating activity compared to that of Aloctin A but Aloctin B had no mitogenic effect on human erythrocytes.

Mitogenic activity

Mitogenic activity was assayed by morphological examination of human lymphocyte transformations detected by Giemsa staining and by [^{3}H]-thymidine incorporation. Aloctin A augmented the number of large lymphocytes with morphologically transformed shapes compared to that treated with Aloctin B and other purified fractions. This degree of transformation is almost the same as that caused by phytohemagglutinin-W, which was tested as a positive control.

Precipitate-forming reactivity with serum proteins

Two human serum proteins that reacted with crude extract of *Aloe* were identified as α2-macroglobulin and α1-antitrypsin by immunoelectrophoresis (Fujita *et al.*, 1978) and Aloctin A reacted with human α2-macroglobulin.

Inhibitory effect on experimental inflammation in rats

Inhibitory effects of Aloctin A on experimental inflammation were examined using adjuvant arthritis and carageneen-induced edema in rats (Saito *et al.*, 1982; Imanishi and Suzuki, 1984). The arthritic syndrome was induced by an intradermal injection of liquid paraffin containing heat-killed *Mycobacterium butyricum* into the plantar surface of the right hind foot. Intraperitoneal administration of Aloctin A effectively suppressed the swelling in the injected foot and in the inflamed (secondary) lesion but oral administration did not affect adjuvant arthritis. Hind paw edema was induced by a subcutaneous injection of carrageenin solution into the right hind foot pads of rats. Intraperitoneal administration of Aloctin A markedly inhibited carrageenin-induced edema.

Modulation of prostaglandin E2 production by rat peritoneal macrophages

Aloctin A was demonstrated to have inhibitory activity of prostaglandin E2 production by activated rat peritoneal macrophages. This study investigated whether the anti-inflammatory activities of Aloctin A, like the activities of non-steroidal anti-inflammatory agents, reflected inhibition of prostaglandin production (Ohuchi *et al.*, 1984) and the effect was compared with that of Con A, WGA, *Pisum sativum* agglutinin and soybean agglutinin. Effects on phagocytosis were also examined by the measurement of uptake of ^{51}Cr-labeled sheep red blood cells by activated rat peritoneal macrophages after treatment with these lectins and Aloctin A inhibited the uptake.

These actions of lectins on inflammatory prostaglandin E2 production and phagocytosis differ among the respective lectins. Therefore, it was speculated that the capacity to inhibit prostaglandin E2 production and phagocytosis of foreign particles does not appear to explain the anti-inflammation activities of Aloctin A.

Effects on gastric secretion and experimental gastric lesions in rats

Effects of Aloctin A on gastric secretion and on acute gastric lesions in rats were examined (Saito *et al.*, 1989); intravenous Aloctin A administration dose-dependently suppressed the secretion of gastric juice, gastric acid and pepsin and the induction of water-immersion stress lesions in pylorus-ligated rats. The administration also suppressed the development of Shay ulcers and indomethacin-induced gastric lesions in rats.

Immunological effects. i Cytotoxic reactivity

Effects of Aloctin A on cytotoxic reactivity were examined using lymphoid cells prepared from spleen or lymph node of Aloctin A-treated mice as effector cells and using ^{51}Cr-labeled tumor cells, such as cell line YAC, a Moloney leukemia virus-induced mouse T (thymus dependent)-lymphoma of A/Sn (mouse strain) origin and P815, a mastocytoma of DBA/2 (inbred mouse substrain) origin, as target cells. Various numbers of effector cells and ^{51}Cr-labeled target cells were combined, incubated and released-^{51}Cr was counted to evaluate lytic activity (Imanishi and Suzuki, 1984). Intravenous administration of Aloctin A augmented cytotoxicity by spleen cells and by peritoneal exudate cells, whereas intraperitoneal administration of Aloctin A augmented cytotoxicity by peritoneal exudate cells but not by spleen cells. The effector cells were speculated to be natural killer cells since augmentation of natural killer activity with lectins had been noted (Imanishi *et al.*, 1986).

Immunological effects. ii Induction of nonspecific cell-mediated cytotoxic reactivity from treated non-immune spleen cells

Culturing of mouse spleen cells with Aloctin A induced cells cytotoxic to syngeneic and allogenic tumor cells *in vitro* (Imanishi and Suzuki, 1986). Aloctin A-induced killer cells could be generated from spleen cells of natural killer cell-deficient beige mice but not from those of T cell-deficient nude mice. IL(interleukin)-2 released from spleen cells in culture was speculated to be closely associate with the generation of killer cells by the results from the assay of IL-2 in culture fluids and from the culturing of mouse spleen cells with IL-2 containing conditioned medium.

Immunological effects. iii Activation of serum complement components

Activation of serum complement C3 by Aloctin A was investigated by the observation of a change in electrophoretic mobility of C3 component subunits after activation. Aloctin A as a lectin was initially recognized to activate the complement C3 activating system via the alternate pathway, suggesting the stimulation of immune systems (Suzuki *et al.*, 1979).

Aloctin B has only a hemagglutinating activity stronger than that of Aloctin A. To date, there have not been any other studies on the biological and pharmacological actions of Aloctin B reported.

Immunological effects. iv Growth inhibition of mouse methylcholanthrene-induced fibrosarcoma

Antitumor effects of Aloctin A were examined by intraperitoneal injection of methyl-cholanthrene-induced fibrosarcoma into the peritoneal cavity of BALB/c mice (Imanishi *et al.*, 1981). Aloctin A was administered intraperitoneally once daily for five days, starting 24 hours after tumor implantation. Antitumor activity was evaluated by the total packed cell volume ratio calculated from collected whole ascites obtained from mice. Aloctin A obviously inhibited the growth of methylcholanthrene-induced fibrosarcoma and administration at a dose of 10 mg/kg/day, for five days, remarkably inhibited growth (p < 0.001).

Whether the antitumor activity was due to the cytotoxicity of Aloctin A for tumor cells or host-mediated effects of Aloctin A was examined; the effect of Aloctin A on growth *in vitro* of methylcholanthrene-induced fibrosarcoma and the other cell lines was examined by ³H-thymidine uptake. Aloctin A had almost no inhibitory effect on the growth of tumor cell lines tested involving methylcholanthrene-induced fibrosarcoma up to a concentration of 200 mg/ml. However, lower concentrations of Aloctin A rather stimulated the growth of some tumor cell lines. The reason for this was not discussed. Host-mediated inhibition mechanisms, e.g. activation of complement C3 via the alternative pathway in the immune system, not by direct cytotoxicity to tumor cells was suggested about mechanisms of the antitumor effect of Aloctin A. Antitumor effects of some other lectins, such as Con A, ricin and abrin were discussed in relation to that of Aloctin A.

ALOE EXTRACTS WHICH AFFECT ON HUMAN NORMAL AND TUMOR CELLS *IN VITRO*

Lectin-like substances of fresh leaves of *Aloe barbadensis* Miller (=*A. vera* Burm.f.) and *Aloe saponaria* (Ait.) Haw. and the commercially 'stabilized' *A. vera* gel were investigated (Winters *et al.*, 1981; Winters, 1993). Leaf extract fractions (Low speed supernatant; SI, High speed supernatant; SII and Highspeed pellet; HP), were prepared by differential centrifugation and examined by immunodiffusion-like precipitation and hemagglutination assay. Fractions of fresh leaf extracts and the commercially 'stabilized' *A. vera* gel had high levels of lectin-like substances and substances in fluid fractions from both fresh leaf sources were found to markedly promote attachment and growth of human normal, but not tumor, cells and to enhance healing of wounded cell monolayers.

Preparation

Fresh leaf homogenate, a mixture of juice and particles approximately 2 mm in size, was centrifuged at 4 °C and the greenish colored particle-free liquid supernatant was collected (SI fraction). Pelleted materials were recentrifuged at high speed at 10 °C. The high speed supernatant (SII fraction) and pellet (HP fraction) were individually collected. SI and SII fractions were dialysed and then concentrated at 4 °C. A commercially 'stabilized' *A. vera* gel was homogenized and separated into fractions as described for the fresh aloe leaf specimens. SI and SII fractions from this aloe source were then concentrated and refrigerated at 4 °C together with all other aloe fractions.

Biological and pharmacological activity. i Hemagglutinating and precipitating activities

Hemagglutination assays were carried out toward human and canine erythrocytes and concentrated SI and SII fractions were found to have the hemagglutinating activity.

Concentrated SI fraction of *A. barbadensis* contained markedly higher amounts of hemagglutination reactive substances than comparable fractions from *A. saponaria* or *A. vera* gel. Human erythrocytes were more sensitive indicators of hemagglutination than canine erythrocytes for tests of aloe fractions.

Precipitation tests were carried out toward human, canine and baboon sera and concentrated SI fractions from all three aloe sources reacted with human and baboon sera. In contrast, none of the aloe fractions reacted in the precipitation tests with canine sera from normal and tumor-bearing adult dogs (Winters, 1981). Concentrated SII from the three aloe specimens did not show immunoprecipitation reactions with any sera.

Biological and pharmacological activity. ii Effects of the aloe extracts on cell attachment, growth and wound healing

Human normal fetal lung and human cervical carcinoma cells in single-cell suspensions were aggregated by fractions of *A. barbadensis* having their high lectin-like activities. A dilution (1:10) of concentrated SI fraction caused a marked enhancement of attachment of human normal fetal lung cells but not human cervical carcinoma cells. There were no differences in cell attachment with the other *A. barbadensis* fractions compared with that of untreated control cells.

Biological and pharmacological activity. iii Enhancement of growth

Fractions SI and HP of *A. barbadensis* markedly enhanced the growth of human normal fetal lung cells treated in suspension and in monolayer cultures. Human cervical carcinoma cells treated in suspension cultures with the two fractions did not grow as well as untreated control cells, while the growth of the cells treated in monolayer cultures did not different from that of untreated control cells.

Biological and pharmacological activity. iv Wound healing model in monolayer culture

The number of cells at the edges of wounds in monolayer human normal fetal lung and human cervical carcinoma cells cultures treated with fraction SI of *A. barbadensis* were observed to be higher than cell densities at the wound edges in other cultures treated with SII and HP fractions and in untreated control cultures. *In vitro* wounded cell monolayer assays were performed as follows. Parallel lanes were scraped on confluent plate cultures while viewed through a dissecting microscope and the border between the monolayer and scraped area was considered wounded. Cell densities at the edges of the wounds, which reflected the rate of cell movement into the wound area, were counted.

Fractions of 'stabilized' *A. vera* gel inhibited attachment of cells treated in suspension and cell detachment in monolayer cultures of human normal fetal lung and human cervical carcinoma cells. Accordingly, these cytotoxic responses prevented the completion of cell attachment and growth experiments with fractions of 'stabilized' *A. vera* gel.

A LECTIN WHICH STIMULATES DEOXYRIBONUCLEIC
ACID SYNTHESIS

A lectin which has a stimulating action of deoxyribonucleic acid synthesis was isolated from *A. arborescens* and characterized (Yagi *et al*., 1985).

Preparation

The supernatant prepared from fresh leaf homogenate was lyophilized to crude extract and the extract was used to isolate a lectin. The extract was applied to a column of Amberlite XAD-2 and eluted with distilled water and the eluate was concentrated under reduced pressure after concentration and lyophilized to a powder. After dialyzation of the powder against water, the dialysate was evaporated to dryness and the non-dialysate was concentrated on a hollow fibre dialyzer concentrator with a rejection limit over 10 kDa in molecular mass. The non-dialysate was lyophilized to a colorless fibrous material and the fibrous material was dissolved in 0.02 M sodium bicarbonate and gel-filtrated on a column of DEAE-cellulofine. After the neutral polysaccharide fraction was eluted with the same buffer, the column was then eluted with 0.3 M sodium chloride and the eluate was lyophilyzed to a powder (glycoprotein fraction) after concentration. The glycoprotein fraction was dissolved in 0.3 M sodium chloride and applied to a Sepharose 6B column which was eluted with the same solution. Two fractions, glycoprotein fraction 1 and glycoprotein fraction 2 were recovered.

Chemical and physical properties

Fraction 1 of glycoprotein is a homogeneous glycoprotein with a molecular mass of 40 kDa containing 34% sugar and 5% ash. The sugar composition was mannose, glucose, galactose, glucosamine, galactosamine and N-acetyl-glucosamine in ratio of 2:2:1:1:4:1 and the amino acid composition was asparagine, threonine, serine, glutamine, glycine, alanine, valine, isoleucine, phenylalanine, ornithine, lysine, arginine and proline. The agglutination by glycoprotein fraction 1 was inhibited by glucose, mannose and galactose. The active component of the glycoprotein is suggested to be the native protein moiety from the result of the abolishment of the stimulation of DNA after heat-treatment of the glycoprotein. Fraction 2 of the glycoprotein showed several bands on SDS-PAGE (sodium dodecyl sulphate-polyacrylamide gel electro-phoresis), indicating that it was non-homogeneous and had no properties as a lectin.

Biological properties. i Hemagglutination and stimulation
of deoxyribonucleic acid synthesis

Fraction 1 of the glycoprotein agglutinated sheep blood cells (1%), whereas Fraction 2 of the glycoprotein did not. Hemagglutinating activity of the fraction was compared to that of phytohemagglutinin P; maximal agglutination by Fraction 1 of glycoprotein 1.25 mg/ml, whereas phytohemagglutinin-P showed maximal agglutination at a concentration of 0.04 mg/ml. The glycoprotein fraction was also shown to stimulate deoxyribonucleic acid synthesis in baby hamster kidney (BHK 21) cells. Heat-treatment of the fraction abolished the stimulation of DNA synthesis, which suggests that the active component is the native protein moiety of glycoprotein. The therapeutic effect of aloe on burns was attributed to the induction of blastmitogenesis by the stimulation of DNA synthesis by the glycoprotein.

ATF1011, A LECTIN AUGMENTING TUMOR-SPECIFIC IMMUNITY

A lectin that did not have hemagglutinating activity or mitogenic activity but which did have agglutinating activity against tumor cells was separated from whole leaves of *A. arborescens* (Yoshimoto *et al.*, 1987). The lectin differed from aloctins which had already been separated from *A. arborescens* and augmented tumor-specific immunity through activation of T cells specific for the lectin.

Preparation

The supernatant solution of homogenates of whole leaves of *A. arborescens* was fractionated with 40% saturation ammonium sulfate as the final concentration. After dialysis of the precipitated fraction, non-dialyzates were dissolved with 50 mM phosphate buffer, pH 7.5 and applied to a series of a DEAE-cellulose column chromatography three times; first elution with 0.4 M NaCl/50 mM phosphate buffer: second elution with a linear gradient of 0 M–0.4 M NaCl/50 mM phosphate buffer; and, third elution with 0 M–0.3 M NaCl/50 mM phosphate buffer. The active fraction was lyophilized.

Chemical properties

ATF 1011 is a glycoprotein containing 0.4–0.7% neutral sugars as mannose and less than 0.3% amino sugars as glucosamine, has isoelectric point 4.3–5.2 and has a molecular mass of about 64 kDa.

Affects on tumor cells. i No direct cytotoxicity of ATF 1011 to the tumor cells

Since Aloctin A was shown to have antitumor effects, antitumor effects of ATF 1011 were also examined using tumor cells such as MM46 and MM102 (mammary tumor cell lines originating from C3H/He, inbred mouse strain), MH134 (hepatoma cell line originating from C3H/He), Meth-A (fibrosarcoma cell line originating from BALB/C, inbred mouse strain), P815 (mastocytoma cell line from DBA/2, inbred mouse substrain) and EL4 (thymoma cell line from C57/BL6, inbred mouse strain).

Effects of ATF 1011 on the growth of NN46, MM102, MH134 and EL4 were examined and all the tumor cell lines tested showed almost the same growth in the presence of ATF 1011 as that of the control culture, indicating that there was no direct cytotoxicity to the tumor cells. However, MM46, MH134 and EL4 did not grow in the presence of Con A (low concentration). Ricin and abrin were also cytotoxic to these tumor cell lines (Fodstad and Pihl, 1978). MM102 was resistant to low concentrations of Con A but growth was suppressed at high concentration.

Effects on immune system of ATF 1011

The following experiments were attempted to determine the relationship to the immune systems of ATF 1011; activation of Thymus helper (Th) cells in antibody production by ATF 1011; augmentation of cytotoxic Thymus(T) cell response by ATF 1011; and, induction of antitumor cytotoxic T cells in syngeneic tumor-bearing mice by intralesionally administered ATF 1011.

These results demonstrated the potential of ATF 1011 as a carrier protein in the immune system. That is, intralesionally administered ATF 1011 binds to the tumor cell membrane and activates T cells specific for this carrier lectin *in situ*, resulting in the augmented induction of systemic antitumor immunity.

A 35 KDA MANNOSE-BINDING LECTIN FROM *ALOE ARBORESCENS* MILLER

A novel lectin having hemagglutinating and mitogenic activities was isolated from leaf skin of *A. arborescens* (Koike *et al.*, 1995a, b). The lectin with a molecular mass of 35 kDa consists of subunits containing 109 amino acid residues. The N-terminal amino acid sequence of the intact subunit showed a homology with that of snowdrop lectin.

Preparation

The homogenates of the leaf skin of *A. arborescens* were filtered and the filtrates were mixed with two-fold cold acetone. The precipitate was collected by centrifugation and lyophilized. The dried powder was dissolved in phosphate-buffered saline and fractionated on a Sephadex G-25 column. Fractions with the lectin activity were precipitated with 80% saturation of ammonium sulfate. The precipitate was collected by centrifugation and dissolved in 5 mM potassium phosphate buffer, pH 8.0. After dialysis the solution was applied to a DEAE 52 column. The active fractions were dialyzed and then lyophilized. The solution of the dried material was applied to a Superdex 75 HR 10/30 column and the fractions with both hemagglutinating and mitogenic activities were combined and used for the experiments.

Chemical properties

The lectin showed a molecular mass of about 35 kDa by gel filtration chromatography and the complete amino acid sequence of the subunit was determined. The subunit consisted of 109 amino acid residues having a molecular mass 12.2 kDa and contained an intrachain disulfide bridge. However, it is unclear whether it consists of three or four subunits, because the molecular mass of the intact lectin was estimated to be roughly 35 kDa. Lectins consisting of three subunits are not general but a leek (*Allium porrum*) lectin being a trimer rather than tetramer of polypeptides with a molecular mass 12.5–13 kDa was reported (Van Damme *et al.*, 1993).

The sequence of the lectin in this study is highly homologous to that of a mannose-binding lectin (105 residues) from snowdrop bulb, belonging to the family Liliaceae as well as *A. arborescens* (Van Damme *et al.*, 1991).

Hemagglutinating activity

The aloe lectin exhibited high agglutinating activity toward trypsinized and glutaral-dehyde-fixed rabbit erythrocytes, whereas it exhibited no activity toward human (type A, B and O) or sheep erythrocytes.

Effects of fifteen sugars (D-mannose, methyl-α-D-mannopyranoside, mannan, mannosamine, D-glucose, L-fucose, lactose, N-acetyl-D-glucosamine, N-acetyl-D-galactosamine, N-acetyl-D-neuraminic acid, maltose, mannitol, fructose, α-methyl-D-glucoside) on the hemagglutinating activity of the aloe lectin were investigated. D-mannose and methyl-α-D-mannopyranoside showed high activity.

Mitogenic acticity

The mitogenic activity of the aloe lectin was examined using BALB/c strain mouse spleen lymphocytes. Cell proliferation was determined by a colorimetric assay which detects the conversion of 3(4,5-dimethylthiazolyl-2)-2,5-diphenyltetrazonium bromide (MTT, Sigma) into the formazan product (blue color) by mitochondrial succinate-dehydrogenase (12, 13).

The aloe lectin showed a mitogenic activity toward mouse lymphocytes. The activity was co-eluted with the lectin activity both on the final two steps of DEAE ion exchange and the subsequent gel filtration chromatographies.

ETHANOL-SOLUBLE EXTRACTS OF *ALOE BARBADENSIS* MILLER

An ethanol extract of aloes was found to have strongly hemagglutinating effects (Bouthet *et al.*, 1996). Lectins or lectin-like substances had been generally prepared as water-soluble substances. However, an ethanol extract of *A. barbadensis* was described to have lectin-like properties, i.e. strong hemagglutinating effects toward human erythrocytes. In this study, the recognition sites of specific sugars with ethanol extract was particularly discussed and identified.

Preparation

A commercially available gel filet of *A. barbadensis* leaves in irradiated, lyophilized form was reconstituted into a solution with double distilled water. The aloe extract was obtained by two subsequent ethanol precipitations at 50% in absolute ethanol. The pellet after centrifugation was resuspended in water and dialyzed against water. Then the extract was lyophilized and reconstituted into solution with water.

Hemagglutinating activity

Hemagglutination activity of the aloe extract was tested toward human erythrocytes. The extract was originally dissolved in water and phosphate-buffered saline (PBS) was used for further dilution to avoid osmotic shock to the erythrocytes, because it was more soluble in water than in the other buffers.

Factors influencing hemagglutination. i Sugars

The hemagglutination of the aloe extract was not blood type dependent and 13 monosaccharide sugars (fructose, galactosamine, galactose, N-acetyl galactosamine, D-fucose, glucosamine, N-acetyl glucosamine, mannose, mannosamine, O-methyl-mannose, muramic acid, N-acetyl mannosamine and L-rhamnose) and four disaccharides (chondrosine, maltose, raffinose and sucrose) were examined for their hemagglutination inhibition. Glucosamine, mannosamine and chondrosine inhibited hemagglutination induced by the aloe extract. The inhibition activity was strongest with glucosamine and weakest with chondrosine.

Factors influencing hemagglutination. ii pH, temperature, trypsin, EDTA

The aloe extract was heat and trypsin resistant (but using bovine pancreatic type I protease) and pH sensitive. Acid treatment (pH 3) of the aloe extract caused a decrease of the hemagglutination activity, whereas high pH (12) treatment did not change the activity. The addition of EDTA to hemagglutination assay caused a decrease of the hemagglutination activity, indicating that Ca^{2+} is required for the hemagglutination.

Protein and polypeptide compositions

SDS-PAGE showed 12 polypeptides of the aloe extract with the following approximate molecular masses: 83, 81, 75, 67, 63, 51, 47, 38, 28, 21, 15 and 12 kDa. Two bands with molecular mass of 15 kDa and 12 kDa seemed to be major components of the aloe extract.

Recognition sites of the specific sugar to the aloe extract

The α-amino at the C2 position and α-hydroxyl at the C4 position of glucosamine were concluded to be critical for binding to the aloe extract. The conclusion was introduced by comparisons of the chemical structures of mannosamine, muramic acid, galactosamine and chondrosine.

LECTINS OF LEAF PULPS OF *ALOE BARBADENSIS* MILLER

Two lectins were partially purified from leaf pulps of *A. vera* (*A. barbadensis*, cultivated in Turkey) and designated as Aloctin I and Aloctin II (Akev and Can, 1999). The lectins agglutinated rabbit erythrocytes but the hemagglutination by Aloctin I was not inhibited by any of the 20 sugars tested. The two lectins did not possess any glycosidase activities.

Chemical properties and sugar-binding specificity

Aloctin I and II were a glycoprotein containing 5% and 4.6% sugars, respectively, and the hemagglutinating activity by Aloctin I was inhibited by N-acetyl-D-galactosamine.

Preparation

The gel was removed from *A. vera* leaves by scraping with a spoon and the remaining pulps were homogenized with PBS. The supernatant (crude leaf pulp extract) of the extract was obtained by centrifugation ($45,700 \times g$) at $2\,°C$ and was fractionated with 50% ammonium sulfate. The precipitate was separated by centrifugation at $2\,°C$, suspended in PBS and dialysed against the same buffer. The dialysate (50% ammonium sulfate fraction) was applied to hydroxyapatite column chromatography. The separation was performed by stepwise elution with phosphate buffer (pH 7) and two protein peaks showing hemagglutinating activity were eluted with 5 mM and 20 mM; the former was named Aloctin I and the latter Aloctin II.

Hemagglutinating activity

Hemagglutinating activity was examined towards rabbit erythrocytes. In comparison with rabbit erythrocytes, the activities of the leaf pulp lectins were tested on human erythrocytes of blood group types A Rh (+), B Rb (+), O Rb (+) and O Rh (−) as well as rat erythrocytes. Neither of the two lectins agglutinated any of the blood groups of human erythrocytes and the activity toward rat erythrocytes was weaker than that toward rabbit erythrocytes.

Factors influencing hemagglutination. i Sugars

Twenty sugars (D(+)-glucose, D(+)-galactose, D(+)-glucosamine HCl, D(+)-galactosamine HCl, N-acetyl-D-glucosamine, N-acetyl-D-galactosamine, D(+)-mannose, D(−)-fructose, D(−)-ribose, L(+)-rhamnose, L(+)-arabinose, D(+)-fucose, D(+)-xylose, D(+)-saccharose, D-maltose, D(+)-cellobiose, D(+)-melezitose monohydrate, D(+)-melibiose monohydrate, D(+)-trehalose dihydrate, D(+)-raffinose pentahydrate) were examined for hemagglutination inhibition; none of the sugars tested inhibited the hemagglutinating activity of Aloctin I up to a 500 mM concentration. Aloctin II was weakly inhibited by N-acetyl-D-galactosamine at 250 mM concentration.

Factors influencing hemagglutination. ii Metal ions, heat treatment

Of the 10 metal cations (Mg^{2+}, Ca^{2+}, Ba^{2+}, Mn^{2+}, Fe^{3+}, Co^{2+}, Hg^{2+}, Al^{3+} [sulfate], Al^{3+} [nitrate], Pb^{2+}) tested, only Al^{3+} salts activated the hemagglutination of the two lectins. Aloctin I was heat stable up to $60\,°C$ and hemagglutinating activity was not completely lost even when heated at $100\,°C$ for 1 hour.

Glycosidase activity

Glycosidase activity, such as α- and β-galactosidase and α- and β-glucosidase activities, were examined to know whether lectins themselves possess glycosidase activity, since studies with some *Leguminosae* seeds showed that α-galactosidase activity was exactly copurified with hemagglutinating activity. The two activities, i.e. the glycosidase and hemagglutinating activities, were separated in different protein peaks obtained through hydroxyapatite chromatography, suggesting that Aloctin I and II does not have glycosidase activity and that the consideration that lectins may in general be plant enzymes (Dey *et al.*, 1986) does not apply to these two aloe lectins.

VERECTIN

Several fractions of glycoprotein in leaf gels of *A. barbadensis* were separated and one fraction demonstrated proliferation-promoting activity on human and hamster cells *in vitro* (Yagi *et al.*, 1997). The glycoprotein was designated as verectin. Although lectin-like activity of the glycoprotein with the promoting activity was not examined, the glycoprotein was described here because it promoted cell growth and resembled the lectin of *A. arborescens* as described by the same authors.

Preparation

Purification was substantially performed according to the same procedures as described in the previous lectin (Yagi *et al.*, 1985). The supernatant of *A. barbadensis* gel was obtained by centrifugation of the homogenate and dialyzed against distilled water. Then the dialysate was evaporated under reduced pressure below 35 °C to produce a colored material. The non-dialysate dissolved in 0.02 M ammonium bicarbonate, pH 7.8, and was applied for further purification by a series of column chromatography of DEAE Sephadex A-25, Sepharose 6B and Sephadex G-50. The adsorbed fraction on DEAE Sephadex A-25 was eluted with a linear gradient of 0.3 M sodium chloride to produce two glycoprotein fractions. An active fraction was isolated from one of the fractions by further chromatographies on Sepharose 6B and Sephadex G-50.

Chemical and physical properties

The active glycoprotein, verectin, contained 82% protein and 11% sugar, showing a single band on polyacrylamide gel electrophoresis (PAGE) and an isoelectric point of pH 6.8. Its molecular mass was 29 kDa on a Sephadex G-50 column and SDS-PAGE provided a single band with a molecular mass of 14 kDa, indicating the composition of two subunits. Deglycosylation of the glycoprotein provided a protein band with a molecular mass of 13 kDa on SDS-PAGE.

Biological activity

Verectin had a proliferation-promoting activity on normal human dermal (NB1RGB) fibroblasts and baby hamster kidney (BHK-21) cells *in vitro*, as described above.

Several other fractions of glycoprotein were separated but they had no proliferation-promoting activity. On the contrary, one of them, a polar, colored glycoprotein fraction which was considered to bind phenolic substances, strongly inhibited the *in vitro* assays. These phenolic components may be responsible for reducing the proliferative effect of lectin-like substances in *A. barbadensis* gel (Hart *et al.*, 1988). A neutral polysaccharide which was an unabsorbed fraction of a DEAE Sephadex A-25 chromatography did not show any growth stimulation.

Verectin antiserum

Verectin antiserum was raised in white rabbits and its specificity against a non-dialysate of aloe gels was examined (Yagi *et al.*, 1998). First, both an immunopreciptin line in an Ouchterlony double immunodiffusion test and immunoprecipitation were formed against *A. barbadensis* but not against those of *A. arborescens* and *A. chinensis*, and second an immunopositive band was detected in the *A. barbadensis* and *A. chinensis* non-dialysate but not in that of *A. arborescens* in immunoblotting. The verectin antiserum could be used to distinguish aloe materials. This discrepancy between the first and second tests was speculated to be caused by different affinities with which the antibody is bound to *A. barbadensis* and *A. chinensis* antigens. One of the reason might be that *A. chinensis* is taxonomically classified as *Aloe vera* var. *chinensis* and is morphologically similar in shape to *A. barbadensis*.

ALOE GLYCOPROTEINS HAVING MITOGENIC ACTIVITY

Glycoproteins having mitogenic activity of *A. arborescens*, *A. barbadensis* and *Aloe africana* Miller were described (Yasuda *et al.*, 1999). Whole aloe leaves were separated into flesh and skin portions and each portion was first extracted with hot water. The residues were first extracted with 0.1 M sodium hydroxide and the secondary residues after the initial 0.1 M-extraction were then extracted with 0.5 M sodium hydroxide again. Mitogenic activity was extracted from the hot water soluble fraction of *A. arborescens* and *A. barbadensis* but in a 0.1 M sodium hydroxide soluble fraction of *A. africana*. All the fractions demonstrating mitogenic activity contained both neutral and amino sugars. None of the fractions tested showed any antitumor effect against sarcoma 180 and their hemagglutinating activity was not described.

General properties of aloe lectins are summarized in Table 5.2.

SUMMARY

Objects of studies

Various species of *Aloe*, *A. barbadensis* and *A. arborescens* have been mostly used as folk medicines. *A. barbadensis* is more popular than *A. arborescens*, because *A. arborescens* has been mainly used in the East, particularly in Japan and Korea. Therefore, studies of aloe lectins have been mainly conducted using these two species of *Aloe*.

During the course of studies on *A. barbadensis*, *A. saponaria* was simultaneously studied for comparison and two fractions, S I and S II, were respectively obtained as the low- and high-speed supernatants of whole leaf and stabilized *A. vera* homogenates using differential centrifugation (Winters *et al.*, 1981). A monomeric or oligomeric major component consisting of 15 and/or 12 kDa subunits was separated from *A. barbadensis* gel filets by ethanol extractin. Aloctins I and II which respectively contain 4.6% and 5.0% sugars were isolated from *A. barbadensis* gels (Akev and Can, 1999). In addition, a 29 kDa glycoprotein with a proliferation-promoting activity (Verectin, sugar contents; 12%) isolated from *A. barbadensis* was also studied because its chemical properties were similar to those of lectins (Yagi *et al.*, 1997; Yagi *et al.*, 1998), although it did not meet the definition of cell-agglutinating lectins.

Table 5.2 Properties of aloe lectins.

Aloe species/Material/Preparation	Particular property/Name designated by author(s)/(Reference)	Chemical property	Hemagglutination	Mitogenic activity	Characteristics
A. arborescens/Whole leaf/Water extraction	Lectin-like substnaces (Fujita et al., 1978)	Fraction containing M.S. higher than 10 kDa			Reacts with α2-macroglobulin and α1-antitrypsin of human and various animal sera
A. arborescens/Whole leaf/Chromatography	Aloctin A (Suzuki et al., 1979)	M.S. 18 kDa consists of α (7,5 kDa) and β (10,5 kDa) subunits linked by a disulfide bond. Contains more than 18% neutral sugars	Human and various animals. No ABO blood group specificity	Human lymphocytes	Reacts with human α2-macro-globulin. Anti-inflammatory effect. Effects on gastric function. Immunological effects (Antitumor effect)
	Aloctin B (Suzuki et al., 1979)	M.S. 24 kDa consists of two subunits (12 kDa). Contains more than 50% neutral sugars	Human and various animals. No ABO blood group specificity		
A. barbadensis and A. saponaria/Whole leaf and 'stabilized' A. vera/Differential centrifugation	Wound healing Low and high speeed supernatants: S I and S II (Winters et al., 1981)		S I and S II; Human and canine. Suger specificity; α-D-glucose, mannose		Reaction with serum; Canine serum did not react aloes. S I reacted with human and baboon. S II did not react with any sera. S I of Aloe barbadensis; Enhancement of attachment and growth, and augmentation of cell density in monolayer culture of human normal fetal lung cells may be relate to wound healing
A. arborescens/Whole leaf/Chromatography	DNA synthesis (Yagi et al., 1985)	M.S. 40 kDa containing 34% sugars	Sheep. Sugar specificity; glucose, mannose, galactose		Stimulation of DNA synthesis may attributed the induction of blastmitogenesis and then to burn cures

Table 5.2 (Continued).

Aloe species/Material/ Preparation	Particular property/ Name designated by author(s)/(Reference)	Chemical property	Hemagglutination	Mitogenic activity	Characteristics
A. arborescens/ Whole leaf/ Chromatography	ATF 1011 (Yoshimoto et al., 1987)	M.S. 64 kDa containing 0.4–0.7% neutral sugars and less than 0.3% amino sugars and having I.P. 4.3–5.2	No hemagglutinating activity		No activity. Agglutinating toward tumor cells. Activation of Th cells in antibody production. Augmentation of cytotoxic T cell response. Induction of antitumor cytotoxic T cells
A. arborescens/Skin/ Chromatography	35 kDa mannose-binding lectin (Koike et al., 1995a)	M.S. 35 kDa assumed to be either a trimeric or tetrameric form consisting of identical subunits with 109 amino acids residues (12.2 kDa)	Rabbit. None for human and sheep. Sugar specificity; D-mannose and methyl-α-D-mannopyranoside	Mouse lymphocytes	N-Terminal amino acid sequence showed homology with that of snow drop lectins
A. barbadensis/ gel filet/Ethanol extraction	Ethanol extracts (Bouthet et al., 1996)	The major components are monomeric or oligomeric forms consisting of M.S.15 kDa and/or 12 kDa subunits	Human. No blood type dependence. Sugar specificity; glucosamine, mannosamine and chondrosine		Critical radicals for binding; α amino at C2 and α hydroxy at C4
A. barbadensis/ Leaf gel/ Chromatography	Verectin (Yagi et al., 1997)	M.S. 29 kDa consists of two subunits (14 kDa). Contains 12% sugars	No activity	No activity	Proliferation-promoting activity on human dermatal fibroblasts and baby hamster kidney cells
A. barbadensis/Gel/ Chromatography	Aloctin I (Akev and Can, 1999) Aloctin II (Akev and Can, 1999)	Glycoprotein containing 5% sugars Glycoprotein containing 4.6% sugars. Sugar specificity; N-acetyl-D-galactosamine.	Rabbit and rat. None for human Rabbit and rat. None for human		Glycosidase activity was separated from Aloctin I Glycosidase activity was separated from Aloctin II

Note
M.S.; Molecular mass.

The following lectins were isolated from whole leaves of *A. arborescens*: an 18 kDa lectin (Aloctin A) containing neutral sugars by 18%; a 24 kDa lectin (Aloctin B) containing neutral sugars by more than 50% (Suzuki *et al.*, 1979); a 40 kDa lectin containing sugar by 34%; and, a 64 kDa lectin (ATF 1011) containing sugars by 0.4–0.7% and amino-sugars by less than 0.3% (Yoshimoto *et al.*, 1987). From the leaf skin of *A. arborescens*, a 35 kDa lectin was also isolated (Koike *et al.*, 1995a).

Non-cell-agglutinating substances with mitogenic activities toward mouse lympho-cytes were also isolated from a low-molecular mass fraction of hot water extracts of the leaf flesh of *A. barbadensis* (Yasuda *et al.*, 1999); from a high-molecular mass fraction of hot water extracts of the leaf skin of *A. arborescens*; from a low-molecular mass fraction of hot water extracts and 0.1 M NaOH extracts of the residue of hot water extracts of the leaf flesh of *A. africana*; and, from a high-molecular mass fraction of hot water extracts and 0.1 M NaOH extracts of the residue of hot water extracts of the leaf skin of *A. africana*.

Agglutinating activity

Human erythrocyte-agglutinating lectins were isolated from ethanol extracts of whole leaves of *A. barbadensis* and gel fillets of stabilized *A. vera*; from low- and high-speed supernatants of whole leaves of *A. saponaria* (Bouthet *et al.*, 1996); from ethanol extracts of gel fillets of *A. barbadensis*; and, from whole leaves of *A. arborescens* (Aloctin A and Aloctin B) (Suzuki *et al.*, 1979). Lectins isolated from whole leaves of *A. saponaria* also agglutinate canine erythrocytes (Winters *et al.*, 1981) and Aloctins A and B also agglu-tinate sheep and rabbit erythrocytes (Suzuki *et al.*, 1979).

Lectins isolated from the leaf gel of *A. barbadensis* (Aloctin I and Aloctin II) (Akev and Can, 1999) and a 35 kDa mannose-binding lection (Koike *et al.*, 1995a) isolated from the leaf skin of *A. arborescens* agglutinate rabbit erythrocytes. Aloctins I and II also agglutinate rat erythrocytes (Akev and Can, 1999).

A 40 kDa lectin isolated from whole leaves of *A. arborescens* agglutinates sheep erythro-cytes (Yagi *et al.*, 1985), while ATF 1011 isolated from whole leaves of *A. arborescens* agglutinates tumor cells (Yoshimoto *et al.*, 1987).

To our knowledge, *A. barbadensis* appears to have many lectins that agglutinate human erythrocytes.

Mitogenic activity

Aloctin A (Suzuki *et al.*, 1979) isolated from whole leaves of *A. arborescens* and a 35 kDa mannose-binding lectin (Koike *et al.*, 1995a) isolated from the leaf skin of *A. arborescens* have mitogenic activities toward lymphocytes. Although verectin (Yagi *et al.*, 1997), which was isolated from the leaf gel of *A. arborescens*, is not a lectin, it has a proliferation-promoting activity that is closely associated with its mitogenic activity. In contrast to *A. barbadensis* that contains many cell-agglutinating lectins, *A. arborescens* appears to have many substances with mitogenic activities.

Modification of the immune system

Using Aloctin A (Suzuki *et al.*, 1979; Imanishi and Suzuki, 1984, 1986) and AFT 1011 (Yoshimoto *et al.*, 1987) isolated from whole leaves of *A. arborescens*, the possibility of cytotoxicity against tumor cells via the modified immune system was evaluated.

Other specific activities

The following actions of aloe lectins were also evaluated: anti-inflammatory and gastric function protective effects of Aloctin A isolated from whole leaves of *A. arborescens* (Saito *et al.*, 1982; Imanishi *et al.*, 1984); the contribution of a 40 kDa lectin to the healing of thermal burn via the blastogenesis induction after stimulation of DNA synthesis (Yagi *et al.*, 1985); and, the wound-healing effects of the whole leaf homogenate supernatant of *A. barbadensis* (Winters *et al.*, 1981).

Inhibition of hemagglutinating activity by sugars

Hemagglutinating activity of ethanol extracts of gel fillets of *A. barbadensis* was inhibited by glucosamine, mannosamine and chondrosine (Bouthet *et al.*, 1996), while hemagglutinating activity of Aloctin II (a lectin containing sugars by 4.6%) isolated from the leaf gel of *A. barbadensis* was similarly inhibited by N-acetyl-D-galactosamine (Akev and Can, 1999). Hemagglutinating activity of a 40 kDa lectin (sugar contents; 34%) isolated from the leaf gel of *A. arborescens* was inhibited by glucose, galactose and mannose (Yagi *et al.*, 1985), while the hemagglutinating activity of a 35 kDa lectin isolated from the leaf skin of *A. arborescens* was inhibited by D-mannose and methyl-α-D-mannopyranoside (Koike *et al.*, 1995a). Although hemagglutinating activities of lectins are generally inhibited by mannose and N-acetyl-D-galactosamine, they are rarely inhibited by glucosamine (Bouthet *et al.*, 1996).

As described above, aloe lectins were isolated from the leaf skin and leaf gel of various *Aloe* species, including commercially available materials, and have been studied from various approaches. Based on their chemical, biological and pharmacological characteristics, the respective lectins or lectin-like substances described in this chapter may be different substances or components.

ACKNOWLEDGMENTS

We would like dedicate this chapter to the late Dr. Keisuke Fujita, the Founding President of the Fujita Health University and one of the pioneers of aloe lectin researches, who died on June 11, 1995.

REFERENCES

Akev, N. and Can, A. (1999) Separation and some properties of *Aloe vera* L. leaf pulp lectins. *Phytotherapy Research*, **13**, 489–493.

Allen, L.W., Svenson, R.H. and Yachnin, S. (1969) Purification of mitogenic proteins derived from *Phaseolus vulgaris*: Isolation of protein and weak phytohemagglutinins possessing mitogenic activity. *Proceedings of the National Academy of Sciences, USA.*, **63**, 334–341.

Bouthet, C.F., Shirf, V.R. and Winters, W.D. (1996) Semi-purification and characterization of haemagglutin substance from *Aloe barbadensis* Miller. *Phytotherapy Research*, **10**, 54–57.

Boyd, W.C. and Shapleigh, E. (1954) Specific precipitating activity of plant agglutinins (Lectins). *Science*, 119, 419.

Dey, P.M., Naik, S. and Pridham, J.B. (1986) *Vicia Faba* α-galactosidase with lectin activity. *Phytochemistry*, 25, 1057–1061.

Etzler, M.E. (1985) Plant lectins: molecular and biological aspects. *Annual Review of Plant Physiology*, 36, 209–34.

Fodstad, O. and Pihl, A. (1978) Effect of ricin and abrin on survival of L1210 leukemic mice and leukemic and normal bonemarrow cells. *International Journal of Cancer*, 22, 558–563.

Franz, H., ed. (1988) *Advances in Lectin Research*. Vol. 1, 187pp. Berlin: Springer-Verlag.

Fujita, K., Iwase, S., Ito, T. and Matsuyama, M. (1958a) Inhibiting effect of chloropromazine on the experimental production of liver cancer. *Nature*, 181, 54.

Fujita, K., Mine, T., Ito, T. and Matsuyama, M. (1958b) Effects of certain compounds related to trypan blue on the experimental production of liver cancer. *Nature*, 181, 1732–1733.

Fujita, K., Ochiai, J., Shimpo, K., Inoue, T. and Murata, T. (1975) Agglutinins of *Aloe arborescens* (Japanese). *Bulletin of the Fujita Medical Society, Supplement*, p. 34.

Fujita, K., Suzuki, I., Ochiai, J., Shinpo, K., Inoue, S. and Saito, H. (1978) Specific reaction of aloe extract with serum proteins of various animals. *Experientia*, 34, 523.

Goldstein, I.J., Hughes, C., Monsigny, M., Osawa, T. and Sharon, N. (1980) What should be called a lectin? *Nature*, 285, 66.

t'Hart, L.A., van Enckevort, P.H., van Dijk, H., Zaat, R., de Silva, K.T.D. and Labadie, R.P. (1988) Two functionally and chemically distinct immunomodulator compounds in the gel of *Aloe vera*. *Journal of Ethnopharmacology*, 23, 61–71.

Imanishi, K., Ishiguro, T., Saito, H. and Suzuki, I. (1981) Pharmacological studies on a plant lectin, Aloctin A. I. Growth inhibition of mouse methlcholanthrene-induced fibrosarcoma (Meth A) in ascites form by Aloctin A. *Experientia*, 37, 1186–1187.

Imanishi, K. and Suzuki, I. (1984) Augmentation of natural cell-mediated cytotoxic reactivity of mouse lymphoid cells by Aloctin A. *International Journal of Immunopharmacology*, 5, 539–543.

Imanishi, K. and Suzuki, I. (1986) Induction of nonspecific cell-mediated cytotoxic reactivity from non-immune spleen cells treated with Aloctin A. *Journal of Immunopharmacology*, 7, 781–787.

Imanishi, K., Tsukuda, K. and Suzuki, I. (1986) Augmentation of lymphokine-activated killer cell activity *in vitro* by Aloctin A. *International Journal of Immunopharmacology*, 8, 855–858.

Imanishi, K., Karasaki, S., Saito, H., Hoshi, G. and Suzuki, I. (1987) Macrophage activation *in vivo* and inhibition of heat-induced hemolysis by anti-inflammatory substance, Aloctin A. *Japanese Journal of Inflammation*, 53, 52–56.

Imanishi, K. (1993) Aloctin A, an active substance of *Aloe arborescens* Miller as an immunomodulator. *Phytotherapy Research*, 7, S30–S22.

Inbar, M. and Sachs, L. (1969) Structural difference in sites on the surface membrane of normal and transformed cells. *Nature*, 223, 710–712.

Kocourek, J. (1986) Historical background. In reference 1, 1–32.

Koike, T., Beppu, H., Kuzuya, H., Maruta, K., Shimpo, K., Suzuki, M., Titani, K. and Fujita, K. (1995a) A 35kDa mannose-binding lectin with hemagglutinating and mitogenic activities from 'Kidachi Aloe' (*Aloe arborescens* Miller var. *natalensis* Berger). *Journal of Biochemistry*, 118, 1205–1210.

Koike, T., Titani, K., Suzuki, M., Beppu, H., Kuzuya, H., Maruta, K., Shimpo, K. and Fujita, K. (1995b) The complete amino acid sequence of a mannose-binding lectin from 'Kidachi Aloe' (*Aloe arborescens* Miller var. *natalensis* Berger). *Biochemical and Biophysical Research Communications*, 214, 163–170.

Liener, I.E., Sharon, N. and Goldstein, I.J., eds (1986) *The Lectins: Properties, Functions and Applications in Biology and Medicine*, 600pp. Orlando: Academic Press.

Lis, H. and Sharon, N. (1986) Lectins as molecules and as tools. *Annual Review of Biochemistry*, 55, 35–67.

Mirelman, D. (1986) *Microbial Lectins and Agglutinins: Properties and Biological Activity*, 443pp. New York: J. Wiley and Sons.

Ohuchi, K., Watanabe, M., Takahashi, E., Turufuji, S., Imanishi, K., Suzuki, I. and Levine, L. (1984) Lectins modulate prostaglandin E2 production by rat peritoneal macrophages. *Agents and Actions*, 15, 419–423.

Olden, K. and Parent, J.B. (1987) *Vertebrate Lectins*, 255pp. New York: Van Nostrand Reinhold.

Saito, H. (1993) Purification of active substances of *Aloe arborescens* Miller and their biological and pharmacological activity. *Phytotherapy Research*, 7, S14–S19.

Saito, H., Imanishi, K. and Okabe, S. (1989) Effects of aloe extracts, Aloctin A, on gastric secretion and on experimental gastric lesuins in rats. *Yakugaku Zasshi*, 109, 335–339.

Saito, H., Ishiguro, T., Imanishi, K. and Suzuki, I. (1982) Pharmacological studies on a plant lectin Aloctin A. II. Inhibitory effect of Aloctin A on experimental models of inflammation in rats. *Japanese Journal of Pharmacology*, 32, 139–142.

Sharon, N. and Lis, H. (1972) Lectins: cell-agglutinating and sugar-specific proteins. *Science*, 177, 949–959.

Sharon, N. and Lis, H. (1987) A century of lectin research (1888–1988). *Trends in Biochemical Science*, 12, 488–91.

Sharon, N. and Lis, H. (1989) *Lectins*, 127pp. London: Chapman and Hall Ltd.

Stillmark, H. (1888) Über Ricin, ein giftiges Ferment aus den Samen von *Ricinus communis L.* und einigen anderen *Euphorbibiaceen. Inaug. Dissertationen.*, Dorpart.

Suzuki, I., Saito, H., Inoue, S., Migita, S. and Takahashi, T. (1979) Purification and characterization of two lectins from *Aloe arborescens* Mill. *Journal of Biochemistry*, 85, 163–171.

Tunis, M. (1964) Agglutinins of kidney bean (*Phaseolus vulgaris*); a new cytoagglutinin distinct from hemagglutinin. *Journal of Immunology*, 92, 864–869.

Van Damme, E.J., Kaku, H., Perini, F., Goldstein, I.J., Peeters, B., Yagi, F., Decock, B. and Peumans, W.J. (1991) Biosynthesis, primary structure and molecular cloning of snowdrop (*Galanthus nivalis L.*) lectin. *European Journal of Biochemistry*, 202, 23–30.

Van Damme, E.J., Smeet, K., Engelborghs, I., Aelbers, H., Barzarini, J., Pusztani, A., van Leuven, F., Goldstein, I.J. and Peumans, W.J. (1993) Cloning and characterization of the lectin cDNA clones from onion, shallot and leek. *Plant Molecular Biology*, 23, 365–376.

Weber, T.H. (1969) Isolation and characterization of a lymphocyte-stimulating leucoagglutinin from red kidney beans (*Phaseolus vulgaris*). *The Scandinavian Journal of Clinical and Laboratory Investigation, Supplement*, 111, 1–80.

Winters, W.D. (1979) Human adenovirus antibody in sera of normal and tumor-bearing dogs *Veterinary Record*, 105, 216–220.

Winters, W.D. (1993) Immunoreactive lectins in leaf gel from *Aloe barbadensis* Miller. *Phytotherapy Research*, 7, S23–S25.

Winters, W.D., Benavides, R. and Clouse, W.J. (1981) Effects of aloe extracts on human normal and tumor cells *in vitro. Economic Botany*, 35, 89–95.

Yagi, A., Egusa, T., Arase, M., Tanabe, M. and Tsuji, H. (1997) Isolation and characterization of the glycoprotein fraction with a proliferation-promoting activity on human and hamster cells *in vitro* from *Aloe vera* gel. *Planta Medica*, 63, 18–21.

Yagi, A., Machii, K., Nishimura, H., Shida, T. and Nishioka, I. (1985) Effect of aloe lectin on deoxyribonucleic acid synthesis in baby hamster kidney cells. *Experientia*, 4, 469–471.

Yagi, A., Tsunoda, M., Egusa, T., Akasaki, K. and Tsuji, H. (1998) Immunochemical distinction of *Aloe vera, A. arborescens* and *A. chinensis* gels. *Planta Medica*, 64, 277–278.

Yasuda, K., Dohgasaki, C. and Nishijima, M. (1999) Mitogenic activity and antitumor activity of *Aloe arborescens* Miller, *Aloe barbadensis* Miller and *Aloe africana* Miller. *Nihon Shokuryou Hozou Kagaku Kaisi*, 25, 201–207.

Yoshimoto, R., Kondoh, N., Isawa, M. and Hamuro, J. (1987) Plant lectin, ATF1011, on the tumor cell surface augments tumor-specific immunity through activation of T cells specific for the lectin. *Cancer Immunology Immunotherapy*, 25, 25–30.

6 Analytical methodology: the gel-analysis of aloe pulp and its derivatives

Yawei Ni and Ian R. Tizard

ABSTRACT

It has been widely believed for several years that many of the beneficial effects of aloe leaf extracts lie in their carbohydrates. The thick fleshy leaves contain both cell wall carbohydrates, such as celluloses and hemicelluloses, as well as storage carbohydrates, such as acetylated mannans, arabinans and arabinogalactans. Like all cells, they also contain many diverse glycoproteins. The acetylated mannan is the primary polysaccharide in the pulp (inner clear portion of the leaf) and has been most widely studied. It has been claimed to possess many therapeutic properties, including immune stimulation. The general structure of the mannan ($\beta1 \rightarrow 4$ linked mannose residues and acetylation) has been well-defined, although many structural features such as degree of acetylation, glucose content, and molecular weight remain to be defined, especially in relation to functional properties, extraction conditions, and harvesting times. There may well be two types of mannans in the pulp, a pure mannan and a glucomannan. There is very little information available on the enzymes used to modify or synthesize the mannan or other aloe carbohydrates. Aloe mannan is indeed a unique polysaccharide. Future studies on the structure-function relationship will certainly yield more insight into its chemical and functional properties.

INTRODUCTION

Aloe vera (L.) Burm.f. has enjoyed a long history as a medicinal plant. The leaf of *A. vera* consists of two parts, the inner clear pulp and the outer green rind. Many of the beneficial effects of this plant have been attributed to the pulp, including both immunostimulation and anti-inflammation. The chemical and biological properties of the pulp have been described in many reviews (Grindlay and Reynolds, 1986; Klein and Penneys, 1988; Kaufman *et al.*, 1989; Haller, 1990; Shelton, 1991; Canigueral and Vila, 1993; Briggs, 1995; Joshi, 1998; Reynolds and Dweck, 1999). The use of aloe pulp extracts in western society is becoming increasingly popular. Currently, it is the basis for many products used in humans and animals for a variety of purposes, including immunostimulation, wound healing, cosmetic uses, and nutraceutical uses. Thus, analyzing and understanding its chemical composition is of great importance, not only to the underlying mechanisms for its functions or biological effects, but also to product standardization and the development of new products. Indeed, methods by which the pulp is processed may vary greatly between different manufacturers. Such variation likely results in a final preparation or a product that has a different chemical make-up and hence a different functional property.

There are over 400 *Aloe* species recognized. Among these, *A. vera* (L.) Burm.f. (= *A. barbadensis,* Miller) is most widely known and used worldwide. However, other species, especially *A. arborescens* Miller, *A. saponaria* (Ait.) Haw. and *A. ferox* Miller have enjoyed regional popularity in Asia and Africa. In the scientific literature, it is these *Aloe* species that have been most widely studied and analyzed, although several other species have also been examined. The chemical composition of the pulp or the liquid gel prepared from them has been described in various reports and reviews (Waller *et al.,* 1978; Grindlay and Reynolds, 1986; Yamaguchi *et al.,* 1993; Reynolds and Dweck, 1999). Various analytical methods have been used to identify and isolate individual compounds. Some examples of the compounds that have been identified in aloe pulp are listed in Table 6.1.

Here we will not concentrate on the quantitative aspect of the components, but instead on the methods that have been used for analyzing them. We have sought to include one or more references for each method used for analyzing the component in the pulp so that the interested reader can find out how it was applied. It is always hoped that when treating a subject from the analytical point of view, a quantitative reference for some constituents in the subject can be established or cited. Unfortunately, such information seems limited with respect to aloe pulp. This is not due to

Table 6.1 Examples of the compounds that have been identified in aloe pulp*.

Carbohydrates	Mannan: 　　Pure mannan 　　Glucomannan 　　Glucogalactomannan Galactan Pectic substance Arabinogalactan Xylan
Proteins	Lectins: 　　Aloctin A (18 kDa) and B (24 kDa) 　　ATF1011 (40 kDa) Enzymes: 　　Phosphoenolpyruvate carboxylase 　　Superoxide dismutase 　　Carboxypeptidase
Lipid	Steroids (cholesterol, campestrol, β-sitosterol) Triterpenoid (lupeol) γ-linolenic acid Arachidonic acid
Small organic compounds	Malic acid or malate Anthraquinones Anthrone C-glycosides (such as aloins A and B) Chromones (such as aloesin) Vitamins Free sugars Free amino acids

Note
* See text for references.

a lack of effort but to variations in the initial processing of pulp or simply to a lack of description of this process and the analytical methods used, which in turn makes data comparison difficult. Of course, further complicating this situation is the natural variation in geographic location, seasonal change, and plant species, etc. However, many efforts have been made to generate such information (Rowe and Parks, 1941; Roboz and Haagen-Smit, 1948; Bouchey and Gjerstad, 1969; Gjerstad, 1971; Robson *et al.*, 1982; Gorloff, 1983; Kodym, 1991; Yaron, 1993; Femenia *et al.*, 1999). Some of the information is presented in Table 6.2.

As is true for all chemical analysis of a living tissue, an understanding of the structure of the tissue is important. Thus, we will start with a description of the plant and the pulp before dealing with the analytical methods.

THE PLANT

Aloes have been well described elsewhere in this book. It is important to reinforce the fact that aloes are xerophytic succulents, adapted to living in areas of low water availability, and characterized by possessing a large volume of water storage tissue. Although not widely discussed in the literature, the pulp of aloes is likely to be the water storage tissue of this plant (Kluge *et al.*, 1979). Another feature of succulents such as the aloes is the possession of crassulacean acid metabolism (CAM), an additional photosynthetic pathway involving malic acid (Kluge and Ting, 1978; Winter and Smith, 1996). In contrast to other plants, CAM plants take up carbon dioxide during the night, which is then fixed by malic acid synthesis. Thus, in the early morning hours, the malic acid content is much higher in the CAM plant tissues. During daytime, the malic acid is decarboxylated and the released carbon dioxide is then converted into carbohydrates. CAM occurs in aloes (Denius and Homan, 1972; Kluge *et al.*, 1979) where it operates

Table 6.2 The amounts of some major components in the aloe pulp.

	Amount (% dry weight)		References
	Intact pulp	*Liquid gel*	
Total polysaccharide	–	10–20	Yaron, 1993
	30	–	Roboz and Haagen-Smit, 1948
Total soluble sugar	16.48 ± 0.18	26.81 ± 0.56	Femenia, 1999
	–	20–30	Yaron, 1993
	6.5	–	Rowe and Park, 1941
	25.5	–	Roboz and Haagen-Smit, 1948
Total protein	7.26 ± 0.33	8.92 ± 0.62	Femenia, 1999
	2.78	–	Roboz and Haagen-Smit, 1948
Total lipid	4.21 ± 0.12	5.13 ± 0.23	Femenia, 1999
	4.76	–	
Malic acid	5.4 ± 0.85 – 8.7 ± 3.0	–	Paez *et al.*, 2000
Ca	5.34 ± 0.14	3.58 ± 0.42	Femenia, 1999
Na	1.98 ± 0.15	3.66 ± 0.07	Femenia, 1999
K	3.06 ± 0.18	4.06 ± 0.21	Femenia, 1999
Ashes	15.37 ± 0.32	23.61 ± 0.71	Femenia, 1999
	13.1	–	Rowe and Parks, 1991
	8.63	–	Roboz and Haagen-Smit, 1948

primarily in the outer green rind where chloroplasts reside (Kluge *et al.*, 1979). The diurnal fluctuation of malic acid content is therefore observed in the rind. Malic acid is also present in the pulp although its content does not show diurnal variation.

THE PULP

The inner part of the leaf (free of green rind) or pulp is a clear, soft, moist, slippery tissue. It has been described using several other terms including mucilage tissue, mucilaginous gel, mucilaginous jelly, inner gel, and leaf parenchyma. Here we will use the term 'pulp.' The term 'gel' is often used, but is not an accurate description because it implies a homogeneous entity. In addition, the term 'gel' is also confusing at times because it is unclear whether it is used to refer to the intact pulp or the viscous liquid gel prepared from it.

The pulp consists of large mesophyll cells (Kluge *et al.*, 1979; Trachtenberg, 1984; Fahn, 1990; Evans, 1996). The vascular bundles are tubular structures located in the pulp, but adjacent to the green rind. The number of these bundles varies, depending on the size of the leaves. They are the conducting system of the leaf responsible for transporting nutrients. The non-viscous yellow liquid that flows freely from freshly cut leaves is derived from the pericyclic cells associated with these vascular bundles.

Cell walls and cell membranes can be observed in the pulp (Kluge *et al.*, 1979; Trachtenberg, 1984), although intact cellular organelles such as nuclei, chloroplasts, and mitochondria are not usually detected. The only organelles that have been observed in the mature pulp mesophyll cells are dilated or degenerated plastids. These may be the source of mucilage polysaccharide during the early stage of mesophyll cell development (Trachtenberg, 1984). Thus, unlike mucilage tissues in other succulents, aloe pulp does not contain chloroplasts and is not involved in CAM (Kluge *et al.*, 1979). This is consistent with the fact that the photosynthetic activity only occurs in the rind where chloroplasts are located.

A viscous clear liquid gel or mucilage is contained within the pulp mesophyll cells. One major component of the liquid gel is a neutral acetylated $\beta 1 \rightarrow 4$ linked mannan. Thus, *A. vera* pulp is also unique in that the polysaccharide that constitutes the mucilage is a neutral polysaccharide (See Chapter 4). Mucilage polysaccharides found in other plants are mostly acidic polysaccharides (Kennedy and White, 1983).

ANALYSIS OF PULP AND ITS COMPONENTS

In analyzing pulp, it is important to realize that we are dealing with material from a living plant whose conditions and properties can be influenced by many factors, including geographical location, seasonal change, and plant genetics (Pierce, 1983; Wang and Strong, 1993; Yaron, 1993). For example, differences in irrigation level can result in significant differences in total soluble sugar and polysaccharide content (Yaron, 1993). Furthermore, there are many different *Aloe* species. Fluctuations in chemical composition should be expected. This situation demands analytical methods that are well standardized and calibrated before one can be sure if any differences or similarities are real or not. Among the components that have been identified so far in aloe pulp, the polysaccharide, especially the mannan, has been most widely studied. In recent years, much effort has also been made to isolate and characterize the pulp proteins, especially the lectins.

Processing of the pulp

Pulp mainly consists of water; the dry matter only accounts for ~1% by weight. The harvesting and processing of the pulp is the first step of all analytical processes. Fresh *A. vera* leaves are usually first allowed to drain off the yellow liquid before the rind is removed with a sharp blade. The clear pulp is homogenized using a blender or a polytron. This homogenized pulp preparation is very viscous and has a pH of 4–4.5. It is often passed through a filter or centrifuge (Waller *et al.*, 1978; Yagi *et al.*, 1984; Yaron, 1993) and the filtrate or centrifugation supernatant is the starting material for various analyses. We will refer to it here as liquid gel preparation. It is often freeze-dried at this stage before being further processed.

Alternatively, water, saline, acetone, ethanol, methanol, or other solvents may be added to the homogenized pulp for extraction before the filtration or centrifugation step (Yagi *et al.*, 1977; Waller *et al.*, 1978; Gowda *et al.*, 1979; t'Hart *et al.*, 1989; Yamaguchi *et al.*, 1993).

Physical appearance

The liquid gel preparation is a viscous solution. Depending on the pore size of the filter or centrifugal force used, the preparation may be clear or very cloudy. The occurrence of a yellowish to brownish color indicates that it has been contaminated by anthranquinone from the rind or that it has not been processed in a timely manner and oxidation of phenols has occurred. Anthraquinones, normally a yellowish color, are air and light sensitive. Upon exposure to air, they will gradually turn pink to brownish. It is difficult to completely eliminate anthraquinones from the pulp preparation as evidenced by Yamaguchi *et al.* (1993).

General analytical methods

Low and high-pressure liquid chromatography (LPLC and HPLC)-based separation techniques are most widely used for fractionation, purification and size determination. LPLC is also referred to as gel permeation or gel filtration chromatography. Since separation is based on the size of the molecules, they are also collectively referred to as size exclusion chromatography. The detectors employed usually are UV/VIS spectrophotometers and/or refractometers. They can also be coupled with mass spectrometry (MS) and light scattering to enhance the analytical power and to permit compounds to be quickly identified.

Ion exchange chromatography is a separation technique based on the charge of molecules. It is useful for separation of charged molecules from the neutral ones. Gas chromatography (GC) is another widely employed technique. The mobile phase is gas, instead of liquid as in LPLC or HPLC. It is suitable for the analysis of volatile compounds. For example, acetyl or methyl groups on a polysaccharide can be measured by this method following saponification. Other commonly used analytical methods include specific rotation, infra-red (IR) and nuclear magnetic resonance (NMR) spectroscopy. These methods can be used to probe the structure of a compound or identify a compound based on the specific signatures of a chemical bond or group present in the compound.

Other analytical methods that have been used for analyzing aloe pulp and its derivatives include element analysis, amino acid composition, ash content, moisture content, and heavy metal content. Description of these methods can be found in many publications, such as AOAC (Official methods of analysis). Many can be performed by automated instruments. Examples of the use of these methods in analyzing aloe pulp and its derivatives can be found in the studies by Waller *et al.* (1979), Yamaguchi *et al.* (1993), and Femenia *et al.* (1999).

Specific analytical methods

Carbohydrates

The basic methods used for carbohydrate analysis have been well described by Roybt (1998). The book by Chaplin and Kennedy (1994) is a good source for a step-by-step description of commonly used methods.

Total carbohydrate

Several chemical assays have been used to determine the total carbohydrate content and the uronic acid content in the various pulp preparations or isolated polysaccharide preparations. The most commonly used method is the phenol-sulfuric acid assay for total carbohydrate content and the *m*-hydroxydiphenyl method for uronic acid content (Blumenkrantz and Asboe-Hansen, 1973; Chaplin and Kennedy, 1994). These assays have been widely used in analyzing the pulp preparations (Paulsen *et al.*, 1978; Radjabi *et al.*, 1983; Yagi *et al.*, 1984; Wozniewski *et al.*, 1990). The total carbohydrate measurement can also be achieved with HPLC and the recently developed HPAE-PAD (high performance anion exchange-pulsed amperometric detector) analysis of individual sugars following hydrolysis. Analysis of the total sugar composition is often desired and serves as a good method of monitoring the consistency of the pulp or products derived from it.

Monosaccharides

For detecting free or soluble monosaccharides, samples are not subjected to hydrolysis. Instead, samples may be directly subjected to HPLC or TLC following fine filtration or dialysis (Roboz and Haagen-Smit, 1948; Yaron, 1993). Femenia *et al.* (1999) used the HPLC-based method for measuring the total soluble sugars in various aloe leaf preparations. Glucose is the dominant monosaccharide found in aloe pulp, accounting for as much as 95% of the total soluble monosaccharides.

Isolation and purification of polysaccharide

Polysaccharides, mainly the mannan, have been isolated from the pulp by various methods including alcohol precipitation and direct fractionation by chromatography (Roboz and Haagen-Smit, 1948; Gowda *et al.*, 1979, 1980; Yagi *et al.*, 1984). Often, extraction is performed directly with the homogenized pulp with hot water or alcohol before or after the filtration or centrifugation step (Yagi *et al.*, 1977; Gowda *et al.*, 1979; t'Hart *et al.*, 1989). It is important to note that different extraction conditions can result in a different polysaccharide preparation (Gowda *et al.*, 1979, 1980). Following the extraction, polysaccharides may be further purified by chromatography. Often gel permeation with Sephadex, Sepharose, or Sephacryl media is used for this purpose (Radjabi *et al.*, 1983; Vilkas and Radjabi-Nassab, 1986; t'Hart *et al.*, 1989). A refractometer is usually used as a detector. If a detector is not available, fractions may be tested by carbohydrate assays to locate the polysaccharide peak. HPLC is more often used for analytical purposes such as size determination and quantification (Ross *et al.*, 1997). Ion exchange chromatography has also been used to analyze *A. vera* polysaccharides (t'Hart *et al.*, 1989). The commonly used media for ion exchange chromatography is DEAE-Sephacel or DEAE-Sepharose.

Determination of molecular weight

The molecular weights of aloe mannans have been determined in most cases by size exclusion chromatography using neutral polysaccharides (such as dextran) with known molecular weights as standards (Paulsen *et al.*, 1978; Gowda *et al.*, 1979; Yagi *et al.*, 1984, 1986; Wozniewski *et al.*, 1990; Ross *et al.*, 1997). Other methods have also been used, including equilibrium ultracentrifugation (Yagi *et al.*, 1977) and osmometry (Mandal and Das, 1980b). The results obtained with these methods are relative and not absolute. Apparently, the mannan isolated from fresh pulp is a very large molecule with a size estimated to be $>1 \times 10^3$ kDa (Paulsen *et al.*, 1978; Wozniewski *et al.*, 1990; Ross, *et al.*, 1997). Use of more advanced methods such as light scattering for molecular weight determination has not been seen in the published reports.

Determination of sugar composition

Sugar composition is determined by GC-MS analysis of sugars derivatized as alditol acetate or trimethylsilyl ether following hydrolysis (Paulsen *et al.*, 1978; Yagi *et al.*, 1984, 1986; Femenia *et al.*, 1999). The alditol acetate method permits analysis of neutral sugars only and the trimethylsilyl ether method allows for analyzing all sugars including uronic acid and aminosugars. These two methods have been widely used for analysis of aloe carbohydrates. Polysaccharides such as mannan are first hydrolyzed with trifluoroacetic acid or hydrochloric acid and individual sugar residues are then derivatized as trimethylsilyl ether or alditol acetate. The derivatized sugars are then analyzed by GC-MS. However, the conditions for acid hydrolysis vary among various reports and the total carbohydrate yield is seldom reported. The hydrolysis conditions determine how efficiently the polysaccharide is degraded into monosaccharides, i.e. the amount of monosaccharides released, which in turn influences the result on the composition. So far, no effort has been made to evaluate the hydrolysis condition for the mannan and its effect on the composition obtained by GC-MS analysis.

Paper partition chromatography (PC) and thin-layer chromatography (TLC) have also been used for analyzing the composition of aloe polysaccharides (Yagi *et al.*, 1977; Mandal and Das, 1980a, 1983). Following separation on cellulose paper or silica gel plate, sugars are detected by development of color after spraying the paper with a mixture of chemical reagents. The advantages of these methods are their simplicity and the requirement of no expensive equipment, but their shortcoming is that the results are somewhat difficult to quantify due to the fact that different sugars yield different colors upon reacting with the detection reagents.

Determination of sugar linkage

Sugar linkage is usually determined by methylation analysis. It has been widely used for determining the glycosidic linkage of aloe polysaccharides, especially the mannans (Paulsen *et al.*, 1978; Gowda *et al.*, 1979, 1980; Yagi *et al.*, 1984). The polysaccharide is commonly methylated by the Hakomori method. Following acid hydrolysis, the methylated monosaccharides are analyzed by GC-MS. The linkage can also be determined using an enzymatic method (see below).

Configuration of the glycosidic linkage

Specific rotation has been widely used to determine the configuration of the glycosidic linkage in aloe polysaccharides, especially the mannan. It is measured using a polarimeter. A polysaccharide produces a positive rotation if it is mostly linked in the α configuration or a negative rotation if in the β configuration. A more accurate method is ^{13}C-NMR spectroscopy. Each anomeric carbon gives a distinct signal on the spectrum. A good example of this is the study carried out by Radjabi-Nassab *et al.* (1984) for determination of the configuration of the linkage of a mannan isolated from *A. vahombe* (=*A. vaombe* Decorse et Poiss.).

Another reliable method is enzymatic degradation. This method is highly specific as long as a well characterized enzyme is available. It has been only occasionally used for aloe polysaccharides (Paulsen *et al.*, 1978; Mandal and Das, 1980a). In both cases, a galactosidase was used to determine the linkage of galactose side-chains. The endo $\beta 1 \rightarrow 4$ mannanases from plant or bacteria are highly specific for the $\beta 1 \rightarrow 4$ linked mannose polymers and may also be used for analyzing the mannan.

Proteins

Total proteins

There are several different methods for measuring total protein content, such as the BCA (bicinchoninic acid) and Lowry techniques. These assays have been used to measure the protein content in isolated polysaccharide preparations (Wozniewski *et al.*, 1990). One needs to be cautious in that these assays may be affected by the presence of other compounds, i.e. they may be suited for purified protein preparation or a purified polysaccharide preparation, but not for preparations containing a mixture of compounds. Alternatively, protein content can be estimated indirectly by measuring the nitrogen content, which is then converted to protein content. This method has been employed by Femenia *et al.* (1999).

Purification of proteins

Proteins are usually purified by gel permeation and ion exchange chromatography. Before this step, proteins may be concentrated from the liquid gel preparation by ammonium sulfate precipitation or lyophilization (Yagi *et al.*, 1987). Individual proteins can be analyzed by SDS-PAGE (sodium dodecyl sulphate-polyacrylamide gel electro-phoresis) coupled with Commassie blue staining. So far, several different proteins have been identified in aloe pulp (Yagi *et al.*, 1986, 1997; Winters, 1993; Winters and Yang, 1996).

Identification of proteins

For identification of the proteins, amino acid composition and N-terminal sequencing can be performed. For further analysis, antibodies may be generated against the protein and the gene for the protein may be cloned and sequenced based on the information from peptide sequencing.

Proteins may also be identified by their function. This is achieved by testing the protein or a preparation containing the protein with a biological or chemical assay such as blood cell agglutination. So far, several lectins have been identified in aloes, many of which are capable of agglutinating red blood cells. However, most of these studies used the whole leaf as the starting materials (Fujita *et al.*, 1978; Suzuki *et al.*, 1979a, 1979b; Yoshimoto *et al.*, 1987; Saito, 1993; Koike *et al.*, 1995). So it is not clear if these lectins are from the pulp or from the rind. Evidence for the presence of lectins in the pulp, however, has been provided by Winters (1993) and Akev and Can (1999).

Yagi *et al.* (1987) identified an enzyme that can degrade bradykinin from a whole leaf extract. Kluge *et al.* (1979) identified a phosphoenolpyruvate carboxylase in the pulp. Very interestingly, this enzyme from the pulp was found to have different molecular weight and kinetic properties from that in the green rind. Several other enzymes (oxidase, amylase, glyoxalase, glutathione peroxidase, superoxide dismutase, and carboxy-peptidase) have also been identified in the pulp (Rowe and Parks, 1941; Norton *et al.*, 1990; Ito *et al.*, 1993; Sabeh *et al.*, 1993, 1996).

There should be many more proteins present in aloe leaves than those which have been identified to date. Among others there should be a series of enzymes involved in the synthesis of the acetylated mannan including a mannosyltransferase. Identification of such enzymes would be important to the understanding of polysaccharide synthesis and modification in aloe leaves.

Lipid

Extraction of lipid components

Lipids can be extracted by diethyl ether; *n*-hexane; chloroform/methanol (Folch method), or *n*-butanol (Waller *et al.*, 1978; Afzal *et al.*, 1991; Yamamoto *et al.*, 1991; Yamaguchi *et al.*, 1993; Kinoshita *et al.*, 1996). In the studies by Yamaguchi *et al.* (1993), the *n*-hexane extract was further extracted with acetone followed by fractionation on a silica gel column.

Analysis of lipid components

Lipids can be first fractionated on a silica gel column and then analyzed by TLC (Waller *et al.*, 1978; Afazal *et al.*, 1991; Yamamoto *et al.*, 1991; Yamaguchi *et al.*, 1993; Kinoshita *et al.*, 1996). Identification or structural determination is obtained by GC-MS and NMR spectroscopy. So far, several lipid compounds, including cholesterol, lupeol, and β-sitosterol, have been identified in aloe pulp. Some of these, (lupeol, β-sitosterol, and campestrol) have been found to have anti-inflammatory activity (Yamamoto *et al.*, 1991; Davis *et al.*, 1994).

Small organic compounds

These compounds include vitamins, amino acids, monosaccharides, many kinds of acids or salts. They can usually be isolated and identified by HPLC and MS. Further confirm-ation can be obtained with IR and NMR.

Glucose is the dominant monosaccharide in the pulp, accounting for as much as 95% of the total soluble monosaccharides (Yaron, 1993; Femenia *et al.*, 1999; Paez *et al.*, 2000).

Malic acid, an acid characteristic of the CAM plants, is present in the pulp although its amount does not show diurnal fluctuations as in the green rind. It can be isolated and detected by HPLC as the so-called 'E peak.' The presence of malic acid can also be determined by ^{1}H-NMR (Diehl and Teichmuller, 1998). The malic acid content in the pulp is estimated to be 5.4–8.7%, depending on growth conditions (Paez *et al.*, 2000).

Anthraquinones derivatives (such as aloins A and B) and chromones (such as aloesin) have been most often detected and identified in the yellowish exudates from the leaf or extracts prepared from rinds or whole leaf (Koshioka *et al.*, 1982; Grindlay and Reynolds, 1986; Holzapfel *et al.*, 1997). However, Okamura *et al.* (1997, 1998) have identified and isolated anthraquinones and chromones from the pulp. Aloins have been identified as the active compound for the purgative effect of aloes.

Activity-guided fractionation and analysis

There has been a long list of biological activities attributed to aloe pulp. Some examples include immunostimulation, anti-inflammation, aspirin-like activity, anti-complement activity, pain inhibition, anti-bacterial, and anti-fungal activity (Reynolds and Dweck, 1999). However, definitive identification of a component in relation to a specific biological activity has not often been successful or pursued. Furthermore, it is quite possible there is more than one compound with the same biological effect, i.e. although one compound has been identified to have a certain biological activity, it may not be the only one with such an activity in the pulp. This is largely due to a lack of a systematic activity-guided fractionation. It can not be emphasized enough that this approach is critical in efforts to identify active ingredients. This approach depends on a fractionation and purification system coupled with a biological assay. Fortunately, assays for a wide range of biological activities have been well established, although in some cases, the assays are performed in small animal models such as mouse. Although such an approach has not been widely used in analyzing aloe pulp constituents, some successful attempts have been made. t'Hart *et al.* (1989) isolated a polysaccharide con-sisting primarily of mannose that had an anti-complementary activity. A glycoprotein that stimulates cell proliferation was isolated by Yagi *et al.* (1997). Anti-inflammatory compounds have been isolated by Yamamoto *et al.* (1991) who used the carrageenan-induced inflammation model in rat as the biological assay.

The full-scale form of this approach is the so-called high-throughput screening, i.e. isolating each group of compounds and then each individual compound from each group and testing them individually against a particular biological indicator at every stage of the fractionation process. In light of increasing interest in aloe constituents and their various biological effect, such a systematic effort seems highly warranted.

The roles of malic acid and mannan in analysis of pulp-based products

There are many different types of aloe pulp-based products. One major type is the so-called 'Aloe gel' or 'Aloe vera gel,' a whole pulp preparation. Its production process has been described by Agarwala (1997). The pulp, after being separated from the rind, is first homogenized. The resulting preparation is then subjected to brief heating (pasteurization) to reduce viscosity and achieve sterilization (Ashleye, 1983). In some cases, this is followed by cellulase treatment. The preparation is then decolorized with activated charcoal and filtered to yield the final product. It is primarily used for cosmetic and nutritional purposes, and is also often used as starting materials for isolating active compounds (McAnalley, 1988, 1990).

There are several parameters that have been used to evaluate and identify aloe pulp-based products, especially those containing the whole pulp such as 'Aloe gel.' These parameters include pH, calcium, magnesium, malic acid, free amino acid, free sugar, total solids, and size (molecular weight) and degree of acetylation of the mannan. Among all these criteria, malic acid and acetylated mannan appear to be the most important and specific, especially for product identification. This is because these two components are the most characteristic of the aloe pulp among all the compounds identified so far.

Malic acid

The presence of malic acid is the result of CAM that occurs in aloe plants. The malic acid is used as a test criterion for identification of aloe products and can be measured using HPLC or NMR. For products with whole pulp or liquid gel, a certain level of malic acid is expected. However, the malic acid is not a highly specific criterion because other succulent plants also have CAM or malic acid. It should be noted that except for being a supplementary indicator for aloes, malic acid is not known to have any significant biological activity.

Mannan

The mannan is the primary polysaccharide in the liquid gel from pulp. It is a partially acetylated $\beta1 \rightarrow 4$ linked polymannose which may also contain a significant amount of glucose. Its structure has been described in detail (see Chapter 4). Although the structural details vary widely among the mannans identified in various *Aloe* species or even in the same *Aloe* species by different investigators, two structural features are highly conserved, the acetylation and the $\beta1 \rightarrow 4$ linkage. It has been shown to be associated with several different biological activities such as immunostimulation, anti-complementary activity, wound healing, and anti-inflammatory activity (Tizard *et al.*, 1989; Marshall *et al.*, 1993; Reynolds and Dweck, 1999). It is also the basis for the viscosity or hydrogel property of the liquid gel prepared from the pulp.

The mannan is inherently unstable and can be rapidly degraded (Yaron, 1991, 1993). The degradation may be caused by an enzyme such as mannanase that may be present in the pulp, by elevated temperature and pH, or by bacterial contamination since bacteria are also a source of mannanase. In some aloe products, the pulp preparation is treated with cellulase. The cellulase preparations, depending on their source and purity, can be contaminated with mannanases that may also degrade the mannan.

Thus, the presence and integrity of this mannan is a good measure of the quality of the product. The mannan, when extracted from fresh gel, generally has a molecular weight greater than one million, as measured by HPLC-based size exclusion analysis. That is, the native mannan is a very large molecule. Ross *et al.* (1997) have found HPLC-based molecular weight analysis of the mannan to be a very useful indicator of quality. The mannan was eluted as a broad peak at 5–7 minutes. In samples that were not well preserved or properly processed, this peak is absent. The method also allows a quantitative estimation of the amount of the mannan present in a product.

Another important role of the mannan is for product identification and prevention of falsification. There are several widely available plant $\beta 1 \rightarrow 4$ linked mannans (galacto-mannan, Konjac mannan, and ivory mannan), but they are not acetylated. Thus, Diehl and Teichmuller (1998) used ^{1}H-NMR to identify the acetylated mannan in the aloe products. This approach allows detection of acetylation, the chemical signature of the acetylated mannan. However, NMR does not measure one important aspect of the molecule's integrity, i.e. its size. In addition, this technology requires expensive instruments. Thus, as an alternative, the size exclusion analysis described by Ross *et al.* (1997) may be combined with acetylation determination by GC-MS. That is, the mannan peak is collected during size exclusion analysis and then subjected to saponification and GC-MS.

CONCLUSION

Aloe pulp is indeed a very unique plant tissue. One aspect of its analysis is to establish the basic chemical composition of the pulp. Another is for detailed structural analysis of various compounds in relation to biological activity. Clearly, the first aspect is not yet mature. This is evidenced by the lack of a consistent approach for the initial pulp processing and adoption of standardized methods. Most descriptions in literature on this aspect are vague; the extent of grinding, filter pore size, and centrifugal force employed is seldom described or their use varied greatly among investigators. Such information is crucial for product development and identification and in particular for the proper employment of technique. It is hoped that in the long run, some consensus over the methods of analyzing the aloe pulp will be generated and some analytical methods can be standardized along with the establishment of a comprehensive and accurate chemical composition of the pulp. As for identifying and analyzing individual compounds, it would be very helpful to place the emphasis on the structure-function correlation in light of the increasing number of biological activities that have been attributed to the pulp.

REFERENCES

Akev, N. and Can, A. (1999) Separation and some properties of *Aloe vera* L. leaf pulp lectins. *Phytotherapy Research*, 13, 489–493.

Afzal, M., Ali, M., Hassan, R.A.H., Sweedan, N. and Dhami, M.S.I. (1991) Identification of some prostanoids in Aloe vera extract. *Planta Medica*, 57, 38–40.

Agarwala, O.P. (1997) Whole leaf aloe gel vs. standard aloe gel. *Drug and Cosmetics industry*, February 22–28.

Ashleye, A.D. (1983) Applying heat during processing the commercial Aloe vera gel. *Erde International*, 1, 40–44.

Blumenkrantz, N. and Asboe-Hansen, G. (1973) New method for quantitative determination of uronic acids. *Analytical Biochemistry*, 54, 484–489.

Bouchey, G.D. and Gjerstad, G. (1969) Chemical studies of *Aloe vera* juice. *Quarterly Journal of Crude Drug Research*, 9, 1445–1453.

Briggs, C. (1995) Herbal medicine: Aloe. *Canadian Pharmaceutical Journal*, 128, 48–50.

Canigueral, S. and Vila, R. (1993) Aloe. *British Journal of Phytotherapy*, 3, 67–75.

Chaplin, M.F. and Kennedy, J.F. (1994) *Carbohydrate analysis: A practical approach*. 2nd edn, Oxford: IRS Press.

Davis, R.H., DiDonato, J.J., Johnson, R.W. and Stewart, C.B. (1994) *Aloe vera*, hydrocortisone, and sterol influence on wound tensile strength and antiinflammation. *Journal of the American Podiatric Medical Association*, 84, 614–621.

Denius, H.R. and Homm, P. (1972) The relation between photosynthesis, respiration, and crassulacean acid metabolism in leaf slices of *Aloe arborescens* Mill. *Plant Physiology*, 49, 873–880.

Diehl, B. and Teichmuller, E.E. (1998) Aloe vera, quality inspection and identification. *Agrofood-industry Hi-Tech.*, 9, 14–16.

Evans, W.C. (1996) *Trease and Evans' Pharmacognosy*. 5th edn, pp. 245–246. London: WB Saunders.

Fahn, A. (1990) *Plant anatomy*. 4th edn, Oxford: Pergamon Press.

Femenia, A., Sanchez, E.S., Simal, S. and Rossello, C. (1999) Compositional features of polysaccharides from Aloe vera (Aloe barbadensis Miller) plant tissues. *Carbohydrate Polymers*, 39, 109–117.

Fujita, K., Suzuki, I., Ochia, J., Shinpo, K., Inoue, S. and Saito, H. (1978) Specific reaction of aloe extract with serum proteins of various animals. *Experientia*, 34, 523–524.

Gjerstad, G. (1971) Chemical studies of *Aloe vera* juice I: amino acid analysis. *Advancing Frontiers of Plant Sciences*, 28, 311–315.

Gorloff, D.R. (1983) Study of organoleptic properties of the exuded mucilage from the Aloe barbadensis leaves. *Erde International*, 1, 46–59.

Gowda, D.C., Neelisidaiah, B. and Anjaneyalu, Y.V. (1979) Structural studies of polysaccharides from *Aloe vera*. *Carbohydrate research*, 72, 201–205.

Gowda, D.C. (1980) Structural studies of polysaccharides from *Aloe saponaria* and *Aloe vanbalenii*. *Carbohydrate Research*, 83, 402–405.

Grindlay, D. and Reynolds, T. (1986) The *Aloe vera* phenomenon: a review of the properties and modern uses of the leaf parenchyma gel. *Journal of Ethnopharmacology*, 16, 117–151.

Haller, J.S. (1990) A drug for all seasons: medical and pharmacological history of aloe. *Bulletin of New York Academy of Medicine*, 66, 647–659.

Holzapfel, C.W., Wessels, P.L., van Wyk, B.-E., Marais, W. and Portwig, M. (1997) Chromones and aloin derivatives from *Aloe broomii, A. africana* and *A. speciosa*. *Phytochemistry*, 45, 97–102.

Ito, S., Teradaria, R., Beppu, H., Obata, M., Nagatsu, T. and Fujita, K. (1993) Properties and pharmacological activity of carboxypeptidase in *Aloe arborescens* Mill. var. *natalensis* Berger. *Phytotherapy Research*, 7, S26–S29.

Joshi, S.P. (1998) Chemical constituents and biological activity of Aloe barbadensis-a review. *Journal of Medicinal and Aromatic Plant Sciences*, 20, 768–773.

Kaufmann, T., Newman, A.R. and Wexler, M.R. (1989) Aloe vera and burn wound healing. *Plastic and Reconstructive Surgery*, 83, 1075–1076.

Kennedy, J.F. and White, C.A. (1983) *Bioactive carbohydrates: In chemistry, biochemistry and biology*. pp. 142–181. New York: Ellis Horwood Ltd.

Kinoshita, K., Koyama, K., Takashashi, K., Noguchi, Y. and Amano, M. (1996) Steroid glucosides from *Aloe barbadensis*. *Journal of Japanese Botany*, 71, 83–86.

Klein, A.D. and Penneys, N.S. (1988) Aloe Vera. *Journal of the American Academy of Dermatology*, 18, 714–720.

Kluge, M., Knapp, I., Kramer, D., Schwerdtner and Ritter, H. (1979) Crassulacean acid metabolism (CAM) in leaves of *Aloe arborescens* Mill: comparative studies of the carbon metabolism of chlorochym and central hydrenchym. *Planta*, 145, 357–363.

Kluge, M. and Ting, I.P. (1978) *Crassulacean acid metabolism*. Berlin: Springer-Verlag.

Kodym, A. (1991) The main chemical components contained in fresh leaves and in a dry extract from three years old *Aloe arborescens* Mill. grown in hothouses. *Pharmazie*, 46, 217–219.

Koike, T., Beppu, H., Kuzuya, H., Maruta, K., Shimpo, K., Suzuki, M., Titani, K. and Fujita, K. (1995) A 35 kDa mannose building lectin with hemagglutinating and mitogenic activities from "Kidachi Aloe" (*Aloe arborescens* Miller var. *natalensis* Berger). *Journal of Biochemistry*, 118, 1205–1210.

Koshioka, M., Koshioka, M., Takino, Y. and Suzuki, M. (1982) Studies on the evaluation of Aloe arborescens Mill. var. *natalensis* Berger and Aloe extract (JPIX). *International Journal of Crude Drug Research*, 20, 53–59.

Mandal, G. and Das, A. (1980a) Structure of the D-galactan isolated from *Aloe barbadensis* Miller. *Carbohydrate research*, 86, 247–257.

Mandal, G. and Das, A. (1980b) Structure of the glucomannan isolated from the leaves of *Aloe barbadensis* Miller. *Carbohydrate Research*, 87, 249–256.

Mandal, G., Ghosh, R. and Das, A. (1983) Characterization of polysaccharides of *Aloe barbadensis* Miller: Part III-structure of an acidic oligosaccharide. *Indian Journal of Chemistry*, 22B, 890–893.

Marshall, G.D., Gibbons, A.S. and Parnell, L.S. (1993) Human cytokines induced by acemannan. *Journal of Allergy and Clinical Immunology*, 91, 295.

McAnalley, B.H. (1988) Process for preparation of aloe products, produced thereby and composition thereof. *US patent* 4, 735, 935.

McAnalley, B.H. (1990) Processes for preparation of aloe products, produced thereby and composition thereof. *US patent* 4, 917, 890.

Norton, S.J., Talesa, V., Yuan, W.J. and Principato, G.B. (1990) Glyoxalase I and glyoxalase II from Aloe vera: purification, characterization and comparison with animal glyoxalases. *Biochemistry International*, 22, 411–418.

Okamura, N., Hine, N., Harada, S., Fujioka, T., Mihashi, K., Nishi, M., Miyahara, K. and Yagi, A. (1997) Diastereomeric *C*-glucosylanthrones of *Aloe vera* leaves. *Phytochemistry*, 45, 1519–1522.

Okamura, N., Hine, N., Tateyama, Y., Nakazawa, M., Fujioka, T., Mihashi, K. and Yagi, A. (1998) Five chromones from *Aloe vera* leaves. *Phytochemistry*, 49, 219–231.

Paez, A., Gebre, G.M., Gonzalez, M.E. and Tschaplinski, T.J. (2000) Growth, soluble carbohydrates, and aloin concentration of *Aloe vera* plants exposed to three irradiance levels. *Environmental & Experimental Botany*, 44, 133–139.

Paulsen, B.S., Fagerheim, E. and Overbye, E. (1978) Structural studies of the polysaccharide from *Aloe plicatilis* Miller. *Carbohydrate Research*, 60, 345–351.

Pierce, R.F. (1983) Comparison between the nutritional contents of the aloe gel from conventionally and hydroponically grown plants. *ERDE International*, 1, 37–38.

Radjabi, F., Amar, C. and Vilkas, E. (1983) Structural studies of the glucomannan from *Aloe vahombe*. *Carbohydrate Research*, 116, 166–170.

Radjabi-Nassab, F., Ramiliarison, C., Monneret, C. and Vilkas, E. (1984) Further studies of the glucomannan from *Aloe vahombe* (Liliaceae). II. Partial hydrolyses and NMR ^{13}C studies. *Biochimie*, 66, 563–567.

Reynolds, T. and Dweck, A.C. (1999) Aloe vera leaf gel: a review update. *J. Ethnopharmacol*, 68, 3–37.

Robson, M.C., Heggers, J.P. and Hagstrom, W.J. (1982) Myth, magic, witchcraft, or fact? *Aloe vera* revisited. *Journal of burn care and rehabilitation*, 3, 157–163.

Roboz, E. and Haagen-Smit, A.J. (1948) A mucilage from Aloe vera. *Journal of the American Chemical Society*, 70, 3248–3249.

Robyt, J.F. (1998) *Essentials of carbohydrate chemistry*. New York: Springer-Verlag.

Ross, S.A., ElSohly, M.A., Wilkins, S.P. (1997) Quantitative analysis of Aloe vera mucilagenous polysaccharides in commercial Aloe vera products. *Journal of AOAC International*, 80, 455–457.

Rowe, T.D. and Parks, L.M. (1941) Phytochemical study of *Aloe vera* leaf. *Journal of the America Pharmaceutical Association*, 30, 262–266.

Sabeh, F., Wright, T. and Norton, S.J. (1993) Purification and characterization of a glutathione peroxidase from the *Aloe vera* plant. *Enzyme protein*, 47, 89–92.

Sabeh, F., Wright, T. and Norton, S.J. (1996) Isozymes of superoxide dismutase from Aloe vera. *Enzyme Protein*, 49, 212–221.

Saito, H. (1993) Purification of Active Substances of *Aloe arborescens* Miller and their Biological and Pharmacological Activity. *Phytotherapy Research*, 7, S14–S19.

Shelton, R.M. (1991) Aloe vera: its chemical and therapeutic properties. *International Journal of Dermatology*, 30, 679–683.

Suzuki, I., Saito, H. and Inoue, S. (1979a) A study of cell agglutination and cap formation on various cells with Aloctin A. *Cell Structure and Function*, 3, 379.

Suzuki, I., Saito, H., Inoue, S., Migita, S. and Takahashi, T. (1979b) Purification and characterization of two lectins from *Aloe aborescens*. *Journal of Biochemistry*, 85, 163–171.

t'Hart, L.A., van den Berg, A.J.J., Kuis, L., van Dijk, H. and Labadie, R.P. (1989) An anti-complementary polysaccharide with immunological adjuvant activity (1990) from the leaf parenchyma gel of aloe vera. *Planta Medica*, 55, 509–512.

Tizard, I., Carpenter, R.H., McAnalley, B.H. and Kemp, M. (1989) The biological activity of mannans and related complex carbohydrates. *Molecular Biotherapy*, 1, 290–296.

Trachtenberg, S. (1984) Cytochemical and morphological evidence for the involvement of the plasma membrane and plastids in mucilage secretion in Aloe arborescens. *Annals of Botany*, 53, 227–236.

Vilkas, E. and Radjabi-Nassab, F. (1986) The glucomannan system from *Aloe vahombe* (Liliaceae). III. Comparative studies on the glucomannan components isolated from the leaves. *Biochimie*, 68, 1123–1127.

Waller, G.R., Mangiafico, S. and Ritchey, C.R. (1978) A chemical investigation of *Aloe barbadensis* Miller. *Proceedings of Oklohoma Academy of Sciences*, 58, 69–76.

Wang, Y.T. and Strong, K.J. (1993) Monitoring physical and chemical properties of freshly harvested field grown *Aloe vera* leaves. A preliminary report. *Phytotherapy Research*, 7, S1–S4.

Winter, K. and Smith, J.A.C. (1996) *Crassulacean acid metabolism*. pp. 1–13. Berlin: Springer-Verlag.

Winters, W.D. (1993) Immunoreactive lectins in leaf gel from *Aloe barbadensis* Miller. *Phytotherapy Research*, 7, S23–S25.

Winters, W.D. and Yang, P.B. (1996) Polypeptides of the three major medicinal aloes. *Phytotherapy Research*, 10, 5736–576.

Wozniewski, T., Blaschek, W. and Franz, G. (1990) Isolation and structure analysis of a glucomannan from the leaves of *Aloe arborescens* Miller. *Carbohydrate Research*, 198, 387–391.

Yagi, A., Makino, K., Nishioka, I. and Kuchino, Y. (1977) Aloe mannan, polysaccharide, from *Aloe arborescens* var. *natalensis*. *Plant Medica*, 31, 17–20.

Yagi, A., Hamada, K., Mihashi, K., Harada, N. and Nishioka, I. (1984) Structure determination of polysaccharides in *Aloe saponaria* (Hill) Haw. (Liliaceae). *Journal of Pharmaceutical Science*, 73, 62–65.

Yagi, A., Harada, N., Shimomura, K. and Nishioka, I. (1987) Bradykinnin-degrading glycoprotein in *Aloe arborescens* var. *natalensis*. *Planta Medica*, 53, 19–21.

Yagi, A., Nishimura, H., Shida, T. and Nishioka, I. (1986) Structure determination of polysaccharides in *Aloe arborescens* var. *natalensis*. *Planta Medica*, 3, 213–218.

Yagi, A., Harada, N., Shimomura, K. and Nishioka, I. (1997) Bradykinin-degrading glycoprotein in *Aloe arborescens* var. *natalensis*. *Planta Medica*, 53, 19–21.

Yagi, A., Egusa, T., Arase, M., Tanabe, M. and Tsuji, H. (1997) Isolation and characterization of the glycoprotein fraction with a proliferation-promoting activity on human and hamster cells in vitro from *Aloe vera* gel. *Planta Medica*, 63, 18–21.

Yamaguchi, I., Mega, N. and Sanada, H. (1993) Components of the gel of *Aloe vera* (L.) Burm. f. *Bioscience, Biotechnology, Biochemistry*, **57**, 1350–1352.

Yamamoto, M., Masui, T., Sugiyama, K., Yokota, M., Nakagomi, K. and Nakazawa, H. (1991) Anti-inflammatory active constituents of *Aloe arborescens* Miller. *Agricultural and Biological Chemistry*, **55**, 1627–1629.

Yaron, A. (1991) *Aloe vera*: chemical and physical properties and stabilization. *Israel Journal of Botany*, **40**, 270.

Yaron, A. (1993) Characterization of *Aloe vera* gel before and after autodegredation, and stabilization of the natural fresh gel. *Phytotherapy Research*, **7**, S11–S13.

Yoshimoto, R., Kondoh, N., Isawa, M. and Hamuro, J. (1987) Plant lectin, ATF1011, on the tumor cell surface augments tumor specific immunity through activation of T cells specific for lectin. *Cancer Immunology Immunotherapy*, **25**, 25–30.

7 Analytical methodology: the exudate

Tom Reynolds

ABSTRACT

Two properties of drug aloes, the bitter leaf exudate from a small number of *Aloe* species, are of interest to pharmacologists. Firstly, they wish to authenticate the plant origin of the material and secondly, they wish to measure the degree of purgation to be expected. The chief purgative compound in the leaf exudate, bitter aloes, is the anthrone *C*-glucoside, barbaloin, although *O*-glycosides and dianthrones, which are not present in all samples, are also active. There are various chemical methods used to detect and quantify barbaloin but these are not sufficiently accurate and reproducible because of the presence of interfering substances. To improve precision chromatographic separation techniques were developed to isolate the active compounds previous to chemical detection and determination. Thin-layer chromatography was convenient and economical but high performance liquid chromatography proved the most precise for a definitive analysis.

Interest in *Aloe* exudate compounds spread from the few drug species to the c. 400 species now described from Africa and Arabia. Many compounds have now been identified and others recognized chromatographically only and it is speculated that in view of the many folk remedies of aloes reported, some of them may have biological activities relevant to human medicine. In addition, the distribution of these compounds among the species determined by the analytical techniques described here provides clues as to aloe chemotaxonomy.

INTRODUCTION

At present the chief therapeutic interest of aloes is focussed on the leaf parenchyma gel and various tests for quality control have been developed (Chapter 8). Although the active ingredients of the gel are not completely defined, assays of the polysaccharide components are already being suggested (Chapters 6, 8). The phenolics in the leaf exudate have received less attention in recent years, although at one time bitter aloes was an acknowledged medicine and is still found in current pharmacopoeias. Other leaf phenolics currently being investigated may well be shown to have biological activities of great value, very different from the original purgation. Bitter aloes have a long history and the purgative agent, which is also the bitter principle, is well known and the analytical techniques are well established.

ANALYSIS OF BITTER ALOES

The dried exudate of a number of *Aloe* species has been described in ancient documents and has been an article of commerce in Europe for centuries (Haller, 1990). *Aloe vera* (L.) Burm.f. seems to have been known to the ancient Egyptians and described in detail by Dioscorides (c. 78 A.D.) (Reynolds, 1966). *Aloe perryi* Baker was valued by Alexander the Great (fourth century B.C.) (Hodge, 1953; Crosswhite and Crosswhite, 1984). *Aloe ferox* Miller was known somewhat later (Reynolds, 1950) following exploration of the Cape (early eighteenth century A.D.). *Aloe arborescens* Miller, prized as an ornamental, is better known medically in Asia, especially Japan, where it is known as Kidachi-aroe (Yagi, this volume, Chapter 14). In addition, many local species unknown to western medicine are used locally in Africa as part of indigenous folk medicine (Morton, 1961; Watt and Breyer-Brandwigk, 1962; Bruce, 1975).

From all this it was necessary to establish some sort of quality control long before the active ingredient was known and pharmacopoeias have included simple visual and colorimetric tests to distinguish the various products offered, which themselves have varied very much in quality (e.g. British Pharmacopoeia; Bisset, 1994; Trease and Evans, 1983; Evans, 2001). More elaborate procedures were used to provide a quantitative estimate of quality (e.g. Morsy, 1983). An attempt to characterize other *Aloe* species by the exudate colour was relevant to folk medicine (Raymer and Raymer, 1995). An analysis based on an accurate quantitation of the active ingredient would obviously be more satisfactory, if time consuming. This substance was discovered in 1851 (Smith and Smith, 1851) and discussions about its properties were still current in 1887 (Brown, 1887). It was at first named aloin and then the compound from *A. vera* (=*A. barbadensis*) was termed barbaloin to distinguish it from similar substances, homonataloin and nataloin, isolated from other species. Barbaloin is still often called aloin, especially by European workers, although 'aloin' sometimes refers to the crude preparation from which barbaloin is crystallized (Ovanovski, 1983). The structure of barbaloin proved difficult to establish. Early work by Leger (1916) was followed by Hauser (1931) who suggested structures which later proved unsatisfactory. Preparation of several significant derivatives by Mühlemann (1952) and Birch and Donovan (1955) led to a satisfactory structural determination (Barnes and Holfeld, 1956; Hay and Haynes, 1956) as the *C*-glucoside of aloe-emodin anthrone.

The way was now open for setting up methods for determining the exact barbaloin content of the various commercial preparations, as well as the crude aloin extracted from them. Three main methods have been reported: the direct chemical, the biological and the chromatographic (Lister and Pride, 1959; Kraus, 1959). An early report suggested four chemical methods: chlorination, determination of pentose after hydrolysis, persulphate oxidation and ferric chloride oxidation (Harders, 1949). A gravimetric method had also been described which was a measure of dry matter in a warm water solution of the drug (Forsdike, 1951). In another procedure, the barbaloin from a warm dilute hydrochloric acid solution was precipitated by calcium hydroxide and recovered by acidification (Lister and Pride, 1959). This procedure was deemed not very satisfactory.

Ferric chloride oxidation in acid solution of the *C*-glycosides, followed by colorimetric determination in alkaline solution of the liberated anthraquinone, has been a method of choice by pharmacologists (e.g. British Pharmacopoeia, 2001; European Pharmacopoeia, 2002; Bhavsar *et al.*, 1991) since it was described by Harders and steadily improved by Auterhoff and Ball (1954), Hörhammer *et al.* (1959) and finally Fairbairn and Simic (1963, 1964). Another analytical method involved breaking down the barbaloin molecule with sodium meta-periodate (0.1%) in ammoniacal (2.5%) solution, followed as before by colorimetric determination of the aloe emodin at 505 nm (Hörhammer *et al.*, 1963b, 1965). This was improved by using a water blank instead of aloe emodin which often proved to contain impurities (Böhme and Kreutzig, 1965). A method said to be specific for barbaloin was a colorimetric reaction (335 nm) with sodium tetraborate, although a preliminary separation on nylon powder was used (Janiak and Böhmert, 1962).

As the properties of barbaloin became known it seemed that a quantitative estimation might be possible by measuring UV absorption at one or more characteristic wavelengths. The ratio of absorbance at 354 nm and 298 nm was used to distinguish barbaloin (Lister and Pride, 1959) but it was considered by these workers and others (Thieme and Diez, 1973) that there were too many interfering substances in a crude drug sample for accurate determination. They advocated a preliminary separation by some sort of chromatographic means. However, the UV absorption method has persisted using a different absorbing wavelength (500 nm) (Fairbairn and Simic, 1963, 1964), giving reasonably consistent results when combined with a chromatographic separation (McCarthy and Mapp, 1970). Again, absorbance at three wavelengths, 269 nm, 295 and 354 nm, was observed, the latter being considered the most reliable, although the measurements were made on purified barbaloin rather than the drug (Ellaithy *et al.*, 1984).

Biological activity assays

Meanwhile there was concern that the purgative activity of an aloes sample might not reside solely in the barbaloin fraction, making a chemical analysis insufficient for pharmacological purposes. Early trials suggested that for senna and cascara the purgative activity was indeed measured by the anthrone *C*-glycoside or anthraquinone *O*-glycoside content and that the anthrone and anthraquinone aglycones were very much less active (Fairbairn, 1949), although there was some activity in the dianthrones (Auterhoff and Scherff, 1960; Fairbairn and Moss, 1970). This activity refers to ingestion by mouth, as the glycosides are thought to be more soluble and also protected by their sugar moieties during their passage through the gut to the large intestine, where the aglycones are liberated and produce purgation (Fairbairn and Moss, 1970). In aloes, however, the aloinosides, *O*-rhamnosides of barbaloin, also have considerable activity (Hörhammer *et al.*, 1963b), so that these need to be taken into account in an analysis intended to reflect purgation (McCarthy and Mapp, 1970; Mapp and McCarthy, 1970). The bioassay commonly used involves measuring the consistency of rat faecal pellets (Lister and Pride, 1959; Mapp and McCarthy, 1970). It is interesting that mice can hydrolyse *O*-glycosides but lack the enzyme to hydrolyse *C*-glycosides, and that rats, as well as man, can hydrolyse both and thus rats prove a suitable test animal (Fairbairn, 1965).

Polarography

An early study indicated that the polarographic signal from barbaloin was sufficient to estimate the compound in drug samples (Stone and Furman, 1947), although the method did not appear to have been followed up (Kraus, 1959).

CHROMATOGRAPHY

As a result of the problems arising from these previously described assays, there came a need for a more precise estimation of the individual chemicals in the exudate, both at the qualitative level to identify and confirm raw material and at the quantitative level to evaluate particular compounds with the view to determining quality and dosage.

An early report of separation of aloe compounds, followed by spectrophotometric determination, used a column of magnesia (magnesium oxide) and celite (diatomaceous earth) (1:3) with quantification by absorbance at 440 nm, although only the anthraquinone aglycones and acid hydrolysates of glycosides were examined (Brody *et al.*, 1950). Later, separation on a perlon (nylon) powder column eluted with aqueous methanol was used (Hörhammer *et al.*, 1959). A summary of these earlier chromatographic and spectrophotometric studies is given by Böhme and Bertram (1955).

Counter current chromatography

The Craig counter-current separation method has been used to separate glycosides from cascara bark, relevent to aloes (Fairbairn and Mital, 1960), although the free anthrone of aloe emodin has already been separated between benzene and aqueous acetic acid (70%) (Böhme and Bertram, 1955). Three solvent systems were found suitable: *iso*-propanol-ether-water = 1:2:2, butanone-water = 4:3, *n*-butanol-ethanol-water = 5:1:4. Elsewhere a method for rhubarb used *n*-propanol-ethyl acetate-aqueous sodium chloride solution (0.05%) = 2:2:4 (Zwaving, 1965). Later, another counter-current procedure using chloroform-methanol-water = 7:13:8 separated the diastereomers of barbaloin (Rauwald, 1982) and this solvent and similar ones were used by this author and co-workers for several years (reviewed, Rauwald, 1990).

Size exclusion chromatography

The separation of dianthrone and anthraquinone glycosides from plant extracts other than aloes on Sephadex LH-20 eluted with 70% methanol was described by Zwaving (1968) and the method has subsequently been used often as a preliminary step in isolation of components from aloe exudates.

Gas chromatography

Barbaloin has been estimated as its trimethylsilyl derivative by gas chromatography-mass spectrometry, but details of quantification are not clear (Paez *et al.*, 2000). An extensive study had been made of the mass spectra of a range of anthraquinone and anthrone derivatives under electron impact, chemical ionization and field desorption conditions (Evans *et al.*, 1979). Steam distillation has been used to extract other volatile substances from *A. arborescens* material. The temperature of distillation had a significant effect on the product. Thus at a higher temperature (100 °C) a major product (32%) was 3-(hydroxymethyl)-furan (Kameoka *et al.*, 1981), whereas at 32 °C this was a minor component (0.2%), (Z)-3-hexenol being prominent (30%) (Umano *et al.*, 1999) Another technique used by the same authors, simultaneous purging and extraction (SPE) using water and dichloromethane, yielded (E)-2-hexenal (46%) as the major product. A somewhat different approach was headspace analysis on the volatiles from various commercial samples (Saccù *et al.*, 2001). Here, *A. vera* gave off a number of simple alcohols, aldehydes and ketones, while terpenes and anisole were more characteristic of 'Kenya aloes.' Analysis of these volatiles is of more interest to the food industry than to the pharmaceutical industry.

Paper chromatography

Paper chromatography using petroleum spirit saturated with 97% methanol separated the aloe anthraquinones, which were verified colorimetrically with magnesium acetate (0.5%) in methanol and quantified by spectrophotometry at an unstated absorption maximum (Mary *et al.*, 1956). The much-used solvent, *n*-butanol-acetic acid-water=4:1:5 (organic phase), successfully separated aloe glycosides (Hagedorn, 1952), while another method used 'butanol acétique' (sic), followed by visual observation under UV light (Paris and Viejo, 1955; Paris and Durand, 1956). This solvent was later defined as *n*-butanol-acetic acid-water=1:1:5 (organic phase) and used to identify different types of drug aloes, with the extra step of observing the colours under UV light before and after fuming with ammonia (Jaminet, 1957).

Returning to four parts of *n*-butanol, this solvent was used to separate and isolate compounds from drug aloes using as a revealing agent, potassium hydroxide (0.5 M) in ethanol, the zones being examined under natural and UV light (Awe *et al.*, 1958; Awe and Kummell, 1962). Using this system it was possible to distinguish Cape aloes (*A. ferox*) from Curaçao or Barbados aloes (*A. vera*) by the presence in the latter of a zone fluorescing light blue under UV (unspecified wavelength) in alkaline conditions, at a characteristic R_f value (0.5) (Jaminet, 1957; Awe *et al.*, 1958). The compound in this zone was named *iso*-barbaloin and characterized much later as 7-hydroxybarbaloin (Rauwald and Voetig, 1982). The same solvent was used to separate barbaloin in cascara (Baumgartner and Leupin, 1961). Then satisfactory separations were obtained by omitting the acetic acid (Lister and Pride, 1959). Meanwhile, a solvent containing water-acetone-benzene=2:1:4 was used to separate anthraquinone and anthrone glycosides and toluene alone to separate anthraquinones (Betts *et al.*, 1958). Elsewhere, *n*-butanol-pyridine-ethyl acetate-water=1:1:1:1 was used (Worthen *et al.*, 1959).

These early methods were reviewed by Kraus (1959) who also determined barbaloin levels by measuring the size of the chromatographic zone. Rhubarb glycosides were successfully separated using *n*-propanol-ethyl acetate-water=4:3:3 (Zwaving, 1965). The method was taken further towards a chemotaxonomic analysis by Durand and Paris (1960) who described a range of chromatographic solvents and revealing reagents, and also introduced a 2-dimensional technique.

Thin-layer chromatography

It was felt that for routine quantitative determinations an increase in precision and a reduction in manipulation time for the preservation of the active compounds was desirable, so the new process of planar chromatography on thin layers of adsorbant was applied to aloe analysis (Gerritsma and Rheede van Oudtshoorn, 1962), following the method introduced for qualitative separations (Stahl and Schorn, 1961). For this purpose, the developing solvent introduced by Teichert *et al.* (1961), chloroform-95% ethanol=3:1, was used with silica gel G plates. The particular zones were eluted with methanol and their contents determined by absorption of UV light at 355 nm. They also included some chemotaxonomic observations, which were later extended to recognize barbaloin in eight *Aloe* species out of 60 screened (McCarthy and Price, 1965). Chemotaxonomy of aloes had been discussed previously by Hegnauer (1963). McCarthy (1968) used this solvent system followed by elution of the zones and estimation by UV spectrophotometry at either 294 nm or 360 nm. The method proved accurate compared to two other established methods (McCarthy and Mapp, 1970) and has been used since (Jansz *et al.*, 1981). An improved solvent system, ethyl acetate-methanol-water = 100:16.5:13.5, was introduced for qualitative recognition of the various drug aloes (Hörhammer *et al.*, 1963a, b). The zones were revealed by UV light or by a colour reaction with Fast Blue B salt, a tetrazotized *o*-dianisidine, zinc chloride complex (0.5% aqueous solution). These methods were used to draw up a key for recognizing varieties of drug aloes (Hörhammer *et al.*, 1965). They also eluted the zones and measured the barbaloin and aloinoside by UV absorption at 360 nm. Similar mixtures were used for separation on paper and thin-layer chromatograms, both qualitatively (Böhme and Kreutzig, 1963) and quantitatively (Böhme and Kreutzig, 1964, 1965). Analysis in the Swiss Pharmacopoeia reverted to the chloroform-ethanol solvent, which used buffered (pH 6) silica gel layers to quantify a range of drug aloes by absorption at 360 nm (Mühlemann and Tatrai, 1967). Further critical studies introduced a 2-dimensional technique using chloroform-95% ethanol-water = 30:15:1 in the first direction and chloroform-95% ethanol = 3:1 in the second (Thieme and Diez, 1973). These authors concluded that spectrophotometric estimation of barbaloin following a chromatographic separation gave results 30% higher than the periodate method without prior separation. The method was used later to measure barbaloin levels in fresh leaf exudates (Chauser-Volfson and Gutterman, 1996).

A chemotaxonomic study of leaf exudates from c. 240 *Aloe* species on silica gel plates using chloroform-ethanol-water = 7:3:1 (lower layer) (Böhme and Kreutzig, 1964) and Fast Blue B salt (Hörhammer *et al.*, 1963a) resulted in the observation of at least 90 coloured zones (Reynolds, 1985). Chemical structures could not be assigned to most of these, so to facilitate discussion a list of codes was made to identify the zones. Thus, O1–24 were orange-staining zones (often chromone derivatives), P1–35 were purple-staining zones and G1–29 usually green or blue. The zones occurring in certain named *Aloe* species were taken as standards so that similar zones from other species could be compared chromatographically by R_f value and colour reaction in UV and visible light and later by HPLC behaviour. This was followed by a 2-dimensional method with the solvent *n*-propanol-di*iso*-propyl ether-water = 7:5:1 for the first direction and the chloroform-ethanol solvent for the second (Reynolds, 1986). Further chemotaxonomic studies using a variety of similar solvents were carried out by Rauwald and co-workers (Rauwald, 1990; Rauwald and Beil, 1993a), Dagne and co-workers (1994) and Van Wyk and co-workers (1995).

Paper electrophoresis

Electrophoretic methods for separating anthraquinones and their derivatives were reported by Siesto and Bartoli (1957) for rhubarb and by Core and Kirch (1968). The review of these by Kraus (1959) suggests the use of a barbitol buffer (pH 8.6) on paper at 400 volts to separate aloe compounds. Later a separation of anthraquinone glycosides and aglycones was described using a sodium borate buffer (pH 9.6) (Chang and

Ferguson, 1974). Elsewhere a number of buffers were used at both high and low voltages to separate similar compounds in rhubarb and senna (Zwaving, 1974).

High performance (pressure) liquid chromatography (HPLC)

Dissatisfaction with the accuracy of colorimetric estimation of barbaloin in drug aloes led to the development of an HPLC method using a reversed-phase process with a silica-based stationary phase (Lichrosorb RP-8, 7 μm), eluted with 45% aqueous methanol, followed by UV detection at 254 nm (Graf and Alexa, 1980). This method was extended by the use of a different solvent, 25% aqueous acetonitrile, which with detection at 295 nm was used to separate barbaloin into its diastereomers (Auterhoff *et al.*, 1980). Previously the method had been used to separate components of fresh exudates of *A. ferox* and *A. arborescens* using a similar column but eluted with an aqueous methanol gradient from 30% to 45% and UV detection at 360 nm (Grün and Franz, 1979). Estimation of barbaloin in Japanese foodstuffs containing *A. arborescens* was achieved by reversed-phase chromatography (YMC A-32) with detection at 293 nm (Yamamoto *et al.*, 1985). Later, quality control of aloe samples for bittering uses was carried out on a Nucleosil 100 C18, 4 μm column. The mobile phase was 16% aqueous acetonitrile for 16 minutes, followed by a gradient to 33% over 25 minutes and then to 60% over 13 minutes, finishing at 100% (Saccù *et al.*, 2001). It has been suggested that because of the uncertainty surrounding the stability of barbaloin during sampling and analysis, that aloesin or aloeresin A would be more reliable markers for aloes in beverages; separation of these substances by HPLC, followed by detection at 220 nm and 360 nm, was demonstrated (Zonta *et al.*, 1995). Alternatively, aloesin and aloeresin E were suggested and determined by HPLC, followed by MS-MS (Dell'Agli *et al.*, 2001).

A quantitative taxonomic survey of the levels of homonataloin in 23 *Aloe* species was carried out on the fresh leaf exudates by a solvent gradient separation, again using a reversed-phase column (Hypersil C22, 5 μm), eluted with aqueous methanol and detection at 295 nm (Beaumont *et al.*, 1984). This was repeated to survey barbaloin levels from 68 *Aloe* species on another reversed-phase column (Spherisorb ODS 5 μm), eluted with aqueous methanol (45% to 60%) and monitored at 375 nm (Groom and Reynolds, 1987). A more elaborate gradient aqueous methanol programme – 25% to 50% over 15 minutes, remaining at 50% for 15 minutes and then rising to 100% over 10 minutes (Lichrospher 100 RP-18, 5 μm), with detection and integration of the peaks at 300 nm, an arbitrary value close to the wavelength maxima for most compounds – was used to determine the composition of 14 historic drug aloes samples (Reynolds, 1994). A qualitative study of a number of drug aloes samples using isocratic separation on a similar column (Nucleosil 7 C18), eluted with 50% methanol and monitored at 360 nm, had already been presented (Rauwald and Beil, 1993b). The method was extended to separate, detect and quantify 18 compounds from two *Aloe* species on a similar column (Nucleosil 5 C18), eluted with a linear gradient of acetonitrile-water-phosphoric acid from 10:89.5:0.5 to 27.5:72:0.5 over 40 minutes and then to 80:19.5:0.5 over 15 minutes, after which the flow was isocratic. Quantification was by peak area at the maximum wavelength for light absorption for each compound compared with standards (Sigler and Rauwald, 1994; Rauwald and Sigler, 1994). Similar systems were used for the separation of ten constituents of four *Aloe* species and two drug samples (Okamura *et al.*, 1996). Another large survey of compounds in 172 *Aloe* species was made using a non-linear gradient of aqueous actetonitrile (10% to 100%)

and detection at two wavelengths, 275 nm and 365 nm (Van Wyk *et al.*, 1995) and this and a system using methanol gradients (30% to 60%) were used in further surveys (e.g. Viljoen *et al.*, 1996; Viljoen *et al.*, 1999, 2001; Viljoen and Van Wyk, 2001).

CONCLUSION

Although no longer considered a purgative drug of choice, bitter aloes still features in the European Pharmacopoeia (2002) and tests are still required to authenticate and quantify commercial samples of the drug, which also finds a place as a bittering agent. Early estimations used various chemical methods, often colorimetric or even spectro-photometric. Even after preliminary purification these were found unsatisfactory because of interfering components. One improvement was to estimate purgative activity directly by a bioassay but this did not distinguish the constituents chemically. Application of some type of chromatographic separation before a chemical or photometric assay led to a great increase in precision and has been adopted for serious quantitative analysis. At the same time, qualitative analysis by chromatography also gave an accurate picture of chemical composition and led to chemotaxonomic studies of the whole genus.

REFERENCES

Auterhoff, H. and Ball, B. (1954) Inhaltsstoffe und Wertbestimmung von *Aloe*-drogen und Zubereitungen. *Arzneimittel-Forschung*, 4, 725–729.

Auterhoff, H. and Scherff, F.C. (1960) Die Dianthrone der pharmazeutisch interessierenden Hydroxyanthrachinone. *Archiv der Pharmazie*, 293, 918–925.

Auterhoff, H., Graf, E., Eurisch, G. and Alexa, M. (1980) Trennung des Aloins in Diastereomere und deren Charakterisierung. *Archiv der Pharmazie*, 313, 113–120.

Awe, W., Auterhoff, H. and Wachsmuth-Melm, C.L. (1958) Beitrage zur papierchromato-graphischen untersuchung von *Aloe*-drogen, *Arzneimittel-Forschung*, 8, 243–245.

Awe, W. and Kümmell, H.-J. (1962) Zum Vorkommen von Aloin in Aloe vera nebst verleichenden Untersuchungen mit einem Frischsaft der Kapaloe (Aloe ferox) und einem daraus hergestellten Trockenextract. *Archiv der Pharmazie*, 295, 819–822.

Barnes, R.A. and Holfeld, W. (1956) The structure of barbaloin and *iso*barbaloin. *Chemistry and Industry*, 873–874.

Baumgartner, R. and Leupin, K. (1961) Über die Inhaltsstoffe der Rinde von Rhamnus purshi-anus (Cascara sagrada Ph. Helv.V). *Pharmaceutica Acta Helvetiae*, 36, 244–267.

Beaumont, J., Reynolds, T. and Vaughan, J.G. (1984) Homonataloin in Alöe species. *Planta Medica*, 50, 505–508.

Betts, T.J., Fairbairn, J.W. and Mital, V.K. (1958) Vegetable purgatives containing anthracene derivatives. *Journal of Pharmacy and Pharmacology*, 10, 436–441.

Bhavsar, G.C., Chauhan, M.G. and Upadhyay, U.M. (1991) Evaluation of Indian aloes. *Indian Journal of Natural Products*, 6, 11–13.

Birch, A.J. and Donovan, F.W. (1955) Barbaloin 1. Some observations on its structure. *Australian Journal of Chemistry*, 8, 523–528.

Bisset, N.G. and Wichtl, M. (2000) *Herbal Drugs and Phytopharmaceuticals* pp. 59–62. Boca Raton etc: CRC Press.

Böhme, H. and Kreutzig, L. (1965) Zur photometrischen Aloinbestimmung nach H.Möhrle. *Archiv der Pharmazie*, 298, 262–271.

Böhme, H. and Bertram, J. (1955) Zur Kenntnis des Aloins aus Kap-Aloe. *Archiv der Pharmazie*, **288**, 510–516.

Böhme, H. and Kreutzig, L. (1964) Über die quantitative Ermittlung von Aloin mittels Papier-und Dünnschichtchromatographie. *Archiv der Pharmazie*, **297**, 681–689.

British Pharmacopoeia (2001) Barbados Aloes. *British Pharmaceutical Society*, I – 84. London.

British Pharmacopoeia (2001) Cape Aloes. *British Pharmaceutical Society*, I – 86. London.

Brody, T.M., Voigt, R.F. and Maher, F.T. (1950) A chromatographic study of the anthraquinone derivatives of Curacao aloe. *Journal of the American Pharmaceutical Association*, **39**, 666–669.

Brown, J.F. (1887) Bitter aloes: a confession of bewilderment. *Pharmaceutical Journal and Transactions*, 678–680.

Bruce, W.G.G. (1975) Medicinal properties in the Aloe. *Excelsa*, 57–68.

Chauser-Volfson, E. and Gutterman, Y. (1996) The barbaloin content and distribution in *Aloe arborescens* leaves according to the leaf part, age, position and season. *Isreal Journal of Plant Sciences*, 44, 289–296.

Core, A.C. and Kirch, E.R. (1958) Separation of some hydroxyanthraquinones by filter paper electrophoresis. *Journal of the American Pharmaceutical Association-Scientific Edition*, 47, 513–515.

Crosswhite, F.S. and Crosswhite, C.D. (1984) Aloe vera, plant symbolism and the threshing floor: light, life and good in our heritage. *Desert Plants*, 6, 46–50.

Dagne, E., Yenesew, A., Asmellash, S., Demissew, S. and Mavi, S. (1994) Anthraquinones, pre-anthraquinones and isoeleutherol in the roots of *Aloe* species. *Phytochemistry*, 35, 401–406.

Dell'Agli, M., Giavarini, F., Galli, G. and Bosisio, E. (2001) Search for alternative markers to aloin and aloe-emodin for aloe identification in beverages. *World Conference on Medicinal and Aromatic Plants, Hungary*, **Section II**, 46.

Durand, M. and Paris, R. (1960) Les Aloès II. – Etude chromatographique des dérivés anthracéniques divers aloès frais. *Annales Pharmaceutiques Francaises*, 18, 846–852.

Elliathy, M.M., Sayed, L. and Bebawt, L. (1984) Spectrophotometric determination of barbaloin. *Spectroscopy Letters*, 17, 743–750.

European Pharmacopoeia Commission (2002) Aloes, barbados. In *European Pharmacopoeia*, 4th edn, p. 607. Strasbourg: Council of Europe.

European Pharmacopoeia Commission (2002) Aloes Cape. In *European Pharmacopoeia*, 4th edn, p. 608. Strasbourg: Council of Europe.

European Pharmacopoeia Commission (2002) Aloes dry extract, standardised. In *European Pharmacopoeia*, 4th edn, p. 609. Strasbourg: Council of Europe.

Evans, F.J., Lee, M.G. and Games, D.E. (1979) Electron impact, chemical ionization and field desorption mass spectra of some anthraquinone and anthrone derivatives of plant origin. *Biomedical Mass Spectrometry*, 6, 374 – 380.

Evans, W.C. (2001) *Trease and Evans Pharmacognosy*, 15th edn. Philadelphia, etc: W.B.Saunders.

Fairbairn, J.W. (1949) The active constituents of the vegetable purgatives containing anthracene derivatives. Part I. Glycosides and aglycones. *Journal of Pharmacy and Pharmacology*, 1, 683–692.

Fairbairn, J.W. and Mital, V.K. (1958) Vegetable purgatives containing anthracene derivatives Part IX. An aloin-like substance in *Rhamnus purshiana* DC. *Journal of Pharmacy and Pharmacology*, 10, 217T–221T.

Fairbairn, J.W. and Moss, J.R. (1970) The relative purgative activities of 1, 8-dihydroxyanthracene derivatives, *Journal of Pharmacy and Pharmacology*, 22, 584–593.

Fairbairn, J.W. and Simic, S. (1960) Vegetable purgatives containing anthracene derivatives Part XI. Further work on the aloin-like substance of *Rhamnus purshiana* DC. *Journal of Pharmacy and Pharmacology*, 12, 45T–51T.

Fairbairn, J.W. and Simic, S. (1964) Estimation of C-glycosides and O-glycosides in cascara (*Rhamnus purshiana* DC., bark) and cascara extract. *Journal of Pharmacy and Pharmacology*, 16, 450–454.

Forsdike, J.L. (1951) The determination of water-soluble extractive in aloes and in dry extracts of cascara sagrada and krameria, *Journal of Pharmacy and Pharmacology*, 3, 351–359.

Gerritsma, K.W. and Van Rheede van Oudtshoorn, M.C.B. (1962) Microanalytische bepaling van aloine. *Pharmaceutisch Weekblad*, 97, 765–775.

Graf, E. and Alexa, M. (1980) Die Bestimmung der Aloine und Aloeresine durch HPLC. *Archiv der Pharmazie*, 313, 285–286.

Groom, O.J. and Reynolds, T. (1987) Barbaloin in *Aloe* species. *Planta Medica*, 53, 345–348.

Grün, M. and Franz, G. (1979) Isolierung zweier stereoisomerer Aloine aus Aloe. *Die Pharmazie*, 34, 669–670.

Hagedorn, P. (1952) Die Papierchromatographie von Urtinkturen. *Deutsche Apotheker-Zeitung*, 92, 985–990.

Haller, J.S. (1990) A drug for all seasons medical and pharmacological history of aloe. *Bulletin of the New York Academy of Medicine*, 66, 647–659.

Harders, C.L. (1949) Curaçao-aloe en aloine. *Pharmaceutisch Weekblad*, 84, 250–258.

Hauser, F. (1931) Ein Beitrag zur Kenntis der Anthraglukoside, speziell des Aloins und des Peristaltins. *Pharmaceutica Acta Helvetiae*, 6, 79–85.

Hay, J.E. and Haynes, L.J. (1956) The aloins. Part I. The structure of barbaloin. *Journal of the Chemical Society*, 3141–3147.

Hegnauer, (1963) *Chemotaxonomie der Pflanzen III Asphodeloideae*, pp. 305–312. Basel and Stuttgart: Birkhäuser Verlag.

Hodge, W.H. (1953) The drug aloes of commerce, with special reference to the cape species. *Economic Botany*, 7, 99–129.

Hörhammer, L., Wagner, H. and Főcking, O. (1959) Eine verbesserte Aloin-Bestimmung. *Pharmazeutische Zeitung*, 104, 1183–1186.

Hörhammer, L., Wagner, H. and Bittner, G. (1963a) Dünnschichtchromatographie von Anthrachinondrogen und ihren Zubereitungen. *Pharmazeutische Zeitung*, 108, 259–262.

Hörhammer, L., Wagner, H. and Bittner, G. (1963b) Vergleichende Untersuchungen an südafrikanischen *Aloe*-Sorten. *Arzneimittel-Forschung*, 13, 537–541.

Hörhammer, L., Wagner, H. and Bittner, G. (1965) Neue methoden im pharmakognostischen unterricht 10. Mitteilung: Unterscheidung handelsblicher aloesorten mittels Dünnschichtchromatographie. *Deutsche Apotheker-Zeitung*, 105, 827–830.

Jaminet, F. (1957) Recherches sur la caractérisationdes extraits végétaux et animaux dans les spécialités pharmaceutiques et les médicaments composés. *Journal de Pharmacie de Belgique*, 12, 87–100.

Janiak, B. and Böhmert, H. (1962) Methode zur quantitativen Bestimmung des Aloins. *Arzneimittel Forschung*, 12, 431–435.

Jansz, E.R., Siva, V. and Ratnayake, D. (1981) The aloin content of local aloe species. *Journal of the National Science Council of Sri Lanka*, 9, 107–109.

Kraus, Lj. (1959) Übersicht der Analytischen Methoden zur Bestimmung von Anthrachinondrogen. *Planta Medica*, 7, 427–446.

Léger, M.E. (1916) Les aloïnes. Première partie. *Annales de Chimie*, 6, 318–338.

Lister, R.E. and Pride, R.R.A. (1959) The characterisation of crystalline and amorphous aloin. *Journal of Pharmacy and Pharmacology*, 11, 278T–282T.

Mapp, R.K. and McCarthy, T.J. (1970) The assessment of purgative principles in aloes. *Planta Medica*, 18, 361–365.

Mary, N.J., Christensen, B.V. and Beal, J.L. (1956) A paper chromatographic study of aloe, aloin and of cascara sagrada. *Journal of the American Pharmaceutical Association*, 45, 229–232.

McCarthy, T.J. and Mapp, R.K. (1970) A comparative investigation of methods used to estimate aloin and related compounds in aloes. *Planta Medica*, 18, 36–43.

McCarthy, T.J. and Price, C.H. (1965) Micro-determination of aloin in some South African aloe species. *Pharmaceutisch Weekblad*, 100, 761–763.

McCarthy, T.J. (1968) The metabolism of anthracene derivatives and organic acids in selected *Aloe* species. *Planta Medica*, 16, 348–356.

Morsy, E.M. (1983) The current official stand of some species of aloe and their derivatives. Reviewing aloe and its derivatives as documented in the USP, USD, NF, BP and FDA. *Erde International*, 1, 7–16.

Morton, J.F. (1961) Folk uses and commercial exploitation of *Aloe* leaf pulp. *Economic Botany*, 15, 311–319.

Mühlemann, H. (1952) Über Anthrachinone und Anthrachinonglykoside. *Parmaceutica Acta Helvetiae*, 27, 17–26.

Mühlemann, H. and Tatrai, O. (1967) Die Wertbestimmung der Anthrachinon-drogen ihrer Präparate und Herstellung der letzteren im Hinblick auf die Ph.Helv. VI. *Parmaceutica Acta Helvetiae*, 42, 717–733.

Okamura, N., Asai, M., Hine, N. and Yagi, A. (1996) High-performance liquid chromatographic determination of phenolic compounds in *Aloe* species. *Journal of Chromatography A*, 746, 225–231.

Ovanoviski, H. (1983) Aloin, *Erde International*, 1, 34–36.

Paez, A., Gebre, G.M., Gonzalez, M.E. and Tschaplinski, T.J. (2000) Growth, soluble carbohydrates and aloin concentration of *Aloe vera* plants exposed to three irradiance levels. *Environmental and Experimental Botany*, 44, 130–133.

Paris, R. and Viejo, J.P. (1955) Identification des drogues simples et contròle des médicaments végétaux par chromatographie sur papier. *Annales Pharmaceutiques Francaises*, 13, 424–429.

Paris, R. and Durand, M. (1956) A propos de l'essai des aloès. Dosage photométrique de l'aloïne. *Annales Pharmaceutiques Francaises*, 14, 755–761.

Rauwald, H.-W. (1982) Präparatve Trennung der diastereomeren Aloine mittels Droplet-Counter-Current-Chromatography (DCCC). *Archiv der Pharmazie*, 315, 769–772.

Rauwald, H.-W. (1990) Naturally occurring quinones and their related reduction forms: analysis and analytical methods. *Pharmazeutische Zeitung Wissenschaft*, 3, 169–181.

Rauwald, H.-W. and Beil, A. (1993a) 5-Hydroxyaloin A in the genus *Aloe* thin layer chromatographic screening and high performance liquid chromatographic determination. *Zeitschrift für Naturforschung*, 48c, 1–4.

Rauwald, H.-W. and Beil, A. (1993b) High-performance liquid chromatographic separation and determination of diastereomeric anthrone-C-glucosyls in Cape aloes. *Journal of Chromatography*, 639, 359–362.

Rauwald, H.-W. and Sigler, A. (1994) Simultaneous dtermination of 18 polyketides typical of *Aloe* by high performance liquid chromatography and photodiode array detection. *Phytochemical Analysis*, 5, 266–270.

Rauwald, H.-W. and Voetig, R. (1982) 7-Hydroxy-Aloin:die Leitsubstanz aus *Aloë barbadensis* in der Ph.Eur.III. *Archiv der Pharmazie*, 315, 477–478.

Raymer, S. and Raymer, D. (1995) Comments on aloe sap. *Ballyia*, 2, 21–23.

Reynolds, G.W. (1950) *The Aloes of South Africa*, pp. 460–468. Johannesburg, South Africa: The Aloes of South Africa Book Fund.

Reynolds, G.W. (1966) *The Aloes of Tropical Africa and Madagascar*, pp. 144–151. Mbabane, Swaziland: The Aloes Book Fund.

Reynolds, T. (1985) Observations on the phytochemistry of the *Aloe* leaf-exudate compounds, *Botanical Journal of the Linnean Society*, 90, 179–199.

Reynolds, T. (1986) A contribution to the phytochemistry of the East African tetraploid shrubby aloes and their diploid allies. *Botanical Journal of the Linnean Society*, 92, 383–392.

Reynolds, T. (1994) A chromatographic examination of some old samples of drug aloes. *Pharmazie*, 49, 524–529.

Saccù, D., Bogoni, P. and Pricida, G. (2001) Aloe exudate:characterization by reversed phase HPLC and headspace GC-MS. *Journal of Agricultural and Food Chemistry*, 49, 4526–4530.

Siesto, A.J. and Bartoli, A. (1957) Papierelektrophoretische Bestimmung von Oxymethyl-anthrachinone in *Rhizome Rhei, Farmaco, Edizione Scientfica*, 12, 934.

Sigler, A. and Rauwald, H.W. (1994a) Tetrahydroanthracenes as markers for subterranean anthranoid metabolism in *Aloe* species. *Journal of Plant Physiology*, 143, 596–600.

Sigler, A. and Rauwald, H.W. (1994b) First proof of anthrone aglycones and diastereomeric anthrone-C-glycosides in flowers and bracts of *Aloe* species. *Biochemical Systematics and Ecology*, 22, 287–290.

Smith, T. and Smith, H. (1851) On aloin: the cathartic principles of aloe. *Monthly Journal of Medical Science*, 12, 127 – 131.

Stahl, E. and Schorn, P.J. (1961) Dünnschicht-Chromatographie hydrophiler Arzneipflanzenauszüge. *Hoppe-Seyler's zeitschrift für Chemie*, 325, 263–274.

Stone, K.G. and Furman, N.H. (1947) Estimation of aloe-emodin and the aloins in Curaçao aloes. *Analytical Chemistry*, 19, 105–107.

Teichert, K., Mutschler, E. and Rochelmeyer, H. (1961) Die plattenchromatographische Untersuchung von Naturstoffgemischen. *Zeitschrift für Analytische Chemie*, 181, 325–331.

Thieme, H. and Diez, V. (1973) Vergleich verschiedener Methoden zur spektrophotometrischen Bestimmung von Aloin in Aloe DAB 7-DDR, *Die Pharmazie*, 28, 331–335.

Trease, G.E. and Evans, W.C. (1983) *Pharmacognosy*, 12th edn, pp. 619–625. London: Ballière Tindall.

Van Wyk, B.-E., Yenesew, A. and Dagne, E. (1995) Chemotaxonomic survey of anthraquinones and pre-anthraquinones in roots of *Aloe* species. *Biochemical Systematics and Ecology*, 23, 267–275.

Viljoen, A.M., Van Wyk, B.-E. and Dagne, E. (1996) The chemotaxonomic value of 10-hydroxyaloin B and its derivatives in *Aloe* species *Asperifoliae* Berger. *Kew Bulletin*, 51, 159 – 168.

Viljoen, A.M., Van Wyk, B.-E. and Newton, L.E. (1999) Plicataloside in *Aloe* – a chemotaxonomic appraisal. *Biochemical Systematics and Ecology*, 27, 507–517.

Viljoen, A.M., Van Wyk, B.-E. and Newton, L.E. (2001) The occurrence and taxonomic distribution of the anthrones aloin, aloinoside and microdontin in *Aloe*. *Biochemical Systematics and Ecology*, 29, 53–67.

Viljoen, A.M. and Van Wyk, B.-E. (2001) A chemotaxonomic and morphological appraisal of *Aloe* series Purpurascentes, *Aloe* section Anguialoe and their hybrid, *Aloe broomii*. *Biochemical Systematics and Ecology*, 29, 621–631.

Watt, J.M. and Breyer-Brandwijk, M.G. (1962) *The Medicinal and Poisonous Plants of South and Eastern Africa*, 2nd edn, pp. 679–687. Edinburgh and London: E. & S. Livingstone Ltd.

Worthen, L.R., Bennet, E. and Czarnecki, R.B. (1959) The chromatographic behaviour of aloin. *Journal of Pharmacy and Pharmacology*, 11, 384.

Yamamoto, M., Ishikawa, M., Masui, T., Nakazawa, H. and Kabasawa, Y. (1985) Liquid chromatographic determination of barbaloin (aloin) in foods. *Journal of the Association of Official Analytical Chemists*, 68, 493–494.

Zonta, F., Bogoni, P., Massoti, P. and Micali, G. (1995) High-performance liquid chromatographic profiles of aloe constituents and determination of aloin in beverages, with reference to the EEC regulation for flavouring substances. *Journal of Chromatography A*, 718, 99–106.

Zwaving, J.H. (1965) Trennung und isolierung der anthrachinonglykoside von *palmatum*, *Planta Medica*, 13, 474–484.

Zwaving, J.H. (1968) Separation of dianthrone glycosides and anthraquinone glycosides in senna and rhubarb on columns of Sephadex LH-20. *Journal of Chromatography*, 35, 562–565.

8 Industrial processing and quality control of *Aloe barbadensis* (*Aloe vera*) gel

Todd A. Waller, Ronald P. Pelley and Faith M. Strickland

ABSTRACT

Aloe barbadensis Miller (=*A. vera* (L.) Burm. f.) is the most common species of *Aloe* used in worldwide commerce and it is the only species of *Aloe* which productively yields gel from fillets. The three most critical factors in the production of *A. barbadensis* gel are (i) proper agronomy; (ii) the control of bacterial proliferation post leaf harvest; and, (iii) avoiding excessive activity of $\beta 1 \rightarrow 4$ glucosidases which will destroy the major glucomannan polysaccharide. Despite its appearance, *Aloe* is not a cactus and it grows best when supplied with an excess of 50 cm of water per year and a slight excess of nitrogen. Because of its somewhat unusual root system, aloes require well-drained soils and do not tolerate deep tillage. The control of bacterial growth during processing should focus on rapid processing of leaves, leaf and environmental sanitization, rapid cooling and processing of gel after filleting, and proper bacterial control. Microbiological analysis of aloe materials should be properly directed toward those organisms commensal with *Aloe*. The enzymes which break down major aloe polysaccharide can come from either the plant itself, be produced by contaminating organisms, or may be fungal enzymes added as processing aids. A lack of understanding of basic enzymology results in the very low level of polysaccharide observed in most of the less costly commercial aloe products. Because of the failure of so many companies in the industry to observe these guidelines and the presence of fraud in the 'Aloe Vera' industry, rigorous quality control must be performed by consumer product manufacturers. Lastly, recent scientific studies indicate that it may be very desirable to strongly control the level of anthraquinones in aloe extracts destined for dermatologic and cosmetic uses.

INTRODUCTION

OVERVIEW – PROCESSING ALOE AND QUALITY CONTROL

Extracts of *Aloe barbadensis*, particularly the gel, are well accepted on a worldwide basis as a premier ingredient in over-the-counter dermal therapeutics and high-end cosmetics.

Abbreviations: Alcohol precipitable hexose – APH; Aloe Research Foundation – ARF; complementary and alternative medicine – CAM; International Aloe Science Council – IASC; methanol precipitable solids – MPS; molecular weight – MW; polysaccharide – Ps; Rio Grande Valley – RGV; tetrahydroxyanthraquinone – THA.

This is because aloe's efficient acceleration of the rate of recovery of skin cells from environmental injury is well documented (see accompanying reviews in this volume by Heggers *et al.* and Strickland *et al.*). *A. barbadensis* and *A. arborescens* Miller extracts are also ingested orally for a variety of reasons. This consumption results in an aggregate world demand in excess of 100,000,000 liters per annum of crude homogenate juice and filleted gel, as estimated by the International Aloe Science Council (IASC). In order to visualize this quantity, one should imagine 5,000 tanker trucks, each of a capacity of 20,000 liters, constituting a queue sufficient to stretch across a major city. Yet despite this volume of production much of the world's supply of aloe juice is still processed without a fundamental understanding of what aloe is, without comprehending the beneficial and deleterious effect of processing and using quality control methods that range from improper to outright fraudulent. Compared to the dairy industry or the citrus juice industries, we lack agreed-upon chemical standards, uniform methods of processing, appropriate microbiology and enforceable regulation. Because of this we are, and could very well remain, a fringe industry experiencing ever decreasing public confidence because the majority of our products fail to perform due to improper processing or outright fraud.

Given the above background, what is the purpose and intended audience of this review? Our first audience is limited to no more than 100 to 200 people. These are our clients, aloe processors including managers in Sales and Marketing, Technical Services, Farm and Processing Line Production and Quality Control. These are people who have direct 'hands on' contact with aloe. We will attempt to influence them, both with our data and by scientific reasoning, to adhere to proper processing and rigorous quality control. Our second audience is much larger, research scientists and purchasers of 'aloe', whether these 'customers' are purchasing agents of large corporations or end consumer product users. This audience has little or no 'hands on' experience with field grown *A. barbadensis* fillets and must generally rely on materials supplied by others. We want to give this audience an appreciation of what authentic aloe gel is like, of what its properties are and what biological activities can be expected from a given preparation.

What this chapter cannot do is to offer simplistic advice and prescriptions. We cannot recommend plant spacing for aloes or irrigation and fertilization regimes. Only experience dictates which parameters can be extrapolated from situation to situation.

The never-ending battle – management versus production/quality control

Many of those who work in the aloe industry, in companies which produce aloe feed stocks, often view their existence as a battle between those who actually produce the product and those who oversee the company. Those who produce the product usually strive to put out a material of superior quality, while senior management are continually concerned with profits. Those who are responsible for keeping the company afloat are confronted with opposing teams – Sales and Marketing, versus Finance, and Production.

If Production and Quality Control are excessively compliant with the pressures of a management unfamiliar with science and technology, then a large quantity of aloe of inferior quality will be marketed at a low price. This company will capture market share for crude aloe because they can sell an abundance of low-priced juice. These sales occur in what is generally a tight market for high quality aloe wherein parameters for this quality are ill-defined. Consumer product manufacturers will accumulate stocks of spoiled raw material and manufacture an end product of inferior quality because there are no good analytical tests for high quality feed stocks. The aloe-containing products initially sell well because of consumer recognition of product label ('with Aloe Vera'). However, consumer resistance eventually develops because of low product efficacy ('degraded or fraudulent aloe doesn't work!'). The consumer product manufacturer then seeks another feedstock vendor. This 'milking' of production by management comprises the classic 'New-Aloe-Company' market cycle that we, as consultants, have seen on many occasions. To date, the only escape from this cycle has been multilevel-direct marketing companies who have integrated aloe production with consumer product manufacture and marking. They feature limited production and command high prices.

In reality, Senior Management and Production/Quality Control cannot remain at war if a company is to prosper or even survive. Survival is dependent on Management becoming science-aware, while Production and Quality Control must be able to put a product of consistent quality out on a reasonable schedule. What this article will attempt to do is to give the general principles for producing quality aloe feed stocks from experience garnered over the last 20 years. Details will vary, however, because soil is different, each region has a different climate, each manufacturing plant has a different microbial flora and every company puts out a different product line.

What is *A. barbadensis* extract?

Twenty years ago, the single greatest problem confronting the aloe industry was that nobody could define in physical and chemical terms what commercial *A. barbadensis* gel actually was. If you do not know, or cannot agree on, the chemical composition of the aloe you are vending, it is virtually impossible to measure it, to establish quality control sigma limits and to therefore consistently produce an aloe product. Furthermore, without a fixed definition of 'aloe', unscrupulous companies could and do, vend almost anything as aloe, undercutting legitimate vendors. This dilemma was partially resolved by two co-operative, inter-related programs in the early to middle 1990s. The first was the Field Study of the International Aloe Science Council (IASC) (Wang and Strong, 1993; Pelley *et al.*, 1993; Wang and Strong, 1995). The second was the Standard Sample program of the Aloe Research Foundation (ARF) (Pelley *et al.*, 1993; Waller *et al.*, 1994; Pelley *et al.*, 1998).

The IASC and the Texas A&M study

The International Aloe Science Council (IASC), established in 1981, is the trade group for the aloe industry. The IASC encompasses all of the major producers of aloe feed stocks and most of the companies which market aloe products directly to consumers. In 1983 the IASC instituted a quality control program (IASC, 1991; Pelley *et al.*, 1993; Waller *et al.*, 1994) for its members, which culminated in certification of the product as authentic aloe gel. However, one of the criticisms of the IASC certification program was that its database was based upon an older analysis (Waller *et al.*, 1978) of aloe gel and represented only a small number of samples.

To strengthen its certification program, the IASC commissioned a multiyear study by the Horticultural Sciences Department of the Texas A&M University, Agricultural Experiment Station in Weslaco Texas. This project, led by Yin-Tung Wang, was to determine the content and variation over time of the major chemical components in

A. barbadensis. The program, which became known as the Texas A&M Study, initially studied leaves from four commercial plantations in the Rio Grande Valley (RGV) of Texas and eventually more than 500 samples were analyzed. The final published database consisted of 336 samples (112 weekly samples from three plantations), analyzed for twelve parameters (Wang and Strong, 1995). Some of the analyses were for general descriptors such as leaf weight, solids content, pH, pulp content and sugar content (Table 8.1, below).

Some of these parameters are important as certification parameters. For example, the IASC has decided that the soluble-solids value is an important indicator that the liquid under consideration is consistent with authentic *A. barbadensis* gel because low values for solids suggest adulteration by dilution. Other parameters, such as leaf weight, pH and reducing sugar, are useful for monitoring agronomy and exploring plant physiology. The metal cations Na^+, K^+, Ca^{++} and Mg^{++} were considered to be key analytical parameters because of their stability to physical/chemical/biological degradation, their stability during processing and the availability of standardized methods for determination. Therefore, early on, calcium and magnesium became IASC certification markers (IASC, 1991; Pelley *et al.*, 1993). Sodium and potassium were added to the Texas A&M study because they were the remaining major metal cations (Table 8.2, below).

Also measured in the Texas A&M study was an organic acid originally referred to as 'E Peak' (Pelley *et al.*, 1993; Waller *et al.*, 1994) and suggested by Wang to be malic acid (Wang and Strong, 1995). Malic acid was a highly effective certification marker because (i) it was degraded readily by aloe-associated bacteria and therefore was an excellent marker for freshness (Pelley *et al.*, 1993; Waller *et al.*, 1994) and (ii) since the structure of 'E Peak' was not widely circulated, it functioned as a method for detecting

Table 8.1 Characteristics of fresh *A. barbadensis* leaves from three producers averaged over a 112 week period[1].

Grower	Leaf weight (g)	pH	Soluble solids (%)	Pulp (%)	Reducing sugar (mg/liter)
1	705±10	4.53+0.01	0.57±0.01	0.12±0.01	1466±87
2	774±7	4.56+0.01	0.62±0.01	0.11±0.01	2391±97
3	465±8	4.57+0.01	0.56±0.01	0.09±0.01	1713±98

Note
1 Modified from Table 1 of Wang and Strong (1995) with permission. All values are mean ± SEM.

Table 8.2 Mineral concentrations of fresh *Aloe barbadensis* gel from three producers averaged over a 112 week period[1].

Grower	Mineral Concentration (mg/liter)			
	Potassium	Sodium	Calcium	Magnesium
1	370+14	195±7	331±10	52±2
2	313+11	182±4	348±7	60±1
3	450+12	181±7	260±10	54±2

Note
1 Modified from Table 2 of Wang and Strong (1995) with permission.

Table 8.3 Physical and Chemical Values for IASC Certified *A. barbadensis* Gel[1].

Test	Average	Acceptable Range
Solids (105 °C Nonvolatile)	0.83 g/dl	0.46 to 1.31 g/dl
Calcium	241 mg/l	98 to 448 mg/l
Magnesium	58 mg/l	23 to 118 mg/l
Malic Acid	2,029 mg/l	818 to 3,427 mg/l
Polysaccharide		Present, exact values not yet finalized

Note

1 Values established by the IASC (2000).

fraud (Pelley *et al.*, 1998). Wang's database should form the basis for interpreting virtually all studies on the commercial aspects of *A. barbadensis*. Our overviews of agronomy, processing, and fraud in the aloe industry are ultimately based on his findings. The Texas A&M study forms the basis for the current IASC parameters, which are shown in Table 8.3 above.

The Aloe Research Foundation and biological activity

During this same 1991–1993 period, another set of standards were being established by the Aloe Research Foundation (ARF). The IASC standards addressed questions of physical and chemical identity of *A. barbadensis* gel but did not address biological activity. This fell to the ARF program. The ARF was established by one of the two major vertically integrated companies in the aloe industry to support biological research on *Aloe sp.* It was hoped that the other companies in the industry would be attracted to support high quality research, which, alas, never happened. At the very beginning it was decided that major factors holding back research on *Aloe sp.* was (i) the lack of standardized biologically active materials (Pelley *et al.*, 1993; Waller *et al.*, 1994) and (ii) concomitant standardized assays for aloe biological activity (for details see Chapter 12). These factors led to the production of the ARF Standard Samples (Table 8.4, below) and their analysis by the Inter University Collaborative Study. The key to the ARF Standard Samples was meticulous processing and freeze-drying from the native gel state.

These materials had a number of biological activities and their fractionation and characterization occupied much of *A. barbadensis* biochemical research during the second half of the decade. These materials were produced by **Process A** (Pelley *et al.*, 1993; Waller *et al.*, 1994), which comprises:

(i)   subjecting leaves harvested only an hour or two in advance to a sanitizing wash and rinse;

(ii)   filleting manually or mechanically, with utmost care to exclude the anthraquinone-rich mesophyll;

(iii) coarsely grinding the filets and removing pulp by passage through a screen with 250 μm openings; and,

(iv)   lyophilizing the depulped gel immediately, without any concentration or further processing.

Table 8.4 Key Analytical Parameters of the Aloe Research Foundation Process A Gel Fillet Standard Samples.

Standard Sample[a]	Conductivity μ Siemens[b]	Ethanol Precipitable Solids[c]	Total Polysaccharide[c]	HPLC Analysis[d]
91A	2,356 μS	30.6%	7.4%	Normal
91D	1,488 μS	23.0%	6.5%	Normal
92A$_2$	2,108 μS	30.2%	8.7%	Normal
92B	2,232 μS	21.6%	5.6%	Normal
92C$_1$	1,054 μS	45.8%	13.5%	Normal
92C$_2$	1,354 μS	38.4%	13.2%	Normal
93A	2,170 μS	40.2%	11.3%	Normal
93B	2,900 μS	47.2%	9.7%	Normal
93D$_1$	2,050 μS	38.0%	6.9%	Normal
93D$_2$	2,350 μS	33.6%	9.1%	Normal
93E$_1$	2,200 μS	37.0%	9.8%	Normal
93E$_2$	2,100 μS	35.6%	8.9%	Normal
Mean	2,030 μS	35.1%	9.2%	
Standard Deviation	502 μS	7.9%	2.5%	
S. E. M.	145 μS	2.3%	0.7%	
Coefficient of Variability	24.7%	22.5%	27.1%	

Notes

a ARF'91A, 91D, 92A$_2$, 92C$_1$ and 92C$_2$ were prepared from leaves at the Aloecrop Lyford Plantation. ARF'92B, and all of the samples from 1993 were prepared from leaves from the Aloecorp de Mexico, Gonzales Plantation. These materials do not appear to differ significantly from each other. ARF'92C$_1$ consisted of leaves from young plants while '92C$_2$ was prepared from older plants. ARF'93C$_1$ and ARF'93E$_1$ were filleted by hand while ARF'93C$_2$ and ARF'93E$_2$ were filleted by machine. Aside from a slight difference in alcohol precipitable solids, there does not appear to be a difference between the materials.

b Solution at 0.62 g/dl (actual 62 mg in 10 ml).

c This method is adapted from Method 988.12 0f the AOAC (AOAC, 2000a). For the purposes of determining crude polysaccharide in aloe, the second precipitation (alkaline Cu(II) reagent can be omitted. 500 mg of lyophilized powder was dissolved and precipitated with 80% EtOH. The solids content (Alcohol Precipitable Solids) of the resolublized pellet was determined by lyophilization. The total hexose content was determined by phenol sulfuric acid assay which yields the values for Total Polysaccharide hexose.

d Determined by the IASC method for analysis of malic acid.

These requirements mean that **Process A** gel approximates the closest possibly to the freshest possible gel. Unfortunately native gel can never be a commercially viable product because it is extremely pseudoplastic and has a low solids content. Since this system was established in 1995, the ARF has maintained a much lower profile.

The consensus composition of matter of *A. barbadensis* gel

If we combine the IASC Texas A&M database and the ARF **Process A** Standard Sample program, the following composition of matter for aloe gel was obtained (Figure 8.1).

The categories used to classify the composition of matter are arbitrarily chosen, based upon the physical properties of the chemical components, which influences their behaviour during processing, the physico-chemical classes revealed by fractionation, and the biological properties found during the ARF Inter University Collaborative Study. The scheme, though arbitrary, does neatly separate the various biological activities and the overall values are in agreement with the semi-quantitative results of Waller *et al.*

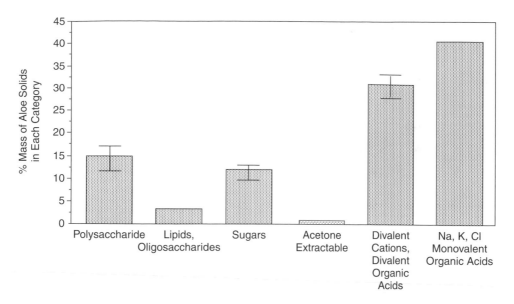

Figure 8.1 A consensus composition of matter for *A. barbadensis* gel in the native state.

(1978). The standard errors indicated above are in remarkably close agreement between the ARF Standard Sample Study and the Texas A&M Study discussed above.

The solids in *Aloe barbadensis* gel are approximately 25% total carbohydrate, comprising c. 11% monosaccharides (95% glucose, 5% fructose), c. 14% polysaccharide and c. 1% oligosaccharide. The monosaccharides (Sugars in Figure 8.1, above) comprise more than 95% of the reducing sugar activity. Wang and Strong (1995) have shown (Figure 6 in the referenced publication) that the reducing sugar content varies greatly from location to location and from time to time. The major polysaccharide of *A. barbadensis* is essentially a block co-polymer (Strickland *et al.*, 1998). In the native state this polysaccharide is so large that it is essentially insoluble. The native polysaccharide consists of linear stretches of an acetylated mannan that activates macrophages, cross-linked with a cytoprotective oligosaccharide that accelerates the recovery from physical injury of epithelial cells (Figure 8.2) (Strickland *et al.*, 1998). In addition to the native, major polysaccharide there are galactans (Mandal and Das, 1980b) and pectins (Mandal *et al.*, 1983). These are relatively minor constituents in undegraded native gel, but they may come to be the dominant polysaccharides when improper processing with cellulase destroys essentially all of the $\beta 1 \rightarrow 4$ linked glucomannan (Pelley *et al.*, 1998).

Approximately 1–3% of solids consist of alcohol-soluble molecules that are unable to pass a cellulose acetate membrane of 5,000 daltons molecular weight cut-off. These materials contain two classes of molecule with two different, and opposing, biological activities. First are the lipophilic free anthraquinones and chromone glycoside esters, some of which are anti-inflammatory. Second, there are also oligosaccharides, including the cytoprotective oligosaccharide (Figure 8.2 below) of Strickland *et al.* (1998). The cytoprotective oligosaccharide is present only when the gel has been processed under certain conditions.

A third category, acetone extractables of lyophilized **Process A** gel, usually comprises about 1% of the solids of aloe gel. It consists of compounds, such as anthraquinone

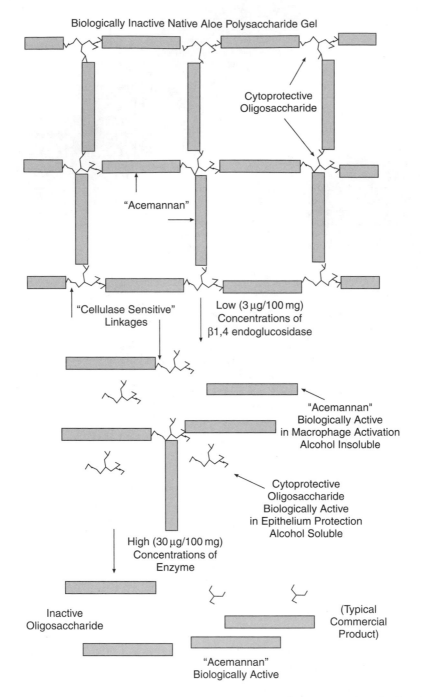

Figure 8.2 Cartoon schematic of the putative structure of the native polysaccharide of *A. barbadensis* and its cleavage by β1 → 4 endoglucosidase. Modified from Sheet 1 of Strickland *et al.* (1998).

and chromone glucosides, readily soluble in acetone or chloroform but sufficiently hydrophilic to pass a cellulose acetate membrane. These molecules are reviewed by Reynolds (Chapter 3). The trihydroxy and tetrahydroxy tricyclic anthraquinone aglycones and anthrone *C*-glycosides, and frequently esterified heterocyclic bicyclic chromones, have been the focus of aloe chemistry for over one hundred years. Among the 30 or so compounds in this group are many biologically active compounds and others which cause color change in commercial aloe.

Finally, approximately 75% of the mass of aloe gel solids consist of monovalent and divalent metal cations, organic acids and chloride. They do comprise most of the solids content of the gel, are little changed during commercial processing and form the basis for the IASC certification system. Among these markers is malic acid, previously known as 'E Peak', which when complexed with calcium can make up 60% of the alcohol-precipitable divalent-cation/diavalent-organic acid fraction (Pelley *et al.*, 1998). Malic acid (Wang and Strong, 1995) is a useful marker for the 'freshness' of aloe gel (Pelley *et al.*, 1993; Waller *et al.*, 1994). This is because malic acid is assimilated by certain cryophilic Gram −ve rods or theromophilic Gram +ve cocci and bacilli commonly associated with *A. barbadensis*, which tend to cause bacterial 'breakthrough' contamination of commercial aloe . Absent or diminished malic acid in authentic aloe indicates that there was extensive bacterial growth at some processing step (Pelley *et al.*, 1993; Waller *et al.*, 1994). The absence of malic acid and extremely low levels of metallic cations raise a strong suspicion that the material under consideration is fraudulent (Pelley *et al.*, 1998).

A summary of processing for *Aloe barbadensis* gel

The rest of this chapter will by and large consider what happens to this composition of matter during the growing and processing of *A. barbadensis*. Processing can be broken down into three basic steps (Figure 8.3). In **Preliminary Processing** the leaves are harvested, scrubbed, the rind is removed to yield the gel fillet, the fillets are lightly ground and the cellulosic pulp is removed. Next, the crude gel is subjected to **Intermediate Processing** to kill bacteria and, if desired, to remove anthraquinones. Intermediate processing yields materials which, with the addition of preservatives, comprise the basic gel products. **Finish Processing** yields preserved liquid products, concentrates and powders.

The ARF created standard samples corresponding to a number of intermediate and final products. In Figure 8.3 these are indicated as ARF **Process A, Process B, Process C, Process D**, and **Process E** materials. These materials are sequentially derived from common starting materials under carefully controlled and monitored conditions using industrial equipment. Study of the **Process A/B/C/D/E** materials has shown how the composition of the material changes during processing.

THE AGRONOMY OF ALOE

One is continually amazed at the hardiness of *Aloe sp*. Given that one is dealing with an organism that is fundamentally frost-sensitive, members of the *Aloaceae* can be found in the most surprising places. Once while on a lecture tour, two of us (RPP and FMS), ascended to the former telegraph station overlooking the harbor of Hobart, Tasmania.

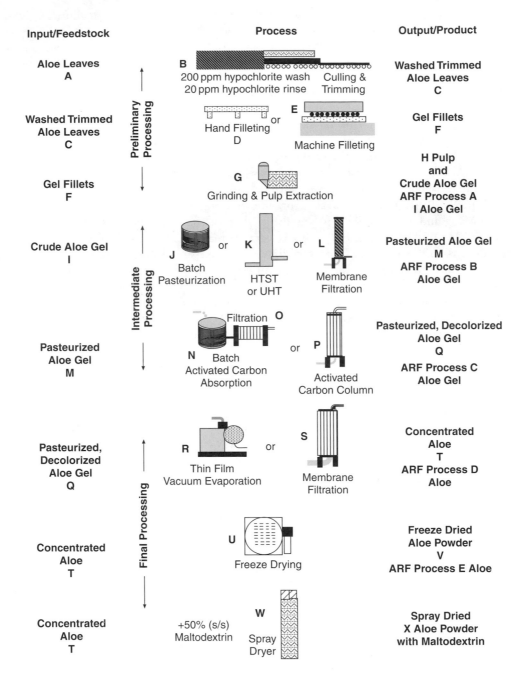

Input/Feedstock	Process	Output/Product
Aloe Leaves A	**B** 200 ppm hypochlorite wash, 20 ppm hypochlorite rinse; Culling & Trimming	Washed Trimmed Aloe Leaves C
Washed Trimmed Aloe Leaves C	Hand Filleting **D** or Machine Filleting **E**	Gel Fillets F
Gel Fillets F	**G** Grinding & Pulp Extraction	H Pulp and Crude Aloe Gel ARF Process A I Aloe Gel
Crude Aloe Gel I	**J** Batch Pasteurization or **K** HTST or UHT or **L** Membrane Filtration	Pasteurized Aloe Gel M ARF Process B Aloe Gel
Pasteurized Aloe Gel M	**N** Batch Activated Carbon Absorption, Filtration **O** or **P** Activated Carbon Column	Pasteurized, Decolorized Aloe Gel Q ARF Process C Aloe Gel
Pasteurized, Decolorized Aloe Gel Q	**R** Thin Film Vacuum Evaporation or **S** Membrane Filtration	Concentrated Aloe T ARF Process D Aloe
Concentrated Aloe T	**U** Freeze Drying	Freeze Dried Aloe Powder V ARF Process E Aloe
Concentrated Aloe T	+50% (s/s) Maltodextrin **W** Spray Dryer	Spray Dried X Aloe Powder with Maltodextrin

Figure 8.3 Steps in commercial processing of *A. barbadensis* gel.

We noticed an aloe plant struggling to maintain a foothold on the cold, rocky, wind-swept summit of the hill. Soon enough we noticed another, and yet another. Soon we realized that the entire hilltop was colonized by aloe plants. These must have been brought more than a century ago, directly from the Cape and planted on this desolate spot for laxative purposes by the staff of the semaphore station and if so were *A. ferox* Miller. However, the endurance of one plant does not equate with the optimal growth conditions for all species or for commercial cultivars.

One of the last things considered by most brokers of aloe feed stocks is agronomy. This consideration is partly because agronomy is probably the last item that consumer product manufacturers care about. However, in our collective experience, there is no item more important than agronomy in determining quantitative yields of gel per hectare, nor more essential to reproducible biological activity. Unfortunately, because we do not yet have chemical assays for everything in the broad spectrum of aloe biological activities, we cannot yet explain the exact agronomic parameters for the production of the best gel.

Water, nitrogen and soil pH

The three most critical factors for the production of aloe gel are the supply of water, the availability of nitrogen, and the acidity of the soil in which the plant grows. Out of the three factors, the most important is water, for although aloe is popularly conceived of as being a desert plant, in reality its growth rate is more highly tied to water availability than to any other factor. Approximately three decades ago it was realized that the anthraquinone content of *Aloe sp.* was dependent on the various inter-related factors (insolation, wind, moisture) which constitute water availability. As recently as 1992, Genet and van Schooten determined that water availability was the key determinant for growth of aloes in the sub-tropics. However, the only systematic study of the interaction of water, macronutrients, and micronutrients in the growth of *A. barbadensis* is the work of Dr Wang at Texas A&M University. Unfortunately, the results of this agronomic study were never published and therefore much of what can be said about the agronomy of aloes remains anecdotal rather than quantitative.

Four general principles can be stated with regard to the cultivation of *A. barbadensis* in the lower RGV. We focus on this locale because that is the location where *A. barbadensis* Miller (cv. RGV) has been cultivated for the longest time. One would therefore expect the agronomy to be well understood. Also, traditionally, this is where most of the aloe gel marketed in the United States is grown. First, the climate of this region is light and quite variable from year to year and severe frosts can happen. Second, adequate rainfall is irregular (average rainfall 26 inches, 66 cm/annum). Droughts (25 cm/annum) lasting up to three years can destroy fledgling plantations. Third, the soil is often low in available nitrogen (<0.15%) and the soil pH is quite high (8.0–8.6). Fourth, due to the soil structure, correcting problems two and three by irrigation and fertilization can lead to salinization. The high calcium content of these soils makes it difficult to adjust the pH to the 6.8–7.6 range and correction of pH will be transient. These variables are interactive with each other and are further interactive with wind and insolation during periods were rain is suboptimal. This means that it is impossible to provide quantitative solutions for a given variable in the absence of controlled site studies. We can, however, provide the following approximate estimates.

Severe frosts occur every six to seven years in the lower RGV (three in the last 20 years; 1983, 1989 and 1991). This means that one risks their profitability if one aloe plants take four to five years (Coats, 1994) to come into full production. They can take this long because in many years the 100 cm of rainfall and 0.25–0.40% nitrogen needed for minimally acceptable growth is not achieved. Higher growth rates (9–12 months to full production) could be achieved with optimal water (200 cm per annum) and nitrogen (0.40–0.50%) inputs at a soil pH of ~7.4. But land and water are expensive in the lower RGV. The practical consequence of these factors is that production of *A. barbadensis* has shifted to areas where land is less expensive, frosts are less frequent, and rainfall is sufficient while water is cheaper. The RGV no longer dominates the world aloe gel market as it did ten years ago. Currently, production is distributed among plantations in Mexico, the Dominican Republic, Venezuela, Guatemala and North Africa. All of these locations feature, above all, an absence of killing frosts. This means that there is not a race to recover the investment before a frost wipes out the crop. Most of these locations are still less than optimal with regard to rainfall and have nitrogen deficient soils. We have no doubts that production of gel in these areas could be improved by such elementary measures as provision of adequate water and nitrogen.

Just as important as water availability is soil composition/porosity and pH. In soils where silt content is significant, drainage will be poor. In the lower RGV rainfall is intermittent and much of the total annual precipitation may occur in three or four deluges. Those locations with poorly draining soils will experience pooling of water on the surface. The roots of *A. barbadensis* are thick and shallow and possess relatively few root hairs. This may explain the predisposition of this plant to root rot. During the RGV droughts during the latter half of the 1990s, it was not uncommon to have episodes of aloe root rot following the occasional torrential rain that broke a year-long drought. Similarly, calcium-rich, high pH soils which drain poorly have a higher than expected problem with salinization. This association makes lowering the soil pH difficult since it exacerbates this salinization.

Spacing and harvesting

We tend to forget, when discussing agronomy, the spacing of plants and harvesting. However with the rising costs of land and water and the need to produce high quality leaves at minimal cost, these factors are becoming increasingly important. We have not uncommonly seen aloes planted at densities as low as 3,000 per acre. Where land is cheap, rainfall plentiful, intercropping is practiced and weed control is achieved by grazing, such a policy makes sense. We have also seen plantations using such practices with aloes planted at densities of 4,500 plants per acre yielding leaf size >800 g. However, to maximize production per acre, densities of up to 6,000 plants per acre can be employed, provided optimal amounts of water are provided and soil nitrogen is maintained in the 0.40–0.50% range. A density of 6,000 + plants/acre is most compatible with drip irrigation and the use of ground-cover poly-film for weed control. This sophisticated agronomy avoids the need for mechanical tillage which can easily disrupt the exceedingly shallow root system of *A. barbadensis*.

The procedure of aloe harvesting properly belongs in the next section since it is done only an hour or so before processing. Also, a major concern in harvesting is using techniques that discourage the proliferation of bacteria, again linking it with processing. However, harvesting is commonly performed by the same field hands that perform the other farming tasks. Harvest supervision is also performed by field managers. Therefore harvesting usually falls under the purview of farm operations.

Harvesting is one of several critical steps in the production of high quality aloe gel. Put another way, improper harvesting is the start of production of poor quality gel. We have taught that a quality control person should always be present during harvesting to make sure two essential steps are observed. First, leaves with tip necrosis should not be harvested for gel production. The necrotic areas of the leaf are where commensal organisms often proliferate. An attempt, during preliminary processing, to trim such areas out, merely contaminates the good gel and the entire processing area. Second, leaves should be treated gently and should be harvested in such a way as to keep the base of the leaf sealed. Puncture of the leaves facilitates entry of bacteria into the gel parenchyma. It is amazing how fast some of these subcuticular organisms can grow when introduced into gel. Enforcing these sanitary precautions will not make quality control popular but control of bacterial proliferation must start in the field.

The message is that proper processing of *A. barbadensis* gel requires a sophisticated farm manager. This person needs to understand soil chemistry. The farm manager must monitor soil nitrogen and pH and adjust soil treatment accordingly. Treatment of the soil can be with conventional sources of nitrogen, or manure in the case of 'organic' fields. We believe that the future of aloes in the lower RGV will utilize drip irrigation and perhaps poly-film weed control. A prudent manager should be prepared to handle these complexities. Finally, the farm manager should eagerly collaborate with Quality Control to enforce the best harvesting methods. Quality aloe gel requires quality agronomy.

Preliminary processing – cleaning, extracting and sanitization

The truck with baskets of aloe leaves or the tractor-drawn trailer with wire 'aloe cages' has just pulled up in front of the plant or farm processing station. The area is full of dust or mud and the leaves are coated with dirt mixed in with exuded aloe sap which is beginning to turn purple from the combination of exposure to air and sunlight. From the other end of the building must come a juice that is essentially sterile. That is the challenge. Failure means that the product will have problems with 'break-through' microbial growth, a cosmetic that turns black, or a biologically inactive product. Preliminary processing is where the battle over aloe quality begins.

Leaf washing

Preliminary processing begins with sanitizing the outer surface of the freshly harvested leaf (Feedstock A in Figure 8.4) and proceeds through removal of the outer rind (Processes D or E) and expression of the gel fillet through a sieving device (Process G) to produce the gel (Product I). Washing begins with freshly harvested leaves. Ideally, leaves can be harvested and washed within two to four hours. However, this requires small batches (less than a ton) and quick transport from the field to either the field processing station or the integrated processing plant. In controlled studies (see Figure 11, H) we have found that if 24 hours elapse between harvesting and washing some biological activity is lost. Delaying washing after harvesting while preserving freshness requires meticulous planning and attention. Leaves must not be bruised and refrigeration should be immediate. Our recommendation is to keep the interval between harvesting leaves and washing and further processing to an hour or less.

Washing can be done in a number of ways (Figure 8.4, Process B). The thoroughness of washing will ideally be a function of how dirty the leaves are. If conditions are chronically muddy, more thorough washing will be required. Similarly, if downstream contamination of gel with soil-associated organisms is a problem, washing should be

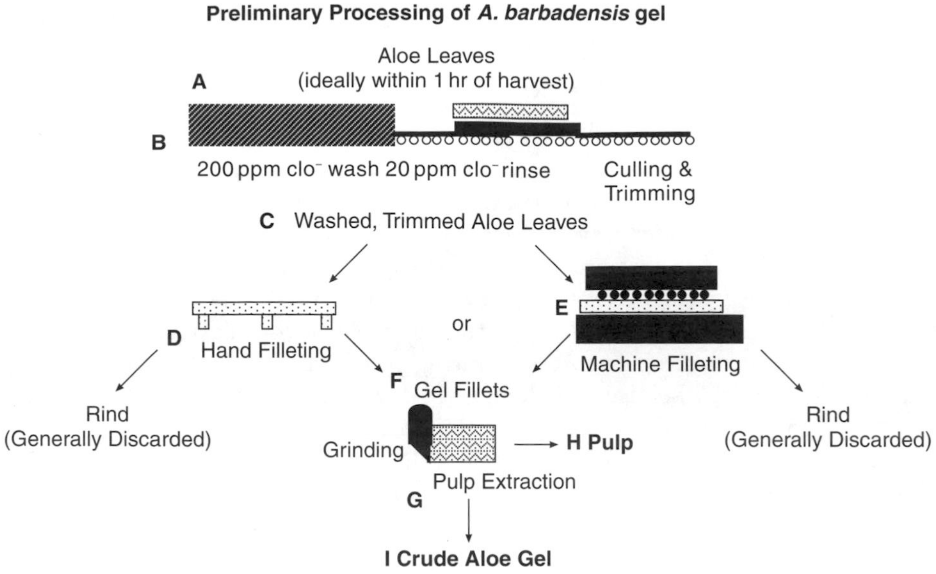

Preliminary Processing of *A. barbadensis* gel

Aloe Leaves
(ideally within 1 hr of harvest)

200 ppm clo⁻ wash 20 ppm clo⁻ rinse   Culling & Trimming

C   Washed, Trimmed Aloe Leaves

or

E

D   Hand Filleting   Machine Filleting

F   Gel Fillets

Rind
(Generally Discarded)   Rind
(Generally Discarded)

Grinding   →   **H Pulp**

Pulp Extraction

G

I Crude Aloe Gel

Figure 8.4 The steps in the initial processing of aloe gel. A: Harvest of leaves and transport to preliminary processing station or integrated processing plant. B: Leaf washing including initial wash in sodium hypochlorite solution (200 ppm), rinsing in 20 ppm hypochlorite solution, trimming off butts and tips, and culling of diseased or damaged leaves. This produces (C), washed and trimmed leaves. Next, the rind is removed, either by hand (D) or mechanically (E), producing gel fillets (F). These fillets are then lightly ground and cellulosic fibers removed by passage through stainless steel screen with 0.25 mm openings (G). This depulping separates pulp (H) from crude gel (I).

oriented toward removing organisms from the outside of the cuticular surface. The schematic diagram above shows the usual sequence of a wash tank with a 200 ppm hypochlorite sanitizing solution, or other suitable sanitizing agent. This feeds a rinsing line via a steel-link conveyor belt. The rinsing portion of the conveyor belt further washes the leaves with a 20 ppm hypochlorite spray. This spraying is followed by culling.

Conditions where mud is a problem can be addressed by increasing the size of the washing tank. A soft, brush scrubber can also be added to the rinsing line to aid in soil removal. However, often the best solution is to add a non-sanitizing pre-wash tank of about half the capacity of the main sanitizing wash tank. This primarily removes mud or adherent sand, sparing the wash tank for its sanitizing function. Although 200 ppm hypochlorite might seem like a lot of oxidative power, it must be remembered that the humic matter in the soil washed from the leaves constitutes a significant input of reducing material. If there is significant mud adherent to the leaves, this hypochlorite oxidizer will quickly be consumed. Thus, under muddy conditions, a prewash tank avoids the necessity of frequent additions of hypochlorite. For monitoring the oxidizing power of the wash tank, a chlorine test kit, designed for use with swimming pools, offers a low cost method to monitor the hypochlorite adequacy of the wash tank.

Downstream from the rinsing assemblage (Figure 8.4, Process B) should be an area where damaged or diseased leaves can be culled out. Culling should be performed by a different crew than that responsible for filleting. This is because the production pressure on the filleting crew is to maximize throughput. If they are also responsible for a step crucial to maintaining quality, they will be pressured to discard fewer leaves in order to maintain speed and increase the amount of product. It is best to have separate groups of workers perform antithetical functions, culling with a goal of maximizing leaf quality versus filleting with a goal of maximizing production of gel fillets. Asking a single group to perform functions with conflicting goals usually results in a compromise wherein quality is sacrificed for throughput.

The product of this process is 'washed leaf' (Figure 8.4, Product C). The overwhelming majority of washed leaves are further processed by either filleting into gel or grinding into whole leaf extract. However, within the last decade, a market has evolved for washed leaves. These are vended for approximately $2.00 per lb (US $4.00/kg) in the fresh, refrigerated state. At first these leaves were sold only in Hispanic specialty stores but at the time of writing they were being increasingly found in major chain grocery stores in the larger cities in Texas and the Northeast USA.

Filleting and depulping

The next step in processing is the production of the gel fillet (Figure 8.4, Product F). The RGV variety of *A. barbadensis* is distinguished from the other cultivars of the species by its large leaves. These have been known to approach 1,300 grams in locations where rainfall is abundant (200 cm per annum), nitrogen is sufficient (0.50% +) and the soil is well drained. These leaves are large enough to be efficiently stripped of their rind to produce gel fillets[1]. In most of the other species of *Aloe*, anthraquinone, chromones, *etc.* are harvested by:-

(i)   incising the leaves and collecting the latex that exudes (as with the South African aloes);

(ii)   harvesting the entire plant including roots, drying it, and either enclosing the ground, whole plant in a capsule or extracting the powder (in the Orient, especially *A. arborescens*);

(iii)   grinding the freshly harvested leaves and processing them to extract the 'gel'[1] (especially with *A. barbadensis or A. arborescens*), also called Whole Leaf Extract (WLE); and,

(iv)   grinding and extracting the rind (a byproduct of fillet production in *A. barbadensis*).

However, it is only *A. barbadensis* leaves that are large enough to produce gel fillet economically. Anatomically, the leaf of *A. barbadensis* can be divided into three zones: the rind, the mesophyll, and the gel[1] (Figure 8.5).

---

1 The use of the term 'gel' is a matter of contention between the aloe industry and the scientific community. Properly described, the scientific community reserves the use of the term gel to those materials with pseudoplasticity. Generally, these materials have significant viscosity in aqueous solution in the range of 0.05 to 0.5 g per 100 ml (0.05 to 0.5% total solids). The IASC defines 'aloe gel' as a water-white extract of *A. barbadensis* without regard to pseudoplasticity. The FDA recognizes the term 'gel' to the meaning usually employed by the scientific community rather than that promulgated by the IASC. We will use the term gel to connote the usual scientific meaning and 'gel' to connote the IASC meaning.

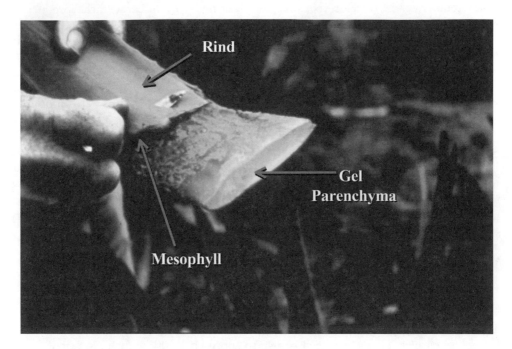

Figure 8.5 Sectioned leaf of the yellow-flowered *A. barbadensis* Miller (cv. RGV). The rind (R) and attached mesophyll (M) are readily stripped away form the gel parenchyma (G). This photograph was taken of leaves at the original 'Hilltop' plantation in Lyford, Texas (*see Colour Plate 5*).

The rind contains multiple layers (see also Sheet 1, Figure 1 in McAnalley '935 patent, 1988). First is the outermost waxy cuticle, which acts as one of the barriers against moisture loss. Just beneath the waxy cuticle lies a region wherein reside the aloe-associated bacteria. Below this level is the chlorophyll-rich rind region where the bulk of photosynthesis occurs. The rind is rich in oxalic acid. Just below the rind and stripping away with it lies the mesophyll. This contains the xylem and phloem vascular bundles. The mesophyll contains the plant's highest concentration of anthraquinones and chromones. We presume that these secondary metabolites function as anti-feedant compounds. When the plant is well hydrated the mesophyll can be easily stripped away with the rind from the gel fillet if attention is paid to detail. Thus the 'rind' fraction (Figure 8.4, 'R' & 'M'), when meticulously prepared, contains both rind and mesophyl elements. This material is second only to the exudate as a source of biologically rich anthraquinones and chromones. However, unlike this exudate, rind is generally discarded as a by-product and is thus low in cost. Furthermore, if properly collected, without excessive exposure to oxygen, light and heat, isolated rind by-product, with mesophyll can be extracted without the tetrahydroxyanthraquinones being converted to red oxidation products or brown/black polyphenolic polymers.

The gel is in the inner parenchyma portion of the *A. barbadensis* leaf (Figure 8.5, G). When properly prepared it contains two elements. One is the liquid portion of the gel (Figure 8.4, Product I, defined by the IASC and ARF parameters) and the other is the cellulose-rich, fibrous, pulp (Figure 8.4, Product H). The ultimate test of a filleting system is its ability to prepare a gel fillet low in mesophyll anthraquinones.

Filleting is accomplished by one of two methods: manual removal of the rind with a knife (Figure 8.4, Process D) and filleting by machine (Figure 4, Process E). In either case, the tip of the leaf is first removed, the butt is trimmed off and the sides of the leaf are trimmed. This process is nicely illustrated by Sheet 1, Figure 2 in the McAnalley '935 patent (1988). Tip removal is usually accomplished using a knife at the culling table (Figure 8.4, Process B) although not usually to the extent that McAnnelley shows. The butt can either be trimmed with a knife at the culling table or by wire at the filleting table. Manual filleting generally takes place on a stainless steel table approximately 1 m wide with raised edges approximately 10 cm high. The length of the table varies depending on how many workstations are desired. Generally each worker, who stand on either side of the table, requires at least a meter of workspace. Table length tends to be in multiples of 2 m accommodating either four, six or eight workers. The flow of cleaned leaves is generally on the head of the table where the rind is removed and discarded to the side. Fillets with exuded pseudoplastic gel are removed from the foot of the table. Trimming and filleting tables are often equipped with stainless steel wire, set up on pegs about 1 cm above the surface. These provide a cutting edge for removal of the butt and trimming of the sides.

After trimming of the tip, butt and sides, the upper and lower surfaces of the leaf are removed. The rind on the flat side of the leaf is then often removed with a knife, discarding the rind. The gel is then scrapped or scooped with the knife away from the rind on the rounder side of the leaf. We have also seen the opposite sequence employed with equal success, discarding the rinds aside and passing the gel fillet down the table.

Mechanical filleting proceeds via a diametrical mechanism (Thompson, 1983). The leaf is placed tip first onto a conveyor belt, which rotates the leaf through 90°. Thus the leaf is carried into the frame of the machine with the broad axis of the leaf at a right angle to the horizontal. A vertical knife then bisects the leaf, splitting it along the broad axis. Almost simultaneously, a set of rollers presses the rind side of the leaf, firmly expressing the gel. The set and tension of the rollers determine how much of the mesophyll is expressed together with the gel. If roller pressure is too high, then the gel will be contaminated with the anthraquinones. If roller pressure is too low, then gel will be discarded with rind and mesophyll. If the reader has difficulty envisioning this system, the drawings of Thompson's '942 patent (1983) should be consulted.

The capacity and throughput of both manual and mechanical filleting is highly dependent on the size and quality of the leaves entering the system. In both systems the yield of gel increases as a power function of leaf size. There is virtually no lower limit to leaf sizes that can be manually filleted. Even the leaves below 100 g from an ornamental *A. barbadensis* houseplant can be hand filleted and are so by millions of households for the treatment of minor burns. However, when leaves less than 250 g are manually filleted, the yield is usually only about 33% (w/w) gel and therefore the output per hour of gel is very low. The smallest leaves that can practically be filleted by machine are approximately 150 g. At this size only about 25% of the wet weight of the leaf can be recovered as gel yielding c.37 g/leaf. Since machine output is determined by leaf per minute throughput, the weight output of gel fillets per minute with small leaves is obviously very poor. For 500 g leaves the recovery rate rises to about 50–55% of leaf weight, yielding 250–275 g of gel per leaf. For 1,000 g leaves the yield of gel approaches

75–80% or 750 g per leaf. Since it takes essentially no more time to fillet a 200 g leaf than it does to fillet a 1 kg leaf there is a disproportionate increase in gel yield with leaf weight. Therefore over the range of leaf weights of 300 to 1,000 g, a three-fold difference in weight, there is a six-fold increase in output (105 g/leaf to 650 g/leaf). Thus, a hidden benefit of proper agronomy is not only a higher gross weight yield per acre but a higher proportionate yield of gel and higher output on the filleting line.

Regardless of whether manual or mechanical filleting is performed, the equipment should be constructed of food-grade, high-quality stainless steel (Grade 316 preferable, 304 is acceptable with a polish finish) with smooth, polished welds. The best guide to filleting machines remains Thompson's 1983 U.S. Patent. Cottrell (1984) advises further modifications, the value of which are uncertain. Filleting machine plans are shown in Sheet 2 of the 1988 McAnalley '95 patent as exemplifying the prior art. As far as we know no one has ever manufactured these machines as their practicality is unclear. The Thompson Company is no longer in business but similar machines are currently being manufactured by Coastal Conveyor (Harlingen, Texas) and International Purchasing and Manufacturing (Harlingen, Texas). Needless to say, despite certain general principles described above, every filleting installation is slightly different depending on the desired throughput, available space and desired capital investment.

The advantages and disadvantages of manual versus mechanical filleting are endlessly debated and will not be resolved in this chapter. Suffice to say, carefully conducted hand filleting produces gel with lower levels of aloin, a quantitative measure of rind and mesophyll contamination, than the best machine fillets. On the other hand, we have seen machine filleted gel of extremely high quality. The most important factor is, 'what does the customer desire and what are they willing to pay?' Filleting technique becomes important only when the producer has mastered agronomy and sanitation, and few producers have mastered these basics.

The filleting process potentially yields two products, gel fillet and rind. Just as there is a small commercial market for whole, washed leaves there is a small commercial market for whole gel fillets (Product F in Figures 8.3 and 8.4). Generally, the fillets are packed into 55 gallon (200 liter) drums, preservatives are added to retard spoilage, and the drums are shipped refrigerated to cosmetic manufacturers who desire minimally modified aloe. Theoretically, the rind could be a rich source for extracting anthraquinones and chromones. However, to our knowledge, very few plantations systematically utilize rind as a feedstock; most rind ends up dumped on compost heaps.

The next and last step in preliminary processing involves removing the cellulosic fibers from the gel fillet. This is accomplished by very coarsely chopping the fillet as with an industrial grade garbage disposal. The coarsely chopped fillet is then passed through a depulper of the type employed in the citrus industry. The aloe industry almost exclusively uses depulpers manufactured by the FMC Corporation (San Jose, California), either the PF 200, MCF 200 or UCF-200A models. The 200 stands for the nominal pore size (the approximate fiber exclusion size in microns). Particles smaller than 200 μm will readily pass through the final screen of the depulper. Furthermore, this sieve size allows the passage of the pseudoplastic gel without shearing the high molecular weight strands and reducing their viscosity. As pore size is decreased, it is more difficult to pass gels through the sieve because of their long-chain molecules. This effect is independent of viscosity but is highly dependent on flow rate and in fact defines the term 'gel'. These effects will be discussed in more detail as we discuss filtration. Removal of fiber yields the crude gel product (Figure 8.4, Product I).

With crude gel, as with the gel fillet (Figure 8.4, Product F), material that is essentially an intermediate can constitute a commercial product. The desirability of such a product is understandable since crude 'aloe gel' is close to the native gel. A small amount of crude gel is sold as such, without removal of bacteria by heat or filtration. Preservatives such as sulfite or benzoate are added in an attempt to prevent bacterial proliferation. Agents such as sorbate, citrate and ascorbate are added in an attempt to prevent oxidation. Unfortunately, these two problems are not easily solved without understanding and attacking the roots of the problems, which are bacteria and anthraquinones. Spoilage severely limits the usefulness of crude gel.

In our scientific studies we have extensively examined crude gel. We solved the problem of stabilization by freeze-drying directly from the crude gel – ARF **Process A**. This requires monopolizing a $650,000 machine for 60 hours to stabilize 100 liters of crude gel worth about $300 on the wholesale market – not a good return on investment. Thus, in our publications we continually emphasize that although ARF **Process A** materials are critical to our understanding of aloe gel in the native state, it is not an economically feasible product. Before proceeding on to Intermediary Processing, we will discuss a second processing pathway from leaf to extract – grinding the entire leaf rather than peeling it.

Slicing, grinding and whole leaf aloe extracts (WLE)

The tradition of preparing gel fillets of aloe leaves goes back approximately 60 years to the early days of the Hilltop Gardens where *A. barbadensis* (cv. RGV) appears to have first been commercially grown. Sometime within the last 30 years and certainly prior to 1989, a second method of processing was developed by grinding the whole leaf followed by extensive treatment with activated charcoal to remove the anthraquinones. This material, termed 'Whole Leaf Extract' or 'Whole Leaf Aloe,' is significantly cheaper than gel fillet for two reasons. First, the labor-intensive and time consuming filleting process is avoided. Second the limitations of gel recovery from small leaves is obviated. By grinding fresh leaves, 'gel' can be efficiently harvested from leaves as small as 100 to 200 g. As the gel fillet seems to have arisen from the RGV Hispanic tradition of filleting out the gel to apply to the skin or put into a tea, so the WLE extract seems to be most closely related to the Oriental tradition of drying the whole aloe plant, especially *A. arborescens*, grinding it into a powder and ingesting the powder as a tonic to enhance vigor.

As alluded to above, although the whole leaf process was adapted to industrial practice long before Coats applied for the predecessor to his '811 patent in 1992, his description of the basic WLE process is apt. In *'Detailed Description of the Preferred Embodiment,'* Coats lays out how the leaves are sanitized and then sliced and ground. His description of slicing and shearing 1000 lbs (about half a ton) of leaves in a 12 inch (c.30 cm) Fritz Mill within a minute or two is in accordance with what we have seen in the industry. We and others call this coarsely chopped material 'guacamole' by analogy with the peeled, deseeded, mashed pulp of avocado. Like real guacamole, the chopped whole aloe leaf will 'spoil' rapidly after preparation with exposure to air unless bacterial proliferation is retarded and oxidation prevented. A very limited amount of *A. barbadensis* gel is sold in this form, using various preservatives and anti-oxidants to retard spoilage. However, if the difficulties of stabilizing crude aloe gel from fillets are significant, the difficulties in preventing spoilage of WLE with preservatives alone are much more extreme. As a consequence the market for 'guacamole' is miniscule.

The next stage of preliminary processing of WLE converts the 'guacamole' to 'gel.' The primary process involved in this is termed 'depulping.' However, this is not as simple a process as described above because the 'guacamole' of fresh, raw WLE has a fiber content several hundred-fold greater than the fiber content of crude gel fillet. As a result it is customary to reduce viscosity and fiber by 'mixing the whole leaf with a cellulose-dissolving compound' (Coats, 1994). Coats recommends the use of Cellulase 4000, an excellent partially purified cellulase from *Trichoderma reesei* at a dose of 20 g per 55 gallons (c.215 liters) of crude WLE. Although Coats does not specify the time and temperature for this process, the 'guacamole' is usually held with stirring at ambient (23–35 °C) until pseudoplasticity is broken (1–2 hours). This treatment certainly reduces the fiber content and viscosity of the WLE and allows it to easily pass through the depulper. However, there are other consequences of this cellulase treatment, as we shall later see. The product of depulping is a yellow liquid with a solids content of 1–2 g per dl and a viscosity little greater than water.

The above example illustrates the advantages of the WLE process – speed and economy. One or two minutes of grinding WLE can be compared with the hour or two it takes six workers to manually fillet a half ton of the same material. Machine filleting takes 15 to 25 minutes to process half a ton, depending on the size of the leaves and the design of the machine. WLE production requires neither the capital investment of the Thompson filleting machine nor the labor expense of hand filleting. The hour or two that the 'guacamole' sits and stirs with the cellulase is the only apparent penalty paid.

The limitations of the WLE process are subtler. The higher bacterial content of WLE means that the methods of removing bacteria must be a hundred to a thousand times harsher than those employed for high quality gel from fillets. WLE is rich in oxalic acid, which is more or less absent from the gel parenchyma and WLE is ten to one hundred times richer in anthraquinones than gel from fillet. This means that anthraquinone removal is mandatory. Lastly, the effects of cellulase are far more profound than Coats ever envisioned. By the time processing of WLE is completed, its polysaccharide has been destroyed and it can no longer rightfully be accorded the term gel.

INTERMEDIATE PROCESSING – BACTERIA AND SPOILAGE

Almost a decade ago, we observed the following pattern during studies of the biological activity of **Process A** crude aloe gel (Strickhand, Waller and Pelley, unpublished results).

Table 8.5 Relationship between the bacterial content of fresh *Aloe barbadensis* gel and its ability to protect the skin immune system from UVB-induced suppression.

Aloe lot	Form	Bacterial content # of organisms/gram of gel	Protective activity
1	**Process A Gel**	<4,000	42–100%
2	**Process A Gel**	10,000	50–100%
3	**Process A Gel**	>100,000	None

This demonstrates that biologically-active aloe material cannot be reliably reproduced unless bacterial proliferation can be controlled prior to processing. Bacteria can only be controlled with an understanding of them and the environment in which they live. Therefore, this section is divided into two parts, the microbiology of aloes and how to control the micro-organisms associated with aloes.

The Microbiology of *Aloe barbadensis*

We should be cognizant of the two following principles:

Endogenous Organisms – In addition to the gel, aloe leaf contains three parts: (i) the exterior surface of the leaf; (ii) the outer (subcuticular) portion of the rind; and, (iii) 'brown tips' and necrotic portions of leaf. Each of these has its own unique bacterial flora well adapted to grow at the particular temperature and pH of that part of the leaf and using the nutrients of that part of the leaf.

Exogenous Organisms – These can be picked up from the environment during processing. Potentially these are of two types: (i) processing plant environmental flora, predominantly micrococci but occasionally fungi; and, (ii) human commensals. The processing plant environment when improperly maintained and sanitized is a rich source of organisms that are able to grow in aloe products. Human commensals are poorly suited for proliferating in native aloes, probably because crude aloe gel has a mild bacteriostatic effect on these organisms.

Everything that follows is derived from these two principles. They apply equally to feed stocks for cosmetics and drinks. However, when manufacturing aloe drinks, a much more detailed understanding of the bacteriology of aloes is essential. This is because the range of bacteriostatic agents available to the beverage industry is greatly restricted compared to what can be used in personal care products. Furthermore, drinks usually contain monosaccharides and/or disaccharides added as sweetners which can offer a rich media in which bacteria may grow. Due to the nature of the process for manufacturing beverages, contamination of even a single barrel (215 liters) of aloe gel could result in spoilage of an entire production run of 4,000 liters of drinks.

Table 8.6 illustrates the range of bacteria found in native **Process A** aloe gel fillets of the highest quality, prior to any processing.

All of these organisms have different growth characteristics and varying sensitivity to heat, sanitizing agents and preservatives. The problems that result will be a combined function of the organism and processing stream breaches in good manufacturing procedures. Therefore, we will examine the major groups of organisms individually.

Specific aloe-associated organisms and their industrial importance – *Micrococcus sp.*

Biochemically, this organism type is known as, in computer biochemical screening, *Streptococcus morbillorum* or *Enterococcus faecium*. Morphologically, they are large Gram +ve coccobacilli in chains. Selected growth traits indicate that these are not *Enterococcus* or *Streptococcus* but probably represent a new species of *Micrococcus* commensal to *Aloe*. Their exact taxonomic status is not of industrial importance. What is important is the realization that certain bacteria have evolved so that they grow well in *Aloe* and poorly

Table 8.6 Organisms on or in *Aloe barbadensis* gel.

Class of organism	Organism	Frequency of isolation[1]	Typical numbers
Gram +ve Cocci[2]	*Streptococcus morbillorum*[3]	5/9	1,000–30,000 CFU/ml
	Enterococcus faecium[3]	3/9	10–55,000 CFU/ml
	Micrococcus species[3]	3/9	500–1,500 CFU/ml
	Staphylococcus hominis	1/9	1,400 CFU/ml
	Streptococcus mitis	1/9	300 CFU/ml
	Staphylococcus auriculoris	1/9	150 CFU/ml
	Staphylococcus cohnii	1/9	20 CFU/ml
	Staphylococcus lentis	1/9	10 CFU/ml
	Staphylococcus carnosus	1/9	50 CFU/ml
Gram −ve Rods[2]	*Enterobacter* species	6/9	10–2,400 CFU/ml
	Klebsiella species	3/9	70–1,800 CFU/ml
	Serratia species	2/9	40–500 CFU/ml
	Cedecea species	3/9	10–60 CFU/ml
Gram +ve Rods[4]	*Bacillus* species	7/9	10–100 CFU/ml
	Diptheroids	1/9	20 CFU/ml

Notes

1 The frequency with which this organism was identified in nine ARF preparations.

2 These organisms were observed only in gel and were not isolated from the surface of the plant.

3 These organisms were identified by biochemical tests. The biochemical tests employed and the computer algorithms used for speciation were developed for the identification of human pathogens. Therefore they are not optimized for identification of plant commensals. All three of these organisms probably belong to a single genus (*Micrococcus*) and may represent different novel species or biotypes of the same species.

4 Concentration of these organisms refer to the gel. These organisms were observed primarily on the surface of the leaf and their presence in gel probably reflects incomplete sanitization.

elsewhere. They are called micrococci not because the organism is small (they are in fact rather large bacteria), but because their colonies are very small on most agar media and can easily be missed if they are not carefully looked for. In particular, when non-selective media such as Nutrient Agar or Plate Count Agar are employed, swarming or heavily encapsulated organisms can overgrow the plate, obscuring micrococci that may be a thousand-fold more numerous. Micrococci are associated with the rind and probably reside just below the waxy cuticle. Oxidative sanitization (e.g. hypochlorite) of the outer leaf surface, which efficiently removes soil organisms like *Bacillus* and *Diptheroides*, does not effectively remove these micrococci. These organisms are the ones most frequently cultured from native high quality aloe gel and they are the organisms present in the highest numbers. When bacteriological breakthrough occurs, these organisms are usually the ones isolated.

Al Davis (now deceased) of Aloe Vera of America was the first to realize that the aloe-associated micrococci were acidophils. He noted that these organisms grow poorly on Nutrient Agar (Difco, 1986; isotonic, pH, 7.3, and no added glucose), which is designed for the isolation of human pathogens, particularly from blood, ascitic fluid or effusions. Plate Count Agar, designed for enumeration of organisms in water, food and milk, has a relatively high pH, (7.0) and is low in glucose (1 g/liter) and salts and does not favor the growth of these organisms either. While exploring other media, Davis noted that aloe-associated organisms grew well in lactobacilli media (20 g of glucose per liter, pH 6.5). He erroneously assumed the organisms were lactobacilli, which was not surprising since his laboratory did not have facilities for biochemical typing.

Later we came to the same conclusion when analyzing bacterial breakthrough in an aloe processing plant. We isolated a set of organisms in enormous numbers (10^8–10^9 per ml) using Columbia CNA agar, a blood agar that retards, by colistin and nalidixic acid, the growth of organisms other than Gram +ve cocci. On CNA the organisms produced small colonies with variable, gamma or weak alpha hemolysis. Morphologically, the organisms were fairly large coccobacilli that grew in chains. Under most conditions, *Micrococcus* sp. did not ferment glucose with evolution of carbon dioxide. Instead they produced lactate which could be detected using the same amino-column HPLC system used to quantitate malate. Once these organisms consumed the available glucose in aloe gel they preferentially used the organic acids, especially malate. The optimum temperature for growth of these organisms was consistent with the ground surface ambient temperature in the RGV, c.35–45 °C. These *Micrococcus* sp. were not killed at 50 °C and the kill curve at 65 °C was significantly lower than for the aloe-associated Gram −ve rods, such as *Cedecea* and *Enterobacter*.

At the ambient temperatures of a processing plant without air conditioning (c. 35–45 °C), *Micrococcus* sp. double their numbers in about 25 minutes. Proliferation in aloe gel can start at 10,000 CFU/ml and increase to more than 10^8 in 24 hours. The growth of these organisms is not inhibited by citrate or sorbate and is only slowed by benzoate. The best preservative to minimize the growth of this organism is sulfite and even sulfite is not completely effective. Unfortunately, the use of sulfite is increasingly restricted in beverages. The presence of micrococci, even in moderate numbers (10^5–10^6) in aloe, has relatively little organoleptic effect.

The detection of the aloe-associated micrococci is highly dependent on proper bacteriology, although the absence of malate and presence of lactate in HPLC profiles of aloes may alert us to the possible presence of micrococci. Not surprisingly, the best liquid media for the propagation of these organisms is autoclaved aloe gel. Columbia colistin, nalidixic acid blood agar, is an expensive medium, generally better purchased in the form of poured plates. Alternative bacteriological media for detection of micrococci, such as KF Enterococcus agar for fecal streptococci in water and Bile Esculin Agar for Group D streptococci, do not support the growth of all aloe-associated *Micrococci* sp. Two media, developed for the food and beverage industry, show preliminary promise (Table 8.7). The first is a modification of Bacto Orange Serum Medium (Difco, 1986), normally used for the detection of lactobacilli and other aciduric organisms causing spoilage of citrus products. The pH of this medium (pH 5.5) more closely approaches the pH of aloe gel (pH 4.5) than any other bacteriological medium and the glucose content (4 g/l) of Orange Agar is congruent with the reducing sugar content of aloe gel (1.86 g/l). It should be noted that the pH of this medium is low enough and there is sufficient glucose to support the growth of fungi. Aloe gel is added to this media to supply any specialized factors necessary for the growth of aloe-associated *Micrococcus* sp.

A second medium, Universal Beer Agar (UBA), has also shown initial promise for identification of these organisms. This medium has proved useful for the isolation of lactobacilli, pediococci, and *Acetobacter* from beer. Integral to the design of UBA is the use of the beverage tested as part of the agar, a principle first employed in milk bacteriology. This principle prompted us to incorporate aloe into our diagnostic media. UBA has a significantly higher pH than Orange agar and a four-fold higher glucose content. At present, it is not clear whether these two agars support growth of a different spectrum of organisms. One problem with these agars is that the low pH of the media and the high content of ions causes significant softening of the agar. This softening becomes

Table 8.7 Modified Agars for Detection of *Micrococcus* sp. in *Aloe barbadensis* materials.

	Orange Serum[1] Agar	Universal Beer[2] Agar
Yeast Extract	3 g	6.1 g
Tryptone	10 g	–
Milk Peptone	–	15 g
Glucose	4 g	16.1 g
Dipotassium Phosphate	2.5 g	0.31 g
Monopotassium Phosphate	–	0.31 g
Salts	–	0.138 g
Orange Serum	Equivalent to 200 ml	–
Dessicated Tomato Juice	–	12.2 g
Agar	17 g	12 g
Water	750 ml	750 ml
pH	5.5	6.3
Aloe barbadensis gel[3]	250 ml	250 ml
Bacto Agar[4] (for pour plates)	13 g	8 g

Notes
1 Based on Difco 0521-01-9.
2 Based on Difco 0856-01-4.
3 Pasteurized and activated charcoal absorbed (ARF Process C or E).
4 Based on Difco 0140-01-0.

particularly severe if the plates are incubated at higher than standard temperatures (35–37 °C) in order to reproduce the temperatures at which aloe-associated bacteria often grow.

There is a long way to go before we can consider this system and the taxonomic status of the organisms established. Are they lactobacilli, streptococci, micrococci or a completely different genus? There are limits to how much morphology and metabolism can tell us about what these organism are, the ultimate answers must come from studies of nucleic acid homology to known organisms. Similarly, our knowledge of these Gram +ve organisms is based on two or three well characterized plantations and processing plants in the RGV. We need to expand this knowledge so that it is more representative of the world-wide distribution of *A. barbadensis*.

The ubiquitous presence and growth characteristics of the aloe-associated micrococci organisms emerge under the selective pressures of a processing plant. The following processing conditions are those wherein micrococci are most likely to cause problems:

(i)   Crude gel is not immediately chilled prior to long distance shipment from plantation to processing plant. Alternatively, leaves are roughly handled during harvesting/washing and then are allowed to sit for 24 hours or more prior to initial processing.

(ii)   Crude gel with a high (>10^8 CFU/ml) bacterial count is processed thereby contaminating processing equipment. Subsequent cleaning and sanitization is less than optimal and micrococci proliferate in the processing plant environment. These micrococci then contaminate the gel being processed.

(iii) During preprocessing of fresh whole leaf, ground homogenates are incubated for more than one hour at ambient temperatures.

Once micrococcal contamination is established in processing equipment, it is difficult to eradicate without complete breakdown and sanitization of the equipment, which can shut a plant down for many days. If finished feedstock product is produced with more than 10^3 micrococci/ml, most beverage preservative systems cannot inhibit their growth. Gel will be received by the consumer product manufacturer with more than 10^6 CFU/ml and upon physical inspection there are few overt signs of contamination. Thus the possibility of micrococcal contamination necessitates rigorous monitoring.

Other aloe-associated organisms

Other gram +ve cocci

Authentic staphylococci and streptococci are isolated inconsistently from *A. barbadensis* gel and, when they are present, they are few in number (Table 8.6). We have not observed them to be present in a contaminated commercial product. Staphylococci and streptococci commensal to humans do not grow well in good quality aloe gel. We have observed, over the years, a great deal of effort being expended to prevent the contamination of aloe with human organisms and orientation of bacteriological surveillance toward human-associated bacteria. These observations took place in processing plants with massive environmental contamination of aloe-associated bacteria, followed by the subsequent contamination of aloe products. The aloe-associated organisms were overlooked because the focus was on human pathogens. We do not wish to see massive outbreaks of human pathogens and we strongly agree with the public health regulations designed to keep human pathogens out of commercial materials, but perhaps the importance of these organisms has been overestimated.

Gram −ve rods

Enterobacter, Klebsiella, Serratia and *Cedecea* are frequently isolated from good quality *A. barbadensis* gel, albeit in lower numbers than the micrococci. They grow quite well on Nutrient Agar or Plate Count Agar. Many *Klebsiella* and *Cedecea* make copious amounts of capsular polysaccharide which allows easily identifiable colonies. The aloe-associated Gram −ve organisms have a significantly high growth rate, with a generation time in the range of two hours in a variety of liquid media at reduced temperature, (4–10 °C) although they also grow well at usual incubator temperature (35 °C). Their growth rate in aloe gel is significant, although not as spectacular as that of the micrococci. The optimal media for culturing the Gram −ve rods is McConkey's EMB agar, but an estimate of their number can be got by comparing the difference in the count between Nutrient Agar and a specialty agar like CNA or Lactobacillus Agar. Many of these organisms ferment glucose with carbon dioxide production. Therefore, they are the organisms that cause foul smelling, bulging drums of aloe feedstocks. Those organisms that produce capsular polysaccharides result in aloe that is 'ropey', 'slimy' or 'soapy'. Contamination with these four organisms generally results in a feedstock that, upon inspection, is immediately recognized as contaminated.

Considering that the aloe-associated Gram −ve organisms have a significant rate of growth at reduced temperatures, it is not surprising that the following has been observed. Gel is produced which at the plantation has a higher than usual content of these four organisms, say 1,000 organisms per ml. The gel is properly chilled prior to transport. However, transport is delayed for one to two days. Alternatively the processing plant delays intermediate gel processing for up to one week. In all cases the gel is stored at low temperature (4–10 °C). Under these conditions the Gram −ve organisms, even with a generation time as short as two hours, can in 24 to 72 hours, grow to bacterial numbers of 10^{6-7} organisms per ml. Once a processing line has been contaminated with these organisms, the same degree of scrupulous clean-up is required as with *Micrococcus*.

Gram +ve rods

These organisms have rarely been found to cause commercial problems. They are normally not found in gel and their presence indicates failure to properly sanitize leaves prior to pre-processing. Both classes of organisms are highly sensitive to pasteurization and neither grow well in aloe. However, if *Bacillus* sporulates these spores are highly resistant to heat, sanitization, and preservatives and can cause similar problems to the Gram −ve rods in processed material. Contamination of feedstocks with these organisms is easy to detect in pre-production quality control testing, they grow extemely fast on Nutrient Agar, and the incidence of problems is very low.

Fungi, molds and yeasts

As Table 8.6 indicates, molds and yeasts are extremely rare in **Process A** ARF aloe. Unfortunately, the same cannot be said of commercial aloe. Fungi from the environment can and do contaminate aloe. They generally enter the processing stream at two points. The first appears to be on leaves if these are stored for a significant period of time between harvesting and pre-processing, particularly if they are refrigerated. The second point of contamination is upon the packaging of feedstocks if care is not taken to keep the processing environment clean and to ensure that drums are sanitized.

Fungal contamination is generally manifest immediately. Upon opening the drums, colonies of mold are seen floating on the surface of the gel. However, in certain circumstances contamination may not be apparent. Culture of gel upon Sabouraud's agar reveals the problem within three to five days. Proper feedstock quality control at the manufacturing facility ensures that no consumer product should ever be manufactured with fungally-contaminated feedstock. We have to admit that we have seen more than one aloe cosmetic product with mold growing on its surface.

Destruction of bacteria and fungi

The following figure outlines the various processes for the reduction of microorganisms in aloe.

One way of estimating the magnitude of the problem of microbial contamination of aloe is to examine the patent literature on stabilization. There are many patents and if any one of them really worked well there would not be so many of them. Examples of such processes are found below:

When reading both the specifications and the claims in detail, one is struck by a lack of enablement. The examples contain not a single real chemical analysis nor any bacterial assay. In the Maret, Cobble and Coats patents, the difference between oxidative effects on anthraquinones and bacterial spoilage is not made clear. There is an impression of

Table 8.8 Technological Developments in 'Stabilization' of *A. barbadensis* gel.

Inventor date	Reference	Claimed technological state of the art
Farkas	1963	U.S. Patent No. 3,103,466 Pasteurization
Farkas	1967	U.S. Patent No. 3,360,511 Pasteurization
Farkas	1968	U.S. Patent No. 3,362,951 Pasteurization
Maret	1975	U.S. Patent No. 3,878,197 Ultraviolet light, ambient temperature
Cobble	1975	U.S. Patent No. 3,892,853 H_2O_2 + heat + antioxidant
Coats	1979	U.S. Patent No. 4,178,372 Cobble + very minor modifications
McAnnelley	1988	U.S. Patent No. 4,735,935 Sterilization by γ or microwave radiation
Coats	1994	U.S. Patent No. 5,356,811 Oxidation + membrane filtration

a lot of bright but unfocused thinking. The practical result of this is that more than 90% of all *A. barbadensis* material processed is pasteurized.

Differing forms of pasteurization remain at the heart of the intermediate processing of aloe. From here on we will consider bacteria alone because mold-laden aloe should not be allowed into the processing plant. When moldy aloe is encountered, the material should be removed with minimal agitation from the area and the manufacturing environment immediately sanitized. Under no circumstances should an attempt be made to process fungally-contaminated aloe. Even killing the vegetative forms does not destroy the spores.

Batch pasteurization in unsealed kettles is the traditional method for reducing bacterial numbers (Figure 8.5, **Process J**). Usually 100 to 2,000 gallons (400 to 8,000 liters) are fed into a steam-jacketed, electrically-heated or gas-fired kettle and the temperature raised to 65 °C (150 °F). Attaining this temperature can take from 15 to 60 minutes depending on the equipment used in the processing facility. In standard pasteurization the material is then held at 65 °C for 15 minutes. With smaller batches the material is then allowed to cool to near ambient by radiation and/or convection. With larger batches, heat exchangers are usually employed to more rapidly reduce the temperature of the gel to the point where it can be further managed. During these processes the kettle is covered, although no attempt is made to exclude oxygen. Sweep agitation is employed in order to speed both the heating and cooling phases. Often decolorization, activated by adsorbtion on charcoal, is performed after the heat is turned off and cooling begins.

However, it is not often that processors intelligently use bacteriology to set the parameters for pasteurization prior to a processing run and the subsequent bacterial quality control. There is insufficient time between the arrival of the gel at the processing plant, and the initiation of intermediate processing to allow for prospective bacteriology. However, if raw gel is routinely cultured quantitatively and a profile of incoming raw material established, then pasteurization can be adjusted to the average and expected bacterial content of the incoming gel. In reality a temperature of 65 °C is seldom used by the industry because it is rare for commercial raw aloe to have less than 100,000 CFU/ml. Even moderately high concentrations of organisms (10^6/ml) can result in insufficient pasteurization under standard conditions (65 °C, 15 minutes). Many companies therefore employ prolonged pasteurization times and elevated temperatures in unsealed kettles in order to ensure the complete killing of bacteria. The results of this will be discussed in the following section.

The numbers of bacteria in Whole Leaf Extracts are so high (minimum observed in raw extract $\geq 35,000$ CFU/ml, usually 500,000 CFU/ml), that standard pasteurization conditions are likely to fail. This is because heat kills a fixed percentage of organisms per unit time and temperature ('the kill curve'). The kill curves are different for different organisms depending on their heat sensitivity. The 'staph' and 'strep' found on human skin and hair have a so-called 'usual' degree of thermal sensitivity and in fact this is how 'usual' is defined. The micrococci are killed less readily by heat than is 'usual' and the aloe-associated Gram −ve rods are killed more readily. However, if a treatment is 99.99% effective in killing, it will sterilize a solution with 10^3 CFU/ml but it will be ineffective on a solution of 10^6 CFU/ml, as 100 organisms will survive. If an organism is less heat tolerant than the average (e.g. the aloe-associated Gram −ve rods) the above treatment will kill 99.9999% of the heat-sensitive organisms and a solution with 10^5 CFU/ml will be sterilized. On the other hand, a heat tolerant organism (e.g. the aloe-associated micrococci) will only suffer a 99% decrease in numbers and bacterial breakthrough may occur in materials with initial counts of 10^3 CFU/ml. In order to reliably process Whole Leaf Extracts with bacterial counts that can routinely exceed 10^6 heat-tolerant micrococcal CFU/ml, a kill ratio of 99.9999% must be achieved. This cannot be done at 65 °C with 15 minutes treatment.

Due to the limitations alluded to above, larger and more modern processing plants are employing high-temperature/short-time (HTST) pasteurization (Figure 8.6, **Process K**). This sealed but non-pressurized system employs a series of heat exchangers to rapidly-within seconds-raise and then lower the temperature of the liquids. In the regeneration cycle, product entering the HTST unit is initially heated in the first stage of the exchanger by product leaving the HTST. It then passes through a heater section where its temperature is raised to 90–95 °C and then proceeds into a set of holding coils where pasteurization occurs. The duration of pasteurization is determined by the length of the coil and the flow rate. Typical pasteurization times are between one to five minutes. The pasteurized product then re-enters the regenerating heat exchanger

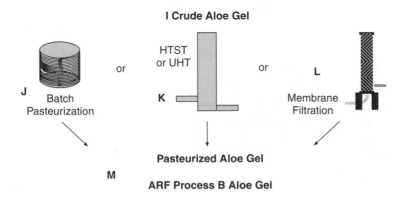

Figure 8.6 Methods for the destruction of microorganisms.

where it is used to raise the temperature of the incoming raw gel to about 50–60 °C while its own temperature decreases to about 35–45 °C. A final heat exchanger cools the leaving gel to the desired temperature with chilled glycol or water. Bacteriologically, the kill efficiency of this HTST process is inherently no better than properly conducted batch processing. However, HTST is capable of handling larger volumes and functions in an automated fashion. HTST can more efficiently reach the high temperatures needed to handle micrococci. However, the caveats inherent in the heavy bacterial loads of the WLE process still apply.

Recently, a modification of HTST has proven successful in the dairy industry. This process is called ultrapasteurization (UHT). In this case, the product stream is not only sealed but is pressurized such that temperatures in excess of 100 °C can be attained. To our knowledge this promising process has not been employed to date on a large scale in the aloe industry.

In the pharmaceutical industry sterilization of liquids by filtration through sub-micron membrane filters has been common for decades, and this process is utilized to a lesser degree by the food and beverage industry. This process (Figure 8.6, **Process L**) has not been widely adopted by the aloe industry, although some small scale production of specialty products employs membrane sterilization. This process is industrially impractical for high quality raw aloe gel because the extreme pseudoplasticity results in unacceptable back pressures. If pseudoplasticity is broken by degradation of the native polysaccharide with enzymes, then aloe liquid can be membrane sterilized, although it is no longer a physical gel. On a laboratory scale, membrane filtration is very commonly used to sterilize aloe. For purposes of HPLC analysis a 25 mm, 0.2 μm, filter can be used to treat the native gel from 1–2 ml of aloe yielding the 100–200 μl of filtrate required for HPLC analysis. However, it is physically very difficult to manually prepare even 10 ml of high quality aloe gel by direct passage through a 0.2 μm filter. The gel must first be 'broken' by passage through multiple filters of decreasing porosity, usually first a glass fiber filter, followed by a 20 μm filter, then 5 μm, then 1.2 μm, prior to final 0.2 μm filtration. Investigators who have not worked with native gel do not appreciate the pseudoplasticity of native gel since commercial aloe lacks this property.

The product

The processes described above yield a very slightly yellow to almost water white liquid. In the case of gel of the highest quality, the material is significantly pseudoplastic. In our studies of the chemistry and biology of *Aloe barbadensis* gel (Strickland and Pelley, Chapter 12), this material, lyophilized without addition of preservatives or further processing and is termed ARF **Process B** material. This material, usually with the addition of preservatives, is marketed as '1:1 Aloe Vera Gel' (IASC, 2001). Ideally when mesophyll contamination is low, pasteurization does not affect the color of aloe gel and commonly pasteurized aloe is marketed without adsorption on activated charcoal. This material is called 'non-decolorized' since it has not been treated (decolorized) with activated carbon. However, because it is difficult to exclude all mesophyll, most non-decolorized aloe gel undergoes some color change after batch pasteurization (see next section). The most commonly employed preservatives to prevent bacterial growth are benzoate, up to 0.1%, and sulfite, up to 0.1% for cosmetics. Sorbate in concentrations up to 0.1% is used to retard the growth of fungi. Anti-oxidants are also added in an attempt to prevent color change. Chief among these are ascorbate, up to 0.1%. It should be noted that sulfite also has anti-oxidant properties. Citrate, up to 0.2%, or other food approved acids, are usually added as a buffer to keep pH in the proper range of less than 4.5, which in itself has a mild bacteriostatic effect. Citrate itself has a slight anti-oxidant effect. In '1:1 Gel' destined for cosmetic use, Germaben II is an excellent anti-microbial preservative. Methyl paraben, propyl paraben, Germall 115 (imazolidinyl urea), either individually or in combination as Germaben II above, are the foundation of cosmetic anti-microbial preservatives.

'Aloe Vera Gel 1:1' was the first product sold by the aloe industry and even today may be the most widely sold feed stock[1]. It is employed in the production of drinks, with the addition of sweeteners and flavors, and in the manufacture of cosmetics. In the Complementary and Alternative Medicine community, it is generally believed to have the best biological activity. Since pasteurization should not theoretically change the composition of matter of a material, the chemical composition of '1:1 Aloe Vera Gel' should be identical to the IASC and ARF standards. In reality the chemistry and bio-logical activity of the commercial materials vary enormously.

The drawbacks of this product are the high cost of shipping a material that is 99% water and the propensity of 'non-decolorized' aloe for color change.

Relationship of bacteriology to overprocessing

It is the firm conclusion of these authors that if aloe contains greater than one million CFU bacteria, it should be discarded and not processed. Table 8.5 illustrates one reason for this opinion – the aloe is likely to have lost biological activity. Furthermore, pro-cessing aloe with massive bacterial contamination is likely to contaminate processing equipment and the processing environment, thereby imperiling future batches. However, many companies will process it anyway, hoping that application of enough heat for a long enough period of time will kill all microorganisms. Consumer product manufacturers need to be able to identify such material so as to avoid incorporating it into their products.

A common procedure is to pasteurize heavily contaminated aloe material at a higher temperature and for a longer period of time than is usual. In batch pasteurization temperatures of 80 °C and times of 30–45 minutes are not uncommon. In HTST, tem-peratures of 95 °C are normally used. When heavy bacterial contamination is suspected, dwell times are increased to 5–19 minutes. If the aloe material is still not sterile after processing, it is run through pasteurization for a second time. This 'reworking' is very often accompanied by additional treatments with activated charcoal in an attempt to remove the results of the caramelization and anthraquinone oxidation that occur during prolonged heating.

The result of this is aloe material that upon organoleptic testing is called 'over pro-cessed.' Despite the use of activated charcoal there is often a yellow-brown color due to products of oxidation that are sufficiently hydrophilic to pass through the charcoal. There may be a tendency for this color to intensify and darken with time. This color

---

1 No precise data on production of the various *A. barbadensis* extracts are available. The IASC makes estimates but admits that these numbers are only guesses. The three largest producers of *A. barbadensis* in the world are privately held and therefore disclose neither profit nor production figures.

Table 8.9 Symptoms and Causes of 'Overprocessed' Aloe.

System	Property/Analyte	Cause
Organoleptic Characteristics	Color	Oxidation of tetrahydroxyanthraquinone to 'red compound' with Subsequent condensation to brown/black polyphenolic material
	Color	Caramelization of glucose and polysaccharide by heat
	Aroma	Loss of aroma due to evaporation of terpenes, etc.
	Taste	Loss of sweet overtones – consumption of glucose by bacteria Production of lactic acid by bacteria
	Taste	'Scorched' taste – oxidation of sugars
Laboratory Findings Glucose	Malic Acid Absent	Absent
	Polysaccharide	Acetylated Glucomannan absent
	Lactic Acid	Present

change problem is particularly common in batch processes rather than HTST. Many of the terpenes and esters that give aloe its distinctive 'woody' aroma have been lost due to evaporation. These are replaced by the odors of oxidized materials reminiscent of burnt sugar. Finally, the taste of the aloe is altered. Although native aloe is somewhat bitter, it also has a sweet overtone, because approximately 10–20% of the solids are glucose. This distinctive taste is lost and the material is very bitter, with the sweetness replaced by with a pronounced 'scorched' taste.

Chemically, overprocessing is associated with changes that we discuss throughout this chapter and in frequent reviews (Pelley *et al.*, 1998; Pelley and Strickland, 2001). These include: (i) a loss of polysaccharide content; (ii) the disappearance of glucose; (iii) the appearance of lactic acid; and, (iv) diminished to absent levels of the organic acid, malate. The normal polysaccharide content of aloe gel is about 6 to 12%, with alcohol precipitable hexose as a percentage of the total solids. 'Over-processing', or mis-processing, generally lowers polysaccharide to 1 to 2% of the total solids due to the uncontrolled activity of endogenous and exogenous β $1 \rightarrow 4$ glucosidases. Specifically, the acetylated glucomannan is hydrolysed and the only polysaccharides that remain are the galactan and pectins (Pelley *et al.*, 1998). This loss of polysaccharide is also observed when Whole Leaf Aloe is exhaustively treated with cellulase during pre-processing to boost the yield of juice. The micrococci that cause most of the bacterial contamination of aloe appear to be very efficient at assimilating malic acid. This results in the loss of the 'E Peak' in the HPLC analysis of aloe.

It is our experience that up to half of the authentic commercial aloe in the market-place displays some of these chemical changes and at least 20% of all products display the complete spectrum of degradative change (Pelley *et al.*, 1998). This does not mean that these materials are not legally, 'authentic' aloe. They have not been adulterated or diluted. However, it means that these 'over processed' materials are not good quality aloe. These materials are simply aloe that has been allowed to rot during processing. This aloe has been 'over-processed', either through a lack of knowledge of proper technique, a failure to adhere to good manufacturing practices, or in order to cover up material that at one time experienced severe bacterial contamination. Proper chemical analysis, which should be done by the feedstock vendor, will inform the knowledgeable consumer product manufacturer of the actual history of the preparation.

Conclusive caveats or how a feedstock user can avoid overprocessed aloe

At the beginning of this chapter we described the production of good quality aloe gel as a conflict between Senior Management, Production and Quality Control. Advantages to Management are sometimes counterbalanced by arguments from Sales and Marketing who will tell the Consumer Product Manufacturer that:

- 'we don't have a quality problem, we haven't had to reject a batch of product in years'
- 'our leaf supplier is so good we don't have to test for bacteria in raw gel'
- 'a bacterial count before pasteurization is not necessary'
- 'sanitation is no problem. Our employees wear hairnets and gloves'
- 'if you have a problem with the product, don't dispose of it, just send it back'
- 'if the final product is clean, the type of original contamination is irrelevant'
- 'glucose and malic acid level tests are not economic'

Ideally, the vendor of an aloe feedstock should be able and willing, as part of the specification sheet, to supply the customer with:

1. **Processing History.** When were the leaves for this batch harvested, how long was transport, how long was preliminary and intermediate processing? Was the product ever reworked?
2. **Bacterial History.** Was microbiology done on the raw gel as well as on the finished product. If bacteria were found, what type of bacteria were they?
3. **Chemical Testing.** What is the glucose, malic acid and lactic acid content of *this* batch?

No aloe supplier is going to give you this willingly. Disposing of batches and laboratory tests cost money. Above all, there is a profound suspicion of science in the aloe industry. The head of one of the two largest aloe companies in the world proudly boasts that he has never spent a dollar on research and development. If we were purchasers of aloe feed stocks we would strongly consider the way we felt about the products we manufacture. If we were proud of our consumer product we would insist upon quality feed stocks and would not put overworked material into our product.

Control of aloe-associated organisms in the processing plant environment

Of all the questions that we are asked, one of the most frequent concerns processing plant sanitation. Since every plant is different we can only make general recommendations. We do, however, have preferences. First of all, we like commercial sanitizing agents that contain quaternary amino compounds (QUATS) with added surfactant. Weekly, this cleaning routine should be interrupted by the use of a different agent to avoid selecting for QUAT-resistant organisms. Our favorite agent for this is 200 ppm sodium hypochlorite solution containing 0.1% detergent. Iodophors are also excellent, though expensive.

Some general principles should be observed. Micro-organisms lodge in obscure places which should therefore be eliminated. Floors should be treated with epoxy resins and walls should be smooth and easily scrubbed down. Stainless steel can be etched by bases which provides a niche for microorganisms. A tidy work area is easier to clean so that heavily contaminated aloe does not lodge; obvious but not always observed. Cleaning in Place (CIP) is effective provided that the aloe material is kept moving. Equipment that is designed for food/dairy/brewing is not designed for handling materials with high bacterial counts (say c.10^8 CFU/ml). Bacteria on contaminated aloes will enter every crack and blind loop not sanitized by CIP. When that happens production time is lost during cleaning. No method of sterilizing aloe will work if downstream equipment is being contaminated with an inoculum of 10^8 organisms/ml.

Intermediate processing – anthraquinones and spoilage

Processing aloe produces color change. At the leaf washing facility, yellow sap exudes from the cut base of the leaf and is washed away. As the sun comes out and the day heats up the yellow exudate turns to red/purple. After pasteurization at a temperature of 65 °C, the vigorously stirred aloe gel turns from light yellow to a 'pretty pink' color in 15 minutes. WLE 'guacamole' fed into the depulper turns to a brown-black. Next to microbiological spoilage, color change is the most perplexing problem with which the production staff is challenged . A decade and a half ago it was thought in the U.S. aloe industry that color change was caused by 'aloin' and that it could be solved by 'decolorizing' using activated charcoal. The reality, of course, is much more complex. Although 'decolorization' works most of the time, its scientific basis for aloe is not completely understood.

The scientific basis of color change in aloe

The U.S. aloe industry has made little or no progress in understanding the chemistry of color change in *A. barbadensis* gel and WLE. The U.S.-dominated aloe vera industry focuses on cosmetic and 'dietary supplement' uses of gel fillets of *A. barbadensis*' which are grown in the sub-tropics of the Americas. European chemists have focussed on the purgative uses of the exudate of plants grown in the African and East Indian tropics. The purgative principle is the *C*-glucoside of aloe emodin anthrone, referred to as barbaloin or in its cruder form as aloin.

Reynolds (1994) established a pair of analytical systems, 2-dimensional Thin-Layer Chromatography (TLC) and C18 water/methanol gradient HPLC and analyzed freshly prepared exudates from five *Aloe* species and 14 museum reference samples. We have in the past used simpler TLC and water/methanol systems of lower resolving power. However, we are able to compare the two systems because the positions of aloesin and barbaloin are known in both systems. In the future, we believe that a higher resolution system, such as that of Reynolds, will be preferable to the simpler systems.

There are four elements in common between the production of the purgative exudate and the color change that occurs in commercial gel and WLE. These are: (i) the vascular elements of the mesophyll; (ii) oxygen; (iii) light; and, (iv) heat. These four factors are present in the environment in which the exudate appears. After incision, yellow juice exudes. Gradually this turns first red, then brown, then black. If the incision is made in the dark, the exudate forms, but remains yellow for several hours. Upon moving the plant into intense sunlight, the gel darkens within a short time. The involvement of the four factors in color change of the gel is evident from observations made during industrial processing, as described above. Also, over a decade ago, a crude test was developed for color change potential. This involved taking 2–3 ml of the gel being processed into a 16×150 mm tube, making the solution basic by the addition of 0.1 ml of 1 M sodium hydroxide and placing the tube for a minute in a boiling water bath, with agitation to aerate the solution. A positive test, the appearance of a pink color, indicated that all of the material with the potential to develop color change, had not yet been removed.

The next advance occurred when we developed a TLC assay for the color change compound. This happened while performing the anthraquinone periodate test (Bohme and Kreutzig, 1963) on the acetone extracts of the low molecular weight dialysates of **Process A** aloe gel. The sample was placed on a TLC plate which was developed, dried and then exposed to ammonia fumes at a temperature of 100 °C. One zone on the TLC plate changed color to the pink shade that developed in the color-change potential test above. Extraction of the pink spot yielded the same visible light adsorption spectrum as observed in the tube test. Using the TLC/ammonia/heat assay, it was possible to partially purify the color change compound using low pressure normal phase silica gel chromatography (Figure 8.7). The color-change-potential material (Figure 8.7, lane 4) is intermediate in its polarity between barbaloin and aloesin. The material isolated was not pure, this chromatographic region being

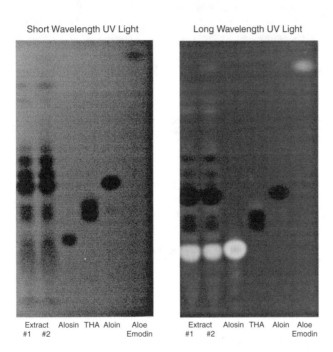

Figure 8.7 Anthraquinones and chromones from *A. barbadensis* gel and representative purified and partially purified compounds. Compounds were placed on Merck silica gel G-60 plates and developed in toluene-ethyl acetate – methanol – water – formic acid = 10:40:12:6:3. Plates were photographed under short (280 nm) or long (350 nm) wavelength UV light without any specific staining. Identity of the standards, aloesin (2 µg), aloin A or barbaloin (10 µg) and aloe emodin (0.2 µg) were confirmed by mass spectroscopy. Other zones contained 100 µg of extract. Extracts were replicate acetone extracts of <5,000 molecular weight cut-off membrane (MWCO) dialysates of ARF'92B Standard Sample Gel (*see Colour Plate 6*).

complex (Reynolds, 1994) The compounds isolated are consistent, upon mass fragmentation, with compounds previously described as 7-hydroxyaloin, 5-hydroxyaloin, 4-hydroxyaloin and their methyl derivatives (Rauwald and Voetig, 1982; Rauwald and Beil, 1993; Graf and Alexa, 1980). At present, we cannot determine which of these tetrahydroxyanthraquinones (THA) is responsible for color change or, indeed, whether any of them actually are the molecules involved in this change. A critical observation in favor of the involvement of 7-, 5- and or 4-hydroxyaloins in color change is the observation that when this change occurs and the products of color change convert into water-insoluble molecules, the content of 7-, 5- and/or 4-hydroxyaloins decrease.

The final understanding of the identity and mechanism of color change of aloe anthraquinones will require further research.

Decolorization with activated charcoal

Figure 8.8 below summarizes the various processes and equipment necessary to remove from aloe the anthraquinones, which contribute to color change. Most activated charcoal adsorption is done in a batch process fashion. Pasteurized aloe gel (Figure 8.8, raw material, M) is run into a mixing tank of 500 to 5,000 gallon capacity (2,000 to 20,000 liters), while activated charcoal (0.05 to 2% w/v) is added with mixing (**Process N**). After 15 to 60 minutes the activated charcoal is removed by filtration (Figure 8.8, **Process O**). Theoretically, the temperature at which adsorption occurs is critical. However, in most aloe industry applications, the temperature of adsorption is determined by convenience. If adsorption is conducted as part of batch pasteurization, then activated charcoal is added immediately after the pasteurization holding period is finished and temperature reduction begins. Filtration is performed when the desired final temperature is attained. Thus adsorption may begin at a temperature of 65 °C and continue for an hour until a temperature of 45 °C is reached. At this point the product is filtered and finished. On the other hand, pasteurized gel that has been stored at 4 °C may be charcoal-treated for an hour at that temperature and then filtered. Obviously the adsorption isotherms for these two processes are very different and the first process might require ten times as much activated carbon as the second process. On the other hand, the first batch may have a much higher color change potential than the second. This is

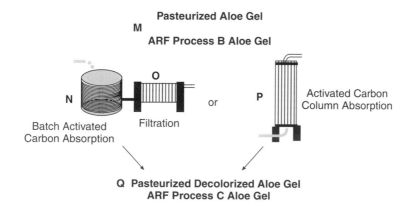

Figure 8.8 Processes involved in the treatment of *A. barbadensis* extracts with activated charcoal.

because the first was processed WLE which had an anthraquinone content 50 times greater than the second material, which is gel carefully hand filleted from large leaves and therefore low in anthraquinones.

The relationship between time of adsorption and activated charcoal concentration is reciprocal. If a more prolonged adsorption period can be tolerated, then lower amounts of charcoal can be used, although the ratio must be maintained between limits. If, on the other hand, it is desired to keep the adsorption period down to 15 minutes, larger quantities must be used. Thus there are four sets of parameters to be optimized for successful adsorption: (i) time, in the range of 15–60 minutes; (ii) temperature: in the range of 4° to 60 °C; (iii) amount of color change potential; and, (iv) amount (0.05 to 2% w/v) and type of activated carbon employed.

Once adsorption is finished, the activated charcoal must be removed. This is generally done with a filter press (Figure 8.8, **Process O**) using paper with a nominal pore size of 20 μm, rather than by sedimentation or centrifugation. The most active carbon systems generally have a small particle size, in order to maximize surface area. When particle size is significantly below 50 μm, it is advisable to employ 0.05 to 0.5% of a clarifying agent such as Celite Filteraide (diatomateous earth) to minimize filter clogging. With a properly designed and operated filtering system, a throughput sufficient to clarify 2,000 to 10,000 liters per hour is feasible.

Technologies encompassing pellicular activated charcoal, ranging from 50–100 μm up to several mm, are widely employed in water treatment facilities. Some of these systems pack the activated charcoal beads in columns through which the process liquid circulates. These systems offer the possibility of continuous processing that is compatible with HTST pasteurization. Furthermore, carbon columns are amenable to regeneration, which may help offset their initial higher capital cost. To our knowledge, columns of activated charcoal (Figure 8.8, **Process P**) have not yet been employed at the production level in the aloe industry.

The product that emerges after activated charcoal treatment (Figure 8.8, **Product Q**) is decolorized '1:1 Aloe vera Gel' according to the IASC nomenclature (IASC, 2001). This 'gel' material makes up almost all of the remainder of the output of non-concentrated aloe 'gels', the remainder being raw aloe gel and pasteurized aloe gel. During the last decade, because of problems with color change and concerns about the carcinogenicity of anthraquinones in Europe, there has been a shift away from natural 'non-decolorized' to 'decolorized gel.' The ARF reference material for this type of product is termed ARF **Process C** 'Decolorized Gel.' This material is universally referred to in the aloe industry as 'Gel.' This *A. barbadensis* material, carefully prepared from fillets, without exogenous cellulase, when treated by adsorption with DARCO activated charcoal and filtered through a 20 μm paper filter press with Celite Filteraide, has pseudoplasticity slightly greater than water. It should not be referred to by the physical-chemical term, gel. The pseudoplasticity is lost, not solely because of the adsorption of polysaccharide on activated charcoal but because of physical entrapment of native polysaccharide into the complex of charcoal, diatomaceous earth, and 20 μm pore size paper filter. In the ARF 1993 Processing Study treatment of *A. barbadensis* gel with activated charcoal resulted in the loss of 19 to 23% of the polysaccharide content, (see Figure 8.11). Furthermore, the nature of the polysaccharide is changed by this process from native polysaccharide-like to Acemannan-like substances (Figure 8.3). There are some suggestions that processes other than entanglement and simple adsorption may be involved.

What is actually adsorbed on to charcoal?

Activated carbon adsorption is the first processing step where gel is intentionally subjected to chemical alteration – all the previous processing steps, washing, filleting, depulping, pasteurization, were extractive or sanitation-related. Charcoal adsorption results in a radical change in the chemical composition. Since the aloe industry regards activated charcoal adsorption as a standard and usual procedure, those molecules removed during this process cannot be regarded as part of the IASC general definition of aloe. This restriction explains the selective bias in the IASC certification standards towards solids, mineral cations and organic acids, since none of these are changed radically by activated charcoal adsorption. Molecules such as barbaloin, aloesin and aloe emodin, long associated with *A. barbadensis*, are not IASC certification analytes because their content changes during legitimate processing. This emphasis on certification parameters that do not change with processing explains why the IASC has grappled for 20 years with polysaccharide tests, because polysaccharide levels and chemistry change with industry-accepted processing. Thus we have diametrically opposed definitions of aloe. The European definition of aloe (e.g. Reynolds, 1994) focuses on the exudate with its mixture of almost 50 chromones and anthraquinones, which are adsorbed on to activated charcoal. The American definition, exemplified by the IASC, focuses on parameters such as solids, salts and organic acids that do not change with processing.

We have developed a different way of looking at what adsorbs on activated charcoal, not what is lost from aloe but what can be eluted from a mixture of carbon and diatomaceous earth. Figure 8.9 illustrates that what goes on activated charcoal can be eluted off. About 6.5% of the dry weight of charcoal with adsorbed material can be eluted using mixtures of organic solvents. We have used mixtures of chlorinated hydrocarbons with acetone or alcohols under refluxing conditions to elute adsorbed aloe materials. The eluate resembles the acetone extract of *A. barbadensis* gel (compare Figure 8.9 with Figure 8.7, lanes 1 and 2). In some cases the eluate resembles the exudates of Reynolds. The complex of 7-, 5- and 4-hydroxyaloins with Rfs between those of aloin and aloesin are greatly reduced in the activated charcoal eluates, as they are in exudate. This may be due to oxidation of some of these THA compounds to the 'red compound(s)'. The red compound(s) are intermediates.

Finally, these extracts of aloe-adsorbed on charcoal/diatomaceous earth contain a significant amount, up to 37%, of hexose. This appears to be a mixture of sugars resulting from the breakdown of polysaccharide trapped on the complex of the adsorbant. In this respect the charcoal eluates further resemble the leaf exudates which are rich in breakdown products of simple and complex sugars.

Decolorization does not always work

The above description give a plausible suggestion as to why aloe undergoes color change. The THA are oxidized to compounds which undergo polyphenolic condensation to a brown/black insoluble residue. Color change can be prevented by removing the THA before they oxidize. On the other hand it is possible that the oxidative method of aloe stabilization (e.g. Cobble, 1975; Coates, 1979; etc.) may prevent color change by converting the THA to insoluble polyphenolics, which are removed during filtration. Whatever the mechanism of 'decolorization' used, experience in the aloe industry

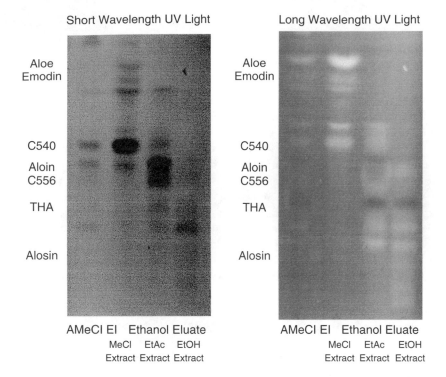

Figure 8.9 TLC analysis of acetone/methylene chloride (AcMeCl) eluates (lane 1) and ethanol eluates (lanes 2–4) of aloe-adsorbed activated charcoal. The ethanol eluate has been fractionated by successive extractions with methylene chloride (MeCl), ethyl acetate (EtAc), and hot ethanol (EtOH). Along the right side of the plate are indicated the Rfs of reference compounds: C540, a chromone ester of mass 540 daltons; C556, a chromone ester of mass 556 d.; THA, tetrahydroxyanthraquinone) (*see Colour Plate 7*).

shows that it does not always work. Treated aloe subsequently subjected to light, air and heat turns brown/black. It may be that this situation reflects partial oxidation. The oxidation product of a THA, the 'red compound(s)', particularly at elevated temperature, may be hydrophilic enough to avoid complete adsorption on activated charcoal. These 'red compound(s)', however, will still be available after 'decolorization' to undergo polyphenolic condensations. Furthermore, process control by analytical chemistry is unknown in the aloe industry. There is no perceived need to know the exact THA content of the aloe which is to be subjected to charcoal adsorption and thus to adjust the amount of activated carbon added. It might be that the color change potential of a given batch of aloe is greatly in excess of the adsorptive capacity of the activated carbon added and that some of this material will remain. However, a final conclusion will await complete characterization of the compounds responsible for color change.

Final processing – concentration and drying

About a decade and a half ago one of the major expenses in the aloe industry was shipping. Pasteurized aloe gel (**Product M**) and pasteurized, decolorized aloe 'gel'

(**Product Q**) are 99% water. As the market for *A. barbadensis* gel expanded, companies who could drastically reduce shipping costs could reduce the ultimate cost of aloe feed stocks to manufacturers by one-third. Given that the primitive state of aloe quality control focused on solids, ions and organic acids, the reduction of costs by one-third meant an economic advantage.

Concentration of aloe gel

An industrial thin film vacuum evaporation system with temperatures maintained at 35–45°C and a throughput usually in the range of several hundred liters per hour results in 'concentrated aloe' (Figure 8.10, **Product T**). This product has a somewhat checkered past. Some of the problems are inherent in the process relating to the composition of 'gel' preparations. Other problems are due to the aloe industry's lack of scientific rigor.

There are three types of physico-chemical problems. Firstly, the removal, in the evaporation process, of terpenes which give fresh aloe extracts their organoleptic 'fresh nose'. Whether this 'ruins' the product or 'improves' it can be debated. However, it is a change in the chemical composition of the product. Secondly, evaporation is the step where a 'darkening' color change frequently occurs. If during earlier processing the 'red compound' has been produced by oxidation of the THAs, the heating that occurs during thin film evaporation may cause a polyphenolic condensation resulting in a brown/black tint, which becomes more pronounced as concentration proceeds.

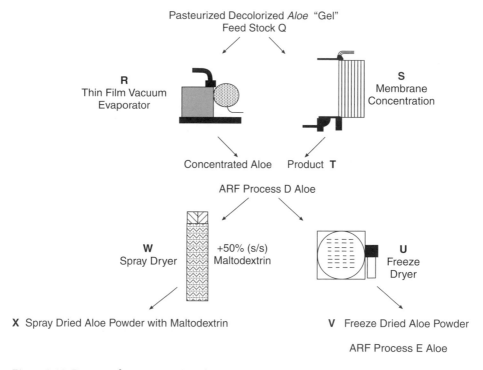

Figure 8.10 Processes for concentrating aloe extracts.

Thirdly, what is probably the most significant chemical change in aloe during concentration is the destruction of polysaccharide that occurs due to the action of exogenous cellulase. In the case of WLE, cellulase was added back at the beginning of production when the viscosity of the 'guacamole' was reduced prior to depulping. The higher operating temperature of the thin film process (45 °C) is close to the temperature optimum for fungal cellulase. If concentration is carried out until a solids content of 15–20% is obtained then 90% of the acetylated glucomannan can be broken down (Pelley *et al.*, 1998, Table 3, Material 4). At this point the only remaining polysaccharides are galactan and pectin. Similar cellulase-induced polysaccharide breakdown problems are encountered when pasteurized aloe gel (**Product M**) or pasteurized, decolorized aloe 'gel' (**Product Q**) are concentrated. In order for thin film vacuum evaporation to proceed efficiently, the material undergoing concentration should have the viscosity of water. Many companies, in order to boost the speed of the vacuum evaporation process, add cellulase to 'break' the viscosity. This results in the loss of considerable polysaccharide (Pelley *et al.*, 1998, Table 2, Materials 1–7). Not all manufacturers employ cellulase so that there is a great deal of variability in polysaccharide content among legitimate samples of concentrated aloe material processed from gel fillets, depending on whether cellulase was employed during the concentration process and at what level and to what degree concentration proceeded.

A wholly separate set of problems, not physico-chemical in nature, are inherent in the uninformed management practices common in the aloe industry. One problem is the refusal to adopt a nomenclature for concentrates in concert with usual food industry practices and the other problem is the tendency to 'push' processes in order to maximize 'throughput/efficiency' at the expense of quality. The aloe industry has long (IASC, 1993) accepted an 'X' nomenclature for concentrates, which assumes that the solids content of aloe gel is 0.5 g/dl. Concentrates are expressed as multiples of this measurement of the solids content. Thus a product concentrated to a solids content of 1.0 g/dl is labelled '2X,' a product concentrated to 2.0 g/dl is '4X,' a product concentrated to 5 g/dl is '10X' and one concentrated to 20 g/dl is '40X.' One consequence of this is that consumer product manufacturers easily and commonly confuse 'X' with solids content. Spray dried aloe, properly produced with 50% matrix is called '100X' which is generally confused with 100% aloe. The system common to the food industry is to label juices based on the Balling refractive index system according to solids content (BRIX). The aloe industry should adopt this and base its nomenclature directly on solids content, otherwise aloe products are viewed with suspicion by companion industries.

A second management-specific problem is the tendency to 'push' processing. By this is meant the tendency to take a process that is working quite satisfactorily and in the interest of cost effectiveness, 'push' it one step further without scientifically monitoring the consequences of the processing change. For example, if aloe gel can be concentrated to a solids content of 20 g/dl rather than 5 g/dl, this will save 60% on shipping costs. This is done without determining if: (i) the further step will cause a significant decrease in polysaccharide content; (ii) certain biological activities will be lost; and, (iii) at this concentration, at shipping temperatures, some of the components will precipitate out of solution. Proceeding to extremes based solely on economic considerations, without resolving the basic biologic and chemical implications of the processing change, is unfortunately all too common in the aloe industry. This has resulted in products that, although legally aloe and not adulterated, have lost key chemical ingredients and biological activities.

It should be noted that there are alternatives to thin film vacuum evaporation for the purpose of concentrating aloe materials. Pressure filtration across membranes (Figure 8.10, **Process S**) can also be used. This is a process somewhat similar to reverse osmosis, used to purify water, but with the opposite outcome-the desired product is retained by the membrane rather than passed by the membrane. This process is used by one company to partially concentrate gel prior to shipping from its third world platntations to its central processing plant. Although the process operates efficiently, there has not been a systematic analysis of the effects on the chemistry and biology of the membrane concentration process.

Freeze-dried or lyophilized aloe

Product T, concentrated aloe, is a major branch point for the production of finished products. A great deal of concentrated aloe liquid is marketed as a feedstock, using the preservative systems discussed above. Most of these liquid concentrates go to the beverage industry. However, some consumer product manufacturers, particularly in the cosmetic industry, prefer a dried product. A dry product has a wide variety of advantages ranging from ease of transportation and storage, to assurance of solids content and in certain cases, better protection against spoilage. Freeze-drying is certainly the most elegant method for stabilizing aloe materials (Figure 8.10, **Process U**). Material with a solids content of 5 to 20 g/dl is poured into stainless steel trays and frozen to low temperature (below $-40\,°C$). The frozen material is then put under a high vacuum (50 milli Tor). Water gradually sublimes from the frozen material which is gradually heated. The rate at which the trays are heated controls the rate at which water sublimes from the frozen blocks and thus the degree of vacuum is controlled. After a cycle time of 36 to 72 hours, most of the water has been removed, a high vacuum is sustained (<25 milli Tor), and the temperature of product remains at ambient (c.30 °C).

This process, employing high vacuum technology and precise temperature control, is obviously very expensive. We have extensively used it in preparation of our ARF controls, which are not commercially feasible products. Because of the expense factor, the commercial employment of freeze-drying or lyophilization operates most efficiently when the feedstock is highly concentrated material. The most economical materials for freeze-drying are the most concentrated, the most oxidized, have the greatest degree of color change and have the most broken down polysaccharide. This is a case where lyophilizing a feedstock with a solids content of 5 g/dl yields a much better product than one made with 30 g/dl feedstock, albeit one that cost about three times more to produce.

Freeze-drying is potentially the method of choice for production of the finest quality finished product to be used in the manufacture of cosmetics. If high quality aloe, prepared from gel fillets and subjected to careful preliminary/intermediate processing is the feedstock for lyophilization, a very high quality product will result. However, it is more economical to feed poor quality, over-processed, over-concentrated color-changed aloe with a high solids content into the process.

Spray-dried aloe gel

This method is potentially an excellent method of concentrating and preserving aloe extracts. However, the reputation of the product was badly tarnished during the early 1990s, because of widespread fraud involving maltodextrin-matrix, spray-dried aloe products. Spray-drying (Figure 8.10, **Process W**) can be a two-step process. First, the aloe concentrate is mixed with a matrix and then sprayed into a stream of hot air, which dries the mix of matrix and aloe. The process begins when matrix is added to the aloe liquid concentrate. Matrix, either the disaccharide lactose or the higher-molecular weight saccharide, maltodextrin, is used to provide a rapidly forming, readily dried nucleus around which the aloe can accrete and dry. The matrix is added at a ratio of about a 1 g matrix/1 g aloe solids. Companies in the aloe industry that use spray-drying seem to prefer maltodextrin, particularly Lodex-10, matrix to lactose matrix. The solution of aloe and the maltodextrin matrix is then pumped to the top of the spray dry tower, which may be up to 10 m in height. The fluid is then sprayed in a downward direction out of a series of nozzles as a fine mist. The tower, which is an enclosed space, has a positive flow of air heated to between 50 to 90 °C. As the mist falls through the hot air down the shaft of the tower, the water evaporates and the aloe solids/matrix dries. The dried product, which consists of tiny granules, falls into a conical collector from which it is continuously removed. There is a dynamic equilibrium between the rate of spraying, the temperature and velocity of the stream of warm air and the physical chemistry of the liquid. Attaining this balance is easy with properly maintained equipment. The freely flowing product is generally white, or if the aloe concentrate has undergone a slight color change, off white and has little apparent aroma. If the maltodextrin employed as matrix has a significant content of lower saccharides, the product may have a very slightly sweet taste.

During the early to mid 1990s, a great deal of spray dried aloe with matrix was sold as 100% aloe which constitutes fraudulent misrepresentation (Pelley *et al.*, 1998). This fraud was possible because the industry promoted, despite all scientific evidence, a test for aloe quality termed the methanol precipitable solids (MPS) test. This MPS test, alleged to specifically measure aloe polysaccharide, yielded roughly comparable values for spray-dried aloe and freeze-dried aloe. Since spray-dried aloe costs about one third as much to make as freeze-dried aloe, the temptation to offer the cheaper product as the more expensive product was overwhelming. Once a vendor discovered that the customer was using the MPS test, a switch was simply made between lyophilized aloe and spray-dried aloe. Eventually, some vendors even sold pure maltodextrin as spray-dried aloe (Pelley *et al.*, 1998). This fraud involved one of the three largest aloe producers in the world and two large aloe brokers. The proper and improper measurement of polysaccharide will be discussed below.

Aside from fraud made possible by ignorance of basic biochemistry, the spray-dry process has much to recommend it. Although the method uses elevated temperatures, the product is exposed to relatively low temperatures due to heat loss during evaporation. Advancements in equipment over the past decade have also allowed aloe to be spray-dried without the use of matrix. Although it would seem that oxidation should be a problem in this technique, prior steps in processing such as pasteurization and concentration have generally already oxidized any oxygen-sensitive compounds in the product. Lastly, spray-drying is not generally employed for products where preservation of the full spectrum of biological activity is paramount. Freeze-drying is used for that purpose.

Consequence of the use of activated charcoal and cellulase

In 1993, after preliminary experiments on processing, the Inter-University Co-operative Group of the ARF decided to address the question as to what the effects of cellulase and activated charcoal adsorption were. Furthermore, both these effects were studied on gel from fillets and from WLE.

The flow chart above illustrates the protocol followed. Three metric tons of leaves were harvested under optimal conditions and divided into three lots. One lot of leaves, A, was transported overnight while refrigerated at 4 °C to a processing plant and ground into WLE. At this point it was divided into two lots and treated (500 kg) or not treated (500 kg) at ambient temperature with cellulase, depulped and then transported refrigerated to a second processing plant. There, each lot was again divided and treated at ambient temperature with 1% (w/v) activated charcoal or alternatively not treated. These four preparations were frozen and then lyophilized. This yielded four WLE products each processed from 250 kg of leaves by one of four processing variants.

A second batch of one metric ton of leaves (Figure 8.11, B) was immediately processed to produce depulped aloe gel. The gel was shipped refrigerated over night to the processing plant. Protocol B was then likewise divided into four streams as before. Lastly, a third protocol (Figure 8.11, C) was established to test the effect of a 24 hour delay in processing upon gel quality. Here, the leaves were stored, refrigerated, for 24 hours after harvesting and washing, and then the gel was produced by filleting. Otherwise, this part of the overall protocol was conducted identically with part B. The experiment consisted of 12 samples comprising four variables, WLE versus gel, cellulase action versus no cellulase, and charcoal treatment versus no charcoal and immediate versus delayed filleting. These 12 samples were then analyzed by a wide variety of chemical and biological tests and fractionated to isolate polysaccharides and other compounds. The lower portion of Figure 8.11 gives the data from selected analyses. An interesting, and complex, biological activity was the ability to protect the skin immune system against suppression by ultraviolet light (see Strickland and Pelley, Chapter 12). However, data are also provided on the classical biological activity of aloe, the amelioration of acute inflammation (Davis, 1986, 1987).

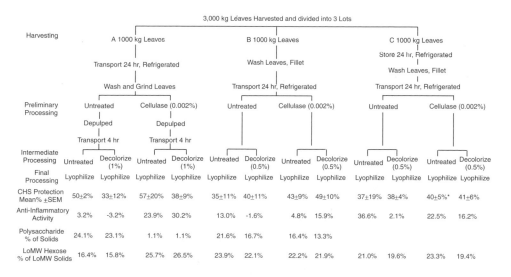

Figure 8.11 Effect of processing upon biological activity and some key chemical parameters.

Protection of the skin immune system varied in a complex fashion. Clean, gently handled leaves stored for 24 hours before filleting caused a borderline decrease in protective activity (42% protection versus 39% protection ±3%). Similarly, this meticulously filleted gel, with a very low anthraquinone content demonstrated only a 3% change in activity after activated charcoal treatment. On the other hand, treatment with cellulase increased protective activity from 37.5% to 43.5%, an increase of 15% where the range of variation was about 1.5% and this lead to further study. It was found that activation of native glucomannan gel by cellulase was followed by subsequent decay of the biological activity upon excessive treatment. There is a similar gross effect with WLE where cellulase action causes an increase of 50 to 57% and 33 to 38% of activity. As stated above, charcoal treatment has little effect on highest quality gel, although treatment of WLE reduces the CHS protective activity from 50–57% to 33–38%, a decrease of almost a half. These data suggest that adsorption removes from WLE one major factor responsible for protection of the skin immune system, a factor that is absent from gel with low anthraquinone content.

Normal processing steps, such as treatment with cellulase or activated charcoal also affect such biological activities as the modulation of acute inflammation. However, the *A. barbadensis* anti-inflammatories are affected by processing in different ways than protection of the CHS response. Thus, treatment of WLE with cellulase strongly activates an anti-inflammatory principle, while absorbing with activated charcoal has no effect. The situation is less clear cut for the anti-inflammatory activity in gel. Native gel, not treated by cellulase or charcoal has good anti-inflammatory activity (13 to 37% inhibition of croton oil-induced swelling), which is lost by adsorption on charcoal (−1.6 to 2.1% inhibition). Thus processing has a profound effect on the biological activities of aloe gel although there is no simple pattern of activity change among different biological activities. Lastly it should be realized that the skin immune system protective activity in Figure 8.11 is based on 14–15 animals per group whereas the anti-inflammatory data is based on only five animals per group.

Chemical parameters change with proper processing. Most notable are the changes that occur with carbohydrates, other than monosaccharides, during processing. Figure 8.11 shows around 11% of total solids as monosaccharides in treated and untreated gels. On the other hand, polysaccharides, which made up on the average 18% of solids, varied between samples. The highest values (27.2% of solids) were found with WLE, which was exposed to neither cellulase nor activated charcoal. Treatment with activated charcoal reduced this to 23.9%. Depulping treatment of WLE with cellulase, even under conditions far milder than usual in the industry, destroyed all of the acetylated glucomannan leaving only the pectins and galactan. Similar effects on polysaccharide were observed with gel. In general, adsorption on activated charcoal caused a drop from about 22% of solids to 17%. Treatment with cellulase under extremely mild conditions reduced gel polysaccharide from about 20.6 to 19% of solids. Again, it should be noted that the time and temperature of cellulase treatment in these experiments are far shorter and lower than is usual in industry. Cellulase treatment increases low molecular weight (<1.6 kd.) saccharides (Figure 8.11, last line), of which 14% are monosaccharide as discussed above. This means that about 2% of untreated WLE and 7–10% of untreated gel can be defined as 'oligosaccharide.' These values increase considerably when WLE is treated with cellulase but not when gel is subjected to much milder cellulase treatment.

It can be seen that processing has a remarkable effect upon all aspects of *A. barbadensis* gel chemistry and biological activity. The activated charcoal adsorption process, used to prevent color change, caused a vast complex of compounds to be adsorbed. Large quantities of cellulase are routinely used for viscosity reduction in WLE process and before evaporative concentration of gel in thin films. Profound biological and chemical changes occur with both the charcoal and cellulase processes. Cleavage and destruction of the cytoprotective oligosaccharide from the native gel is described elsewhere (Strickland and Pelley, Chapter 12). All of these observations are made with considerably cleaner materials, under conditions considerably more mild than those generally employed in industry. Materials processed in industry are often devoid of biological activity because of bacterial proliferation, over pasteurization, excessive use of activated charcoal and addition of large amounts of cellulase followed by prolonged incubation at elevated temperatures.

QUALITY CONTROL USING CHEMICAL SURROGATE MARKERS

We have considered what aloe is, how aloe materials are processed how the products are effected by processing and how the chemistry of aloe materials change during processing. Our focus will shift slightly to describe the chemical tests commonly used to analyze aloe and we will exemplify the types of data these tests yield. None of these analyses directly indicate biological activity but they are surrogate markers for such activity, since when they are low or absent, so is biological activity. These tests are not common for process control in the aloe industry. We will therefore spend most of this section examining typical test data on finished products. This will be oriented toward a description of suitable aloe material, in contrast to bad aloe, and what fraudulent and/or adulterated material looks like. In particular, we will emphasize the usefulness of multiparameter analysis for determining the quality and identity of aloe materials (Pelley *et al.*, 1993; Waller *et al.*, 1994; Pelley *et al.*, 1998). This concept suggests that in material as complex as aloe, a single given analytical system may commonly yield values outside the 2σ limit on a given sample. However, when multiple parameters fall outside their 2σ limits, the sample is either degraded beyond recognition or fraudulent.

Solids and metallic cations

The first quality control parameters used in the aloe industry were the determination of solids and quantitation of calcium and magnesium. The use of solids is obvious and traditional in the plant juice industry. Over 20 years ago it was appreciated that aloe juice was comparatively rich in calcium and magnesium (Waller *et al.*, 1978). Solids are readily determined by simply evaporating at elevated temperature (e.g. 105 °C) to constant weight. If a single analytical test is performed in a processing facility, it is usually for solids. Divalent metal cations, on the other hand, are more complicated to determine, usually by inductively-coupled plasma atomic emission spectroscopy. Although other methods (e.g. ion selective electrodes) could be used for process control, spectroscopy remains the method preferred for IASC certification because of the lack of interference. Subsequent to the Texas A&M study, the monovalent cations have been added to the analytical armamentarium because there is little additional cost to adding K^+ and Na^+, once Ca^{++} and Mg^{++} have been specified.

In general, neither solids nor the content of metallic cations change during processing. If the solids content of the depulped gel is 0.58 g/dl than the pasteurized, decolorized gel is likely to have a solids content of 0.58 g%. This does not mean that there are no losses during processing, but that these losses are of the spillage variety and do not decrease the solids content. On the other hand, final processing involving concentration will radically alter solids content. That is the purpose of evaporative concentration, freeze-drying or spray-drying. As discussed above, the idiosyncratic nomenclature of the aloe industry frequently confuses those not intimately familiar with its less-than-logical reasoning (e.g. '100X' is not 100% pure aloe, which is '200X'). Similarly, enzymatic treatment, heat treatment and treatment with activated charcoal should not remove basic chemical components like mineral salts. The one exception to this may occur during storage, particularly cold storage of '40X' liquid concentrates. An insoluble complex may form between the divalent cations and the divalent organic acids, especially when oxalic acid is present, which then precipitates, removing the calcium from solution. When this precipitate is primarily calcium oxalate, it can be impossible to dissolve unless the pH drops to below 3. On occasion, even calcium malate can be difficult to dissolve. In summary, the solids and metal cations, although not linked to biological activity, are desirable because of their extreme stability; multiparameter deviation of these materials strongly implies dilution or non-ionic adulteration.

Table 8.10 illustrates typical findings of solids content in 13 commercial liquids alleged to be 'Aloe vera.' These liquids were divided, based on multiparameter analysis (Pelley *et al.*, 1998) into two groups corresponding to Samples 1–7, consistent with authentic *A. barbadensis* extracts and Samples 8–13 not consistent with *A. barbadensis*. None of the liquid samples 1–7 had the exact solids content predicted from the label claim, although all fell within the range of acceptability for IASC certification (Table 8.3). As a group their average came close, the absolute deviation expected was only 19%. This degree of variability is consistent with the variability observed in the Texas A&M study (Table 8.2). On the other hand, with samples 8–13, multiparameter analysis indicated a high suspicion of fraud. The solids content of these were different from the label claim (47% discrepancy) and 75% were outside of the IASC certification range (Table 8.3).

As discussed above a sizable discrepancy between the IASC cation distribution and a tested population also raises the possibility of fraud. Table 8.11 illustrates cation values in three groups of commercial materials: (i) those consistent by multiparameter analysis with ARF **Process** *A. barbadensis* gel (#1–3); (ii) those consistent by multiparameter analysis with WLE (#4–6); and, (iii) those consistent with maltodextrin being fraudulently sold as *A. barbadensis* material (#7–10). By reference to the IASC distribution for cations (Table 8.3) all of the ARF Standards and the materials consistent with aloe by multiparameter analysis were within the range of acceptability for calcium and magnesium content. On the other hand, authentic maltodextrin and materials consistent with maltodextrin fell outside the limits of acceptability for IASC certification and as a group were ten times the SEM removed from any of the calcium and magnesium populations in the Texas A&M study (Table 8.2). Thus, just as with the solids content, the abnormal content of mineral cations strongly suggests that the material under consideration is either fraudulent (Table 8.11, 8 and 9) and or so extensively adulterated as to be no longer recognizable as aloe (Table 8.11, 10).

Very few aloe producers or consumer product manufacturers possess an atomic emission spectrophotometer, so these determinations are usually sent out to an independent testing laboratory. Therefore there is frequently a need for a cheap and rapid test to substitute for the measurement of the individual cations. Around the mid 1980s it was

Table 8.10 Characteristics of commercial 'Aloe vera' liquid materials.

		Unfractionated material				Polysaccharide isolated by dialysis				Dialysate	
No.	Source	Discrepancy in solids[a]	Ionic strength[b]	Polysaccharide content[c]	Malic acid[d]	% Retained material[e]	% hexose[f]	UV scan[h]	Glu:Man:Gal percentage[i]	% hexose[f]	Reducing sugar[g]
1	A	+16%	2,484 µS	11.1%	Normal	12.4%	65.2%	0.55	4:90:3	10.7%	13.9%
2	A	−8%	1,799 µS	5.9%	Normal	10.0%	75.9%	0.19	7:81:4	16.8%	14.7%
3	A	−22%	2,108 µS	7.9%	Normal	12.4%	50.7%	0.02	18:81:1	16.3%	11.2%
4	R	NP	2,548 µS	4.5%	Normal	13.1%	66.6%	0.26	18:81:1	10.2%	11.2%
5j	D	+16%	773 µS	2.4%	Low	4.5%	56.3%	0.20	32:51:12	15.6%	N.D.
6j	Q	−10%	1,800 µS	4.7%	Low	3.4%	68.4%	0.65	62:17:8	14.4%	10.3%
7j	Q	+42%	1,830 µS	2.8%	Low	2.1%	71.3%	1.17	69:6:8	16.9%	N.D.
Mean		19%	1,906 µS	5.6%		8.3%	64.9%	0.43	12:83:2	14.4%	12.3%
SEM		5%	223 µS	1.2%		1.8%	3.3%	0.15	4:2:1	1.1%	0.9%
8	N	NP	407 µS	2.6%	Absent	3.3%	38.9%	λ max 250	27:61:3	1.8%	3.0%
9	S	NP	12 µS	15.2%	Absent	10.9%	74.5%	0.06	73:8:0[k]	5.5%	2.7%
10	T	−72%	1,538 µS	3.9%	ABN*	21.9%	27.7%	0.38	68:10:15	3.1%	7.6%
11	T	+60%	3,333 µS	13.9%	ABN*	13.9%	45.5%	2.69	11:79:5	1.6%	5.2%

Table 8.10 (Continued).

Unfractionated material						Polysaccharide isolated by dialysis				Dialysate	
No.	Source	Discrepancy in solids[a]	Ionic strength[b]	Polysaccharide content[c]	Malic acid[d]	% Retained material[e]	% hexose[f]	UV scan[h]	Glu:Man:Gal percentage[i]	% hexose[f]	Reducing sugar[g]
12	U	−45%	636 µS	2.3%	Absent	1.7%	61.2%	1.20	53:47:0	18.1%	15.5%
13	U	+9%	664 µS	2.1%	Absent	2.5%	51.7%	1.64	58:42:0	56.2%	15.2%
Mean		47%	1,098 µS	6.7%		9.0%	49.9%	1.19	48:41:4	14.4%	8.2%
SEM		14%	491 µS	2.5%		3.3%	6.8%	0.47	10:11:2	8.7%	2.4%

Sources are as follows. A is a feedstock manufacturer. D was supplied by a manufacturer's representative and is not A. Q is a broker blending product from feedstock manufacturer A. R is a consumer product manufacturer and Sample 7 is a consumer product. C is a broker. S, T, & U were supplied by consumer product manufacturers.

Notes

a The solids content was determined by freeze-drying 100 ml of the liquid and direct measurement of mass. The discrepancy between the solids found and the label claim was calculated directly using the IASC solids content value in force at that time. For calculation of the means, the absolute values were used. NP: Not Pertainent – either there was no label claim for content, or the claim was for 'aloe' but no concentration was specified.

b Conductivity of a solution with solids content of 0.58 to 0.62 g/dl at ambient laboratory temperature.

c Values expressed are the % of solids that are alcohol precipitable hexose.

d As determined under IASC Conditions. * Extremely abnormal chromatogram.

e Liquids were lyophilized and then dialysed exhaustively until the conductivity of the final dialysate was under 100 µ Siemens. The data are expressed as percentage of mass in dialysate compared to total mass recovered.

f Determined by the phenol sulfuric acid assay for total hexose (Dubois et al., 1956).

g Determined by the neocuprione method (Dygert et al., 1965).

h Extinction coefficient at 280nm of a 0.1% (1 mg/ml) solution in deionized water.

i Retained material was hydrolysed in 6N constant boiling HCl for 10 to 20 minutes at 120 °C. After neutralization, ratio of monosaccharides determined by HPLC on a Dionex PA-1 column with a pulsed amphoteric detector. Statistical values for commercial materials are from the four samples of gel.

j Consistent with WLE, polysaccharide sugar composition is within normal limits: glucose, 54 ± 11%;mannose, 25 ± 13%; galactose, 9 ± 1%.

k Abnormally high in fructose, 17.7%.

Adapted from Pelley et al. (1998) with permission.

Table 8.11 Characteristics of Commercial Powdered Materials of Defined Provenance.

Sample no.[a]	Metal cation content[b]				Unfractionated material				Isolated crude	Polysaccharide
	Na^+	K^+	Ca^{++}	Mg^{++}	Ionic strength[c]	Malic acid[d]	Polysaccharide content[e]	Alcohol ppt'd Solids[f]	% total material[g]	Glu:Man:Gal percentage[h]
1 ARF-Gel	238	95	410	31	2,960 µS	103%	6.22%	34.2%	5.4%	8.4:81.1:1.6
2	188	103	411	30	2,960 µS	124%	6.36%	33.2%	4.8%	3.4:93.5:2.5
3	39	206	274	102	1,870 µS	142%	9.3%	36.8%	7.1%	7.1:89.8:2.1
4 ARF-WLE	128	285	445	51	2,570 µS	186%	0.99%	42.8%	0.3%	19.4:42.8:28.8
5	187	234	229	155	2,820 µS	8.3%	1.85%	36.4%	1.1%	30.4:35.7:25.1
6	69	329	409	70	2,460 µS	92%	1.01%	35.4%	0.3%	17.9:<.1:<.18.6[i]
7 LoDex10	4	1	1	1	30 µS	<5%	51.7%	63.6%	40.6%	91.0:<.1:<.1
8	4	1	2	1	53 µS	<5%	44.6%	49.2%	29.2%	78.2:2.0:<.1[i]
9	8	3	9	1	85 µS	<5%	48.2%	59.6%	37.2%	53.5:8.1:<.1[i]
10	32	53	47	22	618 µS	8.8%	43.2%	53.8%	37.2%	87.6:<.1:<.1[i]

Notes

a The identity of the material and its manufacturer can be obtained by contacting the International Aloe Science Council, Irving, TX.

b Determined at Inchcape Testing Services according to IASC procedure by inductivity-coupled plasma atomic emission spectroscopy. Data expressed are concentrations in mg per liter of a 0.62 mg/ml solids solutions.

c Conductivity of a 0.62 g/dl solution at ambient laboratory temperature. Test conducted at Inchcape Testing Service.

d Conducted according to the IASC protocol. Malic acid data is expressed as a percentage of the ARF Standard Sample currently in use at time of testing. Test conducted at Inchcape Testing Service.

e Assay conducted at Inchcape Testing Service on 500 mg of sample using the published method (Pelley *et al.*, 1993). Values expressed are the % of solids that are alcohol precipitable hexose.

f Alcohol Precipitable Solids. This is the mass of solid precipitated from the 500 mg sample at Inchcape Testing Service determined by lyophilization at UTMB. Data expressed are percentage of precipitated dry mass versus the 500 mg of powder originally tested.

g Lyophilized, alcohol precipitates were exhaustively dialysed. Data expressed are the percentage of retained material verus the 500 mg of powder originally precipitated.

h Retained material was hydrolysed in 6N constant boiling HCl for 10 to 20 minutes at 120 °C. After neutralization, ratio of monosaccharides determined by HPLC on a Dionex PA-1 column with a pulsed amphoteric detector. Data expressed are percentages of individual sugars versus the total sugars identified on HPLC.

i Unusual sugars were identified. Sample 6 was 74% amino sugars; sample 8 was 15.5% fructose; sample 9 was 28.1% fructose; sample 10 was 9.7% fructose.

Adapted from Pelley *et al.* (1998) with permission.

realized that conductivity of 1:1 gel provided a quick spot test for dilutional adulteration. An anecdote from c.1995 (the late Al Davis) stated that diluted feed stocks had low conductivity. At the time we were developing the ARF Standard Samples, Wang was measuring the four major cations in *A. barbadensis* gel. We quickly realized that although individual cations varied greatly, the sum of the cations were remarkably consistent (Pelley *et al.*, 1993). Since conductivity measures the sum of all the ions, it was appreciated that it could provide a cheap and rapid substitute for analysis of the individual ions (Pelley *et al.*, 1993), although it could not replace measurement of individual divalent cations for certification. Since then measurement of conductivity has proved useful in screening for fraudulent 'aloe' (Pelley *et al.*, 1998). Figure 8.12 below illustrates that there is a good correlation between the simple sum of the four Texas A&M analytes and conductivity for the ten samples examined immediately above.

Inspection of Tables 8.10 to 8.12 reveals the utility of conductivity in screening samples alleged to be aloe. Economically, the biggest adulteration/fraud problem is the substitution of maltodextrin for freeze-dried or spray-dried aloe (Tables 8.11 and 8.12). This is one of the areas where a quick conductivity spot test can be extremely useful. There are 11 samples of commercial material consistent with authentic aloe in Tables 8.9 and 8.12 (8.11: 2, 3, 5, 6; 12A: 1–7). Ten of these 11 are within the one σ level (1,528 mS calculated from Table 8.4) and the 11th is within the 2σ level (1,026 mS calculated from Table 8.4). At the opposite end of the spectrum were the 12 commercial samples consisting almost entirely of maltodextrin (Table 8.11: 8–10; Table 12B: 8–16). Ten of the twelve were beyond the 3σ limit (524 mS calculated from Table 8.4) and two (Table 8.11: 8–10; Table 12B: 8–16) approached the 3σ limit. Last are the commercial materials that appear chemically to be spray-dried aloe but whose label clearly claims them to be freeze-dried aloe (Table 12C: 17–19). These all yielded conductivity results exactly consistent with spray-dried aloe, that is a conductivity one-half that

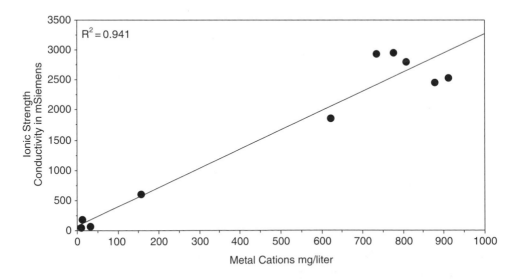

Figure 8.12 Correlation between conductivity and the sum of the metal cations Na^+, K^+, Ca^{++} and Mg^{++} in ten materials of defined provenance. The ratios of ions for the different species were determined by simple addition of the values by ICP in Table 8.11. Ionic strength was determined by conductivity. Reproduced from Pelley *et al.* (1998), by permission.

Table 8.12A Characteristics of Commercial Powdered Materials Consistent with 'Aloe vera'.

No.	Source[a]	Unfractionated material							Material greater than 10,000 MW			Material less than 10,000 MW
		Ionic strength[b]	Polysaccharide content[c]	Malic acid[d]	% Total Material[e]	% hexose[f]	UV Scan[g]	Glu:Man:Gal Percentage[h]	% hexose[f]	Reducing sugar[i]	Ratio	
1	A	1,320 µS	6.6%	Normal	1.9%	75.1%	0.36	6:80:8[h]	27.0%	11.6%	0.43	
2[j]	A	1,860 µS	1.7%	Normal	2.2%	66.5%	1.00	62:17:8[j]	11.4%	10.9%	0.96	
3[j]	A	2,020 µS	3.7%	Normal	1.2%	53.4%	1.24	10:58:21[j]	14.7%	11.5%	0.78	
4	B	2,750 µS	2.1%	Low	5.7%	66.3%	0.90	6:83:5[h]	5.9%	6.3%	1.07	
5	C	2,860 µS	2.2%	Absent	5.4%	75.3%	0.05	6:80:8[h]	14.2%	9.2%	0.65	
6	C	3,830 µS	15.9%	Normal	5.4%	69.3%	0.26	8:81:6[h]	17.1%	10.9%	0.64	
7[j]	D	2,200 µS	9.6%	Low	7.5%	66.5%	0.18	13:70:14[j]	13.3%	5.2%	0.39	
Mean		2,405 µS	6.0%		4.2%	67.5%	0.57	7:81:7[h]	14.8%	9.4%	0.70	
SEM		309 µS	2.0%		0.9%	2.8%	0.18	1:1:1	2.4%	1.0%	0.25	
Coef. Var.		34.0%	87.7%		58%	11.2%	82%	6.3%	43.4%	27.9%	36.0%	

Table 8.12A (Continued).

No.	Source[a]	Unfractionated material							Material greater than 10,000 MW			Material less than 10,000 MW
		Ionic strength[b]	Polysaccharide content[c]	Malic[d] acid	% Total Material[e]	% hexose[f]	UV Scan[g]	Glu:Man:Gal Percentage[h]	% hexose[f]	Reducing sugar[i]	Ratio	
Commercial Maltodextrin		62 µS	47.1%	Absent	58.6%	89.3%	0.01	93:7:0	76.5%	15.3%	0.20	
ARF		2,030 µS	9.2%	Normal	17.3%	73%	1.80	7:85:4	13.8%	15.5%	0.98	
SEM		145 µS	0.7%		1.7%	6%	0.35	1 1 1	1.5%	3.1%		

Notes

a A & C are feedstock manufacturers. B was supplied by consumer product manufacturer and is known to be neither A nor B. D was supplied by manufacturer's representative and is neither A, B, nor C.

b Conductivity of a 0.58 to 0.62 g/dl solution at ambient laboratory temperature. Results for maltodextrin standard are the means for determinations done on four separate occasions. Results for ARF Process A Standard Samples are the means ± SEM for analyses done on twelve different Standard Samples.

c Values expressed are the % of solids that are alcohol precipitable hexose. Results for ARF Process A Standard Samples are the means ± SEM for analyses done on twelve different Standard Samples.

d Conducted according to the IASC protocol. 'E Peak' is graded as normal in quantity, low, very low but still detectable or absent.

e 10 g of powder were dialysed exhaustively using Spectraphor #1 tubing and deionized water. Samples were dialysed for a minimum of three times against at least five volumes of water. In all cases the conductivity of the final dialysate was under 100 µ Siemens. The data are expressed as percentage retained material (mass/mass). Results for ARF Standard Samples are the means ± SEM for analyses done on fifteen different fractionations.

f Determined by the phenol sulfuric acid assay for total hexose (Dubois Assay). Results for hexose content of ARF Standard Samples, high molecular weight fraction are the means ± SEM for analyses done on ten different fractionations. In the case of the low MW fraction, 13 isolations.

g Solutions of isolated polysaccharide, at a concentration of 1 mg/ml were scanned at wavelengths from 200 nm to 800 nm. Data are expressed as OD$_{280\,nm}$. (Mean ± SEM of eight determinations from eight fractionations).

h Retained material was hydrolysed in 6N constant boiling HCl for 10 to 20 minutes at 120 °C. After neutralization, ratio of monosaccharides determined by HPLC on a Dionex PA-1 column with a pulsed amphoteric detector. ARF Standard Sample means are for 15 determinations on separate isolates. Means for commercial materials are for the four samples consistent with gel.

i Determined by the Neocuprione assay for reducing sugar (Nelson Assay). Note: the reducing sugar content of the retained fraction (high molecular weight material) was 5.3 ± 0.3% (Mean ± SEM). This value was based on five fractionations of four ARF Process A Standard Samples of gel. For these Standard Samples the respective value for total hexose content of the low molecular weight fraction was 15.8 ± 1.5%.

j Whole leaf preparation, polysaccharide sugar composition is within normal limits when compared to a limited number of ARF WLE Standard Samples. Statistical values for the three samples alleged to be whole leaf are: glucose, 28 ± 17%; mannose, 48 ± 16%; galactose 14 ± 4%; coefficient of varience, 67%.

Table 8.12B Characteristics of Commercial Powdered Maltodextrin Materials Alleged to be 'Aloe vera'.

No.	Source[a]	Unfractionated material							Material greater than 10,000 MW			Material less than 10,000 MW
		Ionic strength[b]	Polysaccharide content[c]	Malic[d] acid	% total material[e]	% hexose[f]	UV Scan[g]	Glu:Man:Gal Percentage[h]	% hexose[f]	Reducing sugar[j]	Ratio	
8	E	64 µS	54.3%	Absent	50.9%	82.2%	0.02	78:3:1[k]	79.6%	10.2%	0.13	
9	F	395 µS	77.1%	Absent	42.9%	91.9%	0.02	79:4:0[k]	85.0%	12.9%	0.15	
10	G	435 µS	35.7%	Absent	42.5%	88.2%	0.02	93:6:0	60.5%	12.5%	0.21	
11	H	107 µS	29.4%	Absent	48.7%	90.9%	0.02	77:16:0	54.5%	11.5%	0.21	
12	I	144 µS	30.7%	Absent	47.8%	83.4%	0.02	81:3:0[k]	54.8%	11.1%	0.20	
13	J	80 µS	31.9%	Absent	67.4%	82.1%	0.02	83:12:3	73.4%	11.8%	0.16	
14	K	115 µS	28.0%	Absent	64.4%	73.8%	0.05	94:4:0	61.7%	11.3%	0.18	
15	L	270 µS	25.6%	Absent	51.8%	86.8%	0.01	82:18:0	N.D.	N.D.	N.D.	
16	M	690 µS	46.1%	Absent	62.0%	81.0%	0.02	14:70:12	51.3%	11.7%	0.23	
Mean		255 µS	39.9%	Absent	53.2%	85.4%	0.02	83:8:1	65.1%	11.6%	0.18	
SEM		71 µS	5.6%		3.1%	1.9%	0.01	2:2:1	4.5%	0.3%	0.01	
Coef. Var.		83%	42%		17.4%	6.7%	49%	60%	19.0%	7.1%	18.6%	
Commercial Maltodextrin		62 µS	47.1%	our Absent	58.6%	89.3%	0.01	93:7:0	76.5%	15.3%	0.20	
ARF		2,030 µS	9.2%	Normal	17.3%	73%	1.80	7:85:4	13.8%	15.5%	0.98	
SEM		145 µS	0.7%		1.7%	6%	0.35	1:1:1	1.5%	3.1%		

Notes

a E, F & L are feedstock manufacturers. I and M are different brokers. G, H, J, & K were samples supplied by consumer product manufacturers.

k Unusually high fructose content; 8, 14%; 9, 14%; 11, 15.8%. Other footnotes are as per Table 8.12A.

Table 8.12C Characteristics of Maltodextrin-Adulterated Commercial Powdered 'Aloe vera' Materials.

No.	Source[a]	Unfractionated material							Material greater than 10,000 MW			Material less than 10,000 MW
		Ionic strength[b]	Polysaccharide content[c]	Malic[d] acid	% total material[e]	% hexose[f]	UV scan[g]	Glu:Man:Gal percentage[h]	% hexose[f]	Reducing sugar[i]	Ratio	
17	N	1,000 µS	32.8%	Absent	34.2%	84.4%	0.03	79:11:1	44.0%	15.3%	0.35	
18	O	1,000 µS	15.3%	Absent	35.2%	91.0%	0.06	77:3:0	48.3%	13.4%	0.28	
19	P	1,065 µS	25.4%	Absent	36.0%	78.5%	0.09	53:43:0	38.1%	11.2%	0.29	
Mean		1,022 µS	24.5%		35.1%	84.6%	0.06	70:19:1	43.5%	13.3%	0.31	
SEM		22 µS	5.1%		0.5%	3.6%	0.02	8:12:1	3.0%	1.2%	0.02	
Coef. Var.		3.7%	36%		2.6%	7.4%	50%	160%	11.8%	15.4%	12.0%	
Commercial Maltodextrin		62 µS	47.1%	Absent	58.6%	89.3%	0.01	93:7:0	76.5%	15.3%	0.20	
ARF		2,030 µS	9.2%	Normal	17.3%	73%	1.80	7:85:4	13.8%	15.5%	0.98	
SEM		145 µS	0.7%		1.7%	6%	0.35	1:1:1	1.5%	3.1%		

Note
a N & O are brokers. P was supplied by a consumer product manufacturer. Other footnotes are as per Table 8.12A.

expected for pure aloe (c.1,000 µS) at the 2σ limit. Measuring the ionic strength by conductivity gave one of the clearest signals for maltodextrin/aloe fraud and misrepresentation of spray-dried material as freeze-dried.

Liquid samples can be among the most difficult to analyze because of the wide variety of adulterants encountered (Pelley *et al.*, 1998). Among the aloe liquids (1–7) examined in Table 8.10, only one (5) would have been identified by conductivity as suspicious. Among the questionable liquids (8–13) four would have been flagged by conductivity as questionable (8, 9, 12, 13) and the remaining two (10, 11) would be on the borderline of normal.

In conclusion, all of the three basic analytical parameters, total solids, divalent metal cations and ionic strength, have their different places in the laboratory analysis of *A. barbadensis* materials. Solids measured by evaporation and weighing should be measured in every Quality Control laboratory. Because of the cost of the equipment, quantitation of metal cations, including the monovalent cations, will remain a test sent to an outside laboratory as part of preparation for IASC certification. Conductivity meters are cheap (under US$300) and easy to use. The measurement of conductivity should be one of the first things a laboratory does and a conductivity meter should be out on the production floor next to the pH meter and the refractometer.

Malic acid and other organic acids

We have examined the relationship between the composition of matter of *Aloe barbadensis* gel, the microbial physiology of aloe-associated micro-organisms and the HPLC analysis of aloe. This relationship was first suggested in 1993 when we established that an organic acid, at that time termed 'E peak' and now known to be malic acid, was a process-stable, freshness-sensitive analyte (Pelley *et al.*, 1993). Malic acid is universally present in ARF materials (Table 8.4) and the Texas A&M population (Wang and Strong, 1995). However, the content of malic acid varies greatly and therefore the range of acceptable malic acid values for IASC Certification is quite wide (Table 8.3). Malic acid makes an acceptable certification parameter because its content does not vary during standard processing (Pelley *et al.*, 1993). The determination of malic acid is relatively straightforward. For purposes of certification, the IASC has specified an HPLC method (IASC, 1991). This utilizes a 4×250 mm bonded amino column eluted isocratically at 1 ml per minute with a buffer consisting of 70% acetonitrile in 0.05 M phosphate buffer, pH 5.6. The eluate is monitored by UV absorbance at 205 nm. Samples of aloe liquids are adjusted to IASC '1:1' concentration (assuming that the label claim is correct). Currently, dry samples are suspended to 0.83 g% concentration. In all cases samples are filtered through 0.2 µm filter prior to use and 20 µl of sample is injected. Currently, the system is calibrated using chemically pure malic acid. Previously, ARF standard sample was employed for this purpose (Pelley *et al.*, 1998). Figure 8.13 below illustrates typical chromatograms. It should be noted that the AOCA has recently approved a somewhat more complex method for the determination of malic acid in fruit juices (AOAC, 2000b).

Malic acid is a desirable analyte because of its ability to act as a marker for 'freshness'. As Figure 8.13 illustrates, it is the major organic acid in fresh native gel. Bacteria such as bacilli that do not grow well in aloe do not alter the malic acid content. Some of the Gram −ve rods (Panel B) that grow in aloe, assimilate malic acid and produce other organic acid such as lactate. Aloe-associated micrococci (Panel C) are very efficient at

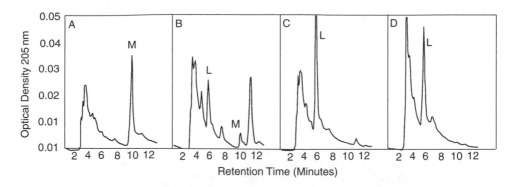

Figure 8.13 Effect of bacterial growth upon the organic acid content of *A. barbadensis* gel. ARF Standard Sample gel incubated with A) *Bacillus* sp., B) *Cedecea* sp., C) *Micrococcus* sp. and D) gel stored without processing or preservation. This figure uses data originally described in Pelley *et al.* (1993) which should be consulted for experimental conditions.

assimilating malic acid and in many cases produce lactate. Aloe gel that has 'rotted' has a pattern quite similar to that of *Micrococcus*-innoculated aloe and although all micrococci produce lactate, all 'spoiled' aloe materials do not contain lactate.

Tables 8.10, 8.11 and 8.12 illustrate some of the patterns of malic acid distribution in commercial aloe materials. Among liquids studied, malic acid was present in all seven of the materials determined by multiparameter analysis to be authentic *A. barbadensis* (Table 8.10: 1–7). It should be noted that in three of these (Table 8.10: 5–7) the levels of malate were low. This is consistent with the suspicion that these materials are WLE. As discussed earlier, WLE inherently has a high bacterial content and control of bacterial proliferation in these preparations is difficult. Liquid materials suspected of severe adulteration or fraud, possessed no detectable malate. Analysis of powdered material yielded similar findings. Of eleven materials consistent with authentic aloe (Table 8.11: 2, 3, 5, 6; Table 8.12A: 1–11) seven (Table 8.11: 2, 3, 6; Table 8.12A: 1–3, 6) had normal levels of malic acid, three (Table 8.11: 5; Table 8.12A: 4, 7) had low but detectable amounts and only one (Table 8.12A: 5) was devoid of malate. Here, there was also an association between commercial materials with low malate and a WLE origin (Table 8.11: 5; Table 8.12A: 7). There were two commercial materials with low to absent malate (Table 8.12A: 4, 5) that were likely to have originated as gel. Commercial materials that were consistent with maltodextrin (Table 8.11: 8–10; Table 8.12B: 8–16) were devoid of malate with the exception of sample 10 in Table 8.11. The possibility that a small amount of aloe is present in this commercial material is strengthened by the conductivity and the presence of metallic cations. Lastly there are three commercial materials that were alleged to be 100% pure aloe but which on examination were more consistent with spray-dried aloe (Table 8.12C). HPLC analysis for malic acid is negative, suggesting that these materials were manufactured with very low quality aloe that had already undergone bacterial spoilage.

In conclusion, after over a decade in use as an IASC certification standard, malic acid is well established as a quality control analyte. No feedstock should vend without the malic acid content on the certificate of analysis, nor should any consumer product manufacturer buy aloe without knowing this. Now that 'E Peak' is known to be malic acid, it is possible that disreputable manufacturers will aduterate with malic acid just as they adulterated with maltodextrin. Until that day arrives, malate will remain at the center of our analytical armamentarium.

Anthraquinones

Anthraquinones have been discussed elsewhere (Chapters 3 and 7) and it is surprising that the aloe gel industry uses essentially no anthraquinone quality control. The IASC does not employ anthraquinones in its certification process because the content of these molecules radically changes during the charcoal absorption of routine processing. Furthermore, the cathartic action of anthraquinones by and large does not enter into the claims promulgated by either the cosmetic or drink industries that utilize *Aloe barbadensis* gel and WLE extracts. Therefore, in the U.S. anthraquinones remain of research interest only.

Although TLC analysis is not strictly quantitative, it is capable of analyzing dozens of samples an hour. We use Merck G-60 silica gel plates, developed with toluene – ethyl acetate – methanol – water – formic acid = 10:40:12:6:3. The only post-development technique used is to heat the plates and expose them to ammonia vapor in order to develop THA zones. For the determination of aloin, aloe emodin and the chromone glycosides and their esters, several HPLC techniques were used, all with C18 reversed phase columns. The simplest separation of aloin and the chromones uses isocratic 70% aqueous methanol. Tests for anthraquinone aglycones uses isocratic 90% aqueous methanol. In order to measure both the aglycones and the glycosides a gradient of 30% to 80% methanol in water is useful. In all cases elution is monitored by UV-absorption. For estimation of single compounds, the λ maximum of that compound is used. When multiple compounds are being analysed, 256 or 280 nm is used as a universal wavelength. Unfortunately, the only reference compounds available in useful purity are aloe emodin or barbaloin (Sigma-Aldrich Co. Ltd).

In the future, measuring the presence and amount of anthraquinones in aloe gel and WLE preparations applied to the skin or orally ingested may become more important. On the beneficial side, we demonstrate in Figure 8.11, that activated charcoal treatment removes a number of biological activities perhaps due to either anthraquinones or chromones. On the regulatory side, Europeans have traditionally closely monitored anthraquinone content in products because of the concern over the potential carcinogenicity of these compounds. This began when Mori *et al.* (1985) found chryazin caused gastrointestinal cancer in rats. These concerns have been extended to aloe emodin (Westendorf *et al.*, 1990; Wolfe *et al.*, 1990; Heidemann *et al.*, 1996). This controversy has passed relatively unnoticed in America. Recently (Strickland *et al.*, 2000) it was found that topically applied aloe emodin in ethanol solution interacts with ultraviolet radiation to cause the development of pigmented skin tumors. Scientifically these observations are extremely important because hitherto it has been very difficult to develop murine models for melanoma (Bardeesy *et al.*, 2000; Berkelhammer *et al.*, 1982; Kelsall and Mintz, 1998; Kusewitt and Ley, 1996; Romerdahl *et al.*, 1989; Takizawa *et al.*, 1985). It is likely that in the future anthraquinones will come under increasing scrutiny.

Polysaccharides

Besides the anthraquinones, polysaccharides are the most studied aloe components. Strickland *et al.* (Chapter 12) critically review the various studies of the structure of aloe

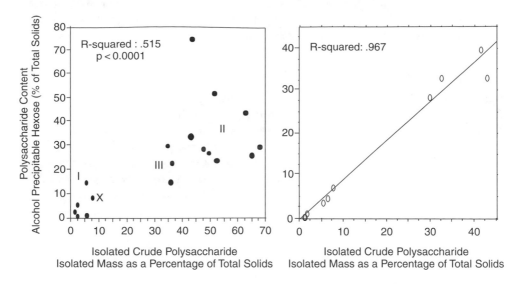

Figure 8.14 Relationship between aloe polysaccharide measured by alcohol precipitable hexose assay (Y ordinate) and polysaccharide isolated by precipitation and dialysis (X ordinate). Data in the left panel are from Study A (Table 8.12) and data in the right panel are from Study B (Table 8.11). Figures are derived from Figures 1 and 4 of Pelley *et al.* (1998).

polysaccharides since the middle of the twentieth century. That chapter also presents our synthesis of the structure and mode of breakdown of the native *A. barbadensis* polysaccharide. We have also examined some aspects by which polysaccharide is intentionally or unsuspectingly broken down during industrial processing. With these findings as background we will briefly review quality control tests currently used in the aloe industry and will examine commercial materials for their polysaccharide type and content.

The first assays for aloe polysaccharides involved the physical isolation of the polysaccharide (Ps Isolation) by dialysis and alcohol precipitation followed by the direct determination of hexose content (Roboz and Haagen-Smit, 1948; Segal *et al.*, 1968; Yagi *et al.*, 1977; Paulsen *et al.*, 1978; Gowda *et al.*, 1979; Solar *et al.*, 1979; Mandal and Das, 1980a, b; Gowda, 1980; Mandal *et al.*, 1983; Hadiabi *et al.*, 1983; Radjabi-Nassab *et al.*, 1984; Yagi *et al.*, 1984, 1986). Although Ps isolation is extremely cumbersome, it unquestionably establishes the amount and identity of the polysaccharide under consideration (Pelley *et al.*, 1998). In the above studies the isolated polysaccharides were then hydrolysed, and the constituent sugars identified and quantified. There have been two notable, related departures from this basic procedure: (i) the 'isolation' of polysaccharide by McAnalley (1988, 1993); and, (ii) the promotion of an unpublished, unverified test for polysaccharides – the methanol precipitable solids (MPS) test. In both cases the flaw was the assumption that everything which precipitates with alcohol is polysaccharide. In reality (e.g. Table 8.11, Polysaccharide Content versus Alcohol Precipitated Solids) polysaccharides isolated by the standard, laborious method above are one-fifth to one-twentieth of the MPS, most of which is a dialyzable complex of metallic cations and organic acids. The MPS test, promoted by a practising physician, defined polysaccharide as that mass which is precipitated from aloe by methanol. McAnalley 'isolated polysaccharide' by alcohol precipitation alone, without the

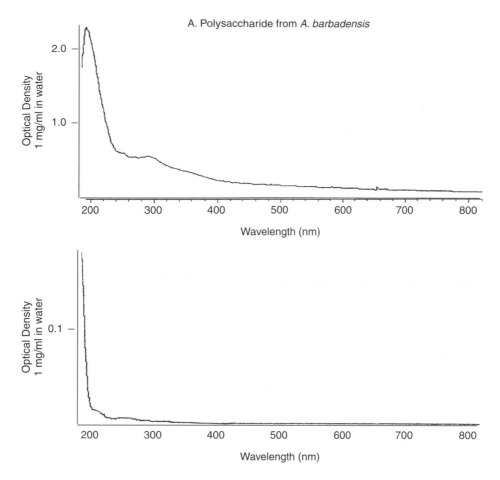

Figure 8.15 Ultraviolet and visible spectra of isolated *A. barbadensis* polysaccharide (A) and commercial maltodextrin (B). These figures are modified from Pelley *et al.* (1998). It should be noted that the scale of Panel A is ten fold greater than the scale of Panel B.

time-consuming dialysis step. This yielded highly impure material, which was distributed to unwitting collaborating biologists (see Strickland and Pelley, 2001). These two assumptions set research back a decade and cost the industry millions by allowing maltodextrin adulterated materials to pass as pure aloe (Pelley *et al.*, 1998).

There have been three legitimate attempts, of increasing scientific sophistication, to quantify aloe polysaccharide: alcohol precipitable hexose or APH (Pelley *et al.*, 1993), size exclusion HPLC (Ross *et al.*, 1997); and, proton NMR (Diehl and Ockels, 1997), each of which has its advantages and disadvantages. McAnalley (1988) first promulgated the use of size exclusion HPLC to study the breakdown of aloe polysaccharides. Ross adapted this technique (without attribution) to employ a column with a somewhat higher molecular weight range, while still monitoring by refractive index, a nonspecific method. The method is further flawed by a lack of validation of quantitation, absence of rigorous standards, lack of specificity controls, and failure to consider that the filtration employed in sample preparation might indeed remove the very highest molecular weight polysaccharide. Despite this lack of knowledge of polysaccharide chemistry in general and a lack of critical thinking about aloes in particular, this study represented a significant advance over McAnalley (1988). With proper attention to controls, both for calibration and specificity, this method may yet achieve its potential for examining commercial materials whose native polysaccharide has broken down. An even greater advance is the work of Diehl and Ockels (1997) using ^{1}H NMR. They appear to focus on the coupling of the mannose 2, 3 and 6 acetyl groups to the polysaccharide backbone and they make an attempt at rough calibration. The major drawback to the method is the cost of the equipment which restricts its use to reference laboratories. In this respect, proton NMR is like FTIR, a useful probe for validating the structure of isolated polysaccharide but not the method of choice for examining large numbers of samples.

Our exploration of aloe polysaccharides extends back a decade to a search for techniques that would allow us to differentiate adulterated aloe from legitimate materials (Pelley *et al.*, 1993). As discussed earlier we had as tools the IASC Certification Parameters and the ability to determine solids, ions and organic acids. Thus our diagnosis of adulteration would not depend on a single parameter. But from the start we felt polysaccharides were important and were willing to spend the time on Ps isolation by classical means, dialysis and alcohol precipitation from every sample we examined. Then the polysaccharides could be examined by definitive tests, FTIR and UV/Vis spectra, gel filtration and most of all, determination of sugar composition. Initially, we performed intertest correlation, rather than just internal calibration, between Ps isolation and our primary, crude assay of polysaccharide, the APH test, which was derived from existing industrial tests (AOAC, 2000a).

These comparisons revealed, not surprisingly, that there was a correlation between the weight of polysaccharide isolated by dialysis and ethanol precipitation and the amount of hexose precipitated by ethanol (APH). The major source of the variability above, particularly apparent in the left panel, is scatter among adulterative maltodextrins. Those with prominent low molecular weight components react differently in the two tests. It should be noted that there is not a correlation between Ps isolation and the results of the methanol-precipitable solids MPS test. A scattergram is not given but simple inspection of the data in Table 8.11 reveals the lack of correlation when APH (Polysaccharide Content) and/or Ps are compared with MPS. These lack of correlations exist because the large mass of the alcohol insoluble complex of organic acids and divalent cations (Figure 8.1, 2nd column from left) is removed by dialysis during polysaccharide isolation and the organic acids/divalent cations do not react in the Dubois assay for hexose (Dubois *et al.*, 1956; Pelley *et al.*, 1998). Although these two principles have been well established in biochemistry for half a century, they were neglected by the aloe industry.

Our confidence in the APH and Ps isolation assays was further enhanced because of the ability to perform various physical and chemical tests upon the isolated polysaccharide and check these with literature values. The first and traditional assay is FTIR spectroscopy, which has been heavily employed by McAnalley (1988, 1995). This relies on the prominence of the acetyl groups on the glucomannan to identify the polysaccharide as being from aloe (maltodextrin is not acetylated). Although this in confirmed by FTIR, this assay is not highly sensitive to maltodextrin adulteration. Somewhat more useful was the UV/VIS spectra (above). Maltodextrin, like most simple dextrans, had very little absorption in the range of 200 to 300 nm.

The strong absorbance at around 200 nm is probably due to the acetyl groups. More puzzling is the absorbance around 280 nm This was not due to contamination of the aloe polysaccharide with protein because hydrolysis and amino acid analysis documented the virtual absence of protein. We suspect that this absorbance is due to coupling of anthraquinones or chromones to the sugars, possibly through a Schiff's base intermediate. The band is uniformly present in ARF Standard Sample polysaccharide. Eight samples had an O.D. of 1.80 ± 0.35 (mean $\pm$ SEM, 1 mg/ml solution, Table 8.12). This 280 nm absorbing material was also generally present in authentic aloe commercial material. The seven powder materials in Table 8.12A had an OD_{280} of 0.57 ± 0.18 (mean $\pm$ SEM) and six of the seven had OD_{280} of 0.18 or higher. On the other hand, authentic malto-dextrin had very low OD_{280} (0.01), as did nine commercial materials alleged to be aloe, consistent with pure maltodextrin by multiparameter analysis (0.02 ± 0.01, mean $\pm$ SEM). Commercial material alleged to be 100% aloe but consistent upon multiparameter analysis with spray-dried aloe with 50% maltodextrin, yielded intermediate results – OD_{280} 0.06 ± 0.02 (mean $\pm$ SEM). Results with liquid samples were less definite, probably because the most prevalent mode of adulteration was with surfactants, which may have significant UV absorbance. This parameter obviously deserves further exploration.

The most important analytical parameter of a polysaccharide is the sugar composition. Ps isolation allowed us to examine sugar composition in the wide variety of commercial materials we studied (Tables 8.10, 8.11 and 8.12). The ARF Standard Samples had a polysaccharide mixture that, upon hydrolysis, yielded a sugar composition (mean $\pm$ SEM) of glucose $7 \pm 1\%$ and mannose $85 \pm 1\%$ (Table 8.12). This approximates to the 1:12 glucose: mannose ratio published in the literature (Gowda *et al.*, 1979; Mandal and Das, 1980a) confirming that we are, indeed, dealing with authentic *A. barbadensis* polysaccharide. The mixture of polysaccharides isolated also contains 4% galactose. This is consistent with a mixture of 23 parts glucomannan polysaccharide to one part galactan polysaccharide. Again, this is in accord with the literature on *A. barbadensis* polysaccharides, which describe a galactan as a minor constituent in the crude polysac-charide (Mandal and Das, 1980b). Commercial maltodextrins are essentially dextrans (i.e. they are predominantly glucose) and vary only slightly in their composition. The maltodextrin standard in Table 8.11 was 91% glucose with <0.1% mannose and <0.1%, galactose, whereas the maltodextrin standard in Table 8.12 was 93% glucose, 7% mannose and 0% galactose.

Lastly among the defined materials are the polysaccharides isolated from WLE treated with cellulase. The three ARF WLE standards had a sugar composition (Table 8.10, note j) of 54% glucose, 25% mannose and 9% galactose, which is consistent with a breakdown of half of the glucomannan. The WLE standard utilized in Table 8.11 on the other hand had a sugar composition of 19% glucose, 43% mannose and 29% galactose which suggests that the mix of polysaccharide had changed from 23 parts glucomannan: one part galactan to two parts glucomannan: one part galactan. This represents a 90% breakdown of the glucomannan. Thus we have three distinct patterns of sugar composition to consider: mannose-rich glucomannan, characteristic of native *A. barbadensis*; glucose-rich dextran, characteristic of maltodextrin adulterant; and, galactose rich galactan which is the limit digest of cellulase cleavage of native *A. barbadensis* polysac-charide.

With these findings in hand we can now examine the polysaccharides of commercial materials purporting to be aloe with regard to polysaccharide content and composition. In Table 8.11 two commercial gel materials of defined provenance are illustrated (2 and 3) that are almost equivalent to ARF Standard Samples (APH, 7.9%; Ps Isolated, 5.9%; 5.2% glucose: 91.7% mannose: 2.3% galactose versus ARF Standard; APH, 6.2%; Ps Isolated, 5.4%; 8% glucose: 81% mannose: 2% galactose). In Table 8.12A four powdered materials (1 and 4–6) are identified that are also very much like ARF Standard gel Samples (APH, 6.7% average; Ps Isolated, 4.6%; 6.5% glucose: 81% mannose: 6.8% galactose versus ARF Standard; APH, 6.2%; Ps Isolated, 5.4%; 8% glucose: 81% mannose: 2% galactose). Finally, four liquid commercial materials (Table 8.10: 1–4) had polysaccharides very much like the ARF standards (APH, 7.5% average; Ps Isolated, 12.0%; 11.8% glucose: 83% mannose: 2.3% galactose versus ARF Standard; APH, 6.2%; Ps Isolated, 5.4%; 8% glucose: 81% mannose: 2% galactose).

A second group of commercial materials were identified whose polysaccharides resembled the mixture of degraded glucomannan and cellulase-resistant galactan characteristic of WLE. In the commercial powdered materials of defined provenance (Table 8.11: 5,6) these had a reduced polysaccharide content (APH, 1.4%; Ps Isolated, 0.7% versus ARF WLE Standard Sample; APH 1.0%; Ps Isolated, 0.3%). The sugar composition was suggestive of a decreased content of glucomannan and an increase in cellulase-resistant galactan (24.2% glucose: 17.8% mannose: 16.8% galactose versus ARF WLE Standard Samples; 19% glucose: 43% mannose: 29% galactose). Similar findings were observed in the study described in Table 8.12A. Commercial powdered materials 2–3 and 7 had reduced levels of polysaccharide (APH, 5.0%; Ps Isolated, 3.6% versus ARF WLE Standard Sample; APH 1.0%; Ps Isolated, 0.3%). There was also a glucomannan-depleted sugar composition suggestive of cellulase digestion (28% glucose: 48% mannose: 14% galactose versus ARF WLE Standard Samples; 19% glucose: 43% mannose: 29% galactose) to an extent where 80% of the glucomannan had broken down. Lastly, among commercial liquid materials there was a subgroup (Table 8.10: 5–7) suggestive of WLE derivation (APH, 3.3%; Ps Isolated, 3.3%; 54% glucose: 24.7% mannose: 9.3% galactose versus ARF WLE Standard Samples; APH 1.0%; Ps Isolated, 0.3%; 19% glucose: 43% mannose: 29% galactose). In summary, the commercial material most resembling authentic aloe material can be divided into two groups based upon their polysaccharide composition, one consistent with a gel origin and one consistent with a WLE origin. Commercial gel has a polysaccharide content about twice as high as commercial WLE. The dominant polysaccharide is glucomannan, which comprises 90+% of all polysaccharides. This glucomannan is characterized by a 1:10 ratio of glucose to mannose. In good quality WLE about two-thirds of the acetylated mannan has been completely broken down by cellulases but the glucomannan still comprises a majority of the polysaccharide.

Analysis of polysaccharide is critically helpful in confirming that misrepresented and adulterated commercial materials are indeed different from authentic aloe materials. In Table 8.11 materials 8–10 have a different polysaccharide content than authentic aloe (8–10; APH, 45.3% APH polysaccharide versus 7.2% polysaccharide in gel and 3.7% polysaccharide in WLE). This polysaccharide content is similar to that observed in maltodextrin (51.7% APH). Similarly the sugar composition (Table 8.10: 8–10) of these materials (73.1% glucose: 3.4% mannose: <0.1% galactose) is much closer to that of commercial maltodextrin (91.0% glucose: <1% mannose: <1% galactose) than it is to either commercial gel (8% glucose: 84% mannose: 4% galactose) or commercial WLE gel (37% glucose: 32% mannose: 13% galactose). A similar pattern is observed with commercial material (Table 8.12B), of less closely defined provenance, more closely identified with maltodextrin than authentic aloe. The polysaccharide content of these

Table 8.13 Sugar Composition of Commercial Materials Consistent by Simple Tests with Authentic 'Aloe vera'.[a]

Group	Polysaccharide content[b]	Sugar composition[c]			Ionic strength[f]
		Glucose	Mannose	Galactose	
Gel[d] n=10	7.2±1.3%	8.3±1.7%	84.0±1.6%	4.1±0.8%	2,453±226 μ Siemens
WLE[e] n=8	3.7±1.0%	37.0±8.5	31.8±9.1%	13.1±2.3%	1,970±212 μ Siemens

Notes

a All values are the Mean ± SEM.

b By alcohol precipitable hexose (APH) Assay.

c After hydrolysis, results from Dionex sugar analyser.

d Samples consist of Table 8.12A: 1, 4–6; Table 8.11: 2, 3; Table 8.10: 1–4.

e Samples consist of Table 8.12A: 2, 3, 7; Table 8.11: 5, 6; Table 8.10: 5–7.

f At a concentration of 0.62 g per dl.

materials (Table 8.12B: 39.9% ± 5.6%) is clearly closer to maltodextrin (47.1%) than to either authentic commercial aloe gel (7.2±1.3%) or commercial aloe WLE (3.7±1.0%). The only example where polysaccharide composition is not consistently altered with adulteration is in liquid commercial materials (Table 8.10: 8–13). This is because commercial materials are not as frequently adulterated with maltodextrin (only one sample of six is consistent with maltodextrin adulteration – Table 8.10: 9). In summary, although the isolation of polysaccharide from commercial material is laborious, there is no method more effective in establishing maltodextrin adulteration of a material that is alleged to be pure aloe.

ACKNOWLEDGEMENTS

This work was supported by grants CA80423 (TAW and RPP) and CA70383 (FMS) from the National Institutes of Health. We wish to acknowledge and thank Lew Sheffield and Bob Davis, our colleagues in the Aloe Research Foundation Inter-University Co-Operative Study and Yin-Tung Wang, our colleague in establishing the chemical basis for aloe quality control. Lastly we wish to thank Gene Hale, Executive Director of the IASC for putting up with all of us for over a decade. He is the one who should be consulted for the details and mechanics of IASC Certification at 414 E. Airport Freeway, Suite 260, Irving TX 75062.

REFERENCES

Aleshkina, J.A. and Rostotskii, B.K. (1957) An aloe emulsion – a new medicinal preparation. *Meditsinkaya Promyshlennost USSR*, **II**, 54–55.

AOAC (2000a) Dextran in raw cane sugar. Robert's copper method, final action, 1990. *AOAC Official Methods of Analysis*, 988.12.

AOAC (2000b) L-Malic/total malic acid ratio in apple juice. Liquid chromatographic method (total malic acid), enzymatic method (L-malic acid), 1997. *AOAC Official Methods of Analysis*, 993.05.

Ashley, F.L. Oloughlin, B.J., Peterson, R., Fernandez, L., Stein, H. and Schwartz, A.N. (1957) The use of Aloe vera in the treatment of thermal and irradiation burns in laboratory animals and humans. *Plastic and Reconstructive Surgery*, 20, 383–396.

Bardeesy, N., Wong, K-K., DePinho, R.A. and Chin, L. (2000) Animal models of melanoma: Recent advances and future prospects. *Advances in Cancer Research*, 79, 123–156.

Berkelhammer, J., Oxenhandler, R.W., Hook, R.R. and Hennessy, J.M. (1982) Development of a new melanoma model in C57BL/6 mice. *Cancer Research*, 42, 3157–3163.

Bohme, H. and Kreutzig, L. (1963) Zur Papier- und Dunnschichtchromatographie von Aloe-drogen *Deutsche Apotheker-Zeitgtung*, 103, 505–508.

Byeon, S.W., Pelley, R.P., Ullrich, S.E., Waller, T.A., Bucana, C.D. and Strickland, F.M. (1998) Mechanism of action of *Aloe barbadensis* extracts in preventing ultraviolet radiation-induced immune suppression I. The role of keratinocyte-derived IL-10. *Journal of Investigative Dermatology*, 110, 811–817.

Coats, B.C. (Dec. 11th, 1979) Hypoallergenic stabilized Aloe vera (sic) gel. *U.S. Patent* No. 4, 178, 372.

Coats, B.C. (Oct. 18th 1994) Method of processing stabilized Aloe vera Gel obtained from the whole Aloe vera leaf. *U.S. Patent* No. 5,356,811.

Cobble, H.H. (Jul. 1st, 1975) Stabilized Aloe vera gel (sic) and preparation of same. *U.S. Patent* No. 3,892,853.

Cottrell, T.A. (Dec. 18th, 1984) Method and apparatus for extracting Aloe vera (sic) gel. *U.S. Patent* No. 4,488,482.

Dalton, T. and Cupp, M.J. (2000) Aloe. In *Forensic Science: Toxicology and Clinical Pharmacology of Herbal Products*, edited by M.J. Cupp, pp. 259–272. Totowa, N.J. : Humana Press.

Davis, R.H., Kabbani, J.M. and Maro, N.P. (1986) *Aloe vera* and inflammation. *Proceedings of the Pennsylvania Academy of Sciences*, 60, 67–70.

Davis, R.H., Rosenthal, K.Y., Cesario, L.R. and Rouw, G.A. (1987) Processed Aloe vera administered topically inhibits inflammation. *Journal of the American Podiatric Medicine Association*, 79, 395–397.

Diehl, B.W.K. and Ockels, W. (1997) ^{3}H-NMR-spectroscopy: The difference between Aloe vera (sic) and 'Aloe falsa'. Privately circulated transcript of talk presented at the 'IASC Work Shop in Cosmetics', May 5th, 1997, Dusseldorf, Germany.

Difco Laboratories. (1986) *Difco Manual*. 10th Edn., Detroit: Difco Laboratories.

Dubois, M., Gilles, K.A., Hamilton, J.K., Rebers, P.A. and Smith, F. (1956) Colorimetric method for determination of sugars and related substances. *Analytical Chemistry*, 28: 350–356.

Ernst, E. (2000) Adverse effects of herbal drugs in dermatology. *British Journal of Dermatology*, 143, 923–929.

Farkas, A. (Dec. 1967) Aloe polysaccharide compositions. *U.S. Patent* No. 3,360,511.

Genet, W.B.M. and van Schooten, C.A.M. (1992) Water requirement of *Aloe vera* in a dry Caribbean climate. *Irrigation Science*, 13, 81–85.

Gowda, D.C., Neelisiddaiah, B. and Anjaneyalu, Y.V. (1979) Structural studies of polysaccharides from *Aloe vera*. *Carbohydrate Research*, 72, 201–205.

Gowda, D.C. (1980) Structural studies of polysaccharides from *Aloe saponaria* and *Aloe vanbalenii*. *Carbohydrate Research*, 83, 402–405.

Graf, E. and Alexa, M. (1980) Uber die Stabilität der diastereomeren Aloien A und B sowie ihr Haupt-Zersetzungs-Produkt 4-Hydroxyaloin. *Planta Medica*, 38, 121–127.

Grindley, D. and Reynolds, T. (1986) The aloe vera phenomenon: A review of the properties and modern uses of the leaf parenchyma gel. *Journal of Ethanopharmacology*, 16, 117–151.

Hadiabi, F., Amar, C. and Vilkas, E. (1983) Structural studies of the glucomannan from *Aloe vahombe*. *Carbohydrate Research*, 116, 166–170.

Heggers, J.P., Pelley, R.P. and Robson, M.C. (1993) Beneficial effects of *Aloe* in wound healing. *Phytotherapy Research*, 7, S48–S52.

Heidemann A., Volkner,W. and Mengs, U. (1996) Genotoxicity of aloe emodin in vitro and in vivo. *Mutation Research*, **367**, 123–133.

IASC, Science and Technical Committee. (1991) Official Certification Program For Aloe Vera. International Aloe Science Council, Inc. 1–22.

Jobst, K.A. (1999) Herbal medicine legislation and registration and stretching the mind: Mental exercise for health? *Journal of Alternative and Complementary Medicine*, **5**, 107–108.

Kelsall, S.R., Mintz, B. (1998) Metastatic cutaneous melanoma promoted by ultraviolet radiation in mice with transgene-initiated low melanoma susceptibility. *Cancer Research*, **58**, 4061–4065.

Kusewitt, D.F. and Ley, R.D. (1996) Animal models of melanoma. *Cancer Survey*, **26**, 35–70.

Lee, C.K., Han, S.S., Shin, Y.K., Chung, M.H., Park, Y.I., Lee, S.K. *et al*. (1999) Prevention of ultraviolet radiation-induced suppression of contact hypersensitivity by Aloe Vera gel components. *International Journal of Immunopharmacology*, **21**:303–310.

Mandal, G. and Das, A. (1980a) Structure of the glucomannan isolated from the leaves of *Aloe barbadensis* Miller. *Carbohydrate Research*, **87**, 249–256 .

Mandal, G. and Das, A. (1980b) Structure of the D-galactan isolated from *Aloe barbadensis* Miller. *Carbohydrate Research*, **86**, 247–257.

Mandal, G., Ghosh, R. and Das, A. (1983) Characterization of polysaccharides of *Aloe barbadensis* Miller: Part III – Structure of an acidic oligosaccharide. *Indian Journal of Chemistry*, **22B**, 890–893.

Manna, S., McAnalley, B.H. and Ammon, H.L. (1993) *Carbohydrate Research*, **243**.

McAnalley, B.H. (Apr. 5th 1988) Process for preparation of Aloe products, products produced thereby and compositions thereof. *U.S. Patent* No. 4,735,935.

Other related patents are *U.S. Patent* Nos. 4,851,224 of Jul. 1989, 4,959,214 of Sept. 1990, and 4,966,892 of Oct. 1990.

The patents subsequent to the '935 patent (the '224, '214, and '892 patents) are *Divisionals and Continuations In Part* of the '935 patent. As such, the scientific content (*The Specification*) is virtually identical to that of the '935 patent but The Claims of these patents differ very significantly from the '935 patent (legally that is what is important).

McAnalley, B.H., Carpenter, R.H. and McDaniel, H.R. (Aug. 15th 1995) Uses of Aloe products. *U.S. Patent* No. 5,441,943.

Other related patents are *U.S. Patent* Nos. 5,118,673, Jun., 1992 and 5,106,616, Apr., 1992.

These patents, although legally linked to the series of patents above as *Divisionals and Coninuations In Part*, have *Specification* sections (the part that contains the data), completely different from the '935 series of patents above. As such, any one of this series of patents could be equally read by scientists seeking information different from the '935 series above.

McIntyre, M. (1999) Protecting the availability of herbal medicine. *Journal of Alternative and Complementary Medicine*, **5**, 109–110.

McIntyre, M. (1999) Alternative licensing for herbal medicine-like products in the European Union. *Journal of Alternative and Complementary Medicine*, **5**, 110–113.

Mori, H., Sugie, S., Niwa, K., Takahashi, M. and Kawai, K. (1985) Induction of intestinal tumors in rats by chryazin. *British Journal of Cancer*, **52**, 781–783.

Paulsen, B.S., Fagerheim, E. and Overbye, E. (1978) Structural studies of the polysaccharides from *Aloe plicatilis* Miller. *Carbohydrate Research*, **60**, 345–351.

Pelley, R.P., Wang, Y-T. and Waller, T.A. (1993) Current status of quality control of *Aloe barbadensis* extracts. *SOFW Journal*. **119**, 255–163.

Pelley, R.P., Martini, W.J., Liu, D.Q., Rachui, S., Li, K-M., Waller, T.A. and Strickland, F.M. (1998) Multiparameter testing of commercial "Aloe Vera" materials and comparison to *Aloe barbadensis* Miller extracts. *Subtropical Plant Science Journal*, **50**, 1–14.

Pelley, R.P. and Strickland, F.M. (2001) Plants, polysaccharides and the treatment and prevention of neoplasia. *Critical Reviews in Oncogenesis* **12**, 1–37(*in press*).

Rauwald, H.-W. and Voetig, R. (1982) *Archive Pharmacology (Weinheim)*, **315**, 477–478.

Rauwald, H.-W. and Beil, A. (1993) *Zeitschrift Naturforschung*, **48c**, 1–4.

Radjabi-Nassab, F., Ramiliarison, C., Monneret, C. and Vilkas, E. (1984) Further studies of the glucomannan from *Aloe vahombe* (liliaceae). II. Partial hydrolyses and NMR ^{13}C studies. *Biochimie*, **66**, 563–567.

Reynolds, T. (1994), A chromatographic examination of some old samples of drug aloes. *Pharmazie*, **49**,524–529.

Roboz, E. and Haagen-Smit, A.J. (1948) A Mucilage from Aloe Vera. *Journal of the American Chemical Society*, **70**, 3248–3249.

Rodriguez-Bigas, M., Cruz, N.I. and Suarez. (1988). Comparative evaluation of aloe vera in the management of burn wounds in guinea pigs. *Plastic and Reconstructive Surgery*, **81**, 386–389.

Romerdahl, C.A., Stephens, L.C., Bucana, C. and Kripke, M.L. (1989) Role of ultraviolet radiation in the induction of melanocytic skin tumors in inbred mice. *Cancer Communications*, **1**, 209–216.

Ross, S.A., ElSohly, M.A. and Wilkins, S.P. (1997) Quantitative analysis of *Aloe vera* (sic) mucilaginous polysaccharide in commercial *Aloe vera* (sic) products. *Journal of AOAC International*, **80**, 455–457.

Rovatti, B. and Brennan, R.J. (1959). Experimental thermal burns. *Industrial Medicine and Surgery*, **28**, 364–368.

Segal et al. (1968) A reinvestigation of the polysaccharide material from *Aloe Vera* Mucilage. *Lloydia* Proceedings, **31**, 423–424.

Solar, S., Zeller, H., Rasolofonirina, N., Coulanges, P., Ralamboranto, L., Andriat-Simahavandy, A.A. et al. (1979) Mis en evidence et etude des proprietes immunostimulantes d'un extrait isole et partiellement purifie a partir d'Aloe vahombe. *Archives Institute Pasteur Madagascar*, **47**, 1–31.

Specter, M. (5th Feb., 2001) Annals of medicine: The outlaw doctor: Cancer researchers used to call him a fraud. What's changed? *The New Yorker*, 48–61.

Strickland, F.M., Pelley, R.P. and Kripke, M.L. (1994) Prevention of ultraviolet radiation-induced suppression of contact and delayed hypersensitivity by *Aloe barbadensis* gel. *Journal of Investigative Dermatology*, **102**, 197–204.

Strickland, F.M. and Kripke, M.L. (1997) Immune response associated with nonmelanoma skin cancer. *Clinics in Plastic Surgery*, **24**, 637–647.

Strickland, F.M., Pelley, R.P. and Kripke, M.L. (Oct. 20th 1998) Cytoprotective oligosaccharide from Aloe preventing damage to the skin immune system by UV radiation. *U.S. Patent* No. 5,824,659.

Strickland, F.M., Darvill, A., Albersheim, P., Eberhard, S., Pauly, M. and Pelley, R.P. (1999) Inhibition of UV-induced immune suppression and Interleukin-10 production by plant polysaccharides. *Photochemistry Photobiology*, **69**, 141–147.

Strickland, F.M., Muller, H.K., Stephens, L.C., Bucana, C.D., Donawho, C.K., Sun, Y. and Pelley, R.P. (2000) Induction of primary cutaneous melanomas in C3H mice by combined treatment with ultraviolet radiation, ethanol and aloe emodin. *Photochemistry Photobiology*, **72**, 407–414.

Takizawa, H., Sato, S., Kitajima, H., Konishi, S., Iwata, K. and Hayashi, Y. (1985) Mouse skin melanoma induced in two stage chemical carcinogenesis with 7,12-dimethylbenz(a)anthracene and croton oil. *Carcinogenesis*, **6**, 921–923.

Tam, A., Pomerantz, J., O'Hagan, R. and Chin, L. (1999) Tetracyclin-regulatable expression of Ras in mouse melanocytes and melanomas. In: *Cancer Biology of the Mutant Mouse: New Methods, New Models, New Insights. Proceedings of the Keystone Symposium*, Abstract B-7.

Thompson, D. (Aug. 2nd, 1983). Aloe vera plant gel separator. *U.S. Patent* No. 4,395,942.

Upton, R. (1999) Traditional chinese medicine and the Dietary Supplement Health and Educational Act. *Journal of Alternative and Complementary Medicine*, **5**, 115–118.

Waller, G.R., Mangiofico, S. and Ritchey, C.R. (1978) A chemical investigation of *Aloe barbadensis* Miller. *Proceedings of the Oklahoma Academy of Sciences*, **58**, 69–76.

Waller, T.A., Strickland, F.M. and Pelley, R.P. (1994) Quality control and biological activity of Aloe barbadensis extracts useful in the cosmetic industry. *In Cosmetics and Toiletries Manufacture Worldwide* pp. 64–80 (ed: Martin Caine), Pub: Aston Publishing Group, London.

Wang, Y-T. and Strong, K.J. (1993) Monitoring physical and chemical properties of freshly harvested field-grown *Aloe vera* (sic) leaves. *Phytotherapy Research*, 7, S1–S4.

Wang, Y-T. and Strong, K.J. (1995) A two-year study monitoring several physical and chemical properties of field-grown *Aloe barbadensis* Miller leaves. *Subtropical Plant Science*, 47, 34–38.

Westendorf, J., Marquardt, H., Poginsky, B., Dominiak, M., Schmidt, J. and Marquardt, H. (1990) Genotoxicity of naturally occurring hydroxyanthraquinones. *Mutation Research*, 240, 1–12.

Williams, M.S., Burk, M., Loprinzi, C.L., Hill, M., Schomberg, P.J., Nearhood, K. et al. (1996) Phase III double-blind evaluation of an aloe vera gel as a prophylatic agent for radiation-induced skin toxicity. *International Journal of Radiation Oncology Biology and Physics*, 36, 345–349.

Wolfle, D., Schmutte, C., Westendorf, J. and Marquardt, H. (1990) Hydroxyanthraquinones as tumor promoters: Enhancement of malignant transformation of C3H mouse fibroblasts and growth stimulation of primary rat hepatocytes. *Cancer Research*, 50, 6540–6544.

Yagi, A., Makino, K., Nishioka, I. and Kuchino, Y. (1977) Aloe mannan, polysaccharide, from *Aloe arborescens* var. *natalensis. Planta Medica*, 31, 17–20.

Yagi, A., Hamada, K., Mihashi, K., Harada, N. and Nishioka, I. (1984) Structure determination of polysaccharides in *Aloe saponaria* (Hill.) Haw. (Liliaceae). *Journal of Pharmaceutical Science*, 73: 62–65.

Yagi, A., Nishimura, H., Shida, T. and Nishioka, I. (1986) Structure determination of polysaccharides in *Aloe arborescens* var. *natalensis. Planta Medica*, 213–218.

Part 3

Therapeutic activity of aloes

9 Healing powers of aloes

Nicola Mascolo, Angelo A. Izzo, Francesca Borrelli, Raffaele Capasso, Giulia Di Carlo, Lidia Sautebin and Francesco Capasso

ABSTRACT

Aloe is a medicinal plant that has maintained its popularity over the course of time. Three distinct preparations of aloe plants are mostly used in a medicinal capacity: aloe latex (=aloe); aloe gel (=aloe vera); and, aloe whole leaf (=aloe extract). Aloe latex is used for its laxative effect; aloe gel is used topically for skin ailments, such as wound healing, psoriasis, genital herpes and internally by oral administration in diabetic and hyperlipidaemic patients and to heal gastric ulcers; and, aloe extract is potentially useful for cancer and AIDS. The use of honey may make the aloe extract therapy palatable and more efficient.

Aloe preparations, especially aloe gel, have been reported to be chemically unstable and may deteriorate over a short time period. In addition, hot water extracts may not contain adequate concentrations of active ingredients and purified fractions may be required in animal studies and clinical trials. Therefore it should be kept in mind that, in some cases, the accuracy of the listed actions may be uncertain and should be verified by further studies.

INTRODUCTION

There are at least 600 known species of *Aloe* (Family *Liliaceae*) (Kawai *et al.*, 1993), many of which have been used as botanical medicines in many countries for thousands of years (Grindlay and Reynolds, 1986; Gjerstad and Riner, 1968; Reynolds and Dweck, 1999; Swanson, 1995). Species of the genus *Aloe* are indigenous to Africa (*A. ferox* Miller, *A. africana* Miller, *A. spicata* Baker, *A. platylepis* Baker, *A. candelabrum* Berger) and Socotra (*A. perryi* Baker, *A. forbesii* Balf.fil.). Some have been introduced in Asia (*A. chinensis* Baker), the Barbados Islands in Central America (*A. barbadensis* Miller, otherwise known as *A. vera* [L.] Burm. or *A. vulgaris* Lamarck) and Europe (*A. arborescens* Miller). *Aloe ferox*, known in commerce as 'Cape aloe,' is easy to hybridise and cultivate in Africa. The term 'Cape aloe,' for accuracy, refers to the dried latex of the leaves of several species of the *Aloe* genus, especially *A. ferox*, and the hybrids and preparations made from them (Blumenthal, 1998). *Aloe barbadensis*, known in commerce as 'Curacao aloe,' was said to be native to northern Africa but was introduced into the Barbados Islands in the seventeenth century and is now cultivated in Florida, USA (Bruneton, 1999). *A. chinensis*, a variety of *A. vera*, was introduced into Curacao from China in 1817 by Anderson. The plant was cultivated in the Barbados Islands until the end of the nineteenth century. Curacao aloe is often called Barbados aloe. Other varieties are cultivated throughout India while some grow wild on the coasts of Bombay, Gujarat, southern Arabia, Madagascar and areas surrounding the Red Sea and the Mediterranean Sea (Morton, 1961; Kapoor, 1990). At the present time the principal areas for production of aloes are South Africa, Venezuela, Haiti, Florida and the Dutch islands of Aruba and Bonaire. The plant grows very well if adequately protected from cold weather; aloes are injured at 2 °C and generally killed at −1 °C.

The genus *Aloe* includes trees (e.g. *A. ferox*: Figure 9.1) of variable height (from 2 to 15 metres), shrubs and herbs (*A. barbadensis*). They are succulent plants with perennial, strong and fibrous roots and numerous (15–30) large, fleshly leaves, carrying spines at the margin. In some species the leaves form a rosette at ground level (*A. perryi*). The flowers are grouped in erect, terminal spikes, and are borne by a floral stalk, which is either unique (*A. vera*) or ramified (*A. ferox*); the corolla is tubular, divided into narrow segments (six) at the mouth and of a red (*A. perryi*), yellow (*A. vera*) or white (*A. speciosa* Baker) colour (Dezani and Guidetti, 1953). In some aloes (*A. arborescens*) the vascular cambium develops with age, initiating extensive secondary growth and allowing considerable lateral expansion (Cotton, 1997).

Aloes resemble to some extent the agave or century plant (*Agave americana* L., Family *Amaryllidaceae*); however, the aloe plant is in flower during the greater part of the year while the agave plant is remarkable for the long interval between its periods of flowering.

Aloe leaves in section show from outer to inner: (i) a cuticularized epidermis; (ii) a parenchyma containing chlorophyll, starch and bundles of needles of calcium oxalate; (iii) large perycyclic cells; and, (iv) a central region (3/5 of the diameter of the leaf) consisting of large parenchymatous cells.

Figure 9.1 Aloe ferox.

Aloe is from the Arabic *alloeh* or the Hebrew *halal* and means a shining, bitter substance (Tyler, 1993); *ferox* is from the Latin and means ferocious or wild; *vera* is from the Latin *verus*, meaning true; *barbadensis* refers to the Barbados Islands; *africana*, or *chinensis*, refers to the habitat of the plant; *spicata* refers to the flowers in spikes. However, the word aloe in pharmacopoeias and formularies means a drug derived from the dried leaf juice. This has always created confusion because the leaves of the genus *Aloe* are the source of two products that are quite different in their chemical composition and their therapeutic properties, aloe latex and aloe gel (Capasso *etal.*, 1998). These two products are obtained from two different specialised cells, latex from pericyclic cells and gel from parenchymatous cells. Therefore the term juice must be avoided, as it could mean either the latex from the pericyclic cells, or the gel after extraction from the leaf. There is also a preparation obtained from the whole leaf (total extract) which contains all the components present in aloe leaves. Aloes contain another medically important part, the leaf epidermis (Imanishi *etal.*, 1981). There are also the aloe wood and the aloe root. The first preparation, so called lignaloe or aloe of the Bible, is a fragrant wood obtained from an entirely different plant that was once used as an incense (Tyler *etal.*, 1993). The second is a synonym of unicort root, a preparation that when dried becomes a valuable bitter tonic (Grieve, 1998).

Apart from aloe wood and aloe root which have nothing to do with the genus *Aloe*, the other preparations have very similar names that are sometimes interchanged. Aloe in one form or another is a common domestic medicine and is the basis of most pharmaceutical preparations.

HISTORY

The topical and internal effects of aloes have been known since ancient times. Nefertite and Cleopatra, two Egyptian queens, used aloes as a beauty aid. The drug was used by Dioscorides to heal skin ailments and haemorrhoids. Aloes were used by Pliny the Elder, Celsus, Galen and other famous physicians to treat wounds and gastrointestinal disturbances, but no mention of aloes was made by either Hippocrates or Theophrastus (Shelton, 1991; Hennessee, 1998). Aloe was largely prescribed by Arabian physicians and was one of the drugs-the others were balsam, scammony, tragacanth and galbanum-recommended to Alfred the Great by Helias, the Patriarch of Jerusalem (Wheelwright, 1974). Aloe's use was first discovered on a Mesopotamian clay tablet dating from 2100 B.C. Later, in 1862, a German egyptologist, George Ebers, discovered that a papyrus found in a sarcophagus near Thebes mentioned at least twelve preparations for preparing aloe to treat external and internal ailments (Atherton, 1997; Hennessee, 1998).

Aloe was considered by the ancient Greeks to be an exclusive production of the island of Socotra, in the Indian Ocean. This is why Alexander the Great, persuaded by Aristotle, his mentor, captured the island of Socotra and sent to it Greek colonists solely to preserve and cultivate the aloe plant (Evans, 1989). The drug was included in the Egyptian Book of Remedies (about 1500 B.C.), as well as in that of the Hebrews, as a laxative and dermatologic preparation. Mesopotamians were also aware of its medicinal properties by that time (Swanson, 1995). Aloe was first reported in Greek literature as a laxative before the first century (Hennessee, 1998). In the first century Dioscorides wrote of its use in treating wounds, chapping, hair loss, genital ulcers, haemorrhoids, boils, mouth irritation and inflammation (Shelton, 1991; Hennessee, 1998). In the seventh century, aloe was also used in the Orient for eczema and sinusitis (Shelton, 1991).

Aloe, when introduced into Europe, was used for constipation and skin ailments and later, in the 1930s, was used to treat radiation burns (Tyler *et al.*, 1981). From the ancient times to the seventeenth century the Socotrine aloe was the only official aloe. This vegetable remedy was imported into Europe by way of the Red Sea and Alexandria; there was a direct trade in aloes between Socotra and some medieval towns like Venice, which by the fifteenth century became the greatest trading centre of Europe. At the end of the seventeenth century it was possible to find aloes from the Barbados in Europe, and later, towards the end of the eighteenth century, the Cape Aloe too. The drug was included as a laxative in the U.S. Pharmacopoeia in 1820, in the Italian Pharmacopoeia in 1892 and in the European Pharmacopoeia in 1969. Today Socotrine aloes are very rare, while Barbados and Cape aloes are the most common.

In addition to its medicinal virtues, aloe was believed to be endowed with power against evil spirits. On this account, it was carefully planted in the neighbourhood of Mecca and hung by Mussulmans who visited the shrine of the Prophet, and hung over doorways as a religious symbol. The Mussulman name of aloe is *saber* and signifies patience, referred to the waiting-time between the burial and the resurrection morning (Grieve, 1998).

ALOE PREPARATIONS

Near the epidermis or outer skin the leaves of aloes contain a row of fibrovascular bundles, the cells of which are much enlarged and filled with a yellow latex. Aloe latex is obtained by cutting the leaf transversely close to the stem and inclining the leaf so that the latex flows out in about six hours (Figure 9.2).

No pressure must be applied otherwise the product will be contaminated with the mucilage present in the inner part of the leaf. The preparation obtained is bitter and yellow; it is concentrated to dryness by evaporation in open kettles, rarely by boiling in a vacuum, until it becomes a shiny mass, like broken glass with a yellow greenish to red-black colour (Capasso and Gaginella, 1997). A slow evaporation carried out either by inappropriate temperature or by spontaneous evaporation gives an opaque mass with a wax-like fracture. The taste is nauseating and bitter and the odour sour, recalling that of rhubarb, apple-tart or iodoform.

Aloes require two or three years standing before they yield their latex. In Africa the latex is collected from the wild plants; in the case of aloe plantations the drug is collected in April/July. Aloe latex (=aloe) contains mainly anthraquinones, cathartic compounds useful in constipation. It is totally different from the aloe gel (=aloe vera), a colourless gelatin obtained from the central portion of aloe leaf. The mucilaginous parenchyma tissue is excised from fresh leaves and immediately utilised for pharmaceutical preparations or lyophilized and kept dry until use. During extraction of the gel it is practically impossible to prevent contamination by the latex as the leaves are cut. On the other hand in intact leaves, anthraquinones may diffuse into gel from the bundle sheath cells. To reduce such contamination the starting material must be from varieties of *Aloe* with a reduced anthraquinone content. Aloe gel is sensitive to heat and light and can quickly deteriorate when exposed at high temperature. It contains mainly mucilage and it is now a familiar ingredient in cosmetics and ointments for skin ailments (Henry, 1979; Reynolds and Dweck, 1999). Some recent observations have proven that the rind and the outer leaf, normally thrown away, contain greater healing components.

Figure 9.2 Aloe ferox leaves: latex collection.

The fresh whole leaf is also cut into small pieces and whipped in order to obtain a homogenous yellowish or reddish material. The preparation (total leaf extract = aloe extract) is bitter and has a very characteristic odour. Finally, it has been reported that the leaf epidermis of *A. arborescens* contains lectin, a product that is able to inhibit the growth of a fibrosarcoma in animals through a host-mediated effect (Imanishi *et al.*, 1981). This could stimulate further research and give rise to a new medicinal preparation.

CHEMICAL CONSTITUENTS

Aloes contain anthraquinone derivatives (10% to 40%) like aloin, mucilage (30%), resinous substances (16% to 63%) like aloesin and aloesone, sugars (about 25%), polysaccharides like acemannan and betamannan, fatty acids and cholesterol, campesterol, β-sistosterol, glycoproteins (aloctins A and B), lectins, a gibberellin-like substance, enzymes such as cyclo-oxygenase and bradykininase, together with other compounds such as lupeol, salicylic acid, urea, cinnamic acid, phenol, sulphur, magnesium lactate, salicylates, and amino acids. Aloin (=barbaloin) is an impure mixture of barbaloin A and barbaloin B, which inter-convert through the anthranol form.

Anthraquinone derivatives possess laxative properties; their content is subject to seasonal variations. Aloins reach a maximum concentration in the dried leaf latex of *A. ferox* in April-July [24.1%] and a minimum concentration in winter [14.8%].

5-hydroxyaloin A, characteristic of Cape aloe, is absent in Curacao aloe. However, studies carried out on plants grown hydroponically under carefully controlled conditions still show these variations; for example, aloin content can vary as much as 80% from one plant to another in the same field. Aloes also contain other healing components such as analgesic and anti-inflammatory agents, (aloctins, cholesterol, campesterol, β-sitosterol, acemannan, salicylates, etc.), immunostimulant agents, (acemannan, lectin, etc.) and antiseptic agents (lupeol, salicylic acid, phenol, sulphur, etc.). The benefits of aloe are, however, due to synergism between compounds; also, it is quite possible that other unidentified co-factors present in aloe may provide for the optimum effects generally encountered.

PHARMACOLOGY AND THERAPEUTIC APPLICATIONS

Constipation

Aloe latex possesses laxative properties and has been used traditionally to treat constipation (Benigni *et al.*, 1962). The old practice of using aloe as a laxative drug is based on its content of anthraquinones like barbaloin, which is metabolised to the laxative aloe-emodin, isobarbaloin and chrysophanic acid (Leung, 1980; Blumenthal, 1998). The term 'aloe' (or 'aloin') refers to a crystalline, concentrated form of the dried aloe latex. In addition, aloe latex contains large amounts of a resinous material. Following oral administration the stomach is quickly reached and the time required for passage into the intestine is determined by stomach content and gastric emptying rate. Glycosides are probably chemically stable in the stomach (pH 1–3) and the sugar moiety prevents their absorption into the upper part of the gastrointestinal tract and subsequent detoxification in the liver, which protects them from breakdown in the intestine before they reach their site of action in the colon and rectum (Breimer and Baars, 1976; van Os, 1976). Once they have reached the large intestine the glycosides behave like pro-drugs, liberating the aglycones (aloe-emodin, rhein-emodin, chyrosophanol, etc.) that act as the laxatives (Figure 9.3).

The metabolism takes place in the colon, where bacterial glycosidases are able to cleave the *C*-glycosidic bond of glycosides (de Witte and Lemli, 1990). Their transformation into the active aglycones is carried out by the *Eubacterium* species of the human intestinal flora (Che *et al.*, 1991). Aloe-emodin is metabolised quickly; therefore its bioavailability is low, perhaps <10%, and the half-life is approximately 48–50 hours. Aloe-emodin and rhein are poorly absorbed. In spite of this, rhein has been recovered in urine, in breast milk, in brown fat, muscle, bones and the gonads, although in very low concentrations. In human volunteers, aloe was as active as aloin. Since aloe contains only about 20% aloin, it looks as though the latter is five times more active than aloes. This means that solubility or resorption differences as influenced by accompanying substances can always play a role. It is also important to remember that the resins in aloes are as active as anthraquinones (Ramstad, 1995); like aloe, the resins require the presence of bile acids in order to act. Aglycones like aloe-emodin and rhein act synergistically (Yagi and Yamauchi, 1999), evoking secretory and motility changes in the gut.

Although there is no doubt that aloe exerts its action on the colonic mucosa, its mechanism of action is still unclear. It is believed that aloe, or its active ingredient aloe-emodin-9-anthrone, acts by disturbing the equilibrium between the absorption of water from the intestinal lumen *via* active sodium transport (Ishii *et al.*, 1990) and the

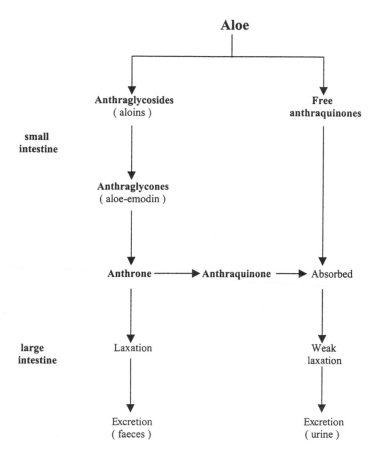

Figure 9.3 An exemplification of aloe constituents metabolism.

secretion of water into the lumen by a prostaglandin-dependent mechanism (Collier *et al.*, 1976; Capasso *et al.*, 1983). Other probable mechanisms could be an increased mucosal permeability of the epithelial cells of the colon, a rise of the level of cAMP in the enterocytes and a stimulation of nerve endings (Canigueral and Vila, 1993; Capasso and Gaginella, 1997; Capasso *et al.*, 2000). Platelet-activating factor (PAF) could also contribute to the laxative effect, as aloe-emodin stimulates the release of PAF in human ileal and colonic mucosa (Tavares *et al.*, 1996) (Table 9.1).

Furthermore, aloe, in contrast to other antraquinones drugs (cascara and senna), inhibits calcium-dependent nitric oxide synthase (NOS) activity in the rat colon, aloin being the active ingredient responsible of this activity (Izzo *et al.*, 1999). Taken in doses of 0.25 mg, aloe causes laxative action after 6–12 hours with loose bowel movements accompanied by abdominal pain. Among the anthraquinone drugs, aloe possesses the most potent action (Table 9.2) and has an effect persisting over several days (Grindlay and Reynolds, 1986). Excessive doses may cause gas and cramps.

These adverse effects, occurring after an overdosage, have reduced its use as a laxative in recent years in some countries while in others it is still widely used. Aloe is also occa-sionally used in association with coleretic and cholagogue drugs to solve atonic

Table 9.1 Mechanisms of action of aloe and its components.

Properties	Mechanisms
Laxative	Increase mucosol permeability. Increase cAMP levels. Increase PAF.
Wound healing	Stimulate fibroblast and connective tissue formation. Stimulate repair process and epitelial growth.
Anti-cancer	Stimulate the production of TNF and cytokines by macrophages. Induce apoptosis.
Anti-inflammatory	Block generation of histamine and bradykinin. Inhibit bradykinin activity. Inhibit eicosanoid formation. Inhibit PMM leucocytes infiltration. Inhibit histamine formation.
Antibacterial	Inhibit the bacterial growth.
Antiviral	Increase the production and function of cytotoxic T cells. Inhibit glycosylation of viral glycoproteins. Inhibit the replication of HIV and HSV-I.
Immunomodulant	Stimulate macrophagy to produce nitric oxide and cytokines. Enhance phagocytosis. Increase the number of circulating monocytes and macrophages. Activate the complement via the alternative pathway.
Antidiabetic	Stimulate release or synthesis of insuline.
Antipsoriatic	Inhibit oxygen consumption of cells. Reduce the size of the intracellular spaces. Induce mitochondrial damage. Retard cell division.

Table 9.2 Difference in the laxative action of anthraquinone drugs.

Drug	Comment
Cascara	Causes a mild laxative action. Its use may generate weak side-effects (griping).
Frangula	Causes a gentle laxative action comparable to cascara. Its use may generate weak side-effects (griping).
Senna	It is not as mild in its action as cascara or frangula. Its use may causes flatulence, cramps, abdominal pain. It is more widely used because it is considered cheaper than other drugs.
Rhubarb	It is more potent that senna, cascara of frangula. Its use may causes flatulence, cramps, abdominal pain.
Aloe	It is the most potent laxative drug. Its use almost always causes abdominal complaints such as flatulence, intestinal griping or colic.

constipation, some cases of acute constipation, dyschesia and before endoscopy of the gastrointestinal tract (Table 9.3). Aloe is not advisable for treating spastic constipation, constipation associated with proctitis or haemorrhoids, in the presence of chronic renal disease and for many other conditions (Table 9.3). Although no clinical evidence seems to have been published, the use of aloe is discouraged to women during menstruation, lactation or in pregnancy (Capasso and Gaginella, 1997; Boon and Smith, 1999). It has been shown that rhein, an active metabolite of aloe-emodin, passes into the breast milk

Table 9.3 Appropriate and inappropriate use of Aloe preparations.

Preparation	*Use*	
	Appropriate	*Inappropriate/controindicated*
Aloe latex	Atonic constipation Acute constipation Dyschesia Endoscopy of gastrointestinal tract	Spastic constipation Chronic constipation Constipation associated with bowel alterations (irritability, proctitis, heamorrhoids, appendicits, etc) Diarrhoea Long-term therapy Obesity Fluid retention Chronic renal disease
Aloe gel	<u>Topical</u> Burns Abrasions Bruises Cuts Herpes simplex Psoriasis <u>Oral</u> Hyperlipidaemia Diabetes Gastric ulcer Mouth ulcer	Known allergy UV radiation
Aloe extract		Cancer Asthma AIDS

but a laxative effect has not been noted in nursing infants (Lang, 1993; Blumenthal, 1998). In addition, although it had been considered abortifacient in folk medicine (Prochnow, 1911), no experimental or clinic evidence for an abortifacient action is found in the literature on aloe (Yago, 1969; Fingl, 1975). Aloe activity is increased when administered with small quantities of soap or alkaline salts. Aloe latex is available in a wide variety of forms, including red and white powders, tablets, capsules, or similar products; it contains standardised amounts of active components along with appropriate directions for use. The use of non-standardised preparations must be avoided because the pharmacological effect could be unpredictable. The laxative aloe is regulated as a drug by the FDA and similar Associations. For the German Commission E, aloe must not be used for more than one or two weeks in self-treatment of constipation. Aloe is commonly available in Italy in combination with: cascara sagrada (Grani di Vals[R]), rhubarb, Gentian (Puntualax[R]); licorice, senna, cascara (Puntuale[R]).

Wound healing properties

Aloe gel is a mucilagenous substance which is 98.5% water (Rowe and Parks, 1941). The mucilage contains predominantly polysaccharides which are partially acetylated glucomannans (Gowda *et al.*, 1979). Aloe gel has been mainly used for the treatment of minor burns, both thermal and from the sun and other skin conditions (Table 9.3) (Hormann and Korting, 1994; Capasso and Grandolini, 1999).

It is incorporated in ointments, creams and lotions and other preparations for external use. Aloe gel is also included in many cosmetics for its emollient properties and anti-ageing effects on the skin (Baruzzi and Rovasti, 1971; Danof, 1987; McKeown, 1987). Several studies have shown that activity of aloe gel varies and that it is attributed to changes in the gel composition during growth or after harvest. On the other hand there is a general agreement that aloe gel used in a fresh state, not from storage, is more able to show therapeutic properties. This variation seems due in part to the instability of its important active ingredients, acemannan and anthraquinone glycosides (Roberts and Travis, 1995; Fujita *et al.*, 1976). It is also important to detect the adulteration of commercial aloe gel powder, as the content of commercial maltodextrin, by HPLC and TLC analysis (Kim *et al.*, 1998).

Aloe gel comes in a variety of forms, such as a pure gel in a natural or decolourised state, or as a liquid concentrate (10- to 40-fold), either natural or decolourised. Aloe gel can be concentrated and dried to produce spray and freeze-dried powders. There are also formulations like Aloe vera oil, separated pulp and preparations labelled 'Aloe vera extracts' which may be highly diluted or 'reconstituted Aloe vera', meaning that the product has been prepared from powder or liquid concentrate (Fox, 1999).

In recent years some interesting studies and reviews have been published (Shelton, 1991; Bunyapraphatsara *et al.*, 1996) and several mechanisms of action for the wound healing properties of aloe gel have been suggested (Table 9.1). Aloe gel may simply act as a protective barrier (Blitz *et al.*, 1963; Ship, 1977) and increase capillary perfusion after local application (Zawacki, 1974; El Zawahry *et al.*, 1973). It may also influence the wound healing process by enhancing collagen turnover in the wound tissue (Chithra *et al.*, 1998). Aloe gel also has impressive topical demulcent properties when applied directly to broken or unbroken skin (Visuthikosoul *et al.*, 1995). However, most of the therapeutic uses of aloe gel can be justified by its anti-inflammatory, immunomodulating and anti-bacterial activities.

Anti-inflammatory activity

The anti-inflammatory activity of aloe gel has been extensively studied in an attempt to find the responsible components and their mechanism of action. For this purpose several models of inflammation (adjuvant arthritis and hind-paw oedema in rats, ear swelling in mice and rabbits, synovial pouch and burn in mice, rats and guinea-pigs) and phlogogen (carrageenan, kaolin, albumin, dextran, gelatin, mustard, croton oil, streptozocin, etc.) have been used (Davis *et al.*, 1986; Davis *et al.*, 1989; Vazquez *et al.*, 1996). Aloe gel contains small amounts of salicylic acid. Other components, such as emodin, emolin and barbaloin, are converted to salicylic acid, perhaps explaining in part the anti-inflammatory activity of aloe gel (Robson *et al.*, 1982).

The anti-inflammatory activity of aloe gel could be due to the inhibition of prosta-noids (Penneys, 1982; Hiroko *et al.*, 1989). Alternative mechanisms are inhibition of polymorphonuclear (PMN) leucocyte infiltration (Davis and Maro, 1989), inhibition of histamine formation by magnesium lactate (Klein and Penneys, 1988), or destruction of bradykininase or inhibition of bradykinin activity (Fujita *et al.*, 1976). However, it has been observed that the vehicles used in commercial aloe gel preparations (mineral oil, petrolatum and aquaphor) may inhibit prostanoid production themselves (Penneys, 1982). Some observations suggest that the presence of small amounts of anthraquinones are necessary for the absorption of aloe gel given orally (Davis *et al.*, 1989) and may enhance its anti-inflammatory effect (Capasso *et al.*, 1998; Grindlay and Reynolds, 1986). One study in support of this hypothesis found that an aloe gel containing anthraquinone was more effective than an anthraquinone-free gel (Davis *et al.*, 1986). On the contrary, another study found that the decolourised (i.e. without anthraquinones) aloe gel is more potent as an anti-inflammatory agent than the colourised form (Davis *et al.*, 1986). However, the same authors also noted that PMN leukocyte infiltration and inflammation are decreased by both colourised and decolourised aloe gel. Therefore the role of anthraquinones on skin ailment management is still confused and further studies are required to clarify the capacity of aloe gel (acemannan) to interact with integrins, heterodimeric cell surface receptors. Integrins play a role in inflammation, permitting inflammatory cells to leave the bloodstream and enter damaged tissues (Capasso *et al.*, 1998). Lastly, aloe gel and hydrocortisone seem to inhibit the inflammatory process in an additive, dose-dependent manner when given concurrently (Davis *et al.*, 1991).

Immunomodulatory action

Aloe gel may also directly stimulate the immune system (Womble and Helderman, 1988) through its active ingredient acemannan. This polysaccharide increases lymphocyte response to alloantigen; it activates macrophages to produce nitric oxide (Karaca *et al.*, 1995) and cytokines (interleukins 1 and 6, interferon, tumor necrosis factor); it enhances phagocytosis (Shida *et al.*, 1985); and, it increases the number of circulating monocytes and macrophages. Aloe gel also causes a local activation of complement at the level of C_3. Recently Qiu *et al.* (2000) have developed a process to activate and stabilise aloe polysaccharide. Modified aloe polysaccharide (MAP) prevents ultraviolet (UV) irradiation-induced immune suppression as determined by contact hypersensitivity response in mice. MAP also inhibits UV irradiation-induced TNF (tumour necrosis factor) release from human epidermoid carcinoma cells. All these results indicate that MAP can be used to reduce the risk of sunlight-related human skin cancer.

Antibacterial activity

Aloe gel (and acemannan) has been shown to have antibacterial activity against *Streptococcus* species, *Enterobacter cloacae, Citrobacter* species, *Serratia marcescens, Klebsiella pneumoniae, Pseudomonas aeruginosa, Staphylococcus aureus* and other bacteria (Heggers *et al.*, 1979). Aloe gel also accelerates the rate of healing, decreases the production of prostanoids and inhibits infection by *Pseudomonas aeruginosa*.

Clinical studies

The first case report on the use of aloe gel for wound healing was published in 1935 (Collins and Collins, 1935). Roentgen dermatitis developed by a woman for depilatory purposes, treated with aloe gel, showed rapid relief. Since then, there are other reports dealing with successful medicinal applications of aloe gel, in roetgen dermatitis, ulcers and telangiectasis (Wright, 1936; Loveman, 1937), palmar eczema and pruritus vulvae (Crewe, 1937), finger abrasions (Barnes, 1947) and thermal burns

(Crewe, 1939). However, the literature on skin ailment management is not always positive (Kaufman *et al.*, 1988; Ashley *et al.*, 1957; Watcher and Wheeland, 1989; Schmidt and Greenspoon, 1993) because in some cases the researcher does not use the fresh drug and standardised preparations. In this way for many years little interest was paid by researchers and clinicians to the medicinal properties of aloe gel. The fact that the active components were little known has also contributed to the concealment of the therapeutic use of aloe gel.

Vogler and Ernst (1999) systematically reviewed the evidence for and against aloe gel clinical effectiveness. Only controlled clinical trials on aloe gel were included in their study. They found two controlled clinical trials related to the topical application of aloe gel for healing wounds and two for the prevention of radiation-induced skin burns. The results of this study indicate that the topical application of aloe gel is not an effective preventative for radiation-induced injuries. Furthermore, whether or not it promotes wound healing is unclear. The essential extractable information from these studies is provided below.

The effects of two different dressings for wound-healing management on full-faced dermabrasion patients was documented by Fulton (1990). He divided the abraded faces of 18 patients suffering from acne vulgaris in half. One side was treated with a standard polyethylene oxide gel wound dressing, while the other side was treated with a polyethylene oxide dressing saturated with aloe gel. He reported that aloe gel reduced oedema, exudate and crusting in two to four days. Reepithelialization was complete to 90% on the aloe side compared with 40–50% on the control side after five days. Overall, wound healing was approximately 72 hours faster at the aloe side. The time interval required for wound healing in a 40-women gynaecological surgery was evaluated by Schmith and Greenspoon (1993) in an open-label study. The authors found that the mean healing time in the conventional care group was significantly shorter than in the aloe gel group (53 versus 83 days). Two randomised clinical trials were reported in one publication by Williams *et al.* (1996). A total of 194 women receiving radiation therapy were treated with aloe gel or placebo gel, twice per day, by self-administration. The severity of the dermatitis was scored weekly during the ten-weeks treatment period both by the patients and by their healthcare providers. No differences between the placebo and the treatment group were noted.

Some clinicians participating in this trial felt that there were fewer skin problems than normally expected. Thus, it was speculated that the inert carrier gel might have had some beneficial effects. A second randomised clinical trial was therefore performed with 108 women (Williams *et al.*, 1996). The only difference compared with the first study was that the control group now received no topical therapy at all. The trial was therefore not blinded. Again, the results did not suggest any benefit of the aloe gel in terms of prevention of radiation-induced dermatitis.

Psoriasis

Anthrones have long been used in medicine as antipsoriatic agents. Their mode of action is not known exactly, although many biological molecules and cell types have been identified as potential targets of anthrones. The antipsoriatic activity of anthrones is probably multimodal: inhibition of the oxygen consumption of cells; a reduction in size of the intracellular spaces and a decrease in ribosomes and mitochondria; interaction with DNA; inhibition of various enzyme systems associated with cell proliferation and inflammation and a redox reaction resulting in mitochondrial damage and destruction of membrane lipids in the psoriatic epidermis, have all been noted (Verhaeren, 1980; Anton and Haag Berrurier, 1980; Friedman, 1980; Muller, 1996). All these mechanisms may in part retard the increased cell division found in the psoriatic epidermis. Syed *et al.* (1996a) randomised 60 patients with mild to moderate chronic psoriasis to receive either an aloe gel or placebo cream. The cream was self-applied three times per day for four weeks. Patients were subsequently followed up for 12 months. The cure rate in the aloe gel group was 83% and only 7% in the placebo group.

Viral diseases

Anthraquinone derivatives have been shown to inhibit several viruses *in vitro*, including herpes simplex of type 1 and type 2, varicella-zoster, pseudo-rabies and influenza (Sydiskis *et al.*, 1991). Acemannan is another compound of aloe extract that has been reported to have antiviral activity and to play a role as an adjuvant treatment in HIV/AIDS (Kahlon *et al.*, 1991a). This polymannan increases *in vitro* the production and function of cytotoxic T cells in a dose-dependent manner. Acemannan in combination with the antiviral agent azidothymidine (AZT) protected the cells from rapid HIV-1 replication, which causes premature cell death (Kahlon *et al.*, 1991b). The combination of acemannan and acyclovir also inhibited HIV-1 replication. The mechanism of action of acemannan's antiviral activity is the inhibition of glycosylation of viral glycoproteins (Kahlon *et al.*, 1991b). Therefore, aloe extract is considered as a possible therapy for AIDS, alone or in association with AZT. It is also able to reduce the dosage of antiviral treatment up to 90%, consequently reducing the side-effects of AZT (Werbach and Murray, 1994).

Dianthrones and other anthraquinone derivatives, including rhein and emodin, have antiviral activity against human cytomegalovirus. However, it is unlikely that systemic antiviral effects would follow from the ingestion of these compounds, due to their low bioavailability. Two randomised clinical trials, conducted by the same research group (Syed *et al.*, 1996b; Syed *et al.*, 1997) indicate that topical application of aloe vera might be effective against the first episodes of genital herpes. In the first study (Syed *et al.*, 1996b), they divided 120 men into three parallel groups. Each group was treated three times daily for two weeks with placebo, aloe gel or aloe cream. The numbers of cured patients were 7.5%, 45% and 70%, respectively. In addition, aloe gel cream showed a shorter mean duration (4.8 days) of healing than aloe gel (7 days) and placebo (14 days). Of the 49 patients healed at the end of this trial period, six had a relapse after 21 months of follow-up. In the second study (Syed *et al.*, 1997), 60 men were randomly divided into two groups (placebo versus aloe gel). The authors reported that the aloe gel cream group had both significantly shorter healing time (4.9 days versus 12 days) and a higher number of cured patients (66.7% versus 6.7%) compared with the placebo group. Of the 22 healed patients, three showed recurrence after 15 months.

Diabetes

Hypoglycaemic action has been studied in an animal (mouse) model of diabetes (Ajabnoor, 1990) and in humans (Ghannam *et al.*, 1986). The mechanism of action for this effect has yet to be determined. It has been hypothesised that aloe may stimulate the release, or synthesis, of insulin from the β-cell of the Isles of Langerhans (Ajabnoor, 1990). Another study has shown that a formula containing aloe vera and a small number of natural agents (*Nigella sativa* L., *Boswella carterii* Birdw., *Commiphora myrrha* Engl. and *Ferula assa-foetida* L.) inhibits gluconeogenesis and lowers blood sugar in an animal model (Al-Awadi *et al.*, 1991).

Two controlled clinical trials suggest that oral administration of aloe gel might be a useful adjunct for lowering blood glucose in patients with diabetes (Yongchaiyudha *et al.*, 1996; Bunyapraphatsara *et al.*, 1996). Yongchaiyudha *et al.* (1996) divided 72 women without drug therapy into two groups. They received one tablespoon of aloe gel or placebo for 42 days. Blood glucose levels subsequently decreased from 250 mg to 141 mg percentage in the experimental group, while controls showed no significant changes. In addition, cholesterol, serum triglycerides, weight and appetite were also monitored. With the exception of triglyceride levels, which fell significantly in the actively treated group, these variables remained unaltered in both groups. This study was neither randomised nor blinded to patient or investigators. The same research team investigated the effects of aloe gel in combination with a standard oral antidiabetic therapy (Bunyapraphatsara *et al.*, 1996). All diabetic patients admitted to this study received 5 mg oral glibenclamide, twice daily. In addition, for the duration of the trial (42 days) they were given either aloe gel or placebo as above. The results show similar decreases in blood glucose and serum triglyceride levels in the actively treated group, as described in the first trial. The same methodological drawbacks apply as to the previous study.

Hyperlipidaemia

Sixty patients with hyperlipidaemia who had not responded to dietary interventions received either 10 ml or 20 ml aloe gel or placebo daily for a period of 12 weeks (Nassif *et al.*, 1993). Blood lipid levels were measured before treatment and after 4, 8 and 12 weeks. Total serum cholesterol decreased by 15.4% and 15.5%, triglycerides by 25.2% and 31.9%, LDL by 18.9% and 18.2%, respectively, in the two groups receiving different doses of aloe gel.

Gastric ulcer

Aloe gel had a prophylactic effect and was also curative if given as a treatment for stress-induced gastric ulceration in rats (Galal *et al.*, 1975; Parmar *et al.*, 1986). A lectin fraction (glycoprotein) from *Aloe arborescens*, aloctin A, had an anti-ulcer effect in rats (Saito *et al.*, 1989), while another high molecular weight fraction, not containing glycoprotein, was very effective in healing ulcers induced by mechanical or chemical stimuli but not those induced by stress (Teradaira *et al.*, 1993). This fraction contained substances with molecular weights between 5,000 and 50,000 Daltons, which were considered to both suppress peptic ulcers and to heal chronic gastric ulcers. In addition, a component from Cape Aloe, named aloe ulcin, suppressed ulcer growth and L-histidine decarboxylase in rats (Yamamoto, 1973). An early clinical study found that oral administration of aloe gel was effective in the treatment of peptic ulcer (Blitz *et al.*, 1963).

Cancer

The whole leaf extract of aloe (=aloe extract) combines aloe gel with aloe latex and aloe epidermis. Aloe extract contains, among other substances, immunomodulatory, mild anti-inflammatory and antitumor mucopolysaccharides, acemannan being the most notable. Mucopolysaccharides are normally produced by the human body until puberty, after which, these substances must be introduced from outside sources. Their deficiency could produce drastic degenerative diseases. Acemannan is able to increase antibody-dependent cellular cytotoxicity and stimulate the proliferation of thymic cells. Acemannan is also effective in the treatment of fibrosarcoma in dogs, cat and mice, in that the survival rate is increased (Manna and McAnalley, 1993; Peng *et al.*, 1991; Harris *et al.*, 1991; Gribel and Pashinki, 1986; Ralamboranto *et al.*, 1982). In addition, polysaccharides from *A. barbadensis, Lentinus edulis* and others (*Ganoderma lucidum, Coriolus versicolor*) have demonstrated anti-genotoxic and antitumor promoting activities in *in vitro* models (Kim *et al.*, 1999). The antitumor effect of acemannan may be due to stimulation of the production of tumor necrosis factor (TNF), interleukin-1 and interferon by macrophages; acemannan is also able to abrogate viral infections in both animals and men (Womble and Helderman, 1988). From the few reports available, it appears that large doses of polysaccharides are necessary to produce immunostimulating and antitumor effects. To achieve excessive amounts of acemannan and consequently of aloe preparation, aloe has been combined with other substances. Aloe vera in combination with squalene and vitamins A and E has been demonstrated to have chemopreventive and curative properties in the prevention and treatment of mouse skin tumors. Aloe vera with vitamin supplementation has been found to be able to reduce the severity of chemical hepatocarcinogenesis in rats (Shamaan *et al.*, 1998).

Aloe extract also contains aloctins, substances which possess many biological activities such as mitogenic activity for lymphocytes, binding of human 22-macroglobulin, and complement activation via the alternative pathway (Suzuki *et al.*, 1979). In addition, aloctin A is considered a promising candidate as an immunomodulator. This substance administered to mice inhibits growth of methylcholanthrene-induced fibrosarcoma and the results have been attributed to the immunomodulatory effect of aloctin A, not to its cytotoxicity (Imanishi *et al.*, 1981).

Anti-inflammatory, immunomodulating and antitumor agents also include anthraquinones. Aloe-emodin is active against P-388 leukemia in mice (Kupchan and Karim, 1976) and selectively inhibits human neuroectodermal tumor cell growth in tissue cultures and in animal models (Pecere *et al.*, 2000). Aloe-emodin does not inhibit the proliferation of normal fibroblasts nor that of hemopoietic progenitor cells. The cytotoxicity mechanism consists of the induction of apoptosis, whereas the selectivity against neuroectodermal tumor cells is due to a specific energy-dependent pathway of compound incorporation. Aloe-emodin is toxic against neuroectodermal tumors with no evidence of acute or chronic toxicity: therefore it shows a favourable therapeutic index. However, others have investigated aloe-emodin as a cytotoxic agent on several tumor cell lines but no significant activity was found (Grimaudo *et al.*, 1997). A stimulatory effect of aloe-emodin on urokinase secretion and colorectal carcinoma cell growth has also been described (Schörkhuber *et al.*, 1998). Antitumor effects are also exhibited by diethylhexylphthalate (DEHP), isolated from *A. vera* but probably as a contaminant. DEHP is a plasticizer and has a potent antileukaemic effect in human cells (Lee *et al.*, 2000) and anti-mutagenic activity in the *Salmonella* mutation assay (Lee *et al.*, 2000). The presence of all these principles might be enough to explain the prophylactic and possible therapeutic effect of aloe extract and its antitumour activity against leukopenia caused by exposure to cobalt 60, sarcoma-180 and Elhrich ascites (Arendarevslii, 1977; Yagi *et al.*, 1977).

Experimental studies have also reported antimetastatic activity of aloe gel in rats and mice (Gribel and Pashinskii, 1986). The importance of platelet aggregation in metastasis is now more widely accepted. Several studies have shown that migrating cells from some cancers induce platelet aggregation by modifying the balance between prostacyclin (PGI_2) and thromboxane (TXA_2). PGI_2 inhibits platelet aggregation while TXA_2 enhances aggregation. Tumors promote platelet aggregation by stimulating the production of TXA_2 and/or inhibiting the production of PGI_2. Aloe gel inhibits metastasis by decreasing TXA_2 and TXB_2 production *in vitro* (Klein and Penneys, 1988) and this could be one of the mechanisms of antimetastatic activity of aloe. Several natural agents that inhibit kinin production or degrade kinins may inhibit kinin-induced angiogenesis. These agents include aloe gel (Klein and Penneys, 1988). Aloe gel and glycoproteins isolated from *A. arborescens* and *A. saponaria* degrade bradykinin *in vitro* and inhibit the formation of histamine *in vitro*. Aloe gel has also demonstrated antiangiogenic activity *in vivo* in the synovial pouch model in mice (Davis *et al.*, 1992). Other studies have shown aloe extract to have an inhibitory effect when used against preneoplastic hepatocellular lesions in rats (Tsuda *et al.*, 1993) and a regression of the pleural tumor in rats has been demonstrated by aloe latex (Corsi *et al.*, 1998). Antitumor effects of aloe may also depend on the ability to augment tumor specific immunity (Yoshimoto *et al.*, 1987). All these findings have encouraged cancer treatment in humans with a preparation as follows: aloe (five years old-fresh leaf), 300 g; honey 500 g; 2 tablespoons of gin, vodka or whisky. The mixture can be left for ten days in a jar, filtered and taken (1 tablespoon one or two times a day for 14 days), or mixed in a blender and then taken as above. Honey increases the palatability of the preparation and could enhance the effect of aloe for its content of caffeic acid phenethyl ester (CAPE), a potent chemopreventive agent useful in combating diseases with a strong inflammatory component, including various types of cancer (Frenkel *et al.*, 1993). A recent clinical study has also shown that concomitant administration of aloe and melatonin enhances the therapeutic result of melatonin given alone in patients with advanced solid tumors such as lung cancer, gastrointestinal cancer, breast cancer or brain glioblastoma, for whom no effective standard anticancer therapies are available (Lissoni *et al.*, 1998). It may be worthwhile to combine melatonin with immunostimulant drug such as aloe, since these may all act together to increase interleukin-2 activity. It has been also demonstrated that aloe latex enhances the activity of 6-fluorouracil and cyclophosphamide (Gribel and Pashinski, 1986). However, until well-designed clinical trials on aloe are conducted, it will not be possible to determine the anticancer activity of the drug with certainty.

Miscellaneous

There is evidence showing the efficacy of aloe extract in chronic bronchial asthmatic patients. The effect of aloe extract seems due to the formation of some prostanoids during dark storage of aloe extract at $4°-30°C$, for a period of three to ten days (Afzal *et al.*, 1991). Some studies also report the effectiveness of aloe gel in increasing the rate of healing after dental procedures (Bovik, 1966; Payne, 1970). The efficacy of a new bioadhesive patch with aloe gel for the treatment of mouth ulcers has recently been evaluated. The results of this study underline the good efficacy and compliance of the patch for the treatment of the aphtous stomatitis (Andriani *et al.*, 2000). On the contrary, a study carried out to evaluate the effectiveness of a medicine containing aloe, silicon dioxide and allantoin on aphthous stomatites indicated a lack of effect of the gel on aphthous ulcers

(Garnick *et al.*, 1998). Rojas *et al.* (1995) have studied the antiparasitic action of an aqueous extract of *A. barbadensis* against an *in vitro* culture of *Trichonomas vaginalis*. The extract inhibited the growth of *T. vaginalis* suggesting its potential use in womens' disturbances. Studies also show that topical and oral administration of aloe preparations in patients with chronic venous leg ulcers may aid healing (Atherton, 1998). It has been also inferred that aloe reduces the growth rate of urinary calcium crystals that contribute to the formation of kidney stones (Marti, 1995). Aloe is considered as a 'panacea' in veterinary medicine although its real efficacy has been questioned (Anderson, 1983). Aloe has been used as a purge for cattle (Crellin and Philpott, 1990) and aloe gel has been used in the treatment of ringworm, allergies, abscesses, fungal infections and different types of inflammation (Northway, 1975). Its use has also been reported to be beneficial in the treatment of thermal burns in dogs (Cera *et al.*, 1980). Its successful use in an extensively burned monkey has also been referred. However, in the absence of larger research studies we must be prudent against generalisations of these therapeutic treatments.

ADVERSE EFFECTS, TOXICITY, DRUG INTERACTION

Aloe latex (laxative use)

Aloe latex may cause abdominal complaints, meteorism, flatulence, cramps and abdominal pain, just like other laxative drugs such as senna, rhubarb, etc., which are digested by colon microflora (Newall *et al.*, 1996). However, because constipation is often associated with abdominal discomfort, the causative role of aloe is not always apparent. Other side-effects include hemorrhoid congestion and coloration of the urine which becomes orange if the pH is acidic, or reddish purple if the pH is alkaline. This is caused by the renal excretion of the hydroxyanthracene derivatives (Boon and Smith, 1999). An overdosage may cause nephritis, vomiting, bloody diarrhoea with mucus and hemorrhagic gastritis (Canigueral and Vila, 1993). Prolonged use or overdosage may result in watery diarrhoea leading to electrolyte imbalance (Figure 9.4). The increased intestinal loss of K^+ can lead to hypokalemia, while Na^+ loss can result in secondary hyperaldosteronism. This can exacerbate renal K^+ excretion, leading to further reduction of colonic motility. This situation results in fatigue, muscular weakness, weight loss, mental disturbances, steatorrhoea, electrocardiographic abnormalities and kidney dysfunction (Figure 9.4) (for ref. see Capasso and Gaginella, 1997).

Hypokalemia, which results from K^+ loss, may potentiate the action of digoxin, a cardiac glycoside. Drugs like thiazide diuretics, corticosteroids and licorice may exacerbate hypokalemia (Dalton and Cupp, 2000). Table 9.4 lists other possible interactions regarding both aloe latex and aloe gel. Damage to surface epithelium and an impairment of function following damage to the autonomic nervous system may also develop (Muller-Lissner, 1993). These changes, however, have not been clearly demonstrated in animals and humans.

The abuse of aloe has been associated with melanosis coli, which consists of a mahogany to dark brown coloration of the intestinal mucosa that begins at the ileocolonic junction and may extend to the rectum (Koskela *et al.*, 1989; Gobel, 1978; Steer and Colin-Jones, 1975; Wittoesch *et al.*, 1958). The morphological basis of melanosis is a pigment, probably lipofuscin, within macrophages of the large intestinal mucosa (Figure 9.5). A correlation between the intake of laxative and melanosis is accepted now

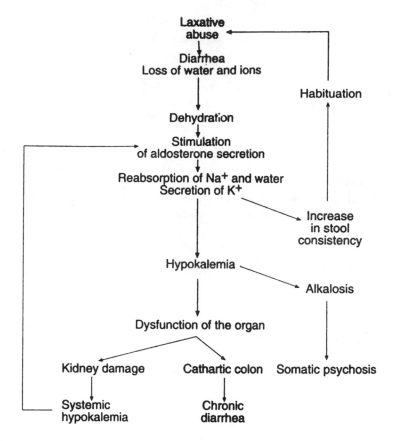

Figure 9.4 Metabolic consequence of laxative abuse (From Capasso and Gaginella, 1997).

but there is no indication that melanosis has any pathophysiological consequences. On the other hand the intestinal mucosa recovers its usual coloration 4–12 months after the intake of the laxative has stopped (Boon and Smith, 1999).

Habituation has not been proved for aloe. Indeed studies in rats suggest that long-term aloe treatment (three to six months) does not induce tolerance in the sense of a reduced laxative effect (Capasso, personal communication).

Anthraquinone derivatives have shown genotoxicity in *Salmonella* assay (Brown and Brown, 1976; Mori *et al*., 1985; Westendorf *et al*., 1990; Muller *et al*., 1996) but the clinical relevance of this experimental result is still not clear. On the other hand, due to the artificial nature of these methods many flavonoids (quercetin, galangin, kaempferol) and antioxidants, including vitamin C, have produced genotoxic effects under these conditions (Mascolo *et al*., 1998).

In recent years the risk of colon cancer has been found to be related to constipation and to use of anthraquinone laxatives (Siegers, 1992; Sonnenberg and Muller, 1993) but epidemiological studies are in disagreement. Jacobs and White (1998), in a recent study, reported that when the colon cancer risks for constipation and laxatives were adjusted for each other, the association with laxatives disappeared, whereas the association with constipation remained strong. Other retrospective data related to laxative use

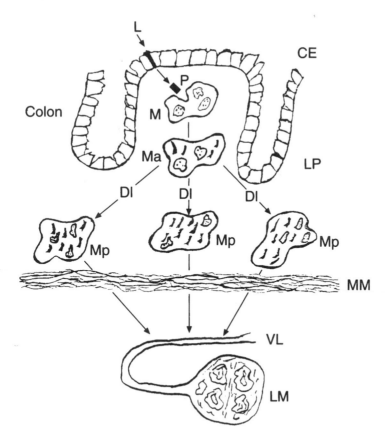

Figure 9.5 Formation of melanosis coli in the mucosa of the colon or rectum. L, laxative; P, cellular fragment; M, macrophage; Ma, macrophage with apoptotic bodies; Mp, pigmented macrophage; Dl, lysosomal digestion; Vl, lymphatic vessel; LM, mesenteric lymphnode; CE, eptithelial cell; LP, lamina propria; MM, muscularis mucosae. The etiology of pseudomelanosis is still unclear. Laxatives could interact with colonic epithelial cells and induce migration of fragments of these cells into the spaces among the crypts. These cellular fragments would be phagosized by macrophages and converted to lipofuscin-like pigments by enzymes present in lysosomes of the macrophages. As a consequence, the epithelium takes on a brown pigmentation. This is reversible upon removing the laxative because the macrophages with pigmentation will migrate towards the lymph-nodes (From Capasso and Gaginella, 1997).

did not show any significant increase of colon cancer incidence (Nakamura *et al.*, 1984; Kune *et al.*, 1988; Kune, 1993). It has also been demonstrated that a positive correlation between melanosis coli, a marker for chronic abuse of anthranoid laxatives, and colon carcinoma exists (Siegers *et al.*, 1993) but this correlation failed in another retrospective study (Nusko *et al.*, 1993). A study carried out *in vivo* shows that aloe given alone for 13 weeks did not induce the development of preneoplastic or neoplastic lesions, or Aberrant Crypt Foci (ACF), and therefore this laxative did not display any tumor initiating activity in the rat colon (Capasso, personal communication). In addition, when aloe was administered with AOM (azoxymethane), a carcinogen agent, it did not produce any significant increase of ACF, and tumors. These results are basically in agreement with previous studies carried out with other anthraquinone drugs

Table 9.4 Aloe – drug interactions.

Preparation	Possible conventional drug interactions	Possible reactions
Aloe latex and anthranoid laxatives	Antidiarrhoeals	Antagonism
	Glucoresins	Synergism
	Cardioactive glycosides	Potassium depletion leading to adverse cardiovascular effects
	Licorice, corticosteroids, Diuretics	Potassium depletion may be exacerbated
	Non steroidal anti-inflammatory drugs	Antagonism
	Vitamins	Reduced absorption of vitamins
	Minerals	Reduced absorption of minerals
Aloe gel	Hydrocortisone	Additive effect
	Antidiabetics	Additive effect
	UV radiation	Photodermatitis

(Mascolo *et al.*, 1999; Borrelli *et al.*, 2001). Recent studies have also shown that anthraquinone compounds (i.e. emodin) selectively block the retardation of oncogene-modulated signal transduction through the inhibition of protein kinase. All these results suggest a sufficient margin of safety for aloe and other anthraquinone drugs when used internally (Capasso and Gaginella, 1997). The FDA has recently proposed to stop the use of aloe, the laxative, until further studies can show its benefit and safety.

Toxicity due to oral and topical use of Aloe vera

Several studies have been made to determine whether aloe causes toxicity in animals or humans in gel form or in its various preparations, pure gel, liquid concentrate, etc. Studies in mice revealed no acute toxicity in therapeutic doses. In high doses, however, a decrease of CNS (central nervous system) activity was noticed. During chronic treatment, there was a decrease in red cell count and significant sperm damage. On the other hand if aloe gel contains very large amounts of aloin, it may cause unfavourable side-effects (see above). In spite of its wound-healing and anti-inflammatory properties, the topical application of aloe gel has been reported in the literature to cause contact and photodermatitis and/or erythema with papulous (Hogan, 1988; Morrow *et al.*, 1980; Shoji, 1982a, b; Nakamura and Kotahima, 1984; Hunter and Frumkin, 1991; Dominguez-Soto, 1992). Aloin can certainly irritate the skin. However it has been recently found that painting aloe emodin on the skin of mice in conjunction with exposure to ultraviolet radiation results in the development of melatonin-containing skin tumors (Strickland *et al.*, 2000). The precise contribution of aloe derivatives to the development of cutaneous melanoma requires further definition. In the meantime it is preferable to avoid local application of aloe gel when the skin is exposed to ultraviolet radiation (Table 9.4). One case of cathartic effect after topical application of aloe gel has also been reported (Boon and Smith, 1999).

Vogler & Ernst (1999), by reviewing ten controlled clinical trials—a total of 803 subjects-reported no withdrawals or serious adverse reactions. Some patients experienced burning after topical application, contact dermatitis and mild itching. All adverse effects were reversible and aloe vera was generally well tolerated.

CONCLUSIONS

Aloe and its preparations have been widely used as a medicine since ancient times. Its beneficial effects are mentioned in Hebers Papyrus and in the writings of famous physicians, from Dioscorides to Galen (for ref. see Heggers, Pelley and Robson, 1993). Aloe has been a common folk medicine in several countries. The whole plant has been used in India as a stomachic, antihelmintic, emmenagogue and purgative; the leaf pulp has also been used for menstrual suppression and the root for colic (Jain and De Filipps, 1991). In China it is a common dermatological remedy and in Mexico it is used to treat minor skin irritations (Grindlay and Reynolds, 1986). Aloe has also been used for centuries for a multiplicity of unrelated human illness: for example to correct kidney ailments, to enhance sexual excitement, to develop the mammary glands, to relieve headaches and reduce fever in children (Morton, 1961; Crosswhite and Crosswhite, 1984). Two uses, however, have persisted for centuries: the main one, internal, has been as a laxative; the second, external, in the treatment of skin injuries. The use of aloe orally, other than for its well-accepted laxative effect, needs furthers studies. Aloe gel can be applied liberally for topical applications. Now a wide range of products are available on the market; however, simply pure aloe gel is sufficient. This product is very sensitive to light and heat, therefore it is removed from the leaf mechanically and preserved, buffered and stabilised immediately. This process may differ from manufacturer to manufacturer, making the quality of aloe gel highly variable (Canigueral and Vila, 1993). This variation in quality must be responsible for some of the negative or inconclusive studies (Grindlay and Reynolds, 1986). In addition aloe gel is unstable and may deteriorate in a short time period, which may also explain some of the conflicting study results.

A recent systematic review (Vogler and Ernst, 1999) suggests that oral administration of aloe gel might be a useful adjunct for lowering blood glucose in diabetic patients as well as for reducing blood lipid levels in patients with hyperlipidaemia. Topical application of aloe vera is not an effective preventative for radiation-induced injuries. It might be effective for genital herpes and psoriasis. Whether it promotes wound healing is unclear. Table 9.5 summarises these results.

Clinical trials are now in progress to provide conclusive evidence not only in these diseases but also in arthritis, gastric ulcer, cancer, AIDS and colitis. However, this is not easy to perform, because there is still confusion about aloe preparations. It is not enough to say that a preparation, if not specified as a latex or gel, is referred to as an extract. Aloe products are classified as drugs while others as dietary supplements. Aloe is also a common ingredient in numerous hand, body and sun lotions, shaving creams, shampoos and personal care-products. This is why people sometimes forget to consider aloe a medicinal drug (Eshinazi, 1999). Aloe is also processed into various flavoured and unflavoured alcoholic drinks (Fernet Branca, etc.) that are ingested with the belief that this bitter liquid (*amarum*) will stimulate the appetite and/or the digestion. There is a general conviction that bitters also have a tonic effect on the body and the term

Table 9.5 Controlled clinical trials of aloe vera (From Vogler and Ernst, 1999, modified).

Reference	Quality score (max 5)*	Design	Sample	Condition treated	Oral vs topical use	Result
Fulton et al., 1990	0	controlled clinical trial	17 patients	wound healing	topical	positive
Schmidt et al., 1991	3	RCT	40 women	wound healing	topical	negative
Syed et al., 1996	4	double blind placebo-controlled two parallel groups	60 subjects	psoriasis	topical	positive
Williams et al., 1996	4	double blind placebo-controlled two parallel groups	194 women	prevention of radiation-induced skin injury in women with breast cancer	topical	negative
Williams et al., 1996	1	randomised two parallel groups	108 women	prevention of radiation-induced skin injury in women with breast cancer	topical	negative
Syed et al., 1996	3	randomised double-blind placebo-controlled three parallel groups	120 men	genital herpes	topical	positive
Syed et al., 1997	4	randomised double-blind placebo-controlled two parallel groups	60 men	genital herpes	topical	positive
Nasiff et al., 1993	abstract	three parallel groups	60 patients	hyperlipidaemia	oral	positive
Yongchaiyudha et al., 1996	1	placebo-controlled single-blind	72 women	diabetes	oral	positive
Bunyapraphatsara et al., 1996	1	placebo-controlled single-blind	72 patients	diabetes (in combination with oral antidiabetics)	oral	positive

Note
* Jadad et al. (1996).

'bitter tonic' is often used. Finally for its very bitter taste aloe has been widely used to discourage nail biting.

REFERENCES

Afzal, M., Ali, M., Hassan, R.A.H., Sweedan, N. and Dhami, M.S.I. (1991) Identification of some prostanoids in Aloe vera extracts. *Planta Medica*, 57, 38–40.

Ajabnoor, M.A. (1990) Effect of aloes on blood glucose levels in normal and alloxan diabetic mice. *Journal of Ethnopharmacology*, 28, 215–220.

Anderson, B.C. (1983) Aloe vera juice: a veterinary medicament? *The Compendium on Continuing Education for Practising Veterinarian*, 5, S364–S368.

Andriani, E., Bugli, T., Aalders, M., Castelli, S., De Luigi, G., Lazzari, N. and Rolli, G.P. (2000) The effectiveness and acceptance of a medical device for the treatment of aphthous stomatitis. Clinical observation in pediatric age. *Minerva Pedriatica*, 52, 15–20.

Anton, R. and Haag Berrurier, M. (1980) Therapeutic use of natural anthraquinone for other than laxative actions. *Pharmacology*, 20, 104–112.

Arendarevslii, L.F. (1977) Factors affecting the efficiency of chemotherapy and recurrence of tumours. Onkologiya, 2, 15–22.

Ashley, F.L., O'Loughlin, B.J., Peterson, R., Fernandez, L., Stein, H. and Schwartz, A.N. (1957) The use of aloe vera in the treatment of thermal and irradiation burns in laboratory animals and humans. *Plastic and Reconstructive Surgery*, 20, 383–396.

Atherton, P. (1997) Aloe vera revised. *British Journal of Phytotherapy*, 4, 176–183.

Atherton, P. (1998) Aloe vera: magic or medicine? *Nursing Standard*, 12, 49–52.

Al-Awadi, F., Fatania, H. and Shamte, U. (1991) The effect of a plant mixture on liver gluconeo-genesis in streptozotocin induced diabetic rats. *Diabetes Research*, 18, 163–168.

Barnes, T.C. (1947) The healing action of extracts of aloe vera leaf on abrasions of human skin. *American Journal of Botany*, 34, 597.

Barnard, D.N., Huffman, J.H., Morris, J.L., Wood, S.G., Hughes B.G. and Sidwell, R.W. (1992) Evaluation of the antiviral activity of anthraquinones, anthrones and anthraquinone derivatives against human cytomagalovirus. *Antiviral Research*, 17, 63–77.

Baruzzi, M.C. and Rovasti, P. (1971) Researches on cutaneous effects of Aloevera L. sap. *Rivista Italiana, Essenze Profumi Piante Officinali Aromi Saponi Cosmetici Aerosol*, 52, 37–39.

Benigni, R., Capra, C. and Cattorini, P.E. (1962) *Piante Medicinali, Chimica, Farmacologia e Terapia*. Vol. I. Milano: Inverni e Della Beffa.

Blitz, J., Smith, J.W. and Gerard, J.R. (1963) Aloe vera gel in peptic ulcer therapy: preliminary report. *Journal of the American Osteopathic Association*, 62, 731–735.

Blumenthal., M. (1998) *The Complete German Commission E monographs: therapeutic guide to herbal medicines*. Austin, Texas: American Botanical Council.

Boik, J. (1996) *Cancer and Natural Medicine*. Princeton, Minnesota, USA: Oregon Medical Press.

Boon, H. and Smith, M. (1999) *The Botanical Pharmacy*. Kingstone: Quarry Health Books Press Inc.

Borrelli, F. and Izzo, A.A. (2000) The plant kingdom as a source of anti-ulcer remedies. *Phytotherapy Research*, 14, 581–591.

Borrelli, F., Mereto, E., Capasso, F., Orsi, P., Sini, D., Izzo, A.A., Massa, B., Boggio and Mascolo, N.M. (2001) Effect of Bisacodyl and Cascara on growth of aberrant crypt foci and malignant tumors in the rat colon. *Life Science*, 69, 1871–1877.

Bovik, E.G. (1966) Aloe vera. Panacea or old wives' tale? *Texas Dental Journal*, 84, 13–16.

Breimer, D.D., Baars, A.J. (1976) Pharmacokinetics and metabolism of anthraquinone laxatives. *Pharmacology*, 14, 30–47.

Brown, J.P. and Brown, R.J. (1976) Mutagenesis by 9,10-anthraquinone derivatives and related compounds in Salmonella thyphimurium. *Mutation Research*, 40, 203–224.

Bruneton, J. (1999) *Pharmacognosy. Phytochemistry. Medicinal Plants.* 2nd edn. Paris: Editions TEC and DOC.

Bunyapraphatsara, N., Jirakulcaiwong, S., Thirawarapan, S. and Manonukul, J. (1996) The efficacy of Aloe vera cream in the treatment of first, second and third degree burns in mice. *Phytomedicine*, 2, 247–251.

Canigueral., S. and Vila, R. (1993) Aloe. *British Journal of Phytotherapy*, 3, 67–75.

Capasso, F., Borrelli, F., Capasso, R., Di Carlo, G., Izzo, A.A., Pinto, L., Mascolo, N., Castaldo, S. and Longo, R. (1998) Aloe and its therapeutic use. *Phytotherapy Research*, 12, S124–S127.

Capasso, F., De Pasquale, R., Grandolini, G. and Mascolo, N. (2000). *Farmacognosia. Farmaci naturali, loro preparazioni ed impiego terapeutico.* Springer-Verlag Italia. Milano, pp 1–557.

Capasso, F. and Gaginella, T.S. (1997) *Laxatives. A Practical Guide.* Milano: Springer-Verlag Italia.

Capasso, F. and Grandolini, G. (1999) *Fitofarmacia. Impiego razionale delle droghe vegetali.* 2nd edn. Milano: Springer-Verlag Italia.

Capasso, F., Mascolo, N., Autore, G. and Duraccio, M.R. (1983) Effect of indomethacin on aloin and 1,8-dioxianthraquinone-induced production of prostaglandins in rat isolated colon. Prostaglandins, 26, 557–562.

Cera, L.M., Heggers, J.P., Robson, M.C. and Hagstrom, W.J. (1980) The therapeutic efficacy of Aloe cera cream (Dermaide Aloe (TM)) in thermal injuries. Two case reports. *Journal of the American Animal Hospital Association*, 16, 768–772.

Che, Q.M., Akao, T., Hattori, M., Kobashi, K. and Namba, T. (1991) Isolation of human intestinal bacterium capable of trasforming barbaloin to aloe-emodin-anthrone. Planta Medica, 57, 15–19.

Chithra, P., Sajithlal., G.B. and Chandrakasan, G. (1998) Influence of Aloe vera on collagen turnover in healing of dermal wounds in rats. *Indian Journal of Experimental Biology*, 36, 896–901.

Collier, H.O.J., MacDonald Gibson, W.J. and Saeed, S.A. (1976) Stimulation of prostaglandin biosynthesis by drugs. Effects in vitro of some drugs affecting gut function. *British Journal of Pharmacology*, 58, 193–199.

Collins, C.E. and Collins, C. (1935) Roentgen dermatitis treated with fresh whole leaf of aloe vera. *American Journal of Roentgenology*, 33, 396–397.

Corsi, M.M., Bertelli, A.A., Gaja, G., Fulgenzi, A. and Ferrero, M.E. (1998) The therapeutic potential of Aloe Vera in tumor-bearing rats. *International Journal of Tissue Reactions*, 20, 115–118.

Cotton, C.M. (1997) *Ethnobotany. Principles and Applications.* Chichester. UK: J. Wiley and Sons.

Crellin, J.K. and Philpott, J. (1990) *Herbal Medicine. Past and Present.* Durham: Duke University Press.

Crewe, J.E. (1937) The external use of aloes. *Minnesota Medicine*, 20, 670–673.

Crewe, J.E. (1939) Aloes in the treatment of burns and scalds. *Minnesota Medicine*, 2, 538–539.

Crosswhite, F.S. and Crosswhite, C.D. (1984) Aloe vera, plant symbolism and the threshing floor. *Desert Plants*, 6, 43–50.

Dalton, T. and Cupp, M.J. (2000) Aloe. In *Toxicology and Clinical Pharmacology of Herbal Products*, edited by Cupp M.J. Totowa, New Jersey: Humana Press Inc.

Danof, I.E. (1987) Aloe in cosmetics – does it do anything? *Cosmetics and Toiletries*, 102, 62–63.

Davis, R.H., Agnew, P.S. and Shapiro, E. (1986) Antiarthritic activity of anthraquinones found in Aloe for podiatric medicine. *Journal of the American Podiatric Medical Association*, 76, 61–66.

Davis, R.H., Kabbani, J.M. and Maro, N.P. (1986) Wound healing and antiinflammatory activity of Aloe vera. *Proceedings of the Pennsylvania Academy of Science*, 60, 79.

Davis, R.H., Leitner, M.G., Russo, J.M. and Bryne M.E. (1989) Anti-inflammatory activity of Aloe vera against a spectrum of irritants. *Journal of the American Podiatric Medical Association*, 79, 263–276.

Davis, R.H. and Maro, N.P. (1989) Aloe vera and gibberellin. Anti-inflammatory activity in diabetes. *Journal of the American Podiatric Medical Association*, 79, 24–26.

Davis, R.H., Parker, W.L. and Murdoch, D.P. (1991) Aloe vera as a biologically active vehicle for hydrocortisone acetate. *Journal of the American Podiatric Medical Association*, 81, 1–9.

Davis, R.H., Stewart, G.J. and Bregman, P.J. (1992) Aloe vera and the inflamed synovial puoch model. *Journal of the American Podiatric Medical Association*, 82, 140–148.

Dezani, S. and Guidetti, E. (1953) *Trattato di Farmacognosia*. Torino: UTET.

Dominguez-Soto, L. (1992) Photodermatitis to Aloe vera. *International Journal of Dermatology*, 31, 372.

Droscoll, J.S., Hazard, G.F., Wood, H.B. and Golding, A. (1974) Structure-antitumor activity relationship among quinone derivatives. *Cancer Chemotherapy Reports*, 4, 1–362.

Eskinazi, D. (1999) *Botanical Medicine*. Larchmont, N.Y.: Mary Ann Liebert Inc.

Evans, W.C. (1989) *Trease and Evans' Pharmacology*, 13th edn, pp. 413–417. London: Boilliere Tindall.

Fingl, E. (1975) Laxatives and cathartics. In *The Pharmacological Basis of Therapeutics*. edited by Goodman L.S. and Gilman A. 5th edn, New York: MacMillan.

Fly, L.B. and Kiem, I. (1963) Tests of Aloe vera for antibiotic activity. *Economic Botany*, 17, 46–49.

Foster, S. and Tyler Varro E. (1999) Tyler's Honest Herbal. 4th ed. New York: The Haworth Herbal Press.

Fox, T.R. (1999) Aloe vera Revered, Mysterious Healer. *Health Foods Business*, 45–46.

Frenkel, K., Wei, H., Bhimani, R., Ye, J., Zadunisky, J.A., Huang, M.T., Ferraro, T., Conney, A.H. and Grunberger, D. (1993) Inhibition of tumor promoter-mediated processes in mouse skin and bovine lens by caffeic acid phenethyl ester. *Cancer Research*, 53, 1255–1261.

Friedman, C.A. (1980) Structure-activity relationships of anthraquinones in some pathological conditions. *Pharmacology*, 20 (Suppl. 1), 113–122.

Fujita, K., Teradaira, R. and Nagatsu, T. (1976) Bradykinase activity of aloe extract. *Biochemical Pharmacology*, 25, 205.

Fulton, J.E. (1990) The stimulation of postdermabrasion wound healing with stabilized Aloe vera gel-polyethylene oxide dressing. *Journal of Dermatologic Surgery and Oncology*, 16, 460–467.

Galal, E.E., Kandil, A. and Hegazy, R. (1975) Aloe vera and gastrogenic ulceration. *Journal of Drug Research*, 7, 73.

Garnick, J.J., Singh, B. and Winkley, G. (1998) Effectiveness of a medicament containing silicon dioxide, aloe and allantoin on aphthous stomatitis. *Oral Surgery, Oral Medicine, Oral Pathology*, 86, 550–556.

Ghannam, N., Kingston, M., Al Meshaal, I.A., Tariq, M., Parman, N.S. and Woodhouse, N. (1986) The antidiabetic effect of aloes: preliminary clinical and experimental observation. *Hormone Research*, 24, 288–294.

Gjerstad, G. and Riner, T.D. (1968) Current status of aloe as cure all. *American Journal of Pharmacy*, 140, 58.

Gobel, D. (1978) Melanosis coli. *Medizinische Klinik*, 73, 519–523.

Gowda, D.C., Neelisiddaiah, B. and Anjaneyalu, Y.V. (1979) Structural studies of polysaccharides from Aloe vera. *Carbohydrate Research*, 72, 201–205.

Gribel, N.V. and Pashinskii, V.C. (1986) Protivometosteticheskie svoistva soka aloe. *Voprosy Onkologii*, 32, 38–40.

Grieve, M. (1998) *A Modern Herbal*. London: Tiger Books International.

Grimaudo, S., Tolomeo, M., Gangitano, R.A., D'Alessandro, N. and Aiello, E. (1997) Effects of highly purified anthraquinoid compounds from Aloe vera on sensitive and multidrug resistant leukemia cells. *Oncology Reports*, 4, 341–343.

Grindlay, D. and Reynolds, T. (1986) The Aloe vera phenomenon: a review of the properties and modern uses of the leaf parenchyma gel. *Journal of Ethnopharmacology*, 16, 117–151.

Harris, C., Pierce, K., King, J., Yates, K.M., Hall, J. and Tizzard, I. (1991) Efficacy of acemannan in treatment of canine and feline spontaneous neoplasms. *Molecular Biotherapy*, 3, 207–213.

Heggers, J.P., Pelley, R.P. and Robson, M.C. (1993) Beneficial effects of Aloe in wound healing. *Phytotherapy Research*, 7, S48–S52.

234   *Nicola Mascolo* et al.

Heggers, J.P., Pineless, G.R. and Robson, M.C. (1979) Dermaide/Aloe vera gel comparison of the antimicrobial effects. *Journal of the American Medical Technologist*, 41, 293–294.

Hennessee, O.M. Some history about aloe vera. Available from: http://www.aloe-vera.com/aloe1.html. Accessed, 1998 Oct 31.

Henry, R. (1979) An updated review of aloe vera. *Cosmetics and Toiletries*, 94, 42–50.

Hiroko, S., Kewnichi, I. and Susumu O. (1989) Effects of aloe extract, aloctin A, on gastric secretion and on experimental gastric lesion in rats. *Yakugaku Zasshi*, 109, 335–339.

Hogan, D.J. (1988) Widespread dermatitis after topical treatment of chronic leg ulcers and statis dermatitis. *Canadian Medical Association Journal*, 138, 336–338.

Hormann, H.P. and Korting, H.C. (1994) Evidence for the efficacy and safety of topical herbal drugs in dermatology: part 1: anti-inflammatory agents. *Phytomedicine*, 1, 161–171.

Hunter, D. and Frumkin, A. (1991) Adverse reactions to vitamin E and Aloe vera preparations after dermabrasion and chemical peel. *Cutis*, 47, 193–196.

Imanishi, K., Ishiguro, T., Saito, H. and Suzuki, I. (1981) Pharmacological studies on a plant lectin, Aloctin A. I. Growth inhibition of mouse methylcholanthrene induced fibrosarcoma (Meth A) in ascites form by Aloctin A. *Experientia*, 37, 1186–1187.

Imanishi, K., Tsukuda, K. and Suzuki, I. (1986) Augmentation of lymphokine-activated killer cell activity *in vitro* by Aloctin A. *International Journal of Immunology*, 8, 855–858.

Ishii, Y., Tanizawa, H. and Takino, Y. (1990) Studies of Aloe. The mechanism of cathartic effect. *Chemical and Pharmaceutical Bulletin (Tokyo)*, 38, 197–200.

Izzo, A.A., Sautebin, L., Borrelli, F., Longo, R. and Capasso, F. (1999) The role of nitric oxide in aloe-induced diarrhoea in the rat. *European Journal of Pharmacology*, 368, 43–48.

Jacobs, E.J. and White, E. (1998) Constipation, laxative use and colon cancer among middle-aged adults. *Epidemiology*, 9, 385–391.

Jain, S.K. and De Filipps, A. (1991) *Medicinal Plants of India*, Volume 1. Algonac, Michigan: USA Reference Publications Inc.

Kahlon, J., Kemp, M.C.X., Carpenter, R.H., McAnalley, B.H., McDaniel, H.R. and Shannon, W.M. (1991a) Inhibition of AIDS virus replication by acemannan *in vitro*. *Molecular Biotherapy*, 3, 127–135.

Kahlon, J., Kemp, M.C.X., Yawei, N., Carpenter, R.H., Shannon, W.M. and McAnalley, B.H. (1991b) *In vitro* evaluation of the synergistic antiviral effects of acemannan in combination with azidothymidine and acylovir. *Molecular Biotherapy*, 3, 214–223.

Kapoor, L.D. (1990) *Handbook of Ayurvedic Medicinal Plants*. Boca Raton, Florida: CRC Press, Inc.

Karaca, K., Sharma, J.M. and Nordgren, R. (1995) Nitric oxide production by chicken macrophages activated by acemannan, a complex carbohydrate extracted from Aloe vera. *International Journal of Immunopharmacology*, 17, 183–188.

Kaufman, T., Kalderon, N., Ullmann, Y. and Berger, J. (1988) Aloe vera gel hindered wound healing of experimental second-degree burns: A quantitative controlled study. *Journal of Burn Care and Rehabilitation*, 9, 156–159.

Kawai, K., Beppu, H., Koika, T., Fujita, K. and Marunauchi, T. (1993). Tissue culture of Aloe arborescens Miller var. *natalensis* Berger. *Phytotherapy Research*, 7, 55–510.

Kim, K.H., Lee, J.G., Kim D.G., Kim, M.K., Park, J.H., Shin, Y.G., Lee, S.K., Jo, T.H. and Oh, S.T. (1998) The development of a new method to detect the adulteration of commercial aloe gel powder. *Archiv der Pharmazie*. Res 21, 514–520.

Kim, H.S., Kacew, S. and Lee, B.M. (1999). *In vitro* chemopreventive effects of plant polysaccharides (Aloe barbadensis Miller, Lentinus edodes, Ganoderma lucidum and Coriolus versicolor). *Carcinogenesis*, 8, 1637–1640.

Klein, A.D. and Penneys, N.S. (1988). Aloe vera. *Journal of the American Academy of Dermatology*, 18, 714–720.

Koskela, E., Kulju, T. and Collan, Y. (1989) Prevalence, distribution and histologic feature in 200 consecutive autopsies at Kuopio University Central Hospital. *Diseases of the Colon and Rectum*, 32, 235–239.

Kune, G.A. (1993) Laxative use not a risk for colorectal cancer. Data from the Melbourne colorectal cancer study. *Zeitschrift fur Gastroenterologisches*, 31, 140–143.

Kune, G.A., Kune, S., Field, B. and Watson, L.F. (1988) The role of chronic constipation diarrhoea and laxative use in the etiology of large bowel cancer. Data from the Melbourne colorectal cancer study. *Diseases of the Colon and Rectum*, 31, 507–512.

Kupchan, S.M. and Karim, A. (1976) Tumor inhibitors. Aloe emodin-antileukemic principle isolated from *Rhammus frangula* L. *Lloydia*, 39, 223–224.

Lang, W. (1993) Pharmokinetic-metabolic studies with ^{14}C-aloe emodin after oral administration to male and female rats. *Pharmacology* (Suppl 1), 110–119.

Lee, K.H., Kim, J.H., Lim, D.S. and Kim, C.H. (2000) Anti-leukaemic and anti-mutagenic effects of di(2-ethylhexyl)phthalate isolated from Aloe vera Linne. *Journal of Pharmacy and Pharmacology*, 52, 593–598.

Leung, A. Y. (1980) *Encyclopedia of common natural ingredients used in food, drugs and cosmetics*. New York, USA: John Wiley and Sons.

Lissoni, P., Giani, L., Zerbini, S., Trabattoni, P. and Rovelli, F. (1998) Biotherapy with the pineal immunomodulating hormone melatonin versus melatonin plus aloe vera in untreatable advanced solid neoplasms. *Natural Immunity*, 16, 27–33.

Loveman, A.B. (1937) Leaf of aloe vera in treatment of roentgen ray ulcers. *Archives of Dermatology and Syphilology*, 36, 838–843.

Manna, S. and McAnalley, B.H. (1993). Determination of the position of the O-acetyl group is a β-(1, 4)-mannan (acemannan) from Aloe barbadensis Miller. *Carbohydrate Research*, 241, 317–319.

Marshall, G.D., Gibbons, A.S. and Parnell, L.S. (1996) Human cytokines induced by acemannan. *Journal of Allergy and Clinical Immunology*, 91, 295–299.

Marti, J.E. (1995) *Alternative Health Medicine Encyclopedic*. Detroit: Visible Ink Press.

Mascolo, N., Capasso, R. and Capasso, F. (1998) Senna. A safe and effective drug. *Phytotherapy Research*, 12, S143–S145.

Mascolo, N., Mereto, E., Borrelli, F., Orsi, P., Sini, D., Izzo, AA., Massa, B., Boggio, M. and Capasso, F. (1999) Does senna extract promote growth of aberrant crypt foci and malignant tumors in rat colon? *Digest Diseases and Sciences*, 44, 2226–2230.

McKeown, E. (1987) Aloe vera. *Cosmetics and Toiletries*, 102, 64–65.

Morrow, D.M., Rapaport, M.J. and Strick, R.A. (1980) Hypersensitivity to aloe. *Archives of Dermatology*, 116, 1064–1065.

Mori, H., Sugie, S., Niwa, K., Takahashi, M. and Kawai, K. (1985) Induction of intestinal tumors in rat by chrysazin. *British Journal of Cancer*, 52, 781–783.

Morton, J.F. (1961) Folk uses and commercial exploitation of Aloe leaf pulp. *Economic Botany*, 15, 311–319.

Muller, K. (1996) Antipsoriatic anthrones: aspects of oxygen radical formation. Challenges and prospects. *General Pharmacology*, 27, 1325–1335.

Muller-Lissner, S. (1993) Laxative-induced damage to the colon. In *Drug Induced Injury to the digestive system*, edited by M. Guslandi and P.C. Braga, pp. 131–142. Berlin Heidelberg New York: Springer-Verlag.

Muller, S.O., Eckert, I., Lutz, W.K. and Stopper, H. (1996) Genotoxicity of the laxative drug components emodin, aloe-emodin and danthron in mammalian cells: Topoisomerase II mediated? *Mutation Research*, 371, 165–173.

Nakamura, G.J., Schneiderman, L.J. and Klauber, M.R. (1984) Colorectal cancer and bowel habits. *Cancer*, 54, 1475–1477.

Nakamura, T. and Kotaijma, S. (1984) Contact dermatitis from Aloe arborescens. *Contact Dermatitis*, 11, 51.

Nassif, H.A., Fajardo, F. and Velez, F. (1993) Effecto del aloe sobre la hiperlipidemia en pacientes refractarios a la dieta. *Revista Cubana de Medicina General Integral*, 9, 43–51.

Newall, C.A., Anderson, L.A. and Phillipson, J.D. (2002) Herbal Medicine. A Guide For Health Care Professionals, 2nd ed. London: The Pharmaceutical Press.

Northway, R.B. (1975) Experimental use of Aloe vera extract in clinical practice. *Veterinary Medicine/Small Animal Clinician*, 70, 89–91.

Nusko, G., Schneider, B., Muller, G., Kusche, J. and Hahn, E.G. (1993) Retrospective study on laxative use and melanosis coli as risk factors for colorectal neoplasma. *Pharmacology*, 47 (Suppl. 1), 234–241.

van Os, F.H.L. (1976) Some aspects of the pharmacology of anthraquinone drugs. *Pharamacology*, 20 (Suppl. 1), 18–29.

Parmar, N.S., Tariq, M., Al-Yahya, M.A., Agreel, A.M. and Al-Said, M.S. (1986) Evaluation of Aloe vera leaf exudate and gel for gastric and duodenal anti-ulcer activity. *Fitoterapia*, 57, 380–383.

Payne, J.M. (1970) Tissue response to Aloe vera gel following peridontal surgery. *MSc Thesis* Baylor College of Dentistry, Dallas, Texas, USA.

Pecere, T., Gazzola, M.V., Mucignat, C., Parolin, C., Vecchia, F.D., Cavaggioni, A., Basso, G., Diaspro, A., Salvato, B., Carli, M. and Palu, G. (2000) Aloe-emodin is a new type of anticancer agent with selective activity against neuroectodermal tumors. *Cancer Research*, 60, 2800–2804.

Peng, S.Y., Norman, J., Curtin, G., Corrier, D., McDaniel, H.R. and Busbee, D. (1991) Decreased mortality of Norman murine sarcoma in mice treated with the immunomodulator acemannan. *Molecular Biotherapy*, 3, 79–87.

Penneys, N.S. (1982) Inhibition of arachidonic acid oxidation in vitro by vehicle components. *Acta Dermato-Venereologica*, 62, 59–61.

Prochnow, L. (1911) Experimentalle Beitrage zur kenntnis der Wirkung von Volk sarbortive. *Archives Internationales de Pharmacodynamie et de Therapie*, 21, 313–319.

Qiu, Z., Jones, K., Wyle, M., Jèa, Q. and Omdorf, S. (2000) Modified Aloe barbadensis polysaccharide with immunoregolatory activity. *Planta Medica*, 66, 152–156.

Ralamboranto, L., Rakotovao, L.H., Le Deaut, J.Y., Chaussoux, D., Salomon, J.C., Fournet, B., Montreuil, J., Rakotonirina-Randriambeloma, P.J., Dulat, C. and Coulanges, P. (1982) Etude des propriétés immunostimulantes d'un extrait isolé et partiellement purifié à partir d'Aloe vahombe. 3. Etude des propriétés antitumorales et contribution à l'étude de la nature chimique du principe actif. *Archives Institut Pasteur*, 50, 227–256.

Ramstad, E. (1995). *Modern Pharmacognosy*. London: McGraw-Hill Book Company, Inc.

Reynolds, T. and Dweck, A.C. (1999) Aloe vera leaf gel: a review update. *Journal of Ethnopharmacology*, 68, 3–37.

Roberts, D.B. and Travis, E.L. (1995) Acemannan-containing wound dressing gel reduces radiation-induced skin reactions in C3H mice. *International Journal of Radiation Oncology Biology Physics*, 32, 1047–1052.

Robson, M.C., Heggers, J.P. and Hagstrom, W.J. (1982) Myth, magic, withchcraft or fact? Aloe vera revisited. *Journal of Burn Care and Rehabilitation*, 3, 157–163.

Rojas, L., Matamoros, M., Garrido, N. and Finlay, C. (1995) The action of an aqueous extract of Aloe barbadensis Miller in an in-vitro culture of Trichomonas vaginalis. *Revista de Cubana Medicina Tropical*, 47, 181–184.

Rowe, T.D. and Parks, L.M. (1941) Phytochemical study of Aloe vera leaf. *Journal of the American Pharmaceutical Association*, 30, 262–266.

Saito, H., Imanishi, K. and Okabe, S. (1989) Effects of aloe extract, Aloctin A on gastric secretion and on experimental gastric lesions in rats. *Yakugalu Zasshi*, 109, 335–339.

Schörkhuber, M., Richter, N., Dutter, A., Sontag, G. and Marian, B. (1998) Effect of anthraquinone-laxatives on the proliferation and urokinase secretion of normal premalignant and malignant colonic epithelial cells. *European Journal of Cancer*, 34, 1091–1098.

Schmidt, J.M. and Greenspoon, J.S. (1993) Aloe vera dermal wound gel is associated with a delay in wound healing. *Obstetrics and Gynaecology*, 78, 115–117.

Shamaan, N.A., Kadir, K.A., Rahmat, A. and Ngah, W.Z. (1998) Vitamin C and aloe vera supplementation protects from chemical hepatocarcinogenesis in the rat. *Nutrition*, 14, 846–852.

Shelton, R.M. (1991) Aloe vera. Its chemical and therapeutic properties. *International Journal of Dermatology*, 30, 679–683.

Shida, T., Yogi, A., Nishimura, H. and Nishioka, I. (1985) Effect of Aloe extract on peripheral phagocytosis in adult bronchial asthma. *Planta Medica*, 51, 273–275.

Ship, A.G. (1977) Is topical Aloe vera plant mucus helpful in burn treatment? *Journal of the American Medical Association*, 238, 1770.

Shoji, A. (1982a) Contact dermatitis to Aloe arborescens. *Contact Dermatitis*, 8, 164–167.

Shoji, A. (1982b) Contact dermatitis to Aloe arborescens. *Chemical and Pharmacuetical Bulletin (Tokyo)*, 36, 4462–4466.

Siegers, C.P. (1992) Anthranoid laxatives and colorectal cancer. *Trends in Pharmacological Science*, 13, 229–231.

Siegers, C.P., von Hertzberg Lottin, E., Otte, M. and Scheider, B. (1993) Anthranoid laxative abuse – a risk for colorectal cancer? *Gut*, 34, 1099–1101.

Sonnenberg, A. and Muller, A.D. (1993) Constipation and cathartics as risk factors of colorectal cancer: a meta-analysis. *Pharmacology*, 47 (Suppl. 1), 224–233.

Steer, H.W. and Colin-Jones, D.G. (1975) Melanosis coli: Studies on the toxic effects of irritant purgatives. *Journal of Pathology*, 115, 199–205.

Strickland, F.M., Muller, H.K., Stephens, L.C., Bucana, C.V.D., Donawho, C.K., Sun, Y. and Pelley, R.P. (2000) Induction of primary cutaneous melanomas in C3H mice by combinet treatment with ultraviolet radiation, ethanol and aloe emodin. *Photochemistry and Photobiology*, 3, 407–414.

Suzuki, I., Saito, H. and Inoue, S. (1979) A study of cell agglutination and cap formation on various cells with Aloctin A. *Cell Structure and Function*, 3, 379.

Swanson, L.N. (1995) Therapeutic value of aloe vera. *US Pharmacy*, 20, 26–35.

Sydiskis, R.J., Owen, D.G., Lohr, J.L., Rosler, K.H. and Blomster, R.N. (1991) Inactivation of enveloped viruses by anthraquinones extracted from plants. *Antimicrobial Agents and Chemotherapy*, 35, 2463–2466.

Syed, T.A., Ahmad, S.A., Holt, A.H., Ahmad, S.H. and Afzal, M. (1996a) Management of psoriasis with Aloe vera extract in a hydrophilic cream: a placebo-controlled, double-blind study. *Tropical Medicine and International Health*, 1, 505–509.

Syed, T.A., Cheema, K.M., Ahmad, S.A. and Holt, A.H. (1996b) Aloe vera extract 0.5% in a hydrophilic cream versus Aloe vera gel for the management of genital herpes in males. A placebo-controlled, double-blind, comparative study. *Journal of the European Academy of Dermatology and Venereology*, 7, 294–295.

Syed, T.A., Afzal, M. and Ahmad, S.A. (1997) Management of genital herpes in men with 0.5% Aloe vera extract in a hydrophylic cream: a placebo-controlled double-blind study. *Journal of Dermatological Treatment*, 8, 99–102.

Tavares, I.A., Mascolo, N., Izzo, A.A. and Capasso, F. (1996) Effects of anthraquinone derivatives on PAF release by human gastrointestinal mucosa in vitro. *Phytotherapy Research*, 10, S20–S21.

Tchou, M.T. (1943) Aloe vera (jelly leeks). *Archives of Dermatology and Syphilology*, 47, 249.

Teradaira, R., Shinzato, M., Beppu, H. and Fujita, K. (1993) Anti-gastric ulcer effects of Aloe arborescens Mill. var. natalensis Berger. *Phytotherapy Research*, 7, S34–S36.

Tsuda, H., Ito, M., Hirono, I., Kawai, K., Beppu, H., Fujita, K. and Nagao, M. (1993) Inhibitory effect of Aloe arborescens Miller var. natalensis Berger (Kidachi aloe) on induction of preneoplastic focal lesions in the rat liver. *Phytotherapy Research*, 7, S43–S47.

Tyler, V.E., Brady, L.Y. and Robbers, J.E. (1981) Pharmacognosy, 8th edn. Philadelphia, USA: Lea and Febiger.

Vazquez, B., Avila, G., Segura, D. and Escalante, B. (1996) Anti-inflammatory activity of extracts from Aloe vera gel. *Journal of Ethnopharmacology*, 55, 69–75.

Verhaeren, E. (1980) Mitochondrial uncoupling activity as a possible base for a laxative and antipsoriatic effect. *Pharmacology*, 20 (Suppl. 1), 43–49.

Visuthikosol, V., Chowchuen, B., Sukwanarat, Y., Sriurairatana, S. and Boonpucknavig, V. (1995) Effect of Aloe vera gel to healing of burn wound: a clinical and histologic study. *Journal of the Medical Association of Thailand*, 78, 403–409.

Vogler, B.K. and Ernst, E. (1999) Aloe vera: a systematic review of its clinical effectiveness. *British Journal of General Practice*, 49, 823–828.

Watcher, M.A. and Wheeland, R.G. (1989) The role of topical agents in the healing of full-thickness wounds. *Journal of Dermatologic Surgery and Oncology*, 15, 1188–1195.

Werbach, M.R. and Murray, M.T. (1994) *Botanical Influences on Illness. A source book of clinical research*. Tarzana, California, USA: Third Line Press.

Westendorf, J., Marquardt, H., Poginsky, B., Dominiak, M., Shmidt, J. and Marquard, H. (1990) Genotoxicity of naturally occurring hydroxyanthraquinones. *Mutation Research*, 340, 1–12.

Wheelwright, G. (1974) Medicinal plants and their history. New York: Dover Pubblication, Inc.

Williams, M.S., Burk, M., Loprinzi, C.L., Hill, M., Shomberg, P.J., Nearhood, K., O'Fallon, J.R., Laurie, J.A., Shanahan, T.G., Moore, R.L., Urias, R.E., Kuske, R.R., Engel, R.E. and Eggleston, W.D. (1996) Phase III double-blind evaluation of an aloe vera gel as a prophylactic agent for radiation-induced skin toxicity. *International Journal of Radiation Oncology, Biology, Physics*, 36, 345–349.

de Witte, P. and Lemli, L. (1990) The metabolism of anthranoid laxatives. *Hepatogastroenterology*, 37, 601–605.

Wittoesch, J.H., Jackman, R.J. and Mc Donald, J.R. (1958) Melanosis coli. General review and a study of 887 cases. *Diseases of the Colon and Rectum*, 1, 172–180.

Womble, D. and Helderman, J.H. (1988) Enhancement of allo-responsiveness of human lynphocytes by acemannan (Carrisyn TM). *International Journal of Immunopharmacology*, 10, 967–974.

Wright, C.S. (1936) Aloe vera in treatment of roentgen ulcers and telangiectasis. *Journal of the American Medical Association*, 106, 1363–1364.

Yagi, A., Makino, K., Nishioka, I. and Kuchino, Y. (1977) Aloe mannan, polysaccharide from *Aloe arborescens* var. *natalensis*. *Planta Medica*, 31, 17–20.

Yagi, T. and Yamauchi, K. (1999) Synergistic effects of anthraquinones on the purgative activity of rhein anthrone in mice. *Journal of Pharmacy and Pharmacology*, 1, 93–95.

Yago, O. (1969) Toxische und kanstische komplikationen durch Gebrauch sogenannter fruchta-breibender Arzneimittel Z. *Geburtshilfe und Gynakologe*, 170, 272–277.

Yamamoto, I. (1973) Aloe ulcin, a new principle of Cape Aloe and gastrointestinal function, especially experimental ulcer in rats. *Journal of the Medical Society of Toho University*, 20, 342–347.

Yongchaiyudha, S., Rungpitarangsi, V., Bunyapraphatsara, N. and Chokechaijaroenporn, O. (1996) Clinical trial in new cases of diabetes mellitus. *Phytomedicine*, 3, 245–258.

Yongchaiyudha, S., Rungpitarangsi, V., Bunyapraphatsara, N. and Cyhokechaijaroenporn, O. (1996) Antidiabetic activity of Aloe vera L. juice. I. Clinica trial in new cases of diabetes mellitus. *Phytomedicine*, 3, 241–243.

Yoshimoto, R., Kondoh, N., Isawa, M. and Hamuro, J. (1987) Plant lectin, ATF1011, on the tumor cell surface augments tumor-specific immunity through activation of T cells specific for the lectin. *Cancer Immunology Immunotherapy*, 25, 25–30.

Zawacki, B.E. (1974) Reversal of capillary stasis and prevention of burns. *Annals of Surgery*, 180, 98–102.

El Zawahry, M.E, Hegazy, M.R. and Helal, M. (1973) Use of aloe in treating leg ulcers and dermatoses. *International Jouirnal of Dermatology*, 12, 68–73.

10 *Aloe vera* in wound healing

Gary D. Motykie, Michael K. Obeng and John P. Heggers

ABSTRACT

Aloe vera gel is a powerful healer that has been successfully employed for millennia. It acts in the manner of a conductor, orchestrating many biologically active ingredients to achieve the goal of wound healing. Aloe can penetrate and anesthetize tissue, it is bactericidal, virucidal, and fungicidal. It possesses anti-inflammatory and immunomodulatory properties and it serves as a stimulant for wound healing, a fuel for proliferating cells and a dressing for open wounds. Although some of the independent fractions of aloe have shown unique and impressive activity by themselves, the number of different substances acting in concert serves to confirm the relative complexity of aloe's actions. *Aloe vera* certainly gives scope to the phrase, 'the whole is more than the sum of its parts.' Since it has been difficult to postulate, separate and isolate one substance that is responsible for aloe's capabilities, many more controlled, scientific studies must be completed before all the secrets associated with the wound-healing abilities of aloe are unlocked. Future research may be directed at further investigation of the gel's ability to stimulate cell growth in tissue culture and its antimicrobial, antifungal, antiviral and anti-inflammatory properties.

INTRODUCTION

Aloe vera is one of nature's most revered therapeutic healing herbs. The beneficial effects of this tropical succulent have spurred recurrent legends about its properties that have persisted from the fourth century B.C. throughout world history (Coats and Ahola, 1979; Cole and Chen, 1943; Gottleib, 1980; Morton, 1961, 1970; Lewis and Elvin-Lewis, 1977). Even so, it has not been until recently that the scientific community has turned greater attention toward this mysterious plant extract (Table 10.1). Scientific interest was initially triggered by a single case report of accelerated wound healing with aloe's use in radiation-induced dermatitis of the scalp (Collins and Collins, 1935). This initial publication in 1935 and three similar cases reported shortly thereafter spurred a great deal of research investigating and detailing *Aloe vera*'s wound-healing abilities (Wright, 1936; Loveman, 1937). In 1937, Crewe noted improvement in burn and scald wounds treated with the topical application of aloe (Crewe, 1939). Mandeville in 1939 and Rowe *et al.* in 1941 observed an increased rate of wound healing in radiation-induced ulcer wounds (Mandeville, 1939; Rowe *et al.*, 1941). Aloe was shown to be effective in the treatment of second-degree thermal burn wounds in 1943 and in the treatment of

Table 10.1 History of Aloe gel in Wound Healing.

Date	Authors	Research
1935	Collins and Collins	Case of radiodermatitis healed with aloe gel.
1937	Crewe	Thermal burns treated with aloe mixed with mineral oil.
1939	Mandeville	Radiation-induced ulcers successfully treated with aloe.
1941	Rowe *et al.*	Increased healing in radiation-induced ulcers in rats with aloe.
1943	Tchou	Successful treatment of second degree thermal and radium burns with aloe.
1947	Barnes	Accelerated healing of abrasions treated with aloe.
1953	Lushbaugh and Hale	Accelerated ulcer healing in rabbits and increased collagen deposition with aloe.
1964	Sjostom *et al.*	Accelerated healing in frostbite wounds treated with aloe.
1980	Raine *et al.*	Increased tissue survival in frostbite wounds treated with aloe.
1981	Winters *et al.*	Fresh aloe promotes human cell growth *in vitro* (15).
1983	McCauley *et al.*	Increased tissue survival in frostbite wounds treated with aloe.
1989	Watcher *et al.*	Accelerated healing of full-thickness thermal burns treated with aloe.
1990	Fulton	Post-dermabrasion wound healing stimulation with *Aloe vera* gel.

dermal abrasions in 1947 (Tchou, 1943; Barnes, 1947). Lushbaugh and Hale in 1953 showed accelerated wound healing and increased collagen deposition in wounds treated with aloe. In 1964, Sjostrom *et al.* showed increased healing in frostbite wounds, and Raine *et al.* in 1980 and McCauley *et al.* in 1983 showed increased tissue survival in frostbite wounds with the application of *Aloe vera* gel. In 1988 and 1989, the topical application of the gel showed improved healing in the treatment of full thickness wounds and in 1990 a commercially prepared aloe product was able to show marked improvement in wound re-epithelialization (Rodriguez-Bigas *et al.*, 1988; Watcher and Wheeland, 1989; Fulton, 1990).

The complete composition of *Aloe vera* gel and each of its components' effects on wound healing is still being unraveled today. Many of the abilities of the gel have already been ascribed to certain of its constituents and it is well known that *Aloe vera* contains many unique biologically active compounds that have wound-healing, anti-inflammatory and antimicrobial activities.

THE EFFECTS OF ALOE VERA GEL ON WOUND HEALING

In order to understand how *Aloe vera* gel effects wound healing, it is necessary to first examine its chemical composition. Working with a single compound makes it simple to develop scientific research to prove or disprove its proposed benefits. This is not the case with *Aloe vera*. Whereas most botanicals have a single ingredient isolated from the parent that is responsible for its observed benefits, the gel is known to contain well over 100 separate ingredients (Rowe and Parks, 1941). In all aloe extracts, quite common substances make up a very large portion of the total solids including organic acids, free sugars (glucose and fructose), potassium, sodium, calcium and magnesium. Also present are fatty acids, sterols, and plant hormones called eicosanoids, terpenes, gibberellins and auxins. Together, these molecules make up 75% of the total solids in the gel.

The other 25% of the aloe extract is composed of primarily polysaccharides. A polysaccharide is a chain of simple sugars such as glucose referred to as a glucan, or mannose referred to as a mannan. Those containing two types of sugars are named accordingly, such as glucomannans or galactomannans. The physiologically active polysaccharides of *Aloe* and the principal component polysaccharide of the gel extract has long been recognized to be the glucomannans (Dan *et al.*, 1951).

Knowing that aloe's effect on wound healing is to do with its complex composition and the ability of its components to act in concert, it is still possible to examine each individual constituent's effect. For more than a century, scientists have attempted to comprehend and analyze the complex interactions involved in the wound-healing process. Simultaneous interactive and mutually-dependent biochemical signaling transpires between the cellular and humoral based elements of the inflammatory response in wound healing. This signaling enables gene activation and suppression as well as organelle and tissue synthesis. It is well known that wound healing progresses through various stages, each of which involves different primary growth factors, inflammatory reactants and proliferating cells.

The topical application of a variety of cytokines and synthetic growth factors to open wounds has been a topic of many recent experiments and the results have revolutionized the understanding of the wound-healing process. The first stage of wound healing is known to involve mainly inflammatory reactants and cell types. A balance must be achieved between destroying foreign bacteria and initiating the wound-healing process. The second stage involves the proliferation of the fibroblast, which is the cell most important for laying the foundation upon which the wound will eventually gain its strength. The fibroblast can be stimulated by various growth factors and requires the availability of essential ingredients in order to produce collagen. Improved local blood flow can increase the delivery of essential ingredients and immune cells. The final stage of wound healing involves the maturation and remodeling of already existing collagen.

In short, *Aloe vera* gel is a modulator. It acts as both an inhibitor and a stimulator. While it can block mediators of inflammation in the immune system, it also stimulates antibody production and wound healing by growth factor-like substances (Heggers *et al.*, 1993; El Zawahry *et al.*, 1961). The gel factor molecule can stimulate fibroblasts to increase collagen and proteoglycan production, increasing wound tensile strength, while inhibiting inflammation and moderators of pain (Shelton, 1991). Consequently, it may be the localized balance of unique substances that bestows aloe with its healing properties. Listed below and in Table 10.2 is the current knowledge on the components of *Aloe vera* that are being investigated for their potential wound-healing properties.

GEL COMPONENTS

Saccharides

Mono- and polysaccharides form about 25% of the solid fraction of the aloe gel. Mannose and glucose are the most significant monosaccharides found in the gel. These sugars most commonly serve as fuels and building blocks. For example, mannose-6-phosphate is required to initiate glycoprotien and glycolipid synthesis in the endoplasmic reticulum of all nucleated cells (Tizard *et al.*, 1989). Optimal nutrition is required for the growth, regulation, reproduction, defense, regeneration and repair during wound healing. In addition, saccharides such as mannose are essential in the golgi apparatus of all cells to complete synthesis of all structural and functional molecules (Campbell *et al.*, 1997). Lastly, the mannose-6-phosphate of *Aloe vera* has been shown to activate the insulin-like growth factor receptor of the fibroblast, stimulating it to increase collagen and proteoglycan synthesis (Danhoff and McAnalley, 1987). This activity has been shown to increase wound tensile strength (Davis *et al.*, 1994a; Davis *et al.*, 1994b).

The polysaccharide component of aloe gel is primarily glucommannans that are comprised of glucose and mannose ($\beta 1 \rightarrow 4$ linked acetylated mannan) (Hirata and Suga, 1977). These polysaccharides, unlike other sugars, are absorbed complete and appear in the bloodstream undigested. Here, they have many activities. It has been very clearly indicated that the immunostimulant and wound-healing activities of *Aloe vera* gel reside, at least in part, in this glucomannan component (Atherton, 1998), for after separation and purification, glucomannans retain immunostimulant, cell-proliferative, and anti-inflammatory activity (Davis *et al.*, 1986). One of the most marked biological activities of mannans is the activation of macrophages and stimulation of T cells. In fact, macrophages are known to possess a specific receptor for mannans that is both

Table 10.2 Chemical Composition of *Aloe vera* gel.

Anthraquinones	Inorganic compounds		Vitamins	
Aloin (=Barbaloin)	Calcium		A	
Isobarbaloin	Sodium		B1	
Anthranol	Magnesium		B2	
Aloetic Acid	Manganese		B6	
Aloe-emodin	Chlorine		C	
Emodin	Zinc		Choline	
Ethereal oil	Copper		Folic Acid	
Resistannol	Chromium		Beta carotene	
	Potassium		Alpha-tocopherol	

Saccharides	Enzymes		Amino acids	
Glucose	Oxidase		Lysine	Aspartic acid
Mannose	Amylase		Valine	Glutamic acid
Glucomannan	Catalase		Leucine	Leucine
Cellulose	Lipase		Proline	Phenylalanine
Aldopentose	Alkaline Phosphatase		Alanin	Isoleucine
	Bradykininase		Tyrosine	Methionine
			Threonine	Histidine
			Glycine	Arginine

Miscellaneous				
Cholesterol	Beta-sitosterol	Lectin-like substance	Eicosanoids	Fatty acids
Triglycerides	Uric acid	Lignins	Terpenes	Lupeol
Campesterol	Gibberellin	Salicylic acid	Auxins	
Steroids				

on the cell surface and intracellularly, and as a result, mannans are potent immuno-stimulants with significant activity against infectious diseases (Lefkowitz *et al.*, 1997; Stahl, 1990).

In 1982, acemannan, aloe's most common glucomannan, was successfully extracted and stabilized from *Aloe vera* gel (Davis, 1997). Acemannan has been well documented to enhance the immune system by stimulating the growth of macrophages, lymphocytes, NK (natural killer) cells, as well as the production of interferon and all known cytokines (Womble and Helderman, 1988; Marshall and Druck, 1993; Kawasaki, 1999). Acemannan has been shown to induce interleukin-1 and prostaglandin E2 production by human peripheral blood cells and the exposure of lymphocytes to ace-mannan has been associated with an increase in antigen expression and NK cell activity (Marshall *et al.*, 1993). These properties have bestowed aloe with significant anti-microbial, anti-fungal and antiviral properties (Bruce, 1967; Lorenzetti *et al.*, 1964; Blumauerova, 1976; Lefkowitz and Lefkowitz, 1999; Stuart *et al.*, 1997; Pulse and Uhlig, 1990).

Sterols

These include campesterol, β-sitosterol and lupeol. These compounds are believed to have anti-inflammatory properties that aid in the coordination of local wound-healing activities (Davis *et al.*, 1988).

Eicosanoids and terpenes

Eicosandoids and terpenes are active in connection with decreasing inflammation in wounds (Davis *et al.*, 1989).

Gibberellin and auxin

The plant hormones gibberellin and auxin have shown an ability to inhibit inflamma-tion and stimulate antibody production and wound healing in a dose-response manner (Davis and Maro, 1989).

Bradykininase

Aloe also contains a bradykininase, which is active in inhibiting inflammation in the wound bed (Fujita *et al.*, 1976). This enzyme is thought to be involved in the initial vasoconstriction seen with the topical application of the gel (Fujita *et al.*, 1979).

Inorganic electrolytes

Sodium, potassium, calcium, magnesium, manganese, copper, zinc, chromium and iron are all found in aloes. Magnesium lactate inhibits histidine decarboxylase, which pre-vents the formation of histidine (Davis *et al.*, 1988). This may partially explain the antipuritic and anti-inflammatory effect of aloe. Also, the remaining electrolytes are essential parts of numerous enzymatic reactions that are important for cell growth, maintenance and wound healing.

Vitamins

Aloe is known to contain numerous vitamins that are vital for the maintenance of cell growth and cellular integrity. In addition, vitamins A, C and E are antioxidants that may be important in reducing inflammation in the wound. Vitamin C is needed in the cross-linking of collagen, which is required for normal wound tensile strength.

Anthraquinones

These phenolic compounds are found in the aloe sap (exudate) and are known to be potent antimicrobial and antiviral agents that possess powerful analgesic effects (Benigni, 1950; Anderson *et al.*, 1991; Saalman *et al.*, 1990; Barnard *et al.*, 1992; Sydiskis *et al.*, 1991). Anthraquinones have also been shown to possess anti-inflammatory and anti-arthritic properties (Atherton, 1998).

Saponins

Saponins comprise approximately 3% of aloe gel and are known to be cleansing agents with antiseptic properties (Atherton, 1998).

Lignin

Lignin is an inert substance that endows aloe preparations with their singular penetrative ability to carry other active ingredients either deep into the skin in order to nourish the dermis or deep into the wound in order to aid in the healing process (Atherton, 1998).

Amino acids

Amino acids are the building blocks of proteins, including enzymes and structural elements that are essential for cell growth and the wound-healing process. Aloe contains 20 of the 22 amino acids required by cells for protein synthesis, including seven of the eight essential amino acids that cannot be synthesized by the cell itself (Gjerstad, 1971).

Mucilage

In 1963, Blitz *et al.* first suggested that aloe may expedite wound healing by serving as a protective barrier due to its occlusive properties (Blitz *et al.*, 1963). The occlusive cover-like qualities of the gel were later attributed to mucilage, which acts as a storage container inside the leaf. It also acts like a sealant or 'bandage' when a plant leaf is injured. It has been shown that dry wounds prevent the migration of cells and disrupt the effects of wound-healing growth factors (Thomas *et al.*, 1998). Aloe gel, acting as a cover, has been shown to keep the wound moist, allowing excellent migration of epidermal and fibroblast cells (Shelton, 1991). In addition, the occlusive cover-like properties of mucilage allow it to serve as an organic bandage or dressing.

Antimicrobial properties

Even today we employ a varied armamentarium of topical substances to control and prevent infection, without a complete understanding of their influence on the process of wound healing. In fact, it has been demonstrated that concentrations of several antimicrobial agents used to control wound infections are in fact toxic to tissue-cultured cells (McCauley *et al.*, 1992). Extracts from *Aloe vera* gel have been shown to have beneficial heterogeneous properties, which include an ability to penetrate tissue, anesthetize tissue, preclude bacterial, fungal and viral growth, and dilate capillaries in order to enhance blood flow to injured tissue (Robson *et al.*, 1982).

Infection plays a major role in influencing the rate of wound healing. In one study, when compared with other topical antimicrobials, the group treated with aloe gel healed significantly faster than the control group and had the shortest half-life overall (Heggers *et al.*, 1995). Topical aloe gel significantly enhanced the rate of wound healing and has also been shown to reverse the wound retardant effects of other topical agents (Heggers *et al.*, 1996; Tizard *et al.*, 1994).

Various studies have provided evidence that aloe retains antimicrobial effects against numerous organisms at concentrations of 60–70% (Table 10.3) (Lorenzetti *et al.*, 1964; Heggers *et al.*, 1979; Fly and Keim, 1963).

In addition, as stated previously, the glucomannan component of *Aloe vera* gel possesses immunostimulant, antifungal and anti-viral activity. It has been shown to stimulate T cells and activate macrophages, causing an enhancement of the respiratory burst and phagocytosis (Lefkowitz *et al.*, 1997). Acemannan has been shown to inhibit virus-induced cell fusion, reduce virus load and suppress the production and release of free virus from acutely infected cells (Kahlon *et al.*, 1991). The ingestion of large mannan molecules has been shown to increase secretion of IL(interleukin)-1, interferon and TNF (tumor necrosis factor) (Marshall and Druck, 1993; Kawasaki, 1999). As a result, aloe is a potent immunostimulant with significant activity against infectious disease. Its lack of significant toxicity, in combination with its macrophage-stimulating activity, makes aloe a prime candidate for further investigation as an immunostimulant in infectious disease and as an antitumor agent (Winters *et al.*, 1981; Bomford and Moreno, 1977; Harris *et al.*, 1991). In fact, acemannan has been used to treat retroviral infections and is currently being studied for its effects in combating the HIV virus (Montaner *et al.*, 1996).

Anti-inflammatory properties

At the time of tissue injury, the eicosanoid cascade is initiated by platelets and the primary mediator of progressive tissue loss is the potent vasoconstrictor, thromboxane

Table 10.3 Bactericidal Effects of 70% concentration of gel.

Gram negative organisms	*Gram positive organisms*
Escherichia coli	*Stapylococcush. Aureus*
Enterobacter sp.	*Streptococcus. Pyogenes*
Citrobacter sp.	*S. agalactiae*
Serratia marcescens	*S.* sp., group D enterococci
Klebsiella sp.	
Pseudomonas aeruginosa	

A_2 (Rosenburg and Gallin, 1999). Aloe acts as a thromboxane A_2 synthetase inhibitor, preventing its production and maintaining a balanced equilibrium between PGE_2 and PGF_{2A} (Heggers and Robson, 1989). Aloe gel's anti-thromboxane properties have been shown to dilate arteries and enhance local blood flow (Davis *et al.*, 1986; Robson *et al.*, 1979, 1980).

Aloe gel can block vasoactive substances responsible for inflammation, can constrict small blood vessels, can block PMN (polymorphonuclear leucocyte) infiltration, can inhibit the production of oxygen free radicals and can dilate capillaries allowing for increased blood supply to damaged tissue (Davis *et al.*, 1987; DelBeccaro *et al.*, 1978; Heggers *et al.*, 1985). These properties have given aloe the ability to reverse progressive tissue necrosis in partially damaged tissue (Zawacki, 1974). The plant hormones called gibberellins and auxins and the group of substances related to aspirin, known as the salicylates, seem to be the mediators of this response (Davis *et al.*, 1991). In addition, there are organic compounds in aloe such as emolin, barbaloin and emodin that can be broken down by the Kolbe reaction into salicylates (t'Hart *et al.*, 1988). The anti-thromboxane effect of these compounds may be due to enzymatic substrate competition. Aloe gel contains an abundance of fatty acids that allow for competitive inhibition of thromboxane production through stereochemical means, while also supplying the necessary nutrient precursors such as triglycerides and cholesterol for the initiation of the remaining arachidonic cascade (t'Hart *et al.*, 1988). This allows for inhibition of inflammatory mediators while supplying the cell with the important constituents to maintain cellular integrity and normal tissue maturation (Brasher *et al.*, 1969; Fujita *et al.*, 1978).

Purified glucomannans have also shown impressive anti-inflammatory activity when separated from other aloe constituents (Davis *et al.*, 1991). This concept has been supported by research in which the administration of mannans prevented arthritic flares in rats (Moreland, 1999). Also, mannose inhibits free radical production by neutrophils, limiting tissue damage (Rest *et al.*, 1988). This is important since the neutrophil is the hallmark cell of inflammation and its presence is crucial to the inflammatory response (Kuby, 1997). Furthermore, polymannose can inhibit the initial step in the migration of neutrophils out of the blood stream and can aid in the clearance of certain pathogens through the presence of cell surface carbohydrate-receptor interactions (Lefkowitz *et al.*, 1999).

REFERENCES

Anderson, D.O., Weber, N.D., Wood, S.G., Hughes, B.G., Murray, B.K. and North, J.A. (1991) *In vitro* virucidal activity of selected anthraquinones and anthraquinone derivatives. *Antiviral Research*, 16, 185–196.

Atherton, P. (1998) *Aloe vera*: magic or medicine? *Nursing standard*, 12, 4952.

Barnard, D.L., Huffman, J.H., Morris, J.L.B., Wood, S.G., Sidwell, R.W. and Hughes, B.G. (1992) Evaluation of the antiviral activity of anthraquinones, anthrones and anthroquinone derivatives against human cytomegalovirus. *Antiviral Research*, 17, 63–77.

Barnes, T.C. (1947) The healing action of extracts of *Aloe vera* leaf on abrasions of human skin. *American Journal of Botany*, 34, 597.

Benigni, R. (1950) Substances with antibiotic action contained in anthraquinonic drugs. *Chemical Abstracts*, 44, 1036.

Blitz, J., Smith, J.W. and Gerard, J.R. (1963) Aloe vera gel in peptic ulcer therapy. *Journal of the American Osteopathic Association*, 63, 731–735.

Blumauerova, C.J., Steinerova, M. and Mateju, N. (1976) Biological activity of hydroxy-quinones and their glucosides toward microorganisms. *Folia Microbiologica*, 21, 54–57.

Bomford, R. and Moreno, C. (1977) Mechanism of anti-tumor effect of glucans and fructosans: a comparison with C.parvum. *British Journal of Cancer*, 36, 41–48.

Brasher, W.J., Zimmeramn, E.R. and Collings, C.K. (1969) The effect of prednisolone, indomethacine and *Aloe vera* gel on tissue culture cells. *Oral Surgery, Oral Medicine and Oral Pathology*, 27, 122–128.

Bruce, W.G. (1967) Investigations of antibacterial activity in the Aloe. *South African Medical Journal*, 41, 984.

Campbell, B.D, Busbee, D.L. and McDaniel, R.H. (1997) Enhancement of immune function using a proprietary complex mixture of glyconutritionals. *Proceedings of the Fisher Institute for Medical Research*, 1, 12–15.

Cheney, R.H. (1970) Aloe drug in Human therapy. *Quarterly Journal of Crude Drug Research*, 10, 1523–1529.

Coats, B.C. and Ahola, R. (1979) *The silent healer. A Modern Study of Aloe vera*, pp. 1–288. Garland, Texas: Bill C. Coats.

Cole, H.N. and Chen, K.K. (1943) Aloe vera in oriental dermatology. *Archives of Dermatology and Syphilology*, 37, 250.

Collins, C.E. and Collins, C. (1935) Roentgen dermatitis treated with fresh whole leaf Aloe vera. *American Journal Roentgenology*, 33, 396.

Crewe, J.E. (1939) Aloes in the treatment of burns and scalds. *Minnesota Medicine*, 2, 538–539.

Dan, D.B., Mitra, M.K. and Wareham, J.F. (1951) Chemical composition of *Aloe* fibre. *Journal of Indian Chemical Science*, 18, 37.

Danhoff, I.E. and McAnalley, B.H. (1987) Stabilized Aloe vera: effect on human skin cells. *Drug and Cosmetic Industry*, 133, 52–55.

Davis, R.H. and Maro, N.P. (1989) Aloe vera and gibberellin: Antiinflammatory activity in diabetes. *Journal of the American Podiatric Medical Association*, 79, 24–26.

Davis, R.H. (1997) *Aloe Vera: A scientific approach*, pp. 1–321. New York: Vantage Press.

Davis, R.H., Di Donato, J.J., Hartman, G.M. and Haas, R.C. (1994) Anti-inflammatory and wound healing activity of a growth substance in *Aloe vera*. *Journal of the American Podiatric Medical Association*, 84, 77–81.

Davis, R.H., Di Donato, J.J., Johnson, R.W.S. and Stewart, C.B. (1994) Aloe vera, hydrocorti-sone, and sterol influence on wound tensile strength and anti-inflammation. *Journal of the American Podiatric Medical Association*, 84, 614–21.

Davis, R.H., Kabbani, J. and Maro, N. (1986) *Aloe vera* and inflammation. *Pennsylvania Academy of Science*, 60, 67.

Davis, R.H., Kabbani, J.M. and Maro, N.P. (1986) Wound healing and anti-inflammatory activity of *Aloe vera*. *Proceedings of the Pennsylvania Academy of Science*, 60, 67–70.

Davis, R.H., Leitner, M.G. and Russo, J.M. (1987) Topical anti-inflammatory activity of Aloe vera as measured by ear swelling. *Journal of the American Podiatric Medical Association*, 77, 610–612.

Davis, R.H., Leitner, M.G. and Russo, J.M. (1988) Aloe vera, a natural approach for treating wounds, edema and pain in diabetes. *Journal of the American Podiatric Medical Association*, 78, 60–68.

Davis, R.H., Parker, W.L., Samson, R.T. and Murdoch, D.P. (1991) Isolation of an active inhibitory system from an extract of Aloe vera. *Journal of the American Podiatric Medical Association*, 81, 258–261.

Davis, R.H., Parker, W.L., Samson, R.T. and Murdoch, D.P. (1991) Isolation of a stimulatory system in an Aloe extract. *Journal of the American Podiatric Medical Association*, 81, 473–478.

Davis, R.H., Rosenthal, K.Y., Cesario, L.R. and Rouw, L.R. (1989) Processed *Aloe vera* administered topically inhibits inflammation. *Journal of the American Podiatric Medical Association*, 79, 397–397.

DelBeccaro, E.J., Heggers, J.P. and Robson, M.C. (1978) Preventing the prostaglandin effect on dermal ischemia in the burn wound. *Surgical Forum*, 29, 603.

El Zawahry, M.E., Hegazy, M.R. and Helal, M. (1961) Use of Aloe in treating leg ulcers and dermatoses. *International Journal of Dermatology*, 12, 68–73.

Fly, L.B. and Keim, I. (1963) Tests of *Aloe vera* for antibiotic activity. *Economic Botany*, 17, 46–48.

Fujita, K., Shasike, I., Teradaira, R., and Beppu, H. (1979) Properties of a carboxypeptidase from aloe. *Biochemical Pharmacology*, 28, 1261–1262.

Fujita, K., Suzuki, I., Ochiai, J., Shimpo, J., Inoue, S. and Saito, H. (1978) Specific reaction of aloe extract with serum proteins of various animals. *Experientia*, 33, 523–524.

Fujita, K., Teradaira, R. and Nagatsu, T. (1976) Bradykininase activity of aloe extract. *Biochemical Pharmacology*, 25, 205.

Fulton, J.E. (1990) The stimulation of post-dermabrasion wound healing with stabilized *Aloe vera* gel-polyethylene oxide dressing. *Journal of Dermatology and Surgical Oncology*, 16, 460–467.

Gjerstad, G. (1971) Chemical studies of *Aloe vera* juice. Amino acid analysis. *Advancing Frontiers of Plant Sciences*, 28, 311–315.

Gottleib, K. (1980) *Aloe vera heals. The scientific Facts*, pp.1–31. Denver, Colorado: Royal Publications.

Harris, C., Pierce, K., King, G., Yates, K.M., Hall, J. and Tizard, I. (1991) Efficacy of acemannan in treatment of canine and feline spontaneous neoplasms. *Molecular Biotherapy*, 3, 207–213.

Heggers, J.P. and Robson, M.C. (1989) Eicosanoids in wound healing. In *Prostaglandins in Clinical Practice*, edited by W.D. Watkins, J.R. Fletcher and D.F. Stubbs, pp. 183–194. New York: Raven Press.

Heggers, J.P., Kucukcelebi, A., Listengarten, D., Stabenau, C.J., Ko, F., Broemeling, L.D., Robson, M.C. and Winters, W.D. (1996) Beneficial effect of Aloe on wound healing in an excisional wound healing model. *Journal of Alternative and Complementary Medicine*, 2, 271–277.

Heggers, J.P., Kucukcelebi, A., Stabenau, C.J., Ko, F., Broemeling, L.D., Robson, M.C. and Winters, W.D. (1995) Wound healing effects of Aloe gel and other topical antibacterial agents on rat skin. *Phytotherapy Research*, 9, 455–457.

Heggers, J.P., Pelley, R.P. and Robson, M.C. (1993) Beneficial effects of Aloe in wound healing. *Phytotherapy Research*, 7, S48–52.

Heggers, J.P., Pineless, G.R. and Robson, M.C. (1979) Dermaide *Aloe/Aloe vera* gel: comparison of the antimicrobial effects. *Journal of American Medical Technologists*, 41, 293–294.

Heggers, J.P., Robson, M.C. and Zachary, L.S. (1985) Thromboxane inhibitors for the prevention of progressive dermal ischemia due to thermal injury. *Journal of Burn Care and Rehabilitation*, 6, 466–468.

Hirata, T. and Suga, T. (1977) Biologically active constituents of leaves and roots of *Aloe arborescens* var. *natalensis*. *Zeitschrift für Naturforschung*, 32, 731–734.

Kahlon, J., Kemp, M.C.X., Carpenter, R.H., McAnnaley, B.H., McDaniel, H.R. and Shannon, W.M. (1991) Inhibition of AIDS virus replication by acemannan. *Molecular Biotherapy*, 3, 127–135.

Kawasaki, T. (1999) Structure and biology of mannan-binding protein, MBP, an important component of innate immunity. *Biochimica et Biophysica Acta*, 1473, 186–195.

Kuby, J. (1997) *Immunology*. 3rd edn. New York: W.H. Freeman and Company.

Lefkowitz, D.L., Gelderman, M.P., Fuhrmann, S.R., Graham, S., Starnes III, J.D., Lefkowitz, S.S., Bollen, A. and Moguilevsky, N. (1999) Neutrophilic myeloperoxidase-macrophage interactions perpetuate chronic inflammation associated with experimental arthritis. *Clinical Immunology*, 91, 145–155.

Lefkowitz, D.L., Lincoln, J.A., Lefkowitz, S.S., Bollen, A. and Moguilevsky, N. (1997) Enhancement of macrophage-mediated bactericidal activity b macrophage-mannose receptor-ligand interaction. *Immunology and Cell Biology*, 75, 136–141.

Lefkowitz, S.S. and Lefkowitz, D.L. (1999) Macrophage candicidal activity of a complete glyconutritional formulation versus Aloe polymannose. *Proceedings of the Fisher Institute for Medical Research*, 1, 5–7.

Lewis, W.H. and Elvin-Lewis, M.P. (1977) *Medical Botany, Plants Affecting Man's Health*, pp. 336–354. New York: John Wiley & Sons.

Lorenzetti, L.J., Salisbury, R., Beal, J. and Baldwin, J.N. (1964) Bacteriostatic property of *Aloe vera*. *Journal of Pharmeutical Science*, 53, 1287.

Loveman, A.B. (1937) *Aloe vera* in treatment of roentgen ray ulcer. *Archives of Dermatology and Syphilology*, 36, 838–843.

Lushbaugh, C.C. and Hale, D.B. (1953) Experimental acute radiodermatitis following beta irradiation: Histopathological study of the mode of action of therapy with *Aloe vera*. *Cancer*, 6, 690–697.

Mandeville, F.B. (1939) *Aloe vera* in the treatment of radiation ulcers of mucous membranes. *Radiology*, 32, 598–599.

Marshall, G.D. and Druck, J.P. (1993) *In vitro* stimulation of NK activity by acemannan. *Journal of Immunology*, 150, 1381.

Marshall, G.D., Gibbons, A.S and Parnell, L.S. (1993) Human cytokines induced by acemannan. *Journal of Allergy and Clinical Immunology*, 91, 295.

McCauley, R.L., Hing, D.N., Robson, M.C. and Heggers, J.P. (1983) Frostbite injuries: A rational approach based on pathophysiology. *The Journal of Trauma* 23, 143–147.

McCauley, R.L., Poole, B., Heggers, J.P., Robson, M.C. and Hemdon, D.N. (1992) Differential *in vitro* toxicity of topical antimicrobial agents to human kerotinocytes. *Journal of Surgical Research*, 52, 276–285.

Montaner, J.S., Gill, J., Singer, J., Raboud, J., Arseneau, R., McLean B.D., Schecter, M.T. and Ruedy, J. (1996) Double-blind placebo-controlled pilot trial of acemannan in advanced human immunodeficiency virus disease. *Journal of Acquired Immune Deficiency Syndromes and Human Retrovirology*, 12, 153–157.

Moreland, L.W. (1999) Inhibitors of tumor necrosis factor for rheumatoid arthritis. *Journal of Rheumatology*, 26, Supplement 57, 7–15.

Morton, J.F. (1961) Folk uses and commercial exploitation of the *Aloe* leaf pulp. *Economic Botany*, 15, 311–317.

Pulse, T.L. and Uhlig, E. (1990) A significant improvement in a clinical pilot study utilizing nutritional supplements, essential fatty acids and stabilized Aloe vera juice in HIV seropositive, ARC and AIDS patients. *Journal of Advancement in Medicine*, 3, 1960–1968.

Raine, T.J., London, M.D., Goluch, K., Heggers, J.P. and Robson, M.C. (1980) Antiprostaglandins and antithromboxanes for treatment of frostbite. *American College of Surgeons Surgical Forum*, 31, 557–559.

Rest, R.F., Farrell, C.F. and Naids, F.L. (1988) Mannose inhibits the human neutrophils oxidative burst. *Journal of Leukocyte Biology*, 43, 158–164.

Robson, M.C., Delbeccaro, E.J. and Heggers, J.P. (1980) Increasing dermal perfusion after burning by decreasing thromboxane production. *Journal of Trauma*, 20, 722–725.

Robson, M.C., DelBeccaro, E.J. and Heggers, J.P. (1979) The effects of prostaglandins on the dermal microcirculation after burning and the inhibition of the effect by specific pharmacological agents. *Plastic and Reconstructive Surgery*, 63, 781.

Robson, M.C., Heggers, J.P. and Hagstrom, W.J. (1982) Myth, Magic, Witchcraft, or Fact? Aloe vera revisited. *Journal of Burn Care and Rehabilitation*, 3, 157–163.

Rodriguez-Bigas, M., Cruz, N.I. and Suarez, A. (1988) A comparative evaluation of Aloe vera in the management of burn wounds in guinea pigs. *Plastic and Reconstructive Surgery*, 81, 386–389.

Rosenburg, H.F. and Gallin, J.I. (1999) *Inflammation. Fundamental Immunology*, pp. 1051–1066. Philadelphia: Lippincott Raven Publishers.

Rowe, T.D. and Parks, L.M. (1941) A phytochemical study of *Aloe vera* leaf. *Journal of the American Pharmeutical Association*, 30, 262–265.

Rowe, T.D., Lovell, B.K. and Parks, L.M. (1941) Further observations on the use of *Aloe vera* leaf in the treatment of third degree x-ray reactions. *Journal of the American Pharmaceutical Association*, 30, 265–269.

Saalman, V., Cannon, D.L., Schinazi, R.F., Eriksson, B.F.H., Chu, C.K., Babu, J.R., Oswald, B.J. and Nasr, M. (1990) Anthraquinones as a new class of antiviral agents against human immunodefieceincy virus. *Antiviral Research*, 13, 265–272.

Shelton, R.M. (1991) Aloe Vera: Its chemical and therapeutic properties. *International Journal of Dermatology*, 30, 679–683.

Sjostrom, B., Weatherly-White, R.C.A. and Paton, B.C. (1964) Experimental studies in cold injury. *Journal of Surgical Research*, 4, 12–16.

Stahl, P.D. (1990) The macrophage mannose receptor: current status. *American Journal of Respiratory Cell and Molecular Biology*, 2, 317–318.

Stuart, R.W., Lefkowitz, D.L., Lincoln, J.A., Howard, K., Gelderman, M.P. and Lefkowitz, S.S. (1997) Upregulation of phagocytosis and candidicidal activity of macrophages exposed to the immunostimulant, acemannan. *International Journal Immunopharmacology*, 19, 75–82.

Sydiskis, R.J., Owen, D.G., Lohr, J.L., Rosler, K.H. and Blomster, R.N. (1991) Inactivation of enveloped viruses by anthraquinones extracted from plants. *Antimicrobial Agents and Chemotherapy*, 35, 2463–2466.

t'Hart, L.A., van Enckevort, P., van Dijk, H., Zaat, R., de Silva, K.T.D. and Labadie, R.P. (1988) Two functionally distinct immunomodulatory compounds in the gel of Aloe vera. *Journal of Ethnopharmacology*, 23, 61.

Tchou, M.T. (1943) *Aloe vera* (jelly leeks). *Archives of Dermatology and Syphilology*, 47, 249.

Thomas, D.R., Goode, P.S., La Master, K. and Tennyson, T. (1998) Acemannan hydrogel dressing versus saline dressing for pressure ulcers: a randomized, controlled trial. *Advances in Wound Care*, 11, 273–276.

Tizard, I.R., Busbee, D., Maxwell, B. and Kemp, M.C. (1994) Effects of acemannan, a complex carbohydrate, on wound healing in rats. *Wounds*, 6, 201–209.

Tizard, I.R., Carpenter, R.H., McNalley, B.H. and Kemp, M.C. (1989) The biological activities of mannans and related complex carbohydrates. *Molecular Biotherapy*, 1, 290–296.

Watcher, M.A. and Wheeland, R.G. (1898) The role of topical agents in the healing of full-thickness wounds. *Journal of Dermatology and Surgical Oncology*, 15, 1188–1195.

Winters, W.D., Benavides, R. and Clouse, W.J. (1981) Effects of aloe extracts on human normal and tumor cells *in vitro*. *Economic Botany*, 35, 89–95.

Womble, D. and Helderman, J.H. (1988) Enhancement of allo-responsiveness of human lymphocytes by acemannan. *International Journal of Immunopharmacology*, 10, 967–974.

Wright, C.S. (1936) Aloe vera in treatment of roentgen ulcers and telangectasis. *Journal of the American Medical Association*, 106, 1363–1364.

Zawacki, B.E. (1974) Reversal of capillary stasis and prevention of necrosis in burns. *Annals of Surgery*, 180, 98–102.

11 *Aloe vera* in thermal and frostbite injuries

*Michael K. Obeng, Gary D. Motykie, Amer Dastgir,
Robert L. McCauley and John P. Heggers*

THERMAL INJURIES: INTRODUCTION

Thermal injury is a major concern in the world. In the United States alone, more than 1.2 million burn casualties are reported each year. Between 50,000–60,000 of these cases require hospitalization as a result of their severity (Brigham and McLoughlin, 1996). Ten percent of those hospitalized will go on to die due to the complications of burns, such as inhalational injury, infections, multiple organ failures, and others (Brigham and McLoughlin, 1996; Bull and Fisher, 1949; Heimbach, 1987; Rodriguez-Bigas *et al.*, 1988). Deaths related to thermal injury are distributed in a bimodal fashion just like any other trauma-related deaths: immediately after the injury or weeks later (Committee on trauma, 1999).

The elderly, young adults and children are often the victims of thermal injury with about 67% of these cases happening at the their homes (Barillo and Goode, 1996; Ryan *et al.*, 1998). It is sad to point out that a substantial amount of these horrendous injuries are as a result of child abuse. Most of these injuries are accident related, however, it is interesting to note that risk factors like low socioeconomic status and unsafe environment have been associated with thermal injuries, making preventive methods worthwhile to mention (Hurley, 1957; Lorenzetti *et al.*, 1964).

The sequelae of burns not only involves disfiguring forms and altered function but also the psychosocial well-being of the individual is significantly altered (Blakeney, 1988). However, with good reconstructive techniques, good rehabilitation, and the availability of psychological support, most of these patients eventually find a sense of fulfillment. With the advances made in burn care, better resuscitation methods, the availability of potent antimicrobials, better wound dressing agents, better attenuation of the hypermetabolic response and improvement in inhalational injury treatment, the survival rate of burn victims has improved dramatically (Alexander *et al.*, 1981; Blakeney *et al.*, 1988; Boswick, 1987; Davis and Maro, 1989; Desai *et al.*, 1990; Desai *et al.*, 1991; Dziewulski, 1992; Heggers *et al.*, 1985; Molnar *et al.*, 1973).

Anatomy and pathophysiology of the burn wound

The burn wound is divided anatomically into a partial versus full thickness wound based on the depth of burn (Figure 11.1). Partial thickness burns are further divided into two subtypes: partial thickness and deep burns.

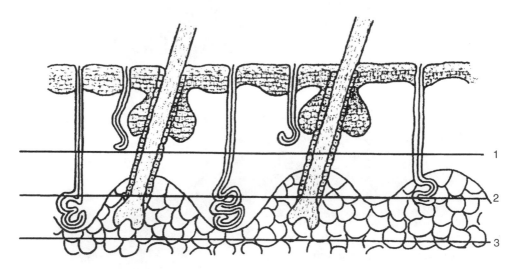

Figure 11.1 Burn wound depth: 1) superficial partial thickness burn; 2) Deep partial thickness burn (superficial tissues injured); 3) Full thickness burn (all tissues injured beyond repair) (Wolf and Herndon, 1999, *Vademecum in Burn Care*. Austin, Texas: Landes Bioscience).

The anatomical terms widely used are described below:

1   *1st degree burns*: Only involves the epidermis and is erythematous and painful in nature. There is usually blistering. The wound blanches under pressure, and sensation is usually intact.
2   *2nd degree burns*: Involves the entire epidermis and the upper portions of the dermis. It is also painful as well. There is some blistering. The wound could be moist or dry. There is no blanching, and sensation is diminished.
3   *3rd degree burns*: Involves the epidermis, the dermis, and all adnexal structures. There is no blistering. It is leathery-like in nature and feels hard to touch. There is charring, and the wound is dry and painless.

As mentioned previously, the partial burns can further be divided into superficial and deep-based on adnexal involvement. They are superficial if most adnexal structures are intact, that is if there is sparing of a significant amount of hair follicles and glands (sebaceous and sweat) and also if a minimum proportion of the dermis is involved. It is deemed deep when only deep adnexal structures are intact. In this case, there is an enormous amount of hair follicles, glands, and a substantial amount of dermis destroyed.

Histologically, the burn wound is classified into three zones deferentially:

1   *Zone of coagulation*: This is the central zone of tissue necrosis, also referred to as the zone of necrosis. Some of the prominent features include but are not limited to denatured proteins, coagulated blood vessels, increase in intracellular sodium as a result of falling sodium-potassium pump, a significant increase in free radicals with subsequent damage to the cell membrane, and finally protein denaturation.

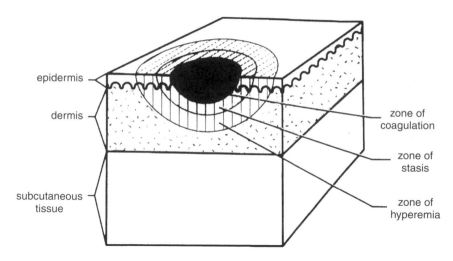

epidermis

dermis

subcutaneous
tissue

zone of
coagulation

zone of
stasis

zone of
hyperemia

Figure 11.2 Burn wound classification in zones. 1) Zone of coagulation; 2) zone of stasis; 3) zone of hyperemia (Modified from Herndon, 2001, *Total Burn Care*. Sidcup, Kent: Harcourt International).

2   *Zone of stasis*: This zone surrounds the central area of the necrosis zone of coagulation. In this zone, there is impaired circulation. It is also known as the zone of ischemia. Tissue in this area can heal without any complication, or the lesion can progress and lead to a greater depth of injury. This zone of ischemia is as a result of thermal injury to the red blood corpuscles rendering them inflexible and unable to enter the microvasculature. Edema occurs over time, as a result. There is an influx of inflammatory mediators including histamine, prostaglandin E (PGE), prostaglandin I (PGI), prostaglandin F_2 (PGF$_2$), interferon, interleukin-1, interleukin-2, monocyte/macrophages, free radicals, tumor necrosis factor, thromboxane A_2 (TXA$_2$), transforming growth factor$_\beta$, platelet-derived growth factor$_\alpha$, PDG$_\beta$, and platelets. This zone is of utmost importance in the initial management of burn wounds.

3   *Zone of hyperemia*: Also know as the zone of inflammation. It surrounds the zone of stasis. Vascular permeability is pronounced in this zone with subsequent edema formation, which if left unattended can lead to hypovolemic shock. This zone generally recovers promptly (Figure 11.2).

FROSTBITE INJURIES: INTRODUCTION

Frostbite is a localized cold-induced injury as a result of improper protection of oneself from the harshness of a cold environment. The first official report of the effects of cold injury was published in 1805 (Wilson and Goldman, 1970). However, these injuries have existed as far back as ancient times, including numerous accounts of soldiers perishing in various wars (Lange and Loewe, 1946; Nakagomi *et al.*, 1985; Orr and Fainer, 1951; Sarhadi *et al.*, 1995; Shelton, 1991; Somboonwong *et al.*, 2000). Some of the signs, symptoms, and sequelae were even described by Hippocrates during his time (Lange and Loewe, 1946; Robson *et al.*, 1982).

Soldiers in military campaigns staged during cold winters have a high propensity to develop frostbite and mountain climbers and explorers are also prone to frostbite injuries (Boswick *et al.*, 1979; Grindlay and Reynolds, 1986; Kulka, 1961; Lange and Loewe, 1946; McCauley *et al.*, 1990). As alluded to earlier, frostbite injury, unlike thermal injury, is more common in people who do not cover themselves appropriately. With the appropriate clothing and accessories, these injuries can be prevented. Some of the risk factors described in the literature include impairment as a result of alcohol, drugs, or other substances, (Klein and Penneys, 1988; Knize, 1977), thereby making the individual unable to clothe themselves adequately. Mental instability has also led to cold-related injury (Klein and Penneys, 1988). Military studies, in addition, have shown that darker colored soldiers are more likely to suffer cold injury in comparison to their white counterparts in similar conditions (Whayne and DeBakey, 1958). Also, individuals from warmer climates are more prone to cold injuries than people from colder climates (Weatherly-White *et al.*, 1964). This observation makes sense, in light of the fact that people from colder climates are accustomed to dressing for colder weather and are physiologically adapted to tolerate such weathers. Furthermore, disease states, such as atherosclerosis, that alter blood flow to vital organs, such as the skin, predispose one to frostbite (Larrey, 1814; Porter *et al.*, 1976). With the advent of proper cold weather clothing and accessories, in addition to sports gear designed for cold weather sports such as skiing and mountain climbing, the incidence of frostbite has decreased compared to historic times.

Anatomy and pathophysiology of frostbite injury

Frostbite injury does not only depend on the ambient temperature and duration of exposure, but also on factors such as: humidity, wind chill, wetness, and the overall physical condition of the individual (Mills, 1973).

Clinically and anatomically, frostbite is divided into four degrees of injury based on the extent of physical injury after freezing and subsequent rewarming.

1   *1st degree frostbite*: Involves the superficial skin with the loss of sensation and erythema. There are firm whitish or yellowish plaques. Even though edema is common, there is no tissue loss.
2   *2nd degree frostbite*: Involves the superficial skin but is a little deeper than 1st degree injuries. There are skin vesiculations with milky or clear fluid within them. Erythema and edema normally surround these blisters (Figure 11.3, Figure 11.4 A–D).
3   *3rd degree frostbite*: Involves the reticular dermis and the dermal vascular plexus. The blisters are deeper, unlike 2nd degree injuries. These blisters frequently contain blood and are red to purplish in color (Figure 11.4 A–D).
4   *4th degree frostbite*: Involves the deeper dermis and adnexal structures. The avascular subcuticular tissue is also affected causing mummification with the involvement of tissues like muscle and bone.

Predicting the extent of injury can be difficult at times, but with good physical examination, the physician should be able to classify the extent of a frostbite injury according to the four historic classifications (Bourne *et al.*, 1986; Knize, 1977; Knize *et al.*, 1969; Kulka, 1956, 1964; Mallonee *et al.*, 1996; Mills, 1964, 1976). A two-division classification system postulated by Mills makes these injuries simpler to classify. Mills'

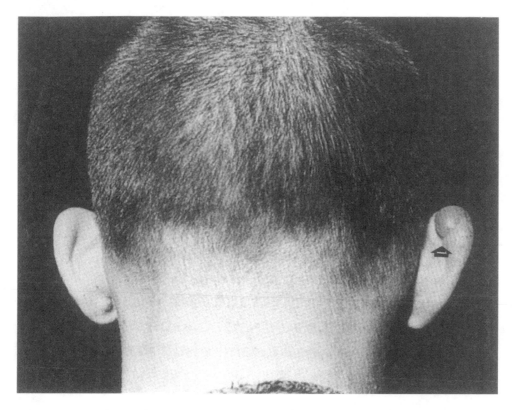

Figure 11.3 Vesiculation of right ear (arrow) with clear fluid characteristic of second-degree frostbite (by permission, Auerbach, 2001, *Wilderness Medicine*, 4th edn. St. Louis, Missouri: Mosby Publishers).

classification is divided into mild and severe (Mills, 1976). The mild injuries are without tissue loss while the severe injuries are associated with tissue loss. Comparing the historic classification to the Mills classification, the 1st and 2nd degree injuries correspond to Mills' mild category, whereas 3rd and 4th degree injuries fall under the severe category. Note, some 3rd degrees can be classified as mild depending on the presentation. Therefore, it is essential to perform a good clinical inspection in order to categorize an injury as mild or severe.

Classically, the frostbite injury has been divided into four phases that may overlap from time to time (Bourne *et al.*, 1986; Kulka, 1956, 1964). These pathologic phases are the pre-freeze phase, free-thaw phase, vascular stasis phase, and late ischemic phase (Kulka, 1956).

1　*Pre-freeze phase*: The temperature of the pre-freeze phase ranges from 3 °C to 10 °C. This phase follows the initial chilling and is before the formation of ice crystals. The changes associated with this phase are a result of the spasm of the capillary vessels, along with the leakage of plasma from the vasculature.
2　*Freeze-thaw phase*: The temperature of this phase is generally below −15 °C during which tissue temperatures reach freezing point. There is actually formation of ice

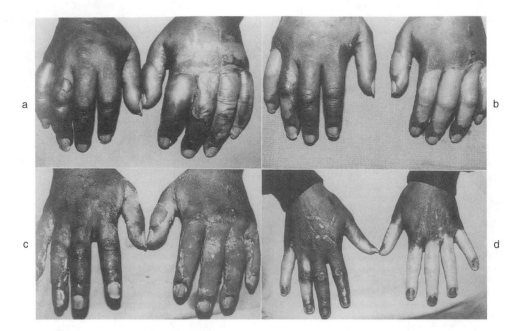

Figure 11.4 A-D A. Acute frostbite with clear fluid in superficial blisters. B. Blisters before treatment but after debridement. C. After 48 hours of therapy with topical *Aloe vera* gel (Dermaide Aloe). D. Sixteen days after injury (by permission, Auerbach, 2001, *Wilderness Medicine*, 4th edn. St. Louis, Missouri: Mosby Publishers).

crystals during this phase causing a severe impairment in circulation and tissue perfusion. This in turn may lead to a precipitous drop in skin temperature at a rate of 1 °C every two minutes.

3   *Vascular stasis phase*: This phase involves the loss of plasma through the vasculature. In addition, there is capillary spasm and dilatation along with the shunting of blood and coagulation due to stasis. This phase is of utmost importance when it comes to the treatment of frostbite with anti-inflammatory agents.

4   *Late ischemic phase*: In this phase, there is ischemic damage and gangrene as a result of thrombosis and arteriovenous shunting. Autonomic dysfunction can ensue as a result. This phase may result in long-term complications and sequelae. In addition, reversibility may be difficult to achieve.

The characteristics of each phase changes with the extent and duration of the injury and also with the velocity of freezing. These injuries are the result of direct and indirect cellular injury (Bulkley, 1983; Herndon *et al.*, 1987; Visuthikosol *et al.*, 1995; Zacarian, 1985). The direct injuries cause changes including extracellular and intracellular ice formation, cell dehydration and shrinking, abnormal intracellular electrolyte concentration, thermal shock, and lipid-protein complex denaturation (Zacarian, 1985). The indirect injuries, in comparison to the direct injuries, are frequently more severe (Visuthikosol *et al.*, 1995). The direct injuries cause more or less progressive dermal ischemia as seen in burns, ultimately leading to tissue loss and subsequent death of the affected extremity or organ (Bulkley, 1983). Robson *et al.*, have determined that the ischemia in frostbite injuries might be due to the same inflammatory mediators that are responsible for the progressive dermal ischemia seen in thermal injuries (Robson *et al.*, 1980). PGE_2, $PGF_{2\alpha}$, and TxA_2 are elevated in frostbite injuries as well as in other eicosanoids, making eicosanoid inhibitors useful in the treatment of frostbite injuries (Robson *et al.*, 1980).

Management of thermal and frostbite injuries

For these injuries to be successfully managed, first we have to understand the anatomy and pathophysiology of the injury. This has already been dealt with. Second, we have to be able to better clinically assess the injury before formulating and undertaking a treatment plan. Clinical judgement is the best available method when it comes to assessing these injuries.

For a thermal injury, the depth of the wound, the size, and the anatomical site of injury should all be determined before any attempts to manage the wound are ensued. Clinical wound inspection, in addition to pinpricking the wound, will help determine the depth of the wound. As discussed previously, the depth of the wound is associated with different qualities such as color and pain. The size of burns are widely estimated using different methods such as:

1   Wallace's 'Rule of Nines'-useful for rapid estimation (Figure 11.5).
2   Lund and Browder Chart-more precise estimation (Figure 11.6).
3   Patient Palm Method (The palm is approximately 1% of the body surface area). Useful in pediatric burns and smaller burns.

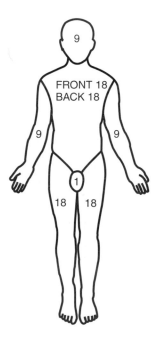

Figure 11.5 Wallace's estimation of burn size 'The Rule of Nines' (Wolf and Herndon, 1999 *Vademecum in Burn Care.* Austin, Texas: Landes Bioscience).

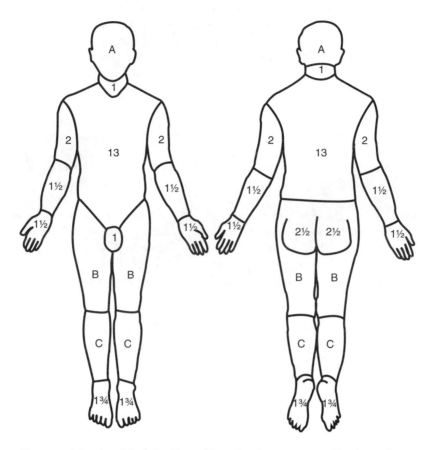

Figure 11.6 Lund and Browder Chart of Burn Size Estimation. (Wolf and Herndon, 1999, *Vademecum in Burn Care*. Austin, Texas: Landes Bioscience).

It is important to examine anatomical sites because certain areas need paramount attention, such as important functional and aesthetic areas including hands, feet, face, eyelids, perineum, genitalia, and joints. Special attention is given to these areas to prevent cosmetic and functional problems as a result of hypertrophic scarring.

For a frostbite injury, the clinical appearance may be deceiving. Most of the time the extremities involved show a frozen appearance. Rapid rewarming almost always produces instant hyperemia, regardless of the severity of the injury (Herndon *et al.*, 1987). Sensation improves after thawing. Assessment of the injury following this initial rewarming phase is imperative. Diagnostic techniques, such as radioisotope scanning using technetium-99m methylene diphosphonate (^{99m}Tc MDP), technetium-99m stannous pyrophosphate, xenon-133 (^{133}Xe), iodine-131 labelled human serum albumin (^{131}I-RISA), angiography, digital plethysmography, and routine x-rays can aid in the estimation of the extent of these injuries. These modalities are also successful in predicting tissue loss and also for estimating the vascular response to vasodilators and identifying tissue boundaries for surgical intervention.

The initial management of these injuries is preceded by the usual initial trauma protocols: airway maintenance, breathing, and circulatory support (Committee on trauma, 1999). Also resuscitation and smoke inhalation injury support are paramount when it comes to thermal injuries. In frostbite injuries, systemic hypothermia support is important to address after the initial resuscitation efforts are carried out. The core temperature should be brought back to at least 34 °C before any further management is undertaken. Unlike thermal injuries, fluid abnormalities are frequently not a problem in frostbite injuries, and thus time should not be wasted in attempts to correct such abnormalities.

After assessing the injury, it is paramount to initiate first aid, by removing the source, which in so doing limits tissue damage. Edema and protein extravasation can also be minimized as a result. Blister management is controversial in both thermal and frostbite injury. In thermal injuries, many experts believe in leaving small blisters thus creating a biological dressing, whereas large blisters are usually removed. In frostbite injuries, the clear and white blisters are debrided just as in burns, whereas the hemorrhagic ones are left intact but should be aspirated to decrease inflammatory agents. Escharatomies are employed in both types of injuries. In thermal injuries, they are reserved for circumferential full thickness burns of the chest, limbs, and digits. Many agents have been used to date in the management of these injuries. Some of the notable agents used to treat burn injuries range from anti-catabolic agents to wound healing agents, such as silvadene, and the potent anti-inflammatory, 'Aloe vera' (Alexander, 1981; Boswick, 1987; Heggersetal, 1987; Reus *et al.*, 1984; Robson *et al.*, 1979; Snider and Porter, 1975; Vaughn, 1980; Zacarian *et al.*, 1970). The same agent, *Aloe vera* gel, is also useful in the treatment of frostbite injuries because of similar pathophysiology in terms of progressive dermal ischemia (Bulkley, 1983; Robson and Heggers, 1981; Zacarian *et al.*, 1970). Some of the other agents that have been used in frostbite injuries include thromboxane inhibitors, like aspirin and ibuprofen, sympathetic blockers, and sympathectomy, as well as IV dextran and intraarterial reserpine, reserving amputation for gangrene (Bouwman *et al.*, 1980; Engrav *et al.*, 1983; Fujita *et al.*, 1976; Kyosola, 1974; Marzella *et al.*, 1989; Muller *et al.*, 1996; Orr and Fainer, 1951; Robson *et al.*, 1982; Simeone, 1960; Visuthikosol *et al.*, 1995).

The use of aloes dates back to as far as 1750 B.C. in Mesopotamia, where clay tablets showed evidence of writings that *Aloe vera* was being used for medicinal purposes. (Coats and Ahola, 1979; Schechter and Sarot, 1968). Books from Egypt in 550 B.C. make mention of aloes in the treatment of skin infection. Discordes, a Greek physician, wrote a book in 74 A.D. in which he wrote that aloes could treat wounds and heal infections of the skin (Coats and Ahola, 1979; Schechter and Sarot, 1968). There are many historical citations about aloes, although much of the information was unproven until recently. Many admit to using *Aloe vera* gel from a houseplant leaf in self-treating small burns and cuts (Coats and Ahola, 1979). The evidence of the successful use of aloes in the treatment of burns is overwhelming. Heggers and others have studied this miraculous plant in much detail with convincing evidence about its beneficial effects (Robson and Heggers, 1981).

The therapeutic properties of aloes are many (Cera *et al.*, 1980; Gottshall *et al.*, 1950; Heggers *et al.*, 1987; Robson *et al.*, 1979; Snider and Porter, 1975). It is important to mention that there are some unforeseen properties and benefits making this miraculous succulent -'the herb of herbs'- a force to be reckoned with. Some of the known and proven properties include:

1 The ability to penetrate the burn wound.
2 The ability to anesthetize the wound thereby providing analgesia, as well as a soothing effect.

3   The bactericidal, viridical, and fungicidal properties.
4   The ability to act as a potent anti-inflammatory agent.
5   The vasodilatory properties, with the ability to dilate capillaries, thereby increasing blood supply to the burn areas.

Many of the beneficial effects of aloes are comparable to those of the well studied and tested eicasanoid inhibitors, which have proven over time to have therapeutic benefit in the prevention of progressive dermal ischemia, after thermal injury, frostbite, and drug injuries (Heggers *et al.*, 1987; Heggers *et al.*, 1993; Lawrence *et al.*, 1994; Muller *et al.*, 1996; Reus *et al.*, 1984; Robson and Heggers, 1981; Robson *et al.*, 1979).

The chemical constituents in aloes that provides its potent therapeutic properties include lignin, lectins, saponin, and gibberellin (Danhof and McAnally, 1983; Gottshall *et al.*, 1950; Schechter and Sarot, 1968). Lignin, a polyphenolic compound, gives aloes their ability to penetrate skin. Lectin, a hemagglutination protein, binds to glycoproteins thereby decreasing inflammation. Gibberellin also acts to decrease inflammation. It acts as a growth hormone in plants but stimulates protein synthesis, unlike steroids (Danhof and McAnally, 1983). Saponin, acts to provide antisepsis, thereby decreasing the microbial load (See Chapter 9).

Overwhelming evidence exists that shows *Aloe vera* to be a potent antimicrobial (Benigni, 1950; Zimmerman and Simms, 1969). Aloes have been effective against deadly bacteria such as *Mycobacterium tuberculosis, Bacillus subtilis, Staphylococcus aureus, Streptococcus pyogenes, Salmonella paratyphi, Streptococcus agalactiae, Klebsiella pneumonae, Pseudomonas aeroginosa, Serratia marcescens, Enterobacter cloacae*, and many more (Benigni, 1950; Bruce, 1967; Golding *et al.*, 1963; Lehr *et al.*, 1991; Zawacki, 1974). In this capacity, it exerts either bactericidal or bacteriostatic effects. The mechanism of the antimicrobial properties of aloes is not well established, but saponin, one of the chemicals of aloe, is credited for these actions.

Many investigators have tried to establish a rationale for the potent anti-inflammatory properties for *Aloe vera*. Robson and Heggers (1981) have shown that salicylate is a by-product of *Aloe vera*, thus contributing to its anti-inflammatory properties. Fujita and colleagues have found a bradykinin-inactivating carboxypeptidase in *Aloe arborescens*, also culminating in its anti-inflammatory properties (Fuhrman and Crissman, 1947; Fujita *et al.*, 1979). Also, magnesium lactate found in aloes is a known inhibitor of the enzyme histidine decarboxylase that catalyzes histamine formation in mast cells from histidine (Kemp and Sibert, 1997). These three postulates best serve to elucidate the anti-inflammatory properties of aloes.

This elucidated anti-inflammatory property explains its benefits in the treatment of thermal injuries, frostbites, and drug injuries which all share a common pathophysiology in terms of the production of inflammatory agents. All three injuries share a common pathway as alluded to earlier, and the end result is the progressive dermal ischemia and eventual loss of tissue (Cera *et al.*, 1982; Heggers *et al.*, 1987; Heggers *et al.*, 1993; Lawrence *et al.*, 1994; Reus *et al.*, 1984; Robson and Heggers, 1981). This loss of tissue is mediated by Thromboxane A_2 (TxA$_2$), a potent vasoconstrictor. *Aloe vera* gel inhibits TxA$_2$ synthetase, and also maintains equilibrium between PGE$_2$ and PGF$_{2\alpha}$, thereby exercising its vasodilatory properties, and eventually causing tissue perfusion and prevention of tissue loss (Cera *et al.*, 1982; Cera *et al.*, 1980; Heggers *et al.*, 1993; Lawrence *et al.*, 1994; Reus *et al.*, 1984; Robson and Heggers, 1981; Vaughn, 1980).

ACKNOWLEDGMENTS

The authors would like to acknowledge the efforts of Mr. Lewis Milutin, Ms. Tina Garcia and Sandy Baxter for their photographic support. The authors would like to recognize Ms. Maness for her input and preparation of a final draft.

REFERENCES

Alexander, J.W., MacMillan, B.G., Law, E. and Kittur, D.S. (1981) Treatment of severe burns with widely meshed skin autograft and meshed skin allograft overlay. *Journal of Trauma*, 21, 433.

Barillo, D.J. and Goode, R. (1996) Fire fatality study. Demographics of fire victims. *Burns*, 22, 85–88.

Benigni, R. (1950) Substances with antibiotic action contained in anthraquinonic drugs. *Chemical Abstracts*, 44, 11036.

Blakeney, P., Meyer, W., Robert, R., Desai, M., Wolf, S.E. and Herndon, D.N. (1998) Long-term psychosocial adaptation of children who survive burns involving 80% or greater total body surface area. *Journal of Trauma*, 44, 625–632.

Boswick, J.A. ed. (1987) *The Art and Science of Burn Care*. Rockville: Aspen Publishers.

Boswick, Jr. J.A., Thompson, J.D. and Jonas, R.A. (1979) The epidemiology of cold injuries. *Surgery Gynecology and Obstetrics*, 149, 326–332.

Bourne, M.H., Piepkorn, M.W., Clayton, F. and Leonard, L.G. (1986) Analysis of microvascular changes in frostbite injury. *Journal of Surgical Research*, 40, 26–35.

Bouwman, D.L. Morrison, S., Lucas, C.E. and Ledgerwood, A.M. (1980) Early sympathetic blockade for frostbite—Is it of value? *Journal of Trauma*, 20, 744–749.

Brigham, P.A. and McLoughlin, E. (1996) Burn incidence and medical care in the United States: Estimates, trends, and data sources. *Journal of Burn Care and Rehabilitation*, 17, 95–107.

Bruce, W.G.G. (1967) Investigations of antibacterial activity in the Aloe. *South African Medical Journal*, 41, 984.

Bulkley, G.B. (1983) The role of oxygen free radicals in human disease processes. *Surgery*, 94, 407.

Bull, J.P. and Fisher, A.J. (1949) A study in mortality in a burn unit: Standards for the evaluation for alternative methods of treatment. *Annals of Surgery*, 130, 160–173.

Cera, L.M., Heggers, J.P., Hagstrom, W.J. and Robson, M.C. (1982) Therapeutic protocol for thermally injured animals and its successful use in an extensively burn Rhesus Monkey. *Journal of the American Animal Hospital Association*, 18, 633–638.

Cera, L.M., Heggers, J.P., Robson, M.C. and Hagstrom, W.J. (1980) The therapeutic efficacy of *Aloe vera* cream (Dermaide Aloe) in thermal injuries: two case reports. *Journal of the American Animal Hospital Association*, 16, 768–772.

Coats, B.C. and Ahola, R. (1979) *Aloe vera the silent healer. A modern study of Aloe vera*, pp.1–288, Garland, Texas: Bill C. Coats.

Committee on Trauma, American College of Surgeons (1999) *Resources for optimal care of the injured patient*.

Danhof, I.E., McAnally, B.H. (1983). Stabilized Aloe vera: Effect on human skin cells. *Drug and Cosmetic Industry*, 133,52–54, 105–106.

Davis, R.H. and Maro, N.P. (1989) Aloe vera and gibberellin. Anti-inflammatory activity in diabetes. *Journal of the American Podiatric Medical Association*, 79, 24–26.

Desai, M.H., Herndon, D.N., Broemeling, L., Barrow, R.E., Nichols, R.J. Jr and Rutan, R.L. (1990) Early burn wound excision significantly reduces blood loss. *Annals of Surgery*, 211, 753–762.

Desai, M.H., Rutan, R.L., Herndon, D.N. (1991) Conservative treatment of scald burns is superior to early excision. *Journal of Burn Care and Rehabilitation*, 12, 482–484.

Dziewulski, P. (1992) Burn would healing. *Burns*, 18, 466–478.

Engrav, L.H., Heimbach, D.M., Reus, J.L., Harnar, T.J. and Marvin, J.A. (1983) Early excision and grafting versus non-operative treatment of burns of indeterminant depth: A randomized prospective study. *Journal of Trauma*, 23, 1001–1004.

Fuhrman, F.A. and Crissman, J.M. (1947) Studies on gangrene following cold injury. Treatment of cold injury by immediate rapid rewarming. *Journal of Clinical Investigation*, 26, 476.

Fujita, K., Ito, S., Teradaira, R. and Beppu, H. (1979) Properties of a carboxypeptidase from aloe. *Biochemical Pharmacology*, 28, 1261–1262.

Fujita, K., Teradaira, R. and Nagatsu, T. (1976) Bradykinase activity of aloe extract. *Biochemical Pharmacology*, 25, 205.

Golding, M.R., Dejong, P., Sawyer, P.N., Hennigar, G.R., Wesolows, S.A. (1963) Protection from early and late sequelae of frostbite by regional sympathectomy: Mechanism of 'cold sensitivity' following frostbite. *Surgery*, 53, 303–308.

Gottshall, R.Y., Jennings, J.C., Weller, L.E., Redemann, C.T., Lucas, E.H. and Sell, H.M. (1950) Antibacterial substances in seed plants active against tubercle bacilli. *American Revue of Tuberculosis*, 62, 475–480.

Grindlay, D. and Reynolds, T. (1986) *The Aloe vera* phenomenon: A review of the properties and modern uses of the leaf parenchyma gel. *Journal of Ethnopharmacology*, 16, 117–151.

Heggers, J.P., Robson, M.C., Manavalen, K., Weingarten, M.D., Carethers, J.M., Boertman, J.A., Smith Jr. D.J. and Sachs, R.J. (1987) Experimental and clinical observations on frostbite. *Annals of Emergency Medicine*, 16, 1056–1062.

Heggers, J.P., Pelley, R.P. and Robson, M.C. (1993) Beneficial effects of Aloe in wound healing. *Phytotherapy Research*, 7, S48–S52.

Heggers, J.P., Robson, M.C. and Zachary, L.S. (1985) Thromboxanes inhibitors for the prevention of progressive dermal ischemia due to the thermal injury. *Journal of Burn Care and Rehabilitation*, 6, 466–468.

Heimbach, D.M. (1987) Early burns excision and grafting. *Surgical Clinics of North America*, 67, 93–107.

Herndon, D.N., Gore, D.C., Cole, M., Desai, M.H., Linares, H., Abston, S., Rutan, T.C., VanOsten, T. and Barrow, R.E. (1987) Determinants of mortality in pediatric patients with greater than 70% full thickness total body surface area treated by early excision and grafting. *Journal of Trauma*, 27, 208–212.

Hurley, L.A. (1957) Angioarchitectural changes associated with rapid rewarming subsequent to freezing injury. *Angiology*, 819.

Kemp, A. and Sibert, J. (1997) Childhood accidents. Epidemiology, trends, and prevention. *Journal of Accident and Emergency Medicine*, 14, 316–320.

Klein, A.D. and Penneys, N. (1988) *Aloe vera. Journal of the American Acadamy of Dermatology*, 18, 714–720.

Knize, D.M. (1977) *Cold injury in reconstructive plastic surgery: General principles. Volume 1*, Philadelphia: W.B. Saunders.

Knize, D.M., Weatherley-White, R.C., Paton, B.C. and Owens, J.C. (1969) Prognostic factors in the management of frostbite. *Journal of Trauma*, 9, 749–759.

Kulka, J.P. (1956) Histopathologic studies in frostbitten rabbits. In *Cold Injury*. edited by M.I. Ferrer, New York:Josiah Macy, Jr Foundation.

Kulka, J.P. (1961) Vasomotor microcirculatory insufficiency: Observations on non-freezing cold injury of the mouse ear. *Angiology*, 12, 491.

Kulka, J.P. (1964) Microcirculatory impairment as a factor in inflammatory tissue damage. *Annals of the New York Academy of Sciencs*, 6, 1018.

Kyosola, K. (1974) Clinical experiences in the management of cold injuries: A study of 110 cases. *Journal of Trauma*, 14, 32–36.

Lange, K. and Loewe, L. (1946) Subcutaneous heparin in the pitkin mastruum for the treatment of experimental human frostbite. *Surgery Gynecology and Obstetrics*, **82**, 256.

Larrey, D.J. (1814) *Memoirs of military surgery. Volume 2*. Baltimore: Joseph Cushing.

Lawrence, W.F., Murphy, R.C., Robson, M.C. and Heggers, J.P. (1994) The detrimental effect of cigarette smoking on flap survival: An experimental study in the rat. *British Journal of Plastic Surgery* 1984; **37**, 216–219.

Lehr, H.A., Guhlmann, A., Nolte, D., Keppler, D. and Messmer, K. (1991) Leukotrienes as mediators in ischemia-reperfusion injury in a microcirculation model in the hamster. *Journal of Clinical Investigation*, **87**, 2036–2041.

Lorenzetti, L.J., Salisbury, R., Beal, J.L. and Baldwin, J.N. (1964) Bacteriostatic property of *Aloe vera*. *Journal of Pharmaceutical Science*, **53**, 1287.

Mallonee, S., Istre, G.R., Rosenberg, M., Reddish-Douglas, M., Jordan, F., Siverstein, P. and Tunell, W. (1996) Surveillance and prevention of residential-fire injuries. *New England Journal of Medicine*, **335**, 27–31.

Marzella, L., Jesudass, R.R., Manson, P.N., Myers, R.A. and Bulkley, G.B. (1989) Morphological characterization of acute injury to vascular endothelium of skin after frostbite. *Plastic and Reconstructive Surgery*, **83**, 67–76.

McCauley, R.L., Heggers, J.P. and Robson, M.C. (1990) Frostbite: Methods to minimize tissue loss. *Postgraduate Medicine*, **88(8)**, 67–68, 73–77.

Mills, W.J. Jr. (1964) Clinical aspects of frostbite injury. In Proceeding of the symposium on arctic medicine and biology. IV Frostbite. Arctic Aeromedical Laboratory. Fort Wainwright, 1964.

Mills, W.J. Jr. (1973) Frostbite A discussion of the problem and a review of an Alaskan experience. *Alaska Medicine*, **15**, 27–47.

Mills, W.J. Jr. (1976) Out in the cold. *Emergency Medicine*, **Jan**, 134. 55. Molnar, G.W., Hughes, A.L., Wilson, O. and Goldman, R.F. (1973) Effect of skin wetting on fingercooling and freezing. *Journal of Applied Physiology*, **35**, 205–207.

Muller, M.J., Nicolai, M., Wiggins, R., Macgill, K. and Herndon, D.N. (1996) Modern treatment of a burn wound. In *Total Burn Care*, edited by D.N. Herndon and J.H. Jones, pp.136–147. Philadelphia:WB Saunders Co.

Nakagomi, K., Oka, S., Tomizuka, N., Yamamoto, M., Masui, T. and Nakazawa, H. (1985) A novel biological activity in Aloe components: Effects on mast cell degranulation and platelet aggregation. *Report of the Fermentation Research Institute*, 23–30.

Orr, K.D. and Fainer, D.C. (1951) *Cold injuries in Korea during winter of 1950–1951*. Fort Knox: Army Medical Research Laboratory.

Orr, K.D. and Fainer, D.C. (1952) Cold injuries in Korea clinic: The winter of 1950–1951. *Medicine* 1952; **31**, 177.

Porter, J.M., Wesche, D.H., Rosch, J. and Baur, G.M. (1976) Intra-arterial sympathetic blockage in the treatment of clinical frostbite. *American Journal of Surgery*, **132**, 625–630.

Reus, W.F., Robson, M.C., Zachary, L. and Heggers, J.P. (1984) Acute effects of tobacco smoking on blood flow in the cutaneous microcirculation. *British Journal of Plastic Surgery*, **37**, 213–215.

Robson, M.C., DelBeccaro, E.J. and Heggers, J.P. (1980) Increasing dermal perfusion after burning by decreasing thromboxane production. *Plastic and Reconstructive Surgery*, **20**, 722–725.

Robson, M.C. and Heggers, J.P. (1981) Evaluation of hand frostbite blister fluid as a clue to pathogenesis. *Journal of Hand Surgery-American volume* 1981;**6**:43–47.

Robson, M.C., Heggers, J.P. and Hagstrom, W.J. (1982) Myth, Magic, Witchcraft or Fact? *Aloe vera* revisited. *Journal of Burn Care and Rehabilitation*, **3**, 157–163.

Robson, M.C., Krizek, T.J. and Wray, R.C. (1979) Care of the thermally injured patient, In *Management of trauma*, Philadelphia: W.B. Saunders.

Rodriguez-Bigas, M., Cruz, N.J. and Suarez, A. (1988) Comparative evaluation of Aloe vera in the management of burn wound in guinea pigs *Plastic and Reconstructive Surgery*, **81**, 386–389.

Ryan, C.M., Schoenfeld, D.A., Thorpe, W.P., Sheridan, R.L., Cassem, E.H. and Tompkins, R.G. (1998) Objective estimates of the probability of death from burn injuries. *New England Journal of Medicine*, **338**, 362–366.

Sarhadi, N.S., Murray, G.D. and Reid, W.H. (1995) Trends in burn admissions in Scotland during 1970–92. *Burns*, **21**, 612–615.

Schechter, D.S. and Sarot, I.A. (1968) Historical accounts of injuries due to cold. *Surgery*, **63**, 527–535.

Shelton, R.M. (1991) Aloe vera. Its chemical and therapeutic properties. *International Journal of Dermatology*, **30**, 679–683.

Simeone, F.A. (1960) Surgical volumes of the history of the United States Army Medical Department in World War II: Cold injury. *Archives of Surgery*, **80**, 296.

Snider, R.L. and Porter, J.M. (1975) Treatment of experimental frostbite with intra-arterial sympathetic blocking drugs. *Surgery*, **77**, 557.

Somboonwong, J., Thanamittramanee, S., Jariyapongskul, A. and Patumraj, S. (2000) Therapeutic effects of *Aloe vera* on cutaneous microcirculation and wound healing in second degree burn model in rats. *Journal of the Medical Association of Thailand*, **83**, 417–425.

Vaughn, P.B. (1980) Local cold injury: Menace to military operations, A review. *Military Medicine*, **145**, 305–311.

Visuthikosol, V., Sukwanarat, Y., Chowchuen, B., Sriurainratana, S. and Boonpuknavig, V. (1995) Effect of *Aloe vera* gel to healing of burn wound: Clinical and histologic study. *Journal of the Medical Association of Thailand*, 403–408.

Weatherly-White, R.C.A., Sjostrom, B. and Paton, B.C. (1964) Experimental studies in cold injury. *Journal of Surgical Research*, **4**, 17.

Whayne, T.J. and DeBakey, M.F. (1958) *Cold injury, ground type*. Washington DC: US Government Printing Office.

Wilson, O. and Goldman, R.F. (1970) Role of air temperature and wind in the time necessary for a finger to freeze. *Journal of Applied Physiology*, **29**, 658–664.

Zacarian, S.A. (1985) Cryogenics: The cryolesion and the pathogenesis of cryonecrosis. In *Cryosurgery for skin and cutaneous disorders*, edited by S.A. Zacarian. St Louis: Mosby.

Zacarian, S.A., Stone, D. and Clater, H. (1970) Effects of cryogenic temperatures in the microcirculation in the golden hamster cheek pouch. *Cryobiology*, **7**, 27–39.

Zawacki, B.E. (1974) Reversal of capillary stasis and prevention of necrosis in burns. *Annals of Surgery*, **180**, 98–102.

Zimmerman, E.R. and Simms, R. (1969) Antibacterial and antifungal *in vitro* properties of *Aloe vera*. Baylor University College of Dentistry, Microbiology Laboratory, Dallas, USA.

12 Plant saccharides and the prevention of sun-induced skin cancer

Faith M. Strickland and Ronald P. Pelley

ABSTRACT

Plants and Fungi have traditionally been the single largest source of lead compounds for the development of therapeutics by the pharmaceutical industry. Currently we are investigating oligosaccharides for the prevention of sun-induced skin cancer. One compound we discovered is a cytoprotective oligosaccharide from *Aloe barbadensis* that is produced by cellulase cleavage from a biologically inactive native polysaccharide precursor. Another is a seed polysaccharide from Tamarind that contains the core repeating unit for xyloglucan hemicelluloses. These oligosaccharides downregulate production of immunosuppressive cytokines by ultraviolet radiation-injured keratinocytes and prevent the ultraviolet-induced suppression of the skin immune system model's cutaneous hypersensitivity and delayed hypersensitivity. We believe these compounds are promising agents for the prevention of environmentally-induced skin cancer.

INTRODUCTION – ALOE, THE IMMUNE RESPONSE AND CANCER

Say the words 'death' and 'disease' and words like AIDS, stroke, heart attacks, and cancer spring to mind. The public perceives that coronary bypass grafts and hypocholesterolemic drugs have greatly reduced heart disease, while anti-reverse transcriptase drugs and protease inhibitors have alleviated HIV infection. Enormous progress has been made in stroke prevention by early, aggressive treatment of essential hypertension. But cancer superficially appears to be breakthrough-resistant (Young, 2000).

Cancer resists breakthroughs because it is not a single disease. Cancer of each organ is essentially a different set of diseases and each organ may have up to a dozen different cancers. Notable progress against a given cancer means only a stepwise attack on cancer in general. Thus progress in cancer prevention and treatment is incremental. For example, the prognosis for breast cancer patients has steadily improved during the last three decades due to earlier detection via mammography, new drugs such as

[**Abbreviations**: *Aloe barbadensis* gel extract – Aloe; Aloe Research Foundation – ARF; biological response modifiers – BRM; complement receptor 3 or mac-1 – CD11b/CD18; complementary and alternative medicine – CAM; minimal erythemal dose – MED; molecular weight – MW; non-melanoma skin cancer – NMSC; skin immune system – SIS; squamous cell carcinoma – SCC.]

Tamoxifen, and better chemotheraputic protocols and adjuvant chemotherapy. Similarly, the rate of increase in incidence of lung cancer in men has finally leveled off, not because of any scientific breakthrough but because tens of millions of men have stopped smoking cigarettes. The Pap (Papanicolaou) Test, the scientific breakthrough vital for the control of squamous carcinoma of the uterine cervix, occurred over three decades ago. What has caused the precipitous drop in the incidence of invasive cervical cancer is the near universal use of this test to identify this cancer in its early, *in situ* stage leading to its extirpation. In fact, of all the major malignancies, only carcinoma of the prostate and skin cancer are still on the rise.

We believe that skin cancer will be the next major family of malignancies to succumb to the broad advances in biomedicine currently under way. As with carcinoma of the cervix, we believe that progress in preventing skin cancer will focus on prevention. Sunscreens will obviously play a key role in prevention by reducing the exposure of the skin to ultraviolet radiation. But as the reader will learn from this chapter, we are also targeting events subsequent to sun exposure, in particular the cancer surveillance system. This is because in addition to preventing sun exposure, it is important to protect the skin immune system so that it can eliminate incipient tumors. This protection is mediated by plant saccharides, typically those in aloe gel, down-regulating the cascade of cytokines that UVB exposure triggers.

The audience we are addressing, as befits this series, is not dermatologists or photobiologists but botanically-oriented scientists. We do presume a knowledge of immunology and will only superficially examine the general role of the immune response in the prevention and treatment of cancer since recent reviews have discussed biological response modifiers (Pelley and Strickland, 2000), immune surveillance (Markiewicz and Gajewski, 1999) and immunotherapy (Pawelec *et al.*, 1999).

Note The name *Aloe barbadensis* Miller, used extensively in this chapter, is now considered to be properly designated as *Aloe vera* (L.) Burm.f. (Newton, 1979) but has been retained here because of its familiarity to many readers.

ULTRAVIOLET LIGHT AND SKIN CANCER – MUTATION AND IMMUNITY

Ultraviolet B (UVB) radiation, tumor suppressor genes and oncogenes

Excessive exposure to UV radiation causes sunburn, premature aging of the skin, and mutations leading to skin cancer (Urbach, 1978). A gradual thinning of the ozone layer, which helps to screen our planet from solar radiation, and an increase in recreational sun exposure in the general population are thought to be significant factors underlying the steady rise in the incidence of all types of skin cancers over the past twenty years (Fears and Scotto, 1983; Green *et al.*, 1985). The non-melanoma types of skin cancer (NMSC), squamous cell carcinoma of the skin and basal cell carcinoma (Weber, 1995), are the most common forms of human neoplasm, representing one-third of all new malignancies diagnosed in the United States (Strom and Yamamura, 1997). The number of new cases of NMSC diagnosed in the United States per year is approaching one million and continues to rise (Boring *et al.*, 1992). Fortunately, NMSC has low mortality although it has considerable morbidity, causing disfigurement, loss of function, and requiring repeated treatments (Weber *et al.*, 1990). Malignant melanoma (Balch *et al.*, 1992), the other form of human skin cancer, has almost one-tenth the incidence of NMSC, 2.5% of all cancer or over 30,000 cases per year (Grin-Jorgensen *et al.*, 1992). However, this neoplasm is feared because of its propensity for rapid metastasis to vital organs such as lung and brain (Fears and Scotto, 1983; Green *et al.*, 1985). Unfortunately, the incidence of melanoma is increasing even faster than that of NMSC.

The evidence linking chronic exposure of the skin to ultraviolet radiation to the development of NMSC is compelling (Kripke, 1974; Kripke, 1979; Fears and Scotto, 1983). Total lifetime exposure to UV has been established as the greatest risk factor for basal cell and squamous cell carcinomas (Urbach, 1978) and numerous murine models exist for the induction and characterization of squamous carcinoma of the skin (Kripke, 1984). The role of UV radiation in the etiology of human melanoma is less clear. Epidemiologic data on melanomas point to periodic exposure to high doses of UV radiation (sunburn) as a risk factor (Green *et al.*, 1985) but attempts to induce melanomas in mice with UV radiation alone have been singularly unsuccessful (Kripke, 1979; Romerdahl *et al.*, 1989; Kusewitt and Ley, 1996). The induction and development of primary melanomas appears to be a complex, multifactorial process that involves non-UV factors, such as predisposing genes, chemicals in the cutaneous environment and the immune response to a much greater degree than NMSC (Camplejohn, 1996; Chin *et al.*, 1997; Langley and Sober, 1997; Strickland *et al.*, 2000).

Integral to the development of skin cancer are mutations and or changes in the expression of a number of tumor suppressor and oncogenes, many of which appear to be due to the effects of UVB radiation (Bardeesy *et al.*, 2000; DePinho, 2000). Oncogenes are genes that regulate cellular metabolism, growth and intercellular communication (Pierceall *et al.*, 1991a; Clark *et al.*, 2000). The first gene implicated in a wide variety of cancers was the oncogene *Ras* which is still felt to play an important role in the development of skin cancer (Chin *et al.*, 1997; Tam *et al.*, 1999). Carcinogens and UV light can induce mutations that cause the over expression of oncogenes, resulting in uncontrolled cell growth (Pierceall *et al.*, 1991a; Bardeesy *et al.*, 2000). Currently oncogenes are not felt to be the primary initiators of skin cancers but appear to play an important role in the growth of these tumors (Chin *et al.*, 1997; Tam *et al.*, 1999; Clark *et al.*, 2000).

Tumor suppressor genes function as guardians of the genome to prevent the replication of cells with damaged DNA (Kuerbitz *et al.*, 1992). Mutations in tumor suppressor genes appear to underlie a goodly proportion of human skin cancers, both NMSC and melanoma (Pierceall *et al.*, 1991b; Nataraj *et al.*, 1995; Zerp *et al.*, 1999). In some cases this suppression of tumorigenesis involves causing the mutated skin cell to commit suicide (Cotton and Spandau, 1997), a process termed apoptosis or programed cell death. For example, after an exposure of a mouse to UVB there is a sequential activation in the epidermis of genes and their products in the p53/p21/bax/bcl-2 pathway which results in a wave of programed cell death, beginning six hours after exposure and reaching a maximum 24 hours after exposure (Ouhtit *et al.*, 2000). When a tumor suppressor gene such as p53 is mutated by UV radiation (Ananthaswamy *et al.*, 1998 and references therein), the tumor suppressor can no longer perform its function of protecting the genome from radiation-induced mutational damage. Such loss of genetic stability, including damage to the telomeric control system, is a major factor in the malignancy of skin cancers (Pai *et al.*, 1999). Although the p53 system is the best characterized tumor suppressor gene system implicated in the induction of skin cancer by UV radiation, recent studies suggest that the $p16^{INK4a}/p19^{ARF}$ system may be equally important in destablizing the genome, particularly under conditions where *Ras* is overexpressed (Chin *et al.*, 1997; Tam *et al.*, 1999). The end effect of mutations in tumor suppressor genes is that the rate of mutational change is radically increased. There is production of aberrant proteins and the expression of genes and their products, that are not usually expressed by cells of that tissue. This releases the mutated cells from normal growth controls and permits their unrestrained growth.

Is there such a thing as immunity to cancer?

Recently, Markiewicz and Gajewski (1999) reviewed the history of immune surveillance beginning with the discoveries of the 1960s, through the disappointments of the 1980s and the new hopes of the 1990s. First, immunologists realized that cancers contain tumor specific antigens which are developmental and oncogenic markers. On occasion, tumors contained viral antigens. However, attempts in the 1980s to develop cancer vaccines or to upregulate the immune response with BCG or Interleukin-2 were singularly unsuccessful. These failures led many to doubt the existence of immune surveillance (Stutman, 1974, 1979; Klein and Boon, 1993) but, as experience with immunotherapy of cancer increased, it became clear that human neoplastic disease is much more complex than any mouse model. The evidence for protective immune involvement in cancer is summarized by the reviews cited in Table 12.1 below. To paraphrase Markiewicz and Gajewski, we now know that in humans: (i) immunosuppression increases the risk of cancer, particularly skin cancer; and, (ii) immune defects are often found in patients with cancer, above and beyond those predicted from the immunosuppressive effect of treatment with cytotoxic anti-cancer drugs.

Lastly, (iii) the most successful methods of immunization with tumor antigens and low molecular weight immunopotentiating agents have yielded positive clinical results in a significant number of patients with melanoma and renal cell carcinoma and some patients with adenocarcinoma of the breast. Thus, there is little doubt that immune surveillance does exist and plays a role in cancer immunity.

Table 12.1 Evidence in favor of immune surveillance.

Cancer incidence is increased in immunosuppressed individuals	
HIV Infection	
Viral associated cancer	Brockmeyer and Barthel, 1998
Melanoma and lung cancer	Smith *et al.*, 1998
Recipients of organ transplants	Birkeland *et al.*, 1995
	Leigh *et al.*, 1996
	Schreiber, 1999
Ultraviolet light and skin cancer	Strickland and Kripke, 1997
Cancer patients have depressed immune responsiveness	
Escape from tumor immunity	Pawelec *et al.*, 1997
Skin cancer and suppressor T cells	Strickland and Kripke, 1997
Alterations in T cell response	Pawelec *et al.*, 1999
	Markiewicz and Gajewski, 1999
	Hadden, 1999
Immunological intervention benefits some cancer patients	
Melanoma and renal cell carcinoma	Pawelec *et al.*, 1999
Breast cancer	Hadden, 1999

UVB, the skin immune system and skin cancer

In addition to ultraviolet light's mutagenic effects, UVB promotes skin cancer by suppressing immune surveillance (reviewed by Strickland and Kripke, 1997). In laboratory animals, radiation at UVB wavelengths (280–320 nm) impairs the ability of the immune system to reject highly antigenic UV-induced skin cancers and respond to allergens and infectious organisms (van der Leun and Tevini, 1991; Kripke, 1984; Denkins *et al.*, 1989; Jeevan *et al.*, 1992). Additionally, subcarcinogenic doses of UV radiation contribute to the growth of skin cancers by generating antigen-specific suppressor T cells. These regulatory lymphocytes recognize UV-specific antigens on the tumors and prevent rejection of the tumor. Since their description in 1982 by Fisher and Kripke, this system of antigen-specific suppressor T cells, induced by UV radiation, has become the premier system for studying clinically-relevant immune suppression in cancer. Table 12.2 below summarizes two decades of progress in this area.

The induction of immune suppression by UV radiation can be broken down into the following discrete steps: (i) damage to keratinocytes; (ii) alteration of antigen presentation; (iii) induction of suppressor T cells; and, (iv) the effector function of suppressor T cells, directing the immune response away from induction and elicitation of delayed type hypersensitivity and cutaneous hypersensitivity. UV radiation directly and indirectly alters many of the immune mechanisms that recognize and control the growth of cutaneous neoplasms. Despite the many differences between mice and humans, findings in animal NMSC and melanoma models provide new insights into the immunobiology of human skin cancer and suggest new lines of investigation and new approaches to the prevention and immunotherapy of cutaneous malignancies.

Table 12.2 UV Radiation induces cells that suppress skin cancer immunosurveillance.

Investigators	Finding
Toews *et al.*, 1980	Critical role of Langerhans cells in inducing downregulation.
Noonan *et al.*, 1981	Suppression correlates with development of tumors.
Fisher and Kripke, 1982	First description of suppressor T cells in UV system.
Sauder *et al.*, 1981	Epidermal cell are actively involved in suppression.
Elmets *et al.*, 1983	Central role of T cells in suppression.
Schwarz *et al.*, 1986	Cytokines are involved in inducing suppression.
Ullrich *et al.*, 1986	Involvement of multiple pathways.
Okamoto and Kripke, 1987	Different pathways for effector and suppressor arms of response.
Cruz *et al.*, 1989	Use of purified cell populations for induction of suppression.
Welsh and Kripke, 1990	Role of epidermal dendritic T cells.
Glass *et al.*, 1990	Suppression is seen in all strains of mice.
Simon *et al.*, 1991	Langerhans cell's function changes with UV treatment.
Simon *et al.*, 1994	Th1 subpopulation of T cells is affected by UV.
Bucana *et al.*, 1994	Ultrastructure of antigen presenting cells in UV suppression.
Muller *et al.*, 1995	Alterations of antigen presenting cells.
Saijo *et al.*, 1995	Antigen presenting function and suppressor cell induction are separable.
Shreedhar *et al.*, 1998	Characterization of suppressor T cells.
Strickland *et al.*, 1999	Cytoprotective oligosaccharide prevent induction of suppressor T cells.

A strategy for preventing sun-induced skin cancer

Current approaches for skin cancer prevention focus on sun avoidance, protective clothing, and/or the use of sunscreens. Chemical sunscreens can reduce the numbers of p53 mutations under experimental conditions (Ananthaswamy *et al.*, 1997). Sunscreens do reduce suppression of the cutaneous immune response but they appear to be more effective at preventing erythema than in protecting the immune responses (Bestak *et al.*, 1995; Pathak *et al.*, 1991; Wolf *et al.*, 1993; Wolf and Kripke, 1996). Since sunscreens are quite effective in preventing painful sunburn, they have the potential to extend a person's time in the sun. Extended exposure time can lead to inadequate protection of the cutaneous immune response that rejects newly neoplastic cells. Thus there is a need to develop consumer-acceptable, post-sun-exposure agents that are effective at reversing UV-induced damage.

Knowledge of the mechanisms involved in UV-induced carcinogenesis has been useful in designing therapeutic agents to prevent skin cancer. For example when UV radiation damages DNA and enzymes which accelerate the removal of UV-induced lesions in DNA and prevent suppression of T cell-mediated immune responses. The use of these agents is extremely costly and may be feasible only for diseases of DNA repair (Kripke *et al.*, 1992; Vink *et al.*, 1996). Compounds such as green tea polyphenolics, retinoids, and ascorbic acid have been investigated for their potential use as anti-oxidant adjuncts to sunscreens (Mukhtar and Ahmad, 1999) but while preventing free radical-induced damage, these agents are ineffective once cellular damage has occurred. We have pursued the opposite approach by taking a widely accepted therapeutic agent for cutaneous injury, *Aloe barbadensis* gel, and elucidating its mechanism of action.

BIOLOGIC EFFECTS OF CRUDE *A. BARBADENSIS* EXTRACTS ON EPITHELIAL TISSUES

Extracts from a number of *Aloe* species are widely regarded as having therapeutic dermatologic properties useful in the treatment of sunburn and mild thermal injury (Heggers *et al.*, 1993). These properties seem to be separate from aloe's laxative activity, anti-gastrointestinal ulcer activity and from efficacy in the treatment of severe thermal burns and wound-healing activity (Pelley and Heggers, unpublished observations). The popular recognition of aloe as a dermatologic has led to the widespread incorporation of aloe extracts in healthcare and cosmetic products.

Development of Aloe Research Foundation (ARF) standard samples

Scientific evidence for aloe's efficacy is limited and studies using commercial 'Aloe vera' have been extremely difficult to reproduce. Probably the best example of this is in the area of the treatment of radiation dermatitis. During the 1930s there were reports of treating radiation-induced skin lesions with crude extracts of *Aloe vera* (reviewed in Grindlay and Reynolds, 1986). Subsequent publications were divided in their findings with the more rigorous clinical investigations failing to show a beneficial effect (Aleshkina and Rostotskii, 1957; Ashley *et al.*, 1957; Rovatti and Brennan, 1959; Rodriguez-Bigas *et al.*, 1988; Williams *et al.*, 1996). One factor in this lack of repro-ducibility is undoubtedly the fraud, adulteration of feed stocks, and misrepresentation that is currently widespread in the 'Aloe vera' industry (Pelley *et al.*, 1998). Active ingredients are also commonly destroyed by routine industrial processing (Waller *et al.*, 1994; Pelley *et al.*, 1998, and most recently Waller *et al.* in this volume). Given the lack of knowledge of proper processing on the part of academic investigators, the lack of scientific expertise on the part of commercial entrepreneurs, and the difficulty of obtaining fresh, undegraded aloe gel, it is remarkable that any positive results were attained at all during this era.

We began our studies by developing standardized *Aloe barbadensis* gel materials (Aloe Research Foundation Standard Samples) with uniform chemical and biological properties (Pelley *et al.*, 1993; Waller *et al.*, 1994). The ARF Process 'A' materials used in most experiments with unfractionated aloe represent depulped aloe gel fillets produced under conditions of rigorous sanitation not usual to the industry and lyophilized extremely rapidly, within hours of harvest. ARF materials (Processes B, C, D and E) were also produced approximating to the various processes, pasteurization, filtration, absorption with activated charcoal, treatment with cellulase, and concentration by rising/falling thin film evaporation, employed in the industry. The ARF Process A material corresponds to no commercial product in current existence and its production is not economically feasible. These ARF materials were produced at several commercial sites by Todd A. Waller and R.P. Pelley, chemically characterized by RPP and distributed to three groups of investigators who characterized the biological activities therein. The first was the late Robert H. Davis at the University of Pennsylvania's College of Podiatric Medicine who was, for many years, one of the few investigators publishing controlled experiments on biological properties of *A. barbadensis* in peer reviewed journals (Davis *et al.*, 1986; Davis *et al.*, 1987). Davis examined the anti-inflammatory activities of ARF materials in the phorbal ester-induced foot pad swelling assay. A second investigator was at the University of Wisconsin in Madison. Dr Sheffield's laboratory was working on mammary gland ductal epithelium and its response to physical injury and injury by activated phagocytes. Lastly, the laboratory of F.M. Strickland at the University of Texas, M.D. Anderson Cancer clinic, examined the effect of UVB radiation on the skin immune system of mice. The ability to concordantly assay materials of known provenance and chemically defined composition with highly standardized assays has led us to realize that: (i) multiple biologically active substances are present in crude extracts of A. *barbadensis*; (ii) these materials are differentially labile: and, (iii) the biologically active molecules are variably present in authentic commercial materials. We further realized that the biologically active molecules were not being measured by existing methods of commercial chemical detection. For a more complete discussion of these concepts the reader is referred to the accompanying article (Waller *et al.*, Chapter 8, this volume).

Effects of crude *A. barbadensis* gel on numbers and morphology of dendritic T cells and Langerhans cells in UV exposed mice

UVB irradiation decreases the number of immune cells in the skin (Toews *et al.*, 1980; Lynch *et al.*, 1981). In the epidermis, the outer most of the three layers of the skin, these immune cells are of two types (Table 12.3). One type is the Langerhans cells, a cell that processes antigen into a form in which it can trigger the immune response. The second type is the dendritic thymus-derived lymphocyte. The skin dendritic T cell recognizes processed antigen and regulates the immune response. Low doses of UV

Table 12.3 Crude *A. barbadensis* ARF Process A gel protects skin immune system cells from UV radiation damage.

	Number of cells in epidermis as a percentage of control				
	Aquaphor Vehicle	*Aloe + Vehicle*	*0.4 kJ/m^2 UV*	*UV + Vehicle*	*UV + Aloe*
Langerhans Cells	99%	106%	19%	22%	41%
Dendritic T Cells	84%	74%	16%	16%	29%

Notes

Values summarize data from Table I in Strickland *et al.* (1994). The number of cells per mm^2 in untreated mice were: Langerhans cells, 839 ± 70 (mean $\pm$ S.E.M.); Dendritic T cells, 839 ± 70. The SEM for experimental values averaged less than 10% of the mean in unsuppressed groups and ~15% in UV suppressed groups. Suppression by UV radiation was statistically significant as was protection by aloe.

radiation (about half the dose that causes sunburn or 0.5 M.E.D.) injure these epidermal Langerhans cells and epidermal dendritic T cells .

Even at as low a dose as 400 Joules/m^2 daily for four days, UVB radiation causes a disappearance of ~80% of the ATPase$^+$, Ia$^+$ Langerhans cells in the skin's epidermis. Morphologically, the remaining antigen- processing Langerhans cells appeared to be damaged with pale staining of specific markers and blunted dendritic processes. UV irradiation reduced epidermal dendritic T cells to an almost similar extent. We demonstrated that the specially prepared ARF extracts of aloe gel prevent UV-induced immune suppression of T cell-mediated immune responses in mice (Strickland *et al.*, 1994). Aloe treatment doubled the number of Langerhans cells and dendritic T cells observed after UV irradiation. The Langerhans cells remaining in the skin had a morphology more closely approximating normal. An equivalent degree of morphologic protection was not observed for dendritic epithelial T cells. The degree of morphological protection of the epithelial cells of the skin immune response was only 30–40% of normal compared to almost complete restoration of skin immune system function. This is consistent with our overall view of cutaneous immune regulation, wherein the cellular potential for responsiveness is usually in excess and the limiting factor on the magnitude of the immune response is determined by the induction of negative immunoregulatory influences.

Effects of *A. barbadensis* gel on local suppression of cutaneous contact hypersensitivity (CHS)

One of the ways that UV radiation contributes to the growth of skin cancer is by suppressing T lymphocyte-mediated immune responses that would otherwise reject the newly transformed skin cells. Models developed to better understand the mechanism of UV-induced immune suppression show that very low doses of UV can suppress responses to model antigens such as contact allergens and infectious organisms. Our studies have used both of the common models for cutaneous T cell-mediated-immune function; contact hypersensitivity (CHS) to allergens and delayed-type hypersensitivity (DTH) to antigens of infectious organisms. The murine CHS responses to hapten allergens such as the organic chemical compounds dinitrofluorobenze (DNFB) or fluorescein isothiocyanate (FITC) are primarily an epidermal reaction which can be suppressed by suberythemal doses of UV radiation (Toews *et al.*, 1980; Lynch *et al.*, 1981; Okamoto and Kripke, 1987). At these low UVB doses and also at higher doses, suppressor T cells are induced (Okamoto and Kripke, 1987; Elmets *et al.*, 1983; Noonan *et al.*, 1984; Ullrich *et al.*, 1990).

First we tested ARF aloe materials to see if they were biological response modifiers (BRM). That is would they, in the absence of UVB, either suppress or enhance the skin immune response? *A. barbadensis* extract in a variety of doses or an Aquaphor vehicle was applied to the shaved abdominal skin of mice and the treated skin was sensitized with FITC. One week later, CHS was tested by challenging a distant site (ear) and measuring the swelling that was caused by the influx of immune inflammatory cells in response to the allergen. Aloe was without either positive or negative effect beyond that of vehicle alone in this system. Thus, topically applied aloe extracts are neither immuno-suppressive nor immunopotentiating. This differentiates our system from BRM's such as the immuno-regulatory polysaccharides acemannan or β glucans (reviewed in Pelley and Strickland, 2000 and below).

We studied the effect of low dose (400 Joules/m^2 daily, for four days) UVB radiation upon the local CHS reaction to FITC (Strickland *et al.*, 1994). Aloe extracts were applied each day within ~1 hour after UVB injury at three doses (16.7 mg/ml, 5 mg/ml and 1.7 mg/ml). These concentrations (Table 12.4, below) respectively correspond to a ten-fold range centered about the solids content (5 mg/ml or 0.5 g/dl) of native aloe gel. The maximum dose of aloe corresponds to the highest concentration of high quality *A. barbadensis* gel that can be achieved, due to the high psuedoplasticity of undegraded native polysaccharide. The data are measured as ear swelling and are expressed as percentage suppression using the following formula:

$$\% \text{ Suppression} = 1.00 - \frac{\mu\text{m swelling, UV experimental}}{\mu\text{m swelling, positive control}} \times 100$$

Formula 1 Calculation of suppression of skin immune response.

In the case of CHS, the positive control is the swelling in the ears of C3H mice not exposed to UV radiation, sensitized with either DNFB or FITC and challenged with the appropriate hapten. Typical positive responses for CHS are 70 to 150 μm (or 7 to 15×10^{-2} mm).

Table 12.4 below illustrates the result of a typical experiment.

Exposure of abdominal skin to 400 Joules/m^2 UVB daily for four days followed by sensitization on the belly and subsequent ear challenge results in a CHS ear swelling response upon challenge approximately half that of unirradiated, sensitized, challenged

Table 12.4 Crude *A. barbadensis* ARF Process A gel protects the local cutaneous hypersensitivity immune response from low dose (cumulative 1,600 Joules/m^2) UV radiation.

	Percentage decrease from homologous control (Groups of ten mice each)				
	UV No Treatment	*UV+ Vehicle*	*UV+ 1.67% Aloe*	*UV+ 0.5% Aloe*	*UV+ 0.17% Aloe*
400 Joules/m^2, daily, 4 days	60%	52%	13%	9%	12%
p versus homologous control	<0.012	<0.012	NS	NS	NS

Note

Values summarize data from Figure 1 in Strickland *et al.* (1994). NS is not significantly different, indicates that no significant immune suppression is occurring. It should be noted that in the legend to that Figure 1, there is a printer's error and the amount of challenging FITC is given as 10 ml rather than 10 μl.

Table 12.5 Crude *A. barbadensis* ARF Process A gel protects the local cutaneous hypersensitivity immune response from low dose (2,000 Joules/m^2) UV radiation.

	Percentage decrease from homologous control (Groups of 15 mice each)		
	UV No Treatment	UV+ Vehicle	UV+ 0.5% Aloe
2000 Joules/m^2	67%	60%	13%
p versus homologous control	<0.001	<0.001	NS

Note

Values summarize data from Figure 4 in Strickland *et al.* (1994). NS is non-significant, indicating that no significant immune suppression is occurring. The experiment was performed on groups of five mice each and was repeated three times. Analysis of variance revealed that with three repeats, experiment to experiment variability was non-significant.

controls (UVB, ~4–5 mm×10^{-2} ear swelling versus ~9–10 mm×10^{-2} in unirradiated controls). Treatment with aloe after UVB injury almost completely prevented the UVB-induced suppression of CHS induction. Swelling was ~8 mm×10^{-2}, 87 to 91% of that usually observed.

The amount of daily UV radiation, 400 Joules/m^2, in the experiment above is quite small, not quite the minimal erythemal dose (MED) for the C3H mouse. We therefore also tested the protective ability of crude aloe materials at a dose (2,000 Joules/m^2) just sufficient to cause a minimal sunburn (Table 12.5 above).

The effect is essentially identical to that described in Table 12.4. UV radiation decreases the T cell-mediated CHS response by about one half. Treatment of the skin with ARF Process A gel immediately after injury restores the immune response to where there is no longer statistically significant suppression by UV. Interestingly, we almost never observe a protective effect where the UV versus UV+aloe values are numerically identical, even through there is not a statistical difference between them.

Recently, our findings with crude materials have been verified by other investigators (Lee, 1999). Aloe's ability to prevent some of the consequences of UV-induced injury make it a potentially useful adjunct to sunscreens.

Effects of *A. barbadensis* gel on suppression of cutaneous hypersensitivity – local versus systemic actions

Aloe, applied topically, may act locally by penetrating the epidermis and preventing immunosuppression or it may penetrate and have a systematic action. To distinguish between these two possibilities the abdominal skin of mice was treated with a single 2,000 Joules/m^2 dose of UVB. Various sites, non-irradiated back versus irradiated abdomen, were then treated with vehicle or 0.5% ARF Process A gel. Three days after injury and treatment, the mice were immunized on the abdominal skin with FITC. One week later their sensitization status was determined via challenge on the ear. Table 12.6 gives the results.

Aloe gel protected the CHS response from UV suppression only when it was applied to the injured skin. This finding indicates that although the gel is being absorbed

Table 12.6 Crude *A. barbadensis* ARF Process A gel protects the cutaneous hypersensitivity immune response from low dose (2,000 Joules/m^2) UV radiation only when the site of injury is treated.

	Percentage decrease from homologous control (Groups of five mice each)			
	UV *No Treatment*	*UV+* *Vehicle to* *abdomen*	*UV+* *0.5% Aloe* *to abdomen*	*UV+* *0.5% Aloe* *to back*
2000 Joules/m^2 to abdominal skin	59%	53%	29%	55%

Note

Values summarize data from Table IV in Strickland *et al.* (1994). Significance of site to site treatment effect, back versus abdomen, p = 0.03.

Table 12.7 Crude *A. barbadensis* ARF Process A gel prevents systemic suppression of the cutaneous hypersensitivity immune response by high dose (10,000 Joules/m^2) UV radiation.

	Percentage decrease from homologous control (Groups of five mice each)		
	UV *No Treatment*	*UV+* *Vehicle to back*	*UV+* *1.67% Aloe to back*
5000 Joules/m^2 to back skin	71%	55%	0%

Note

Values summarize data from Figure V in Strickland *et al.* (1994). Significant suppression was observed in both the UV exposed/untreated group and the UV exposed/vehicle treated group, p = 0.01 by ANOVA. There was no suppression in the UV exposed/aloe treated group.

locally, it is not absorbed in sufficient quantities to treat the immune suppression being induced on the opposite side of the animal.

If a sufficiently high dose of UVB irradiation, a solid sunburn dose, is given, there is a suppression of the ability to induce a CHS response throughout the animal (Noonan *et al.*, 1981, 1984). For example, if enough UVB is given to the skin of the back, and painting hapten on non-UV-exposed abdominal skin of the mouse will not sensitize it, there is systemic suppression of CHS responses. We tested whether aloe extract would be protective against a sunburn-range dose of UV. The backs of mice were exposed to 10,000 Joules/m^2 UV radiation which is about five times the MED of the C3H mouse. The back skin was then treated with the highest dose of ARF Process A gel or the Aquaphor vehicle and the experiment completed as above.

Even at this high dose, one that if repeated is capable of inducing skin cancer in this strain of mice, the gel was capable of restoring the CHS response (Table 12.7).

Effects of *A. barbadensis* gel on systemic suppression of delayed type cutaneous hypersensitivity (DTH)

Delayed type hypersensitivity (DTH) is another type of cell-mediated cutaneous immune response. It primarily involves the dermis although in severe reactions it can also involve the epidermis. It is generally directed against protein antigens such as alloantigen (Major histocompatibility complex (MHC) antigen) or infectious agents

Table 12.8 Crude *A. barbadensis* ARF Process A gel prevents systemic suppression of the DTH immune response by high dose (5,000 Joules/m^2) UV radiation.

	Percentage decrease from homologous control (Groups of five mice each)	
	UV *No Treatment*	*UV+* *1.67% Aloe to back*
5000 Joules/m^2 to back skin	52%	0%

Note
Values summarize data from Figure VI in Strickland *et al.* (1994). Significance of treatment effect p < 0.01.

(e.g. the yeast *Candida albicans*). UVB suppresses both CHS and DTH responses. Suppression of DTH is almost never local but usually systemic (Ullrich, 1986; Denkins *et al.*, 1989; Ullrich *et al.*, 1990). One of the reasons we embarked upon studies of aloe as a protective agent against UVB-induced suppression of the skin immune system was the hope that differential effects on DTH and CHS might help us to dissect the role of DTH and CHS in tumor immunity. This hope has come to fruition (Byeon *et al.*, 1998: Strickland *et al.*, in press and in this chapter).

Systemic suppression of DTH immunity is generally measured by exposing one site, usually the abdomen or back, to a high (5,000 Joules/m^2) dose of UVB, immunizing subcutaneously and challenging in the footpad. This large dose of UVB radiation usually systemically inhibits sensitization so that the subsequent injection of challenge antigen yields only 50% or less of the expected swelling response (see Table 12.8). Again, in this system we first tested whether aloe extracts, cutaneously applied, were immuno-potentiating or immunosuppressive. Once again, such 'nonspecific' BRM effects were not found (Strickland *et al.*, 1994). Therefore the efficacy of aloe was tested by exposing the backs of mice to 5,000 Joules/m^2 UVB and then immediately applying a high dose (16.7 mg/ml) of Process A ARF aloe gel (Strickland *et al.*, 1994). Three days later the mice were immunized subcutaneously with killed *Candida* organisms and ten days later they were challenged by injection of *Candida* antigen into the hind foot pads with subsequent measurement of foot pad swelling.

The aloe extract completely prevented the UVB-induced systemic suppression of DTH responses (Table 12.8). In other experiments, materials from aloe prevented the suppression of the DTH response to alloantigen with efficacy similar to that above. These encouraging results led us to undertake the costly and time consuming experiments to determine if aloe is protective against carcinogenic regimens of UV radiation.

Mechanism of action of crude *Aloe barbadensis* extracts in UVB-damaged skin

The crude extracts employed in these studies were known to contain a mixture of chemically distinct compounds interacting with multiple, distinct mechanisms. However, studies with aloe aimed at elucidating the mechanism(s) of action have none the less been performed (Strickland *et al.*, 1994; Byeon *et al.*, 1998). First, since aloe is traditionally associated with UV absorbing anthraquinones, we asked whether the extracts

Table 12.9 Relationship between time of UV irradiation and efficacy of 0.5% *A. barbadensis* in preserving CHS responses.

Time Sequence	Percentage decrease from homologous control (Groups of five mice each)				
	UV + Vehicle	Aloe 24 hours UV	UV 0 hours Aloe	UV 24 hours Aloe	UV 48 hours Aloe
UV Dose – 2,000 Joules/m^2	70%	76%	46%	28%	72%
p versus homologous control	0.004	0.006	0.02	NS	0.004

Note

Values summarize data from Table III in Strickland *et al.* (1994). NS is non-significant, indicating that no significant immune suppression is occurring.

were acting as sunscreens. Therefore, mice were exposed to a single dose of 2,000 Joules/m^2 of UVB and treated at various times before or after UVB injury.

Aloe applied 24 hours before injury had no effect. Aloe applied immediately after UV (0 Time) reduced suppression by one-third and aloe applied 24 hours after injury reduced suppression by two-thirds. By 48 hours after injury the train of events leading to suppression of the CHS response was set fully in motion and aloe was without efficacy (Table 12.9). This effectively ruled out the possibility that aloe works as a sunscreen and in fact suggested that it interferes with the cascade of inflammatory cytokines released by injured keratinocytes.

Anti-inflammatory agents have been described (Davis *et al.*, 1986, 1987) in *Aloe barbadensis* and sunburn inflammation blocking has been reported to prevent immune suppression by UV (Anderson *et al.*, 1992; Reeve *et al.*, 1995). Therefore, we determined if the immune protective effects were due to an anti-inflammatory action. Although the erythema component of sunburn can be difficult to determine in pigmented C3H mice, the edema component of sunburn can be readily quantitated by measuring ear swelling (Cole *et al.*, 1989). We administered a moderate sunburn (5,000 Joules/m^2 or about 2.5 MED) to C3H mice and treated with aloe or vehicle control.

The ARF Process A aloe gel extract reduced UVB-induced edema (Strickland *et al.*, 1994) of the ears by only 17.5%, a non-significant effect (Table 12.10). This negligible anti-inflammatory activity stands in marked contrast to the almost complete protective effect observed with functional assays of UV-induced damage to the skin immune system.

Table 12.10 Crude *A. barbadensis* ARF Process A gel has little effect upon edema induced by high dose (5,000 Joules/m^2) UV radiation.

Treatment	UV-induced increase in ear thickness – Mean ± S.D. mm × 10^{-2} (Groups of five mice each)		
	24 hours	48 hours	72 hours
UVB	0.7 ± 0.7	7.9 ± 2.1	5.5 ± 1.4
UVB + Vehicle	1.2 ± 0.8	8.0 ± 5.0	7.6 ± 5.0
UVB + 0.5% aloe	1.2 ± 0.8	8.0 ± 5.0	7.6 ± 5.0

Note

Values summarize data from Table II in Strickland *et al.* (1994). There are no statistically significant differences due to treatment effects.

Damage to DNA and its repair are crucial events in carcinogenesis, and increased DNA repair is one of the few previously available theraputic interventions for post-exposure skin protection. Therefore we investigated whether aloe extracts increased the rate of DNA repair (Strickland *et al.*, 1994). The number of cyclobutyl-pyrimidine dimers in UVB treated skin were quantitated using the endonuclease sensitive-site assay and alkaline agarose gels (Yarosh and Ye, 1990). Irradiation with 5,000–10,000 Joules/m^2 resulted in 40 to 56 dimers per 10^6 base pairs, depending on the dose. Even the highest dose of aloe employed (16.7 mg/ml) had no effect upon the numbers of cyclobutyl-pyrimidine dimers. Thus the therapeutic action of aloe takes place at a step down-stream from DNA damage and repair.

CHS and DTH are differentially protected from UV-induced suppression by process A crude aloe gel

The studies above established that crude aloe extracts can protect the skin immune system from suppression by UV radiation and explored some aspects of its mechanism of action (Strickland *et al.*, 1994). We next sought to dissect the various aspects of the skin immune system using crude materials (Byeon *et al.*, 1998). In particular we asked 'is protection of CHS and DTH responses mediated by the same or different agents in aloe gel?' Differential responses establish that CHS and DTH are quite mechanistically different. Aloe offers the opportunity to determine which response, DTH or CHS, is most important in skin tumor immune surveillance.

One way of accomplishing this with crude material is to determine if the agent protecting CHS has the same stability as the agent protecting DTH. Bioassay data obtained from three lots of crude, lyophilized aloe gel (ARF91A, ARF94B, and ARF94G) were analysed for their stability in protecting CHS and DTH immunity. The responses to antigens vary from experiment to experiment to some degree, thus necessitating the use of the normalizing Formula 1 in order to reduce experiment to experiment variability. When experiments are conducted over a long period of time, the degree to which suppression occurs may also begin to vary significantly. Therefore it is necessary to normalize using the slightly more complicated Formula 2, below, in order to factor out that source of variability.

$$\%\text{Protection} \times 100 = \frac{\mu\text{m swelling irradiated aloe treated} - \mu\text{m swelling suppressed control}}{\mu\text{m swelling positive control} - \mu\text{m swelling suppressed control}}$$

Formula 2 Calculation of protection of skin immune response.
The positive control is the aloe-treated, unirradiated pair for each of the UV irradiated, aloe-treated groups. The suppressed control consists of UV exposed animals either without treatment or treated with control vehicle. This process of double normalization to both positive and negative controls allows comparisons between experiments separated by some time with minimal experiment to experiment variability.

In groups of UV irradiated mice treated with phosphate buffered saline (PBS) alone, a 50–80% reduction in CHS and DTH responses is observed compared to their unexposed, matching controls. This degree of response represents 0% restoration while the response of unirradiated, aloe-treated sensitized positive controls was set as 100%.

The results presented in Figure 12.1, show that treatment of UV-irradiated skin with aloe extract prevented suppression of the CHS response to hapten to a variable degree.

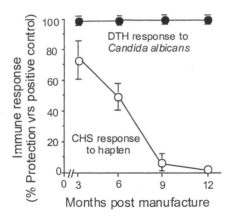

Figure 12.1 Stability of CHS and DTH-protective activities in aloe gel extract.
The effect of Process A *A. barbadensis* gel extracts on UV-induced suppression of contact and delayed type hypersensitivity responses was measured using three lots (ARF91A, ARF94B and ARF94G) of Process 'A' gel at various times after manufacture. For CHS assays ventral skin of groups of five mice was exposed to $2\,kJ/m^2$ UVB, the irradiated skin was treated with 5 mg/ml gel or saline, three days later, UV-irradiated and unirradiated control mice were sensitized with hapten and challenged five days later. For the DTH response dorsal skin of mice was exposed to $5\,kJ/m^2$ UVB radiation, treated with aloe or PBS, UV-irradiated and unirradiated control animals were sensitized three days later with 2×10^7 formalin fixed *Candida albicans* cells *s.c.* and their DTH response was elicited ten days later. The aloe extracts were stored as a lyophilized powder at $-20\,°C$ until use. The data are normalized according to Formula 1 and are the mean ± SEM. Modified from Figure 1 of Byeon *et al.* (1998).

The protection afforded by the three different lots of gel varied. For example, three months after manufacture, application of ARF91G gel to UV-irradiated animals provided only 43% restoration of their CHS response while the ARF91A lot of gel extract completely restored the CHS response. The activity of ARF91B was intermediate between these two values. The levels of protection provided by the gel were maximal at a dose of 5 mg/ml (w:v) and could not be improved by increasing the dose used (not shown). The activity of all three lots of gel extract decayed with time, despite their storage as a lyophilized powder. After nine months, none of the extracts prevented UV-induced suppression of CHS responses to hapten (Figure 12.1). Other lots of 'Process A' gel extract gave similar results except that the levels of CHS protection ranged from 30 to 100% and longevity of CHS-protective activity ranged from three to nine months after manufacture. Commercially prepared (non-Process A) gels from the same source were uniformly inactive even when tested within one month after manufacture.

Systemic suppression of DTH responses to *C. albicans* was measured in mice whose shaved dorsal skin was exposed to $5\,kJ/m^2$ of UVB and treated with the same lot of gel extract used for CHS-protection studies above. In contrast to the partial protection of the CHS responses to hapten, the gel completely prevented systemic suppression of DTH to *C. albicans*. The immunoprotective activity of all three lots of lyophilized extract remained unchanged after 12 months of storage (Figure 12.1). At each time point at which the gel was tested, the experiment was repeated at least once in order to confirm the results. Restoration of immunity was not due to non-specific immunostimulation by the gel, since neither CHS nor DTH responses were affected in unirradiated control animals at any dose tested (Strickland *et al.*, 1994,). Also, the structurally unrelated polysaccharide, methycellulose, failed to protect CHS and DTH responses against UV-induced suppression.

The different decay rates observed for CHS and DTH-protective activity in crude gel suggested that these activities are mediated by different factors. Additional evidence for the presence of distinct factors was obtained by performing a dose response. From $0\,\mu g$ to $5,000\,\mu g$ aloe gel in PBS was applied to unirradiated controls and to the UV-irradiated skin of mice immediately after exposure. Three days later the animals were sensitized with hapten through UV-irradiated skin (local CHS model) or injected with formalin-fixed *C. albicans* cells (systemic DTH model). The data, presented in Figure 12.2 show that protection of CHS responses against suppression by UV radiation was mediated by only the highest dose ($5,000\,\mu g$) of gel. In contrast, as little as $10\,\mu g$ gel completely protected DTH responses.

These experiments were performed using ARF Standard aloe gel four months after manufacture. The levels of protection were consistent with those observed in Figure 12.1 for gel of that age. Similar results were observed using lots ARF 94G, ARF 94B of crude gel, and an oligosaccharide-enriched fraction prepared from cellulase-treated ARF 94G aloe polysaccharide by hollow fiber ultrafiltration (see Section IIIB). Taken together, the results indicate that protection of CHS and DTH immune responses from suppression by UV radiation is mediated by at least two separate factors in crude aloe gel.

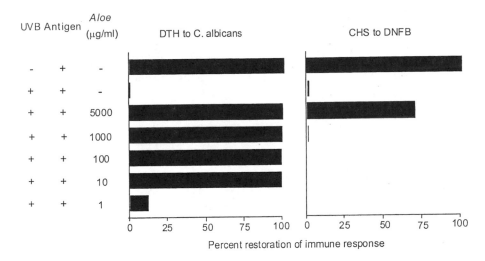

Figure 12.2 Protection of CHS and DTH responses from suppression by UV radiation requires different doses of *Aloe barbadensis* gel.
Groups of five mice were exposed to $2\,kJ/m^2$ (CHS) or $5\,kJ/m^2$ (DTH) UVB radiation followed immediately by topical applicaiton of $1\,\mu g$ to $5\,mg$ of aloe gel in saline. Three days later the mice were sensitized with DNFB or *C. albicans* and challenged. The data, from three experiments using five animals each are expressed as percent protection using Formula 2 and are the Mean$\pm$SEM. Modified from Figure 1 of Byeon *et al.* (1998).

Conclusions from studies with crude aloe gel

Crude, Process A, ARF gel extract was capable of protecting the epidermal immune DTH and CHS systems from UVB-induced damage. This protection does not involve absorbing the UV radiation in a sunscreen-like manner or by altering DNA damage or repair. Although these lots of Process A crude gel had the anti-inflammatory activity (R.H. Davis, unpublished observations) classically ascribed to it (Davis *et al.*, 1986, 1987), this anti-inflammatory activity did not prevent the UVB-induced edema of sunburn. The cytoprotective activities and phagocyte activating activities in these materials were complex and will be presented later. The likeliest step, therefore, at which aloe extracts act is the immune cytokine cascade (reviewed below). However, before further applications could be pursued or mechanisms explored it was necessary to characterize the active ingredient.

THE STRUCTURE OF ISOLATED, PURIFIED POLYSACCHARIDES FROM VARIOUS SPECIES OF ALOE AND THEIR PROBABLE STRUCTURE IN THE PLANT

In the literature on biological activity and quality control of aloe materials many investigators insist that the molecule they are studying is the only biologically active molecule in aloe. Organic chemists tend to focus on the anthraquinones and chromones, biochemists on the proteins, and immunologists on the polysaccharides. The authors and our associates are indebted to Robert H. Davis, a physiologist working for many years on *Aloe barbadensis* extracts and applying countless biological assays to crude extracts, who was, perhaps, the first to suggest that what we were observing in aloe was the combination of several different active moieties of differing chemical structure, operating with different mechanisms and combining to produce a given effect. Davis' paradigm has certainly proved to be true in the case of the effect of aloe on the interaction of UVB and the skin immune system. We suspect that in this system three factors are interactively operating, all chemically distinct and all probably using three different mechanisms. Here we will discuss only one group of molecules, glucomannan (see Figure 12.3), the family comprising the native aloe polysaccharide, and the two fragments that are cleaved from it, the epithelial-protective oligosaccharide and the macrophage-activating acetylated mannan. This saccharide focus does not mean that the cytoprotective oligosaccharide which we will subsequently describe, is the only biologically active molecule in *A. barbadensis* nor is it the only one with effect on skin. Similarly, the macrophage-activating acetylated glucomannan polysaccharide we will immediately describe does not subsume all aloe biological activity. We have simply chosen at present to focus on this family of interesting molecules. There may be at least two other non-saccharide families of molecules which are also important in protecting the skin against environmental injury.

Aloe barbadensis polysaccharides

There are at least three polysaccharides readily apparent in commercially feasible extracts of *A. barbadensis*, an acetylated glucomannan, a galactan, and a pectin (for review see Pelley and Strickland, 2000). The dominant polysaccharide of aloe gel is the glucomannan

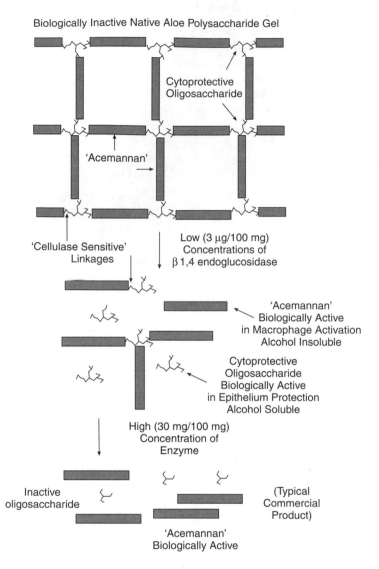

Figure 12.3 Cartoon of the putative structure of the native polysaccharide of *A. barbadensis* and its cleavage by β 1 → 4 endoglucosidase. Modified from Sheet 1 of Strickland *et al.* (1998).

classically described by Roboz and Haagen-Smit (1948). Another glucomannan was also examined although in fairly degraded material, by Segal *et al.* (1968). Gowda *et al.* (1979) were the first to determine the structure of this major aloe polysaccharide, purified by dialysis and then alcohol precipitation. These investigators further fractionated the precipitated polysaccharides based on a gradient of alcohol concentrations. They found a glucose to mannose ratio of 1:4.5 in the more lightly acetylated (9.25% m/m), less alcohol-soluble fraction. The more highly alcohol-soluble fractions had a somewhat higher mannose content (glucose:mannose 1:13.5 to 1:19) and were somewhat more heavily acetylated (10.3 to 17.2% m/m). In all cases a molecular weight of >200 kDa was assigned based on total exclusion from G-200 chromatographic gel. They describe the material initially isolated as a jelly. After further purification they describe a material that, although of a molecular weight beyond the resolving capacity of their analytical method, was relatively non psuedoplastic, since it could be subjected to column chromatography. This is the first published instance we have been able to find wherein the investigators noted that the native gel was of an extremely high molecular weight and that with time and purification it broke down to a non-psuedoplastic state.

All three of the above investigators stated that they were studying '*Aloe vera.*' Unfortunately, we have no way of absolutely determining exactly what plant they used. The species under investigation is important since there are significant differences in the polysaccharides from different species, first noted by Mandal and Das (1980b). Commercial plantings in the western hemisphere refer to *Aloe barbadensis*, strictly an outdated name for *A. vera*, while cultivation of any other species is extremely rare. Leaves described by Roboz and Haagen-Smit (400 g, 45% yield of gel) are consistent in our experience only with our strain *A. barbadensis*. In the case of Gowda *et al.* (1979) the source was simply stated as 'locally available *Aloe vera*' and insufficient detail is given of the isolation to allow us to speculate as to the actual species or variety. Mandal and Das (1980b), working in West Bengal (eastern India), commenting on the work of Gowda *et al.* (1979) working in Mysore (southern India), noted a difference between the material they were working with and the material Gowda *et al.* (1979) were working with. Mandal and Das noted that these differences were potentially important for the type of polysaccharide isolated and we agree.

Analysis of a glucomannan purified by alcohol precipitation followed by precipitation with Fehling's solution yielded a polysaccharide with a glucose to mannose ratio of 1:20.6 and suggested a repeating subunit of 3.3 kDa (Mandal and Das, 1980a). The average molecular weight determined by osmometry was 15 kDa. This group also purified and characterized two other polysaccharides from *A. barbadensis*(sic), one of which is a galactan (Mandal and Das, 1980b) and the other of which is a pectin (Mandal *et al.*, 1983). We agree with their findings (Pelley *et al.*, 1998). We find that the glucomannan greatly predominates in the freshly harvested gel, perhaps comprising >90% of all polysaccharide. However, this fresh gel is subject to degradation by β 1-4 endoglucosidases, often termed 'cellulase.' Some of this activity appears to be endogenous. It is fairly well known throughout the industry that loss of psuedoplasticity proceeds with time even in gel to which no exogenous 'cellulase' has been added and which is devoid of bacterial or fungal contamination. More often, in commercial materials, the acetylated glucomannan is broken down by glycosidases of proliferating, contaminating microorganisms (Pelley *et al.*, 1993; Waller *et al.*, 1994) or by fungal 'cellulase' exogeneously added to decrease viscosity and thereby increase yields (Coats, 1994). Often the glucomannan is broken down to such an extent that the galactan becomes the predominate polysaccharide (Pelley *et al.*, 1998). Thus, it not uncommon for commercial aloe materials to have a polysaccharide content very different from those classically described in the scientific literature (Pelley *et al.*, 1998).

Polysaccharides from other *Aloe* sp.

Polysaccharides have also been isolated and characterized from other *Aloe* species (*A. arborescens* Miller, *A. plicatilis*(L.) Miller, *A. vahombe*(sic) Decorse et Poisson, *A. vanbalenii* Pillans and *A. saponaria*(Ait.) Haw). The first of these to be characterized, in fact the first aloe polysaccharide to be subjected to detailed examination, was the mannan of *A. arborescens* (Yagi *et al.*, 1977). These investigators dialysed the gel, precipitated the protein with chloroform and isolated the polysaccharide by repeated cycles of precipitation with acetone. Analysis of the isolated polysaccharide by hydrolysis and paper chromatography revealed mannose to be the predominant sugar. Spectroscopy suggested that the sugars were β-linked and saponification revealed that the mannan was acetylated. The molecular weight, determined by ultracentrifugation, was 15 kDa. Lastly, these investigators were the first to determine a biological activity for an aloe polysaccharide. Ten injections of either 5 or 100 mg of *A. arborescens* mannan produced a 38 to 48% reduction in the growth rate of Sarcoma 180. This is a classical Biological Respose Modifier (BRM) assay – stimulation of immune responses to tumors (Pelley and Strickland, 2000), although as we have noted, not an assay with strong predictive power.

In 1978 Paulsen *et al.* reported the isolation of a glucomannan from *A. plicatilis* gel by dialysis. This polysaccharide had a glucose:mannose ratio of 1:2.8 and a molecular weight by gel filtration upon Sepharose 4B of at least 1,200 kDa. Reducing sugar analysis indicated that the polysaccharide was essentially linear and almost all sugars bore at least one acetyl group.

The polysaccharides of *Aloe saponaria* and *A. vanbalenii* were isolated by alcohol precipitation and dialysis and chemically characterized by Gowda (1980). The predominant polysaccharide of *A. saponaria* was a linear mannan with negligible amounts of glucose and a 20.7% (m/m) degree of acetylation. The predominant polysaccharide of *A. vanbalenii* was also a pure mannan with significant (19.5% m/m) acetylation. Neither molecular weight nor biological activity were reported.

Aloe vahombe has been described as being a BRM (Solar *et al.*, 1979). Radjabi-Nassab *et al.* (1984) isolated the polysaccharide by ethanol precipitation and gel filtration using Sephadex G-100. The 1→4 linked polysaccharide, which eluted in the void volume upon gel filtration (molecular weight ≥100 kDa) had a glucose:mannose ratio of 1:2.6 and 33% of the glucose residues were acetylated. In these respects, this polysaccharide more closely resembles that found in *A. plicatilis* than it does *A. barbadensis*. Further studies of the *A. vahombe* polysaccharide by Radjabi-Nassab *et al.* (1984) confirmed that the basic 1→4 linkage was of the β configuration and that cellulobiose-like units were rare.

More recently, two mannans from *A. saponaria* were isolated by dialysis, size exclusion chromatography and ion exchange chromatography (Yagi, 1984). *A. saponaria* (As) mannan-1, isolated from material harvested in September 1980 consisted of a linear 1→4 linked, acetylated (18% m/m) polysaccharide composed exclusively of mannose residues. Molecular weight based on gel permeation chromatography was 15 kDa. As-mannan-2 isolated from material harvested in December 1980 was similar to mannan-1 excepting that its molecular weight was 66 kDa, a 'trace' amount of glucose was present, the degree of acetylation was lower (10% m/m) and branching was evident. Biological activity of mannan-1 after parenteral administration was evidenced by inhibition of carrageenin-induced edema. Later, neutral polysaccharide was isolated by dialysis, chromatography on DEAE cellulofine and gel permeation on Sepharose 6B (Yagi, 1986). Three polysaccharides were observed which differed in their molecular weight and structure. Some confusion exists in this publication because the positions at which polysaccharides A, B and C elute from the preparative gel filtration column (A largest, B midsized and C smallest) do not correspond to the molecular weights described for the polysaccharides in the text (A, MW 15 kDa; B, 30 kDa; and C, 40 kDa). Assuming that the assignments in the text are correct and that Figure 2 is mislabelled, the conclusions are as follows.

Polysaccharide A (MW 15 kDa) is present in the largest quantity and it is dextran-like ($1 \rightarrow 6$ linked glucose). Polysaccharide C (MW 40 kDa) is present in second largest amounts and it is an acetylated (10% m/m) mannan of $\beta 1 \rightarrow 4$ linkage. Present in trace amounts was an arabinogalactan (Polysaccharide B) of intermediate (30 kDa) molecular weight. Polysaccharide C, which corresponds to the polysaccharide described in 1977 with the exception of its molecular weight, 15 kDa in 1977 and 40 kDa in 1986, was biologically active in that it enhanced phagocytosis and promoted the reduction of NBT dye by human leukocytes. This finding suggested that their polysaccharide was a classical BRM having both anti-tumor effect, described in 1977 and activating effect on phagocytes.

Conclusions about the β-glucomannans of *Aloe* sp.

Findings over the period 1948–1986 established the structure and biological activity of some aloe polysaccharides. They are variably-acetylated, predominantly-linear polymers of mannose with a $\beta 1 \rightarrow 4$ linkage. In some cases, they have a significant glucose content. There appear to be two forms described. One is a highly linear mannose-rich form of molecular weight ~15 kDa and the other is of higher molecular weight, generally indeterminant because of technological limitations perhaps related to degradation. These polysaccharides are immunostimulants upon parenteral injection, acting by phagocyte activation. The parallels between this system and the immunostimulatory $\beta 1 \rightarrow 3$ glucans (Ross *et al.*, 1999; Pelley and Strickland, 2000) which use the CD11b/CD18 receptor are striking. Usually when considering receptor-ligand specificity, the difference between a $\beta 1 \rightarrow 3$ linkage and $\beta 1 \rightarrow 4$ linkage is so great that one would not even consider the possibility of cross-reactivity. None the less, given the similarities between the two systems it will be interesting to see if the acetylated glucomannans of the various *Aloe* sp. use the Mac(macrophage)-1, (CD11b/CD18) receptor.

Aloe polysaccharides, Acemannan®, Carrisyn® and the patent literature

McAnalley at Carrington Laboratories used the findings of the late 1970s and early 1980s to develop an industrial-scale process to isolate *A. barbadensis* polysaccharide using alcohol precipitation. Recently we have reviewed the history of their Acemannan® and Carrisyn® products (Pelley and Strickland, 2000). The remaining legacy of that story is the numerous polysaccharide patents that have been filed. The shortcomings of the two series of Carrington Laboratories' patents can be addressed (see McAnalley, 1988; McAnalley *et al.*, 1995 and attached bibliographic notes). First, it was hasty to claim the $\beta 1 \rightarrow 4$ glucomanan structure since it had been previously well described in the literature as we have seen above. Second, the biological activities were well established in the industry and the BRM nature of the polysaccharides were taught as prior art by numerous scientists, cited above. Third, they ascribed all biological activity in aloe to Acemannan® and Carrisyn® polysaccharide. Fourth, the primary analytical tool they employed, FTIR, is extremely inefficient in establishing the purity of complex carbohydrates. Therefore, they consistently overestimate the purity of their materials and the process they developed produced only a crude product. Lastly, although they did teach that the freshness of the raw material was important and breakdown of polysaccharide was to be avoided, they failed to realize that the material they were starting with was already highly broken down. Thus they misappraised the structure of the *A. barbadensis* glucomannan polysaccharide as it exists in the living plant. As we subsequently success-fully pointed out (Strickland *et al.*, 1998), the glucomannan is not a homogeneous linear polysaccharide of about a million molecular weight but, rather, a block copolymer gel.

McAnalley and Danhoff implied that anything which precipitated from crude aloe extracts with alcohol is Acemannan® polysaccharide. In reality, aloe extracts are plant juices and contain very significant amounts of mixed salts of sodium/potassium/calcium malate or oxalate which precipitate with alcohol. Depending on the type of aloe product examined, these mixed organic acid precipitates may comprise up to 50% of the mass of the precipitated 'polysaccharide' (Pelley *et al.*, 1998). This is why in the literature discussed above, polysaccharide is generally purified by dialysis first and then alcohol precipitation afterwards, occasionally purified by precipitation first and dialysis later but never purified by precipitation with alcohol alone. The consequence of this mis-understanding was that many in the industry used the 'Methanol-Precipitable Solids' method as a specific test to measure 'Aloe vera' polysaccharide. Maltodextrin, indeed, was offered as 'Aloe vera' before proper analytical chemistry put a stop to this fraud, adulteration and misrepresentation (Pelley *et al.*, 1998).

'Acemannan' structure and molecular weight by process and analysis

The molecular weight and the structure of the saccharide lies at the heart of the aloe polysaccharide problem. The problem arises because the polysaccharide was never purified to chemical homogeneity, and the appropriate molecular weight analysis performed before biological testing. In U.S. Patent No. 4,735,935 patent McAnalley claims that all acetylated $\beta 1 \rightarrow 4$ mannans from the disaccharide (degree of polymer-ization n = 2) to the high linear polymers (n = 50,000) were found and that the polymer is at least 80% mannose. This implies a linear structure, while the native polysaccharide is obviously a gel.

In the examples given in the various patents the molecular weights of the various materials exemplified vary greatly. By way of process, in the '935, patent, Example 1 (columns 20–21) specifies that the product is produced by ultrafiltration, which removes undesirable compounds of less than 10 kDa (nominal) and retains the polysac-charide, presumably of nominal molecular weight >10 kDa. It is further specified that this retained fraction can be further fractionated by passage through an ultrafilter of nominal molecular weight cutoff 50 kDa wherein desired material is obtained. This would define Acemannan® as having a molecular weight of 10–50 kDa. Example 27 of the '935 patent illustrates molecular weight as determined by high pressure liquid chromatography using a 7.5 × 300 mm Beckman Spherogel TSK 2000 column. Unfor-tunately, detection of eluted material was by nonspecific (refractometry) methods so that the peaks are not defined. Furthermore, the chromatograms, which are crucial to evaluating the data, are not shown. From the tabulation of calibration it is apparent that analytical precision is obtained only in the molecular weight range of 40 to 9 kDa. Three classes of materials were exemplified with molecular weights of >80 kDa, >10 kDa and <1 kDa. From the summary (65, lines 28–30; 'cleave the function groups and glycosidic bonds ($\beta(1 \rightarrow 4)$)) thus reducing or eliminating its activity.') and Claim 1a ('substantially non-degradable'), it is apparent that materials in the third chromato-graphic region are undesirable. By reference to the published scientific literature reviewed above, it is highly probable that the first described materials (50, line 59

Table 12.11 Molecular weight distribution of Acemannan®.

| | Percentage of material in class (range) | | |
	Fraction #1 (>80 kDa)	Fraction #2 (>10 kDa)	Fraction #3 (undesired)
Experimental	47%	18%	35%
Preparations n=6	(12–60%)	(15–21%)	(24–67%)
Manufactured	28%	15%	57%
Preparations n=2	(19–37%)	(7–24%)	(56–57%)

'Fraction #1, MW > 80 kDa.') corresponds to the material described by Gowda *et al.* (1979) and that the second material (line 60, MW > 10 kDa.) corresponds to the material of molecular weight 15 kDa described by Mandal and Das (1980). The third material consists of undesired material of molecular weight < 10 kDa. This undesirable low molecular weight material was removed by dialysis or by solvent precipitation of the desired higher molecular weight polysaccharide, in the investigations by Gowda *et al.* (1979) and Mandal and Das (1980) immediately above. By these examples, Acemannan® could be a 40 kDa linear polysaccharide or a >80 kDa polysaccharide.

With all of these assumptions in mind, the examples of McAnalley can be understood as follows. There are two polysaccharides in Acemannan®. One is of molecular weight greater than 80 kDa, although the actual molecular weight cannot be more precisely determined because it is beyond the analytical range of the method. The second polysaccharide is of molecular weight of perhaps 12 kDa. There is a third component of undesired lower molecular weight contaminants and breakdown products. The content in Acemannan® of these materials varied greatly in Example 27 of the '935 patent as is seen in Table 12.11 below.

As interpreted by McAnalley this data implies that the process described is relatively uncontrolled in its enzymatic breakdown until a subunit size of ~10–15 kDa is reached after which breakdown occurs to very small fragments. Our findings are in agreement with this. It is then infered in Example 27, Example 31 and Example 32 that the process of monitoring the breakdown of Acemannan® can be achieved by IR spectroscopy. However, this method cannot measure the molecular weight distribution. What appears to be happening is that the polysaccharide is being broken down prior to alcohol precipitation. The <5 kDa saccharide fragments are remaining in the supernatant but the mixed mineral cation/organic acid coprecipitate is still precipitating with what little polysaccharide is left. This means that the resulting FTIR pattern shows that the Acemannan® product is increasingly contaminated with alcohol-insoluble malate and oxalate salts. The same phenomenon can be observed with broken-down commercial material. Example 5 of US Patent 4,851,224 is concerned with, among other things, the isolation, purification and characterization of the Carrisyn® Acemannan® polysaccharide. This section of the '224 patent has errors similar to those referred to above. With these in mind, the data can be interpreted as being consistent with the Acemannan® product consisting of a Material 'A' of molecular weight >100,000 and a Material 'B' having a molecular weight 'greater than 10,000 but less than 100,000 daltons.' Inspection of the figures, in fact, suggest a molecular weight of approximately 12 kDa for Material 'B'. It is further stated that 'the sum of peaks A and B constitute the active fractions' although the nature of the activity is unspecified and unexemplified. This data, taken together with the Summary (42, lines 21–31) indicates that 'A' (>100 kDa) is active and upon decomposition converts to 'B' (~12 kDa) and subsequently to inactive 'C' (dialysable, alcohol-soluble, MW < 10 kDa). The conclusion is that the Acemannan® is breaking down to a ~12 kDa fragment and from there to inactive fragments, but it is not known from what it is breaking down. Furthermore the assumption is that breakdown is to a single entity, fragments of the acetylated mannan.

Consensus summary of polysaccharide structure

There appears to be a consensus between the publications in the scientific literature and the patents reviewed above, concerning the structure of aloe polysaccharides. There exists in aloe an essentially linear, acetylated polysaccharide of discrete molecular weight between 12 kDa and 15 kDa. In the case of *A. arborescens* (Yagi *et al.*, 1977), *A. barbadensis* (Mandal and Das, 1980) and *A. saponaria* (Yagi *et al.*, 1984) it has the following structure:-

$$(2,3\text{-acetyl mannose, } \beta 1 \rightarrow 4, 6\text{-acetyl mannose})_{28-35}$$

Structure 1 The acetylated mannan disaccharide repeating unit (MW 450) and its polymeric polysaccharide, Acemannan®. With a polymer molecular weight to 12.6 to 15.7 kDa, there would be approximately 28 to 35 repeating units.

This is the structure later claimed by McAnalley (1988). There also exists a higher polymer of this structure, possibly linked together through 1,6 linkages:-

$$\{(2,3\text{-acetyl mannose, } \beta 1 \rightarrow 4, 6\text{-acetyl mannose})_{28-35}\}_n$$

Structure 2 The polysaccharide formed by linking together Acemannan® units into a higher molecular weight form but not crosslinked into a gel, n ~8.

where n is at least 8 and probably generally much larger. This structure 2 has been reported for *A. barbadensis* (Gowda, 1979), *A. saponaria* (Yagi *et al.*, 1984) and *A. arborescens* (Yagi *et al.*, 1986) and also claimed by McAnalley (1988). Mannans of chemical composition similar to the above have been described for *A. saponaria and A. vanbalenii* (Gowda, 1980) although it was not stated whether they were present as Structure 1 or Structure 2. Glucomannans have been described for *A. plicatilis* (Paulsen *et al.*, 1978) and *A. vahombe* (Radjabi *et al.*, 1983) with structures similar to Structure 2 excepting that glucose substitutes for some of the mannose residues.

A novel model for native polysaccharide structure

Structure 2, however, fails to account for three physico-chemical findings. First, how do we account for the 5 to 10% glucose that is so consistently found in the major *A. barbadensis* polysaccharide? Second, how do we account for the psuedoplastic gel observed at such an extremely low (0.5 g%) concentration? Third, why is there such sensitivity of this gel to the viscosity-reducing activity of 'cellulase,' given the relative resistance of acetylated mannan to the reducing sugar generating ability of this enzyme? These shortcomings of Structure 2 were fairly easy to ignore because almost no investigators had access to Process A type material. Rather they were working with material that mostly had the composition of Structure 1 or low multimers thereof.

Fractionation of the ARF Process A gel revealed its unusual physical properties and the correlation of changes in these properties with biological activity. We have claimed (Strickland *et al.*, 1998) that the structure (Figure 12.3) of the native glucomannan gel from *A. barbadensis* consists of linear portions of Structure 1, above, crosslinked into a gel by oligosaccharide linkers. We believe that these linking oligosaccharides may have 1,4,6 mannose branch points and that the $\beta 1 \rightarrow 4$ glycopyranosides adjacent to the branch points are unacetylated and thereby considerably more cellulase-sensitive than Acemannan®. Thus, cleavage proceeds most rapidly by clipping adjacent to the branch points and liberating linear acetylated mannan polysaccharides of molecular weight ~10–15 kDa, branched unacetylated oligosaccharides of molecular weight <5 kDa, and mixtures of the two.

Biological activity for suppressing inflammation and increasing the immune response has been claimed for both Structure 1 and Structure 2 by Yagi *et al.* (1977, 1984, 1986) by activating phagocytic cells. McAnalley (1988) claims an increase of the rate of wound healing by increasing the growth of fibroblasts and phagocytic activity. We agree that acetylated polysaccharide of molecular weight >10 kDa is capable of activating phagocytic cells (Strickland *et al.*, 1998, Figure 9). However, we do not think that the wound-healing and epithelial cell stimulating activity in aloe is due to a linear acetylated $\beta 1 \rightarrow 4$ mannan. We have demonstrated that *A. barbadensis* oligosaccharides, <5 kDa in molecular weight, are protective from UV radiation of the skin immune system. We suspect that the weak fibroblast activity noted by McAnalley above is due to a cytoprotective oligosaccharide contaminating his impure materials. The family of cytoprotective oligosacchrides appears to act, in the case of UV, by down-regulating Interleukin-10 production by injured keratinocytes (Byeon *et al.*, 1998; Strickland, 1999).

CYTOPROTECTIVE OLIGOSACCHARIDE AND INTERLEUKIN-10

When the Aloe Research Foundation originated ten years ago, the initial task laid before the Scientific Advisory Board was to assess the biological activities of *Aloe* sp. Most of us would have been content to assess the biologic activities in the best commercial 'Aloe vera' materials. B. William Lee, President of Aloecorp, challenged the Scientific Advisory Board to do better and determine the biological activities in the 'freshest possible gel' regardless of the commercial feasibility of the material. One individual, Todd A. Waller, was able to produce material to these standards, cutting and processing leaves first from the Lyford plantation of Aloecorp in January 1991 and later from other sites. Todd's ARF Process A material (ARF'91A) laid the foundation for all the work that followed which revolutionized our understanding of the saccharides of aloe.

Properties of fresh native *A. barbadensis* glucomannan polysaccharide

Our discovery of the biological activity of the plant oligosaccharides arose during a systematic study of multiple biological activities in crude *A. barbadensis* extracts (Waller *et al.*, 1994). This ARF Process A gel material was filleted with considerable attention to sanitation from prime leaves, depulped by passage though a 250 µm commercial citrus depulper, and lyophilized (Waller *et al.*, 1994; Pelley *et al.*, 1998). Physico-chemically the polysaccharide in this Process A material was of a very high molecular weight.

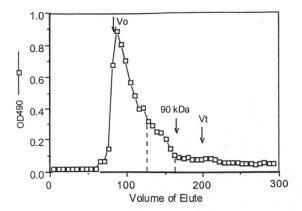

Figure 12.4 Molecular weight distribution of polysaccharide isolated from Process A *A. barbadensis* gel as determined by gel filtration on Sepharose 4B.

Modified from Sheet 5, Figure 4C of Strickland *et al.* (1998). 10 g of Process A gel is suspended in 200 ml of water and dialysed across Spectraphor #1 tubing (MW cutoff 6–8 kDa) versus 4 liters water at 4 °C for 24 hours. This process is repeated four times until the conductivity of the retained material is <100 mSiemens. The desalted polysaccharide solution is then clarified by centrifugation for ~20 kG for at least 15 minutes and the supernatant is further purified by precipitation at 80% v/v ethanol in the cold. Yields of polysaccharide from gel are in the range of 9 to 12% of solids. This purified polysaccharide was analysed by applying 30 mg in a volume of 3 ml to a 2.5 × 40 cm column of Sepharose 4B and eluting with 0.0125% sodium azide at a flow rate of ~10 ml/hour. The eluate is analysed by the Dubois *et al.* (1956) phenol sulfuric acid method for hexose and the results expressed as absorbance at 490 nm. The three arrows represent respectively, the Vo of the column wherein molecules with MW > 2,000 kDa are excluded from the beads, the elution volume of a marker with a 90 kDa molecular weight and the Vt or total volume of the column. Three pools of eluted material, indicated by the dashed lines, are lyophilized and their hexose content determined. This profile is representative of three runs, each performed with polysaccharide isolated from a different Process A ARF Standard Sample. For more detailed information on this method consult the companion chapter by Waller *et al.* in this volume.

Analysis by Sepharose 4B gel filtration of the isolated polysaccharide gives a single peak with a long and variable tail (Figure 12.4). At least $60 \pm 3\%$ of polysaccharide has a MW of 2,000,000 kDa or above and $83 \pm 3\%$ has a MW above 80 kDa. However, up to 50% of the polysaccharide applied to the top of the column did not elute from the column, presumably because it was too large to easily flow through the bed of beads. However, it could be recovered by removing the top several cm of beads and washing the polysaccharide from them. Thus more than 90% of our ARF Process A gel polysaccharide is of a molecular weight at least ten-fold higher than anything previously described in the literature, excepting the allusion to insolublity in Gowda *et al.* (1979), referred to above. By comparison with the findings of McAnalley in Table 12.9, 10% of ARF Process A gel has a MW of 80 kDA or lower, whereas 53% of experimental Acemannan® and 72% of commercial Acemannan® has a molecular weight of 80 kDa or lower.

On the other hand, the sugar analysis of the polysaccharide isolated from Process A gel (Pelley *et al.*, 1998) gives a composition ($85 \pm 1\%$ mannose, $7 \pm 1\%$ mannose and

Table 12.12 Role of time and cellulase dose in the activation and decay of the CHS response protective activity of *A. barbadensis* ARF Process A gel.

Time after manufacture	Percent protection of CHS from suppression by 2000 Joules/m² UV radiation (number of mice in group)			
	Concentration of cellulase (g. per 215 liters)			
	None	0.5–1.5	2.0–2.5	5.0
3 Weeks	12±6% (10)	16±6% (20)	22±6% (15)	24±9% (10)
4 Weeks	21±5% (10)	27±16% (5)	53±10% (15)	48±9% (10)
5 Weeks	32±10% (10)	46±8% (15)	62±8% (18)	55±18% (10)
6 Weeks	5±10% (10)	31±9% (20)	19±6% (15)	3±7% (10)

Notes
Values summarize data from Table 1 in Strickland *et al.* (1998). Percent protection is calculated according to Formula 2. It should be noted that due to printer's errors, there are multiple shift errors in this table as originally published. The statistical analysis of this data is complex and the interested reader is advised to consult Table 2 in the original publication.

$4\pm1\%$ galactose, means $\pm$SEM, analysis of 15 different isolations of polysaccharide from separate ARF lots) which is in accord with the sugar analyses in the older literature. Thus, in chemical terms we are dealing with the same polysaccharide with which everyone else is dealing. In physical terms, however, it is obvious that this material is substantially undegraded. We found (Strickland *et al.*, 1998) that when fresh Process A gel was first harvested, it was biologically inactive in protecting the CHS immune response against UV injury. Table 12.12 illustrates the composite results from three preparations analysed over a period. Without addition of exogenous 'cellulase' (Table 12.12, 'None' column) there was a marginally significant treatment effect (Prep A, $p=0.026$, Prep B, $p=0.006$, Prep C, $p > 0.05$) although there was a significant 'time' effect (Prep A, $p=0.0001$, Prep B, $p=0.003$, Prep C, $p > 0.0015$). Thus, although there was very little biological activity, something happened over time that was far greater than could be accounted for by experiment to experiment variation.

The surprising absence of therapeutic effect with extremely clean, extremely fresh gel was also seen in the classic phagocytic cell activation BRM assay with aloe. Figure 12.5 illustrates experiments by Dr Sheffield at the University of Wisconsin using materials we supplied. ARF Process A gel, in the absence of 'cellulase' (0 Cellulase) had only a meagre ability to activate macrophages. There was no significant activity at $1\,\mu g/ml$, marginal activity at $10\,\mu g/ml$, and substantial activity only at $100\,\mu g/ml$. These results, and other measurements of activities traditionally ascribed to aloe extracts suggested that the ultra-clean Process A gel with the major polysaccharide in the native form was missing a major factor necessary to activate the extracts. 'Cellulase' treatment is necessary to activate native *A. barbadensis* gel for certain biological activities. We suspected that it was only as the Process A aloe gel was stored over time that it became biologically active (Table 12.12, 'None' column; '4 Weeks', $21\pm5\%$ Protection; '5 Weeks', $32+10\%$) although with prolonged storage, the protective activity eventually decayed

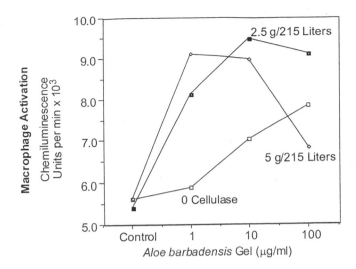

Figure 12.5 Activation of macrophage cultures by incubation with crude, *A. barbadensis* Process A gel processed in the absence of exogenous 'cellulase' or processed with various concentrations (2.5 g per 215 liters or 5 g per 215 liters) of *T. reesei* cellulase. Data of Lewis Sheffield from Sheet 3 (Figure 3) of Strickland *et al.* (1998).

(Table 12.12, 'None' column; '6 Weeks', $5 \pm 10\%$). However, the statistical significance of this effect was weak (Strickland *et al.*, 1998, Table 2, Significance of Treatment Effect, p=0.026, p=0.06, p not significant). However, at the same time, we were studying the process of concentrating *A. barbadensis* gel extracts by a rising-falling thin-film evaporation process that required the addition of 'cellulase' for mechanical efficiency. We noticed that 'cellulase' altered the biological effect in that system (Strickland *et al.*, 1998, Example 1). Therefore we set up specific experiments to determine the effects of cellulase upon activation and decay or CHS protective activity (Table 12.12 above). The use of cellulase accelerated the activation and the decay process in a dose-related fashion. There was now (Strickland *et al.*, 1998, Table 2) a strong statistical therapeutic effect (Prep A, p=0.001 to p=0.0001; Prep B, p=0.005 to p=0.0001; Prep C, p=0.003).

This activation of biological activity was accompanied by physico-chemical changes in the gel polysaccharide. Native Process A gel, prior to being activated, has significant psuedoplasticity, that is, physically it is a gel and its polysaccharide is entirely in a very high molecular weight form (Figure 12.4 above). The process of activation is associated with loss of psuedoplasticity, cleavage of the polysaccharide and a shift in the size exclusion chromatograph to a lower molecular weight profile (Figure 12.6).

Finally, excessive use of cellulase results in the destruction of biological activity. In the parallel experiments with Dr Sheffield referred to above (Figure 12.5), there was also a 'cellulase' effect in his *in vitro* phagocyte activation system. Thus we began to think in terms of a 'cellulase-sensitive' factor that protected the skin immune system against UV injury. We need to stress that this 'cellulase-sensitive' UV-protective activity is not the only epithelial protective activity encountered. There are other, 'cellulase-insensitive' factors with different physico-chemical properties and different mechanisms

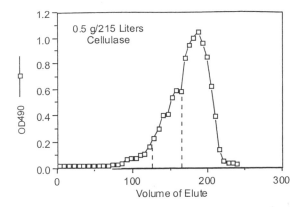

Figure 12.6 Crude Process A *A. barbadensis* gel was incubated with 0.5 g per 215 liters cellulase, the polysaccharide isolated by dialysis /ethanol precipitation and molecular weight distribution determined by gel filtration on Sepharose 4B.

Modified from Sheet 5, Figure 4D of Strickland *et al.* (1998). This material was from the same lot of gel as illustrated in Figure 4 except that it was incubated with 'cellulase'. Subsequent to incubation, the polysaccharide was isolated and analysed by the same method as in Figure 4.

of action. A second factor can be removed by common commercial processing steps. There may even be a third factor, distinct from either of the two above.

Studies on cellulase cleavage of purified native polysaccharide

Investigations to determine the nature of the skin immune system (SIS) UVB-protective and macrophage stimulating molecules revealed a polysaccharide was purified from Process A gel under conditions (see flow chart Figure 5, sheet 7 in Strickland *et al.*, 1998) which resulted in minimal alteration of the gel as demonstrated by the polysaccharide molecular weight distribution analysis in Figure 12.7A below.

The purified polysaccharide was then incubated with various concentrations of partially purified *T. reesei* 'cellulase' and the reaction terminated by alcohol addition, separating the alcohol-soluble oligosaccharides (see flow chart Figure 6, sheet 8 in Strickland *et al.*, 1998; Byeon *et al.*, 1998; Strickland, 1999). This treatment progressively reduced psuedoplasticity (Figure 12.8) and altered the molecular weight distribution profile of the polysaccharide. Instead of observing only the ultra-high molecular weight form of the native polysaccharide, some material appeared that resembled Acemannan® (Figure 12.7B, region eluting between 160 to 200 ml). All the applied polysaccharide was able to migrate through the column. However, the degree of breakdown of polysaccharide was not as extreme as that which occurs in industrial type processing with even the lowest dosage of 'cellulase' (compare Figure 12.7B to Figure 12.6). This observation, that pseudoplasticity is 'broken' and biological activity is generated more readily than Acemannan® structure is digested, has prompted us to propose that there is a highly 'cellulase' sensitive site near the cross-links in the native polysaccharide gel (Figure 12.3).

Studies by Dr Sheffield indicated that the ethanol-insoluble polysaccharide fraction cleaved from native polysaccharide by 1 µg, 3 µg and 10 µg 'cellulase' per 100 mg

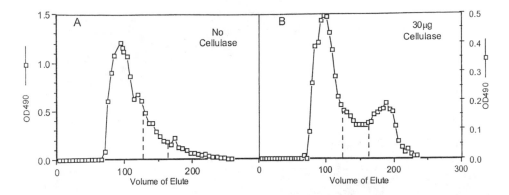

Figure 12.7 Molecular weight distribution of, A. Polysaccharide purified from Process A *A. barbadensis* gel and incubated in buffer and B. Purified polysaccharide incubated for 2.5 hours with 30 µg cellulase per 100 mg polysaccharide.
Modified from Figure 8, Sheets 10 and 11 of Strickland *et al.* (1998) and conducted according to the protocol of Figure 6, Sheet 8. The chromatographs are representative of two runs each, conducted with materials from different preparations. After incubation and prior to chromatography, the polysaccharide was concentrated by precipitation with ethanol.

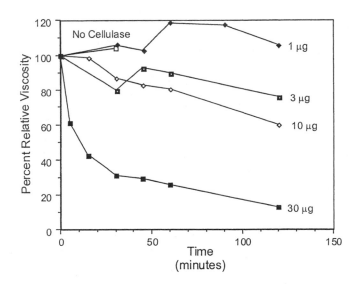

Figure 12.8 Change in relative viscosity of polysaccharide purified from Process A *A. barbadensis* gel and incubated with *T. reesei* cellulase.
Modified from Figure 7, sheet 9 of Strickland *et al.* (1998) and conducted according to the protocol of Figure 6, Sheet 8 using 100 mg portions of polysaccharide.

polysaccharide, contained the phagocyte activating activity (Strickland *et al.*, 1998, Figure 9). The ethanol soluble oligosaccharide fraction contained the CHS protective activity (Strickland *et al.*, 1998).

Native polysaccharide incubated in the absence of cellulase failed to generate activity protective of CHS from suppression by UVB (Table 12.13). Protection was achieved by

Table 12.13 CHS-UVB protective activity of oligosaccharides cleaved from *A. barbadensis* polysaccharide.

Oligosaccharide concentration	Percent protection of CHS from suppression by 2000 Joules/m UV radiation		
	Cellulase/polysaccharide concentration		
	None	3 µg/100 mg	30 µg/100 mg
0.5% (# of Mice)	3±4% (5)	−22±3% (20)	4±8% (15)
0.05% (# of Mice)	32±15% (4)	17±6% (5)	11±7% (5)
0.005% (# of Mice)	10±11% (5)	81±15% (4)	0±8% (5)

Note

Values summarize data from Table 7 in Strickland *et al.* (1998). Percent protection is calculated according to Formula 2. Statistical significance of protection, 3 µg cellulase, 0.05%, p = 0.05; 0.005%, p = 0.015.

oligosaccharides generated by moderate (3 µg/100 mg) 'cellulase' cleavage of native polysaccharide, albeit this factor demonstrated significant suppression at supraoptimal doses. Cleavage of polysaccharide with excessive amounts of cellulase (30 µg/100 mg) resulted in inactive preparations. A similar SIS protective activity was observed with assays utilizing DTH (Byeon *et al.*, 1998, Table I).

Other parallel studies by Dr Sheffield led us to name this novel oligosaccharide activity the cytoprotective oligosaccharide (L. Sheffield, unpublished observations). It was first noted that crude extracts of *A. barbadensis* gel accelerated heat shock recovery, that is the re-entry of cultured cells into the cell cycle after thermal injury. This is measured as DNA synthesis with time of murine mammary epithelial cells in 45°C temperature shift cultures. Our cleavage oligosaccharides were also active in this heat shock protection system. Thus these oligosaccharides appear to have a generalized function in accelerating the recovery of epithelial cells from physical injury – they are cytoprotective oligosaccharides.

In conclusion these studies revealed that aloe oligosaccharides, cleaved from native polysaccharide by 'cellulase,' can protect CHS responses from UV-induced suppression. Aspects of these basic findings have been confirmed by other laboratories (Lee *et al.*, 1999). This cytoprotection by cleavage oligosaccharides led us to explore if other plant complex carbohydrates, associated with a novel mechanism of response to cellular injury, could protect the cutaneous immune responses (Strickland *et al.*, 1999).

Oligosaccharins and other therapeutic complex carbohydrates

At present, our best chemically-defined SIS therapeutic saccharide is the Tamarind-seed xyloglucan polymer made from the nonasaccharide repeating unit (Structure 3) (Strickland *et al.*, 1999).

Structure C

Tamarind Xyloglucan Nine sugar 'monomeric' repeating unit

☐ Glucose   ☐ β 1-4 ☐ β 1-4 ☐ β 1-4 ☐

△ Xylose   α 1-6   α 1-6   α 1-6
                          △   △   △
○ Galactose
                                   α 1-2   α 1-2
                                     ○   ○

Structure 3 The core xyloglucan repeating unit of the Tamarind polysaccharide. Modified from Pelley and Strickland (2000).

This repeating subunit of the Tamarind polysaccharide is an analogue of the classical xyloglucan 'Oligosaccharin' plant growth regulator (reviewed in Albersheim *et al.*, 1992; Darvill *et al.*, 1992). Our finding that a biologically active SIS oligosaccharide in aloe was generated by cleavage of a precursor plant polysaccharide was reminiscent of this biologically active 'Oligosaccharin' generated by cleavage of precursor plant polysaccharide. This led us to test oligosaccharides and their analogues for activity in protecting the skin immune system from UVB-induced injury. Plants utilize the structural complexities of carbohydrates to specifically regulate important physiological processes such as growth, organ formation, and responses to injuries caused by mechanical action or pathogens. A xyloglucan hemicellulose isolated from cultured sycamore cells with an oligosaccharide repeating unit (Figure 6 in Bauer *et al.*, 1973) resembled Structure 3 except that the sycamore xyloglucan oligosaccharide had fucose attached to some of the galactoses that are terminal in the Tamarind repeating unit. Bauer *et al.* (1973) realized at this time the close structural relationship between the cell wall xyloglucan hemicellulose repeating unit oligosaccharide and the Tamarind-seed xyloglucan repeating unit (Structure C above), originally described by Kooiman (1961). Other investigators confirmed the importance of the xyloglucan hemicellulose repeating unit and noted variations to this motif (Mori *et al.*, 1980).

The next advance came in 1984 when York *et al.* (1984) found that the sycamore xyloglucan oligosaccharide could regulate auxin mediated cell wall elongation, a discovery soon to be repeated in other laboratories (McDougall and Fry, 1988). Nanogram quantities of the cell wall hemicellulose saccharides galacturonan and xyloglucan, elicit plant defense responses, trigger cell elongation, and regulate development (Albersheim *et al.*, 1992; Darvill *et al.*, 1992). These discoveries of biological activity of xlyoglucan led to extensive studies of the structure of various xyloglucan oligosaccharides (York *et al.*, 1988; York *et al.*, 1990; Kiefer *et al.*, 1990; Hisamatsu *et al.*, 1992; York *et al.*, 1993) in order to establish structure/function relationships. Other investigators have examined the three-dimensional conformation of the higher order members of the repeating unit series (Gidley *et al.*, 1991), and the fucose containing xyloglucan with the highest biological activity has been systhesized (Sakai *et al.*, 1990). Because of the limitations on the amount of highly purified material available from tissue culture of sycamore cells, enzymatic fragments of Tamarind polysaccharide have been developed for structural study (York *et al.*, 1990; York *et al.*, 1993). Molecular genetics has even been used to confirm some of the crucial structure/function relationships in plants

(Zablackis *et al.*, 1996). Thus, although the aloe oligosaccharides protective of the skin immune response are not yet purified and it may be several years until the structure/function relationship is established, it is possible to begin establishing structure/function relationships using the already characterized family of 'Oligosaccharins.'

Defined saccharides and protection of UVB-injured cutaneous immune responses

On examining a number of oligo- and polysaccharides of defined structure for their effect, in mice, of protection against injury by UV light (Strickland *et al.*, 1999), the most striking finding was the ability of purified, intact Tamarind-seed polysaccharide to protect the DTH immune response from suppression by UVB. This polysaccharide displayed a 10^6-fold activity over crude aloe oligosaccharide (Figure 12.9). This degree of activity is not predicted from theory and represents a 1,000-fold greater *specific activity* than aloe oligosaccharide.

We have found other oligo- and polysaccharides with SIS protective activity but no others with specific activities greater than our aloe lead compound. Since this structure has not been elucidated, structure/function relationships cannot be discussed. However, these findings raise the possibility that there are parallel signaling pathways in plants and animals (Bergey *et al.*, 1996; Stone and Walker, 1995). The production of antibodies against oligosaccharides should help to dissect this complexity (see, e.g. Puhlmann *et al.*, 1994).

Equally important to these positive responses are the negative responses obtained. Also, BRM is a nonspecific effect wherein almost any compound triggers a response. Aloe oligosaccharides are labile so that the CHS-protective effect of the gel declines with storage (Byeon *et al.*, 1998, Figure 1) and an excess of cellulase destroys the biological

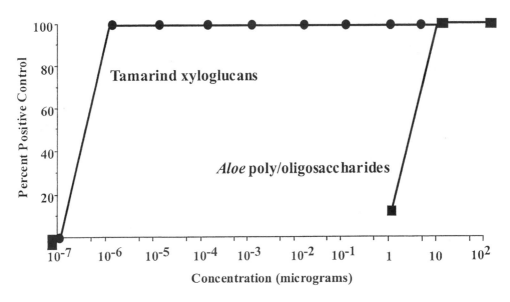

Figure 12.9 Comparative Dose Response curves of DTH-UVB Protection of *A. barbadensis* cleavage oligosaccharide (squares) and Tamarind polysaccharide (circles). Modified from Figure 3, of Strickland *et al.* (1999).

Table 12.14 Role of interleukin-10 in UV radiation-induced suppression of the Skin Immune Response.

Investigators	Finding
Schwarz *et al.*, 1986.	Cytokines are involved in UV-induced suppression.
Kim *et al.*, 1990	Immune suppressive cytokines are produced by epidermal cells.
Ullrich *et al.*, 1990	Immune suppression by cytokines fromUV-injured keratinocytes.
Rivas and Ullrich, 1992	Interleukin-10 production is obligatory for suppression.
Schwarz and Luger, 1992	Review of early work on cytokines and UV-induced suppression.
Enk *et al.*, 1993	Interleukin-10 inhibits antigen presentation by Langerhans cells.
Ullrich, 1994	Interleukin-10 is produced by keratinocytes and alters antigen presentation.
Kang *et al.*, 1994	Epidermal macrophages are major sources of UV-induced IL-10.
Yagi *et al.*, 1996	Th2 T cells produce IL-10 after UV exposure.
Nishigori *et al.*, 1996	Interleukin-10 production is triggered by UV-induced DNA damage.
Niizeki and Streilein, 1997	Interleukin-10 is major cytokine effecting suppression.
Byeon *et al.*, 1998	Downregulation of IL-10 by aloe CPO reduces suppression.
Strickland *et al.*, 1999	Downregulation of IL-10 by Tamarind polysaccharide reduces suppression.

activity of the cytoprotective oligosaccharide (Strickland *et al.*, 1998). Furthermore, structurally unrelated polysaccharides are inactive in the DTH UVB protective system. There is a lack of reactivity in dextrans ($\alpha 1 \rightarrow 4$ and $\alpha 1 \rightarrow 6$ glucose rich molecules) and various analogues of cellulose ($\beta 1 \rightarrow 4$ glucose) whose hydrophobicity is similar to acetylated mannan because of methylation (Figure 2, Strickland, 1999).

Therapeutic saccharides and the role of cytokines

Studies with crude aloe extracts were capable of yielding some mechanistic information. Aloe was not a sunscreen but acted at a step after DNA damage and had a cytoprotective effect on Langerhans cells and epidermal dendritic T cells (Strickland *et al.*, 1994). With the advent of better defined sacchride preparations (Byeon *et al.*, 1998; Strickland *et al.*, 1999) more detailed mechanistic studies become feasible. One of the advantages of using the mouse immune response to model antigens and its suppression by UV radiation is that much is known about the mechanisms involved. Since UV light does not penetrate into deeper layers of skin, systemic suppression of DTH responses are mediated by soluble factors (Ullrich, 1994).

High doses of UV induce systemic suppression DTH responses to proteins, allogenic cells and infectious organisms such as *C. albicans* by affecting different populations of antigen presenting cells (Table 12.14) (see e.g. Saijo *et al.*, 1995, 1996). Suppression also appears to involve the action of soluble agents such as prostaglandin-E2, *cis*-urocanic acid, and the cytokines TNF (tumor necrosis factor)-α, IL (interleukin)-1, IL-10 and IL-12 (Chung *et al.*, 1986; Robertson *et al.*, 1987; Araneo *et al.*, 1989; Vermeer and Steilein, 1990; Schmitt *et al.*, 1995). At times it appears that all 30 interleukins, growth factors and mediators of inflammation that have been described as being produced by keratinocytes (Kondo, 1999) are involved in UVB-induced suppression. Of all of these cytokines, the one that has attracted the most interest is IL-10.

UV-induced, keratinocyte-derived IL-10 systemically suppresses DTH responses to alloantigen indirectly by down-regulating antigen presenting cell functions and triggering the formation of antigen-specific suppressor T cells. Neutralizing anti-IL10 antibody or absorption of IL-10 from the supernatants of UV-irradiated keratinocyte cultures removes their *in vivo* suppressive activity. Besides lymphocytes and keratinocytes, CD11b+ (Mac1+) macrophages, infiltrating the dermis as a part of the inflammatory response in UV irradiated skin are also sources of IL-10. The role of IL-10 in suppression of CHS immunity is less clear since injection of the cytokine blocks the induction of CHS responses in mice but treatment of UV-irradiated mice with anti-IL-10 antibody fails to restore their CHS responses. Crude aloe gel and aloe oligosaccharides reduce the production of IL-10 by UV-irradiated epidermal cells *in vitro* and *in vivo* (Byeon *et al.*, 1998).

Normal murine epidermis has little IL-10 and treatment with aloe oligosaccharides has little effect upon this baseline expression (Figure 12.10, left, top two panels). High dose UV treatment of murine skin induces, four days later, high levels of epidermal IL-10. Immediate treatment of the injured epidermis with aloe oligosaccharides markedly reduces UV-induced expression of IL-10. Similarily (right panels), treatment of the UV irradiated, transformed murine epidermal cell line, Pam (pulmonary alveolar macrophage) 212 with oligosaccharides cleaved from purified aloe polysaccharides reduces IL-10 production by the cells. This down regulation of IL-10 production prevents immuno-suppressive activity of supernatants and blocks the generation of suppressor cells when injected into unirradiated mice (Byeon *et al.*, 1998, Table III). The immunoprotective activity of the oligosaccharides from aloe is not a feature of oligosaccharides in general since neither commercially prepared dextran nor cellulose prevented UV-induced immune suppression.

Similar effects upon UVB-induced expression of IL-10 were observed using Tamarind polysaccharide (Strickland *et al.*, 1999). Treatment of high dose UVB-injured skin prevented the accumulation of IL-10 (Figure 12.11).

We believe that this system offers optimal possibilities for elucidating the structural function relationships of monomer and polymeric oligosaccharides.

CONCLUSION AND FUTURE DIRECTIONS

First, our studies have shown that there are multiple mechanisms operating in UVB-induced suppression of the skin immune system. The wider spectrum of purified, structurally-defined therapeutic ligands we are developing is likely to reveal further complexity of a detailed mechanism. Second, we expect that, together with our collaborators at the University of Georgia, we shall further define the complexity of mechanism of this UVB suppression. This will lead us to the discovery of further novel compounds in the same way as our prior work on the mechanism of aloe led to the discovery of the immune-protective activity of the Tamarind saccharides. Third, our mechanistic studies on the respective roles of CHS, DTH and suppressor cells help us to determine which test best correlates with UVB facilitation of tumor growth. This knowledge will set the stage for selection of the best surrogate for use in future clinical investigations of the efficacy of therapeutic saccharides in humans exposed to UVB. Understanding the separate contributions of CHS, DTH and suppressor T cells to immunity to skin cancer in the context of preventive regimens will further the prevention and treatment of these, the most common human malignancies.

C3H mouse   PAM 212

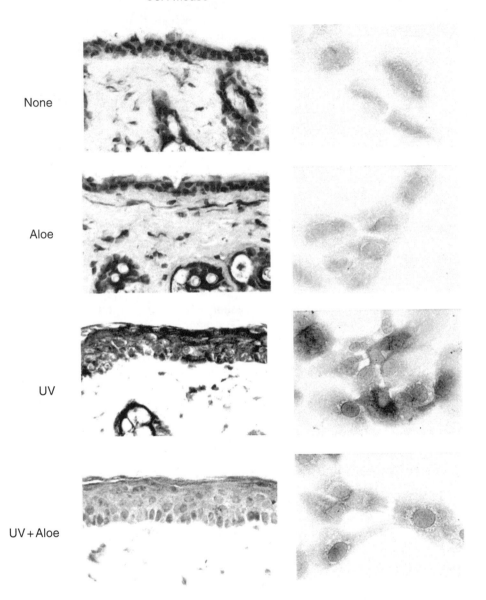

None

Aloe

UV

UV+Aloe

Figure 12.10 Oligosaccharides cleaved from native *A. barbadensis* polysaccharide decrease IL-10 production by UV irradiated murine keratinocytes. From Beyeon *et al.* (1998). Left Panel, Mice were irradiated with 15,000 J/m² UV and treated topically with 500 mg per ml oligosaccharide. Four days later skin was removed and cryosections stained for IL-10 by immunohisto-chemistry. Right Panel, subconfluent cultures of PAM 212 cells were irradiated with 300 J/m² UV and were then treated for 1 hr with 10 mg per ml oligosaccharide. Cells were washed and IL-10 was detected 24 hrs later by immunohistochemistry (*see Colour Plate 8*).

15 kJ/m² UV (no 1° Ab)

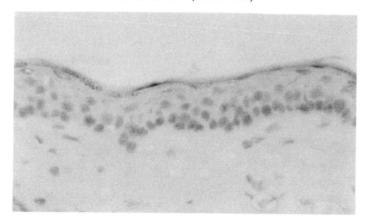

15 kJ/m² UV Only

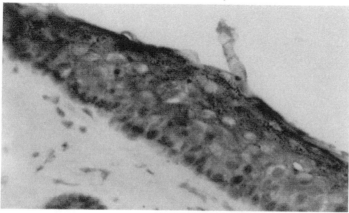

15 kJ/m² UV + Tamarind (1 μg)

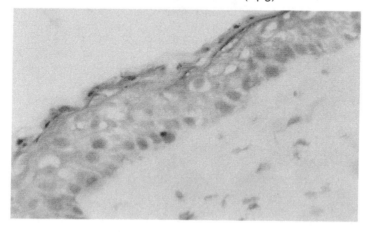

Figure 12.11 Tamarind polysaccharides decrease IL-10 production by UV irradiated murine keratinocytes. From Strikland *et al.* (1999). Mice were irradiated with 15,000 J/m² UV and treated topically with Tamarind polysaccharide. Four days later skin was removed and cryosections stained for IL-10 by immunohistochemistry (*see Colour Plate 9*).

ACKNOWLEDGEMENTS

This work was supported by grants CA70383 (FMS) and CA80423 (RPP) from the
National Institutes of Health. We also wish to acknowledge and thank our colleagues,
Todd Waller, Lew Sheffield and Bob Davis, in the Aloe Research Foundation Inter-
University Co-operative Study. We are also grateful to our collaborator in explorations
of saccharide structure, Alan Darvill at the Center for Complex Carbohydrates, The
University of Georgia.

REFERENCES

Albersheim, P., Darvill, A., Augur, C., Cheong, J-J., Eberhard, S. and Hahn, M.G. *et al.*
(1992) Oligosaccharins: Oligosaccharide regulatory molecules. *Accounts of Chemical Research*,
25, 77–83.

Aleshkina, J.A. and Rostotskii, B.K. (1957) An aloe emulsion – a new medicinal preparation.
Meditsinkaya Promyshlennost USSR, II, 54–55.

Ananthaswamy, H.N., Fourtanier, A., Evans, R.L., Sylvie, T., Medaisko, C., Ullrich, S.E. and
Kripke, M.L. (1998) p53 mutations in hairless mouse skin tumors induced by a solar simulator.
Photochemistry Photobiology, 67, 227–232.

Ananthaswamy, H.N., Laughlin, S.M., Cox, P., Evans, R.L., Ullrich, S.E. and Kripke, M.L.
(1997) Sunlight and skin cancer: inhibition of p53 mutations in UV-irradiated mouse skin
by sunscreens. *Nature Medicine*, 3, 510–514.

Anderson *et al.* (1992) Ultraviolet B dose-dependent inflammation in humans: a reflectance
spectroscopic and laser doppler flowmetric study using topical pharmacologic antagonists on
irradiated skin. *Photodermatology Photoimmunology & Photomedicine*, 9, 17–23.

Araneo, B., Dowell, T., Moon, H. and Daynes, R. (1989) Regulation of murine lymphokine
production *in vivo*: ultraviolet radiation exposure depresses IL-2 and enhances IL-4 production by T cells through an IL-1-dependent mechanism. *Journal of Immunology*, 143, 1737.

Ashley, F.L. Oloughlin, B.J., Peterson, R., Fernandez, L., Stein, H. and Schwartz, A.N. (1957).
The use of Aloe vera in the treatment of thermal and irradiation burns in laboratory animals
and humans. *Plastic and Reconstructive Surgery*, 20, 383–396.

Balch, C.M., Houghton, A.N., Milton, G.W., Sober, A.J. and Soong, S-J. (eds.) (1992)
Cutaneous Melanoma, 2nd edn. Philadelphia, J.B. Lippincott Co.

Bardeesy, N., Wong, K-K., DePinho, R.A. and Chin, L. (2000) Animal models of melanoma:
Recent advances and future prospects. *Advances Cancer Research*, 79, 123–156.

Bauer, W.D., Talmadge, K.W., Keegstra, K. and Albersheim, P. (1973) The structure of plant
cell walls. II. The hemicellulose of the walls of suspension-cultured sycamore cells. *Plant
Physiology*, 51, 174–187.

Bergey, D.R., Howe, G.A. and Ryan, C.A. (1996) Polypeptide signaling for plant defensive
genes exhibit analogies to defense signaling in animals. *Proceedings of the National Academy of
Sciences, USA*, 93, 12056–12058.

Bestak, R., Barneston, St C.R., Nearn, M.R. and Halliday, G.M. (1995) Sunscreen protection
of contact hypersensitivity responses from chronic solar simulated ultraviolet irradiation
correlates with the absorption spectrum of the sunscreen. *Journal of Investigative Dermatology*,
105, 345–340.

Birkeland, S.A., Storm, H.H., Lamm, L.U., Barlow, L., Blohme, L., Forsberg, B., Eklund, B.,
Fjelborg, O., Friedburg, M. and Frodin, L. *et al.* (1995) Cancer risk after renal transplantation
in the Nordic countries, 1964–1986. *International Journal Cancer*, 60, 183–189.

Boring, C.C., Squires, T.S. and Tong, T. (1992) Cancer Statistics. *CA, a Cancer Journal for
Clinicians*, 42, 19.

Brockmeyer, N. and Barthel, B. (1998) Clinical manifestations and therapies of AIDS associated tumors. *European Journal of Medical Research*, **3**, 127–147.

Bucana, C.D., Tang, J.M., Dunner, K., Strickland, F.M. and Kripke, M.L. (1994) Phenotypic and ultrastructural properties of antigen presenting cells involved in contact sensitization of normal and UV-irradiated mice. *Journal of Investigative Dermatology*, **102**, 928–933.

Byeon, S.W., Pelley, R.P., Ullrich, S.E., Waller, T.A., Bucana, C.D. and Strickland, F.M. (1998) Mechanism of action of *Aloe barbadensis* extracts in preventing ultraviolet radiation-induced immune suppression I. The role of keratinocyte-derived IL-10. *Journal of Investigative Dermatology*, **110**, 811–817.

Camplejohn, R.S. (1996) DNA damage and repair in melanoma and non-melanoma skin cancer. *Cancer Survey*, **26**, 193–206.

Chin, L., Pomerantz, J., Polsky, D., Jacobson, M.C., Cohen, C., Cordon-Cardo, C., Horner II, J.W. and DePinho, R.A. (1997) Cooperative effects of *INK4a* and *ras* in melanoma susceptibility *in vivo*. *Genes & Development*, **11**, 2822–2834.

Chung, H.T., Burnhan, D.K., Robertson, B., Roberts, L.K. and Daynes, R.A. (1986) Involvement of prostaglandins in the immune alterations caused by the exposure of mice to ultraviolet radiation. *Journal of Immunology*, **137**, 2478–2484.

Clark, E.A., Golub, T.R., Lander, E.S. and Hynes, R.O. (2000) Genomic analysis of metastasis reveals an essential role for RhoC. *Nature*, **406**, 536–540.

Coats, B.C. (Oct. 18th 1994) Method of processing stabilized Aloe vera gel obtained from the whole Aloe vera leaf. *U.S. Patent* No. 5, 356, 811.

Cole, C.A., Forbes, P.D. and Ludwigsen, K. (1989) Sunscreen testing using the mouse ear model. *Photodermatology*, **6**, 131–136.

Cotton, J. and Spandau, D.F. (1997) Ultraviolet B-radiation dose influences the induction of apoptosis, and p53 in human keratinocytes. *Radiation Research*, **147**, 148–155.

Cruz, P.D. Jr., Nixon-Fulton, J., Tigelaar, R.E. and Bergstresser, P.R. (1989) Disparate effects of in vitro low-dose UVB irradiation on intravenous immunization with purified epidermal cell subpopulations for the induction of contact hypersensitivity. *Journal of Investigative Dermatology*, **92**, 160–165.

Darvill, A., Augur, C., Bergmann, C., Carlson, R.W., Cheong, J., Eberhard, S., Hahn, M.G., Lo, F.M., Marfa, V. and Meyer, B. (1992) Oligosaccharins – oligosaccharides that regulate growth, development and defence responses in plants. *Glycobiology*, **2**, 181–198.

Davis, R.H., Rosenthal, K.J., Cesario, L.R. and Rouw, G.A. (1989) Processed *Aloe vera* administered topically inhibits inflammation. *Journal of the American Podiatric Medical Association*, **79**, 395–397.

Davis, R.H., Kabbani, J.M. and Maro, N.P. (1986) *Aloe vera* and inflammation. *Proceedings of the Pennsylvania Academy of Science*, **60**, 67–70.

Denkins, Y., Fidler, I.J. and Kripke, M.L. (1989) Exposure of mice to UVB radiation suppresses delayed hyper-sensitivity to *Candida albicans*. *Photochemistry Photobiology*, **49**, 615–619.

DePinho, R.A. (2000) The age of cancer. *Nature*, **408**, 248–254.

Dubois, M., Gilles, K.A., Hamilton, J.K., Rebers, P.A. and Smith, F. (1956) Colorimetric method for determination of sugars and related substances. *Analytical Chemistry*, **28**, 350–356.

Early Breast Cancer Trialists' Collaborative Group. (1992) Systemic treatment of early breast cancer by hormonal, cytotoxic or immune therapy. *The Lancet*, **339**, 71–85.

Elmets, C.A., Bergstresser, P.R., Tigelaar, R.E., Wood, P.J. and Streilein, J.W. (1983) Analysis of the mechanism of unresponsiveness produced by haptens painted on skin exposed to low dose ultraviolet radiation. *Journal of Experimental Medicine*, **158**, 781–794.

Enk, A.H., Angeloni, V.L., Udey, M.C. and Katz, S.I. (1993) Inhibition of Langerhans cell antigen-presenting function by IL-10: A role for IL-10 in induction of tolerance. *Journal of Immunology*, **151**, 2390–2398.

Fears, T.R. and Scotto, J. (1983) Estimating increases in skin cancer morbidity due to increases in ultraviolet radiation exposure. *Cancer Investigation*, **1**, 119–126.

Fisher, M.S. and Kripke, M.L. (1982) Suppressor T lymphocytes control the development of primary skin cancers in ultraviolet-irradiated mice. *Science*, 216, 1133–1134.

Gidley, M.J., Lillford, P.J., Rowlands, D.W., Lang, P., Dentini, M., Crescenzi, V., Edwards, M., Fanutti, C. and Reid, J.S.G. (1991) Structure and solution properties of tamarind-seed polysaccharide. *Carbohydrate Research*, 214, 299–314.

Glass, M.J., Bergstresser, P.R., Tigelaar, R.E. and Streilein, J.W. (1990) UVB radiation and DNFB skin painting induce suppressor cells universally in mice. *Journal of Investigative Dermatology*, 94, 273–278.

Gowda, D.C., Neelisiddaiah, B. and Anjaneyalu, Y.V. (1979) Structural studies of polysaccharides from *Aloe vera*. *Carbohydrate Research*, 72, 201–205.

Gowda, D.C. (1980) Structural studies of polysaccharides from *Aloe saponaria* and *Aloe vanbalenii*. *Carbohydrate Research*, 83, 402–405.

Green, A., Siskind, V., Bain, C. and Alexander, J. (1985) Sunburn and malignant melanoma. *British Journal of Cancer*, 51, 393–397.

Grin-Jorgensen, C.M., Rigel, D.S. and Friedman, R.J. (1992) The worldwide incidence of malignant melanoma. *Cutaneous Melanoma*, 2nd edn, edited by Balch, C.M., A.N. Houghton, G.W. Milton, A.J. Sober and S-J Soong. Philadelphia, J.B. Lippincott Co.

Grindlay, D. and Reynolds, T. (1986) The Aloe vera phenomenon: A review of the properties and modern uses of the leaf parenchyma gel. *Journal of Ethanopharmacology*, 16, 117–151.

Hadden, J.W. (1999) Review article: The Immunology and immunotherapy of breast cancer: An update. *International Journal of Immunopharmacology*, 21, 79–101.

Heggers, J.P., Pelley, R.P. and Robson, M.C. (1993) Beneficial effects of *Aloe* in wound healing. *Phytotherapy Research*, 7, S48–S52.

Hisamatsu, M., York, W.S., Darvill, A.G. and Albersheim, P. (1992) Characterization of seven xyloglucan oligosaccharides containing from seventeen to twenty glycosyl residues. *Carbohydrate Research*, 227, 45–71.

Jeevan, A., Evans, R., Brown, E.L. and Kripke, M.L. (1992) Effects of local ultraviolet irradiation on infections of mice with *Candida albicans, Mycobacterium bovis* BCG and *Schistosoma mansomi*. *Journal of Investigative Dermatology*, 99, 59–64.

Kang, K., Hammerberg, C., Meunier, L. and Cooper, K.D. (1994) CD11b[+] macrophages that infiltrate human epidermis after *in vivo* ultraviolet exposure potently produce IL-10 and represent the major secretory source of epidermal IL-10 protein. *Journal of Immunology*, 153, 5256–5264.

Kiefer, L.L., York, W.S., Darvill, A.G. and Albersheim, P. (1990) Structural characterization of an arabinose-containing heptadecasaccharide enzymically isolated from sycamore extracellular xyloglucan. *Carbohydrate Research*, 197, 139–158.

Kim, T.Y., Kripke, M.L. and Ullrich, S.E. (1990) Immunosuppression by factors released from UV-irradiated epidermal cells: selective effects on the generation of contact and delayed hypersensitivity after exposure to UVA or UVB radiation. *Journal of Investigative Dermatology*, 94, 26–32.

Klein, G. and Boon, T. (1993) Tumor immunology: Present perspectives. *Current Opinions in Immunology*, 5, 697–692.

Kondo, S. (1999) The roles of keratinocyte-derived cytokines in the epidermis and their possible responses to UVA radiation. *Journal of Investigative Dermatology, Symposium Proceedings*, 4, 177–183.

Kooiman, P. (1961) The constitution of *Tamarindus*-amyloid. *Recueil Travaux Chimiques des Pays-Bas*, 80, 849–865.

Kripke, M.L. (1974) Antigenicity of murine skin tumors induced by ultraviolet light. *Journal of the National Cancer Institute*, 53, 1333–1336.

Kripke, M.L. (1979) Speculations on the role of ultraviolet radiation in the development of malignant melanoma. *Journal of the National Cancer Institute*, 63, 541–548.

Kripke, M.L. (1984) Immunological unresponsiveness induced by ultraviolet radiation. *Immunological Reviews*, 80, 87–102.

Kripke, M.L., Cox, P.A., Alas, L.G. and Yarosh, D.B. (1992) Pyrimidine dimers in DNA initiate systemic immunosuppression in UV-irradiated mice. *Proceedings of the National Academy of Sciences, USA*, **89**, 7516–7520.

Kuerbitz, S.J., Plunkett, B.S., Walsh, W.V. and Kastan, M.B. (1992) Wild-type p53 is a cell cycle checkpoint determinant following irradiation. *Proceedings of the National Academy of Sciences, USA*, **89**, 7491–7495.

Kusewitt, D.F. and R.D. Ley (1996) Animal models of melanoma. *Cancer Survey*, **26**, 35–70.

Langley, R.G.B. and Sober, A.J. (1997) A clinical review of the evidence for the role of ultraviolet radiation in the etiology of cutaneous melanoma. *Cancer Investigations*, **15**, 561–567.

Lee, C.K., Han, S.S., Shin, Y.K., Chung, M.H., Park, Y.I., Lee, S.K. and Kim, Y.S. (1999) Prevention of ultraviolet radiation-induced suppression of contact hypersensitivity by *Aloe vera* gel components. *International Journal of Immunopharmacology*, **21**, 303–310.

Leigh, I.M., Newton-Bishop, J.A. and Kripke, M.L. (eds) (1996) *Cancer Surveys*, vol. 26.: *Skin Cancer*. Cold Springs Harbor, NY. Cold Springs Harbor Laboratory Press.

Lynch, D.H., Gurish, M.F. and Daynes, R.A. (1981) Relationship between epidermal Langerhans cell density, ATPase activity and the induction of contact hypersensitivity. *Journal of Immunology*, **126**, 1892–1897.

Mandal, G. and Das, A. (1980a) Structure of the glucomannan isolated from the leaves of *Aloe barbadensis* Miller. *Carbohydrate Research*, **87**, 249–256.

Mandal, G. and Das, A. (1980b) Structure of the D-galactan isolated from *Aloe barbadensis* Miller. *Carbohydrate Research*, **86**, 247–257.

Mandal, G., Ghosh, R. and Das, A. (1983) Characterization of polysaccharides of *Aloe barbadensis* Miller: Part III – Structure of an acidic oligosaccharide. *Indian Journal of Chemistry*, **22B**, 890–893.

Manna, S., McAnalley, B.H. and Ammon, H.L. (1993) 2, 3, 4-Tri-*O*-mannopyranose, an artifact produced during carbohydrate analysis. A total synthesis of 2,3,5-tri-*O*-acetyl-1, 6-anhydro-β-D-mannofuranose. *Carbohydrate Research*, **243**, 11–27.

Markiewicz, M.A. and Gajewski, T.F. (1999) The immune system as anti-tumor sentinel: Molecular requirements for an anti-tumor immune response. *Critical Reviews in Oncogenesis*, **10**, 247–260.

McAnalley, B.H. (Apr. 5th 1988) Process for preparation of Aloe products, products produced thereby and compositions thereof. *U.S. Patent* No. 4,735,935.

Other related patents are *U.S. Patent* Nos. 4,851,224 of Jul. 1989, 4,959,214 of Sept. 1990, and 4,966,892 of Oct. 1990.

The patents subsequent to the '935 patent (the '224, '214, and '892 patents) are *Divisionals* and *Continuations In Part* of the '935 patent. As such, the scientific content (*The Specification*) is virtually identical to that of the '935 patent but The Claims of these patents differ very significantly from the '935 patent (legally that is what is important).

McAnalley, B.H., Carpenter, R.H. and McDaniel, H.R. (Aug. 15th 1995) Uses of Aloe products. *U.S. Patent* No. 5,441,943.

Other related patents are *U.S. Patent* Nos. 5,118,673, Jun. 1992 and 5,106,616, Apr. 1992.

These patents, although legally linked to the series of patents above as *Divisionals* and *Coninuations In Part*, have *Specification* sections (the part that contains the data), completely different from the '935 series of patents above. As such, any one of this series of patents could be equally read by scientists seeking information different from the '935 series above.

McDougall, G.J. and Fry, S.C. (1988) Inhibition of auxin-stimulated growth of pea stem segments by a specific nonasaccharide of xyloglucan. *Planta*, **175**, 412–416.

Mori, M., Eda, S. and Kato, K. (1980) Structural investigation of the arabinoxyloglucan from *Nicotiana tabacum. Carbohydrate Research*, **84**, 125–135.

Mukhtar, H. and Ahmad, N. (1999) Green tea in chemoprevention of cancer. *Toxicologic Science*, **52(S)**, 111–117.

Muller, H.K., Bucana, C.D., Kripke, M.L., Cox, P.A., Saijo, S. and Strickland, F.M. (1995) Ultraviolet irradiation of murine skin alters cluster formation between lymph node dendritic cells and specific T lymphocytes. *Cellular Immunology*, 157, 263–276.

Nataraj, A.J., Trent, J. and Ananthaswamy, H.N. (1995) p53 gene mutations and photocarcinogenesis. *Photochemistry Photobiology*, 62, 218–230.

Newton, L.E. (1979) In defence of the name *Aloe vera*. *The Cactus and Succulent Journal of Great Britain*, 41, 29–30.

Niizeki, H. and Streilein, J.W. (1997) Hapten-specific tolerance induced by acute, low-dose ultraviolet B radiation of skin is mediated via interleukin-10. *Journal of Investigative Dermatology*, 109, 25.

Nishigori, C., Yarosh, D.B., Ullrich, S.E., Vink, A.A., Bucana, C.D., Roza, L. and Kripke, M.L. (1996) Evidence that DNA damage triggers Interleukin-10 cytokine production in UV-irradiated murine keratinocytes. *Proceedings of the National Academy of Sciences, USA*, 93, 10354–10357.

Noonan, F.P., Bucana, C., Sauder, D.N., DeFabo, E.C. (1984) Mechanism of systemic immune suppression by UV radiation in vivo. II. The UV effects on number and morphology of epidermal Langerhans cells and the UV-induced suppression of contact hypersensitivity have different wavelength dependencies. *Journal of Immunology*, 132, 2408–2416.

Noonan, F.P., DeFabo, E.C. and Kripke, M.L. (1981) Suppression of contact hypersensitivity by UV radiation and its relationship to UV-induced suppression of tumor immunity. *Photochemistry Photobiology*, 34, 683–689.

Okamoto, H. and Kripke, M.L. (1987) Effector and suppressor circuits of the immune response are activated in vivo by different mechanisms. *Proceedings of the National Academy of Sciences, USA*, 84, 3841–3845.

Ouhtit, A., Muller, H.K., Davis, D.W., Ullrich, S.E., McConkey, D. and Ananthaswamy, H.N. (2000) Temporal events in skin injury and the early adaptive responses in ultraviolet-irradiated mouse skin. *American Journal of Pathology*, 156, 201–207.

Pai, S.A., Cheung, M.C.P., Romsdahl, M.M., Multani, A.S. and Pathak, S. (1999) Can genetic instability be studied at a single chromosome level in cancer cells? Evidence from human melanoma cells. *Cancer Genetics, Cytogenetics*, 109, 51–57.

Pathak, MA. (1991) Sunscreens: Principles of photoprotection. In *Pharmacology of the Skin*, edited by H. Mukhtar, p229. Boca Raton, Florida: CRC Press.

Paulsen, B.S., Fagerheim, E. and Overbye, E. (1978) Structural studies of the polysaccharides from *Aloe plicatilis* Miller. *Carbohydrate Research*, 60, 345–351.

Pawelec, G., Zeuthen, J. and Kiessling, R. (1997) Escape from host-anti-tumor immunity. *Critical Reviews in Oncogenesis*, 8, 111–145.

Pawelec, G., Rees, R.C., Kiessling, R., Madrigal, A., Dodi, A., Baxevanis, C., Gambercorti-Passareni, C., Masucci, G. and Zeuthen, J. (1999) Cells and cytokines in immunotherapy and gene therapy of cancer. *Critical Reviews in Oncogenesis*, 10, 83–127.

Pelley, R.P., Wang, Y-T. and Waller, T.A. (1993) Current status of quality control of *Aloe barbadensis* extracts. *SOFW Journal*. 119, 255–163.

Pelley, R.P., Martini, W.J., Liu, D.Q., Rachui, S., Li, K.-M., Waller, T.A. and Strickland, F.M. (1998) Multiparameter testing of commercial 'Aloe vera' materials and comparison to *Aloe barbadensis* Miller extracts. *Subtropical Plant Science Journal*, 50, 1–14.

Pelley, R.P. and Strickland, F.M. (2001) Plants, polysaccharides and the treatment and prevention of neoplasia. *Critical Reviews in Oncogenesis* 12, 1–37 (*in press*).

Pierceall, W.E., Goldberg, L.H., Tainsky, M.A., Mukhopadhyay, T. and Ananthaswamy, H.N. (1991a) *Ras* gene mutation and amplification in human nonmelanoma skin cancers. *Molecular Carcinogenesis*, 4, 196–202.

Pierceall, W.E., Mukhopadhyay, T., Goldberg, L.H., and Ananthaswamy, H.N. (1991b). Mutations in p53 tumor suppressor gene in human squamous cell carcinomas. *Molecular Carcinogesis*, 4, 445–449.

Puhlmann, J., Bucheli, M.J., Dunning, N., Albersheim, P., Darvill, A.G. and Hahn, M.G. (1994) Generation of monoclonal antibodies against plant cell-wall polysaccharides. I. Characterizaztion of a monoclonal antibody to a terminal α (1-2)-linked fucosyl-containing epitope. *Plant Physiology*, 104, 699–710.

Radjabi, F., Amar, C. and Vilkas, E. (1983) Structural studies of the glucomannan from *Aloe vahombe*. *Carbohydrate Research*, 116, 166–170.

Radjabi-Nassab, F., Ramiliarison, C., Monneret, C. and Vilkas, E. (1984) Further studies of the glucomannan from *Aloe vahombe* (liliaceae). II. Partial hydrolyses and NMR 13C studies. *Biochimie*, 66, 563–567.

Reeve *et al.* (1995) The protective effect of indomethacin on photocarcinogenesis in hairless mice. *Cancer Letters*, 95, 213–219.

Rivas, J. and Ullrich, S.E. (1992) Systemic suppression of delayed-type hypersensitivity by supernatants from UV-irradiated keratinocytes: An essential role for keratinocyte-derived interleukin-10. *Journal of Immunology*, 149, 3865–3871.

Robertson, B., Gahring, L., Newton, R. and Daynes., R.A. (1987) In vivo administration of IL-1 to normal mice depresses their capacity to elicit contact hypersensitivity responses: Prostaglandins are involved in this modification of immune function. *Journal of Investigative Dermatology*, 88, 380–387.

Roboz, E. and Haagen-Smit, A.J. A mucilage from Aloe vera. *Journal of the American Chemical Society*, 70, 3248–3249.

Rodriguez-Bigas, M., Cruz, N.I. and Suarez. (1988). Comparative evaluation of Aloe vera in the management of burn wounds in guinea pigs. *Plastic and Reconstructive Surgery*, 81, 386–389.

Romerdahl, C.A., Stephens, L.C., Bucana, C. and Kripke, M.L. (1989) Role of ultraviolet radiation in the induction of melanocytic skin tumors in inbred mice. *Cancer Communications*, 1, 209–216.

Ross, G.D., Vetvicka, V., Yan, J., Xia, Y. and Vetvickova, J. (1999) Therapeutic intervention with complement and beta-glucan in cancer. *Immunopharmacology*, 42, 61–74.

Rovatti, B. and Brennan, R.J. (1959). Experimental thermal burns. *Industrial Medicine and Surgery*, 28, 364–368.

Sakai, K., Nakahara, Y. and Ogawa, T. (1990) Total synthesis of nonasaccharide repeating unit of plant cell wall xyloglucan: An endogenous hormone which regulates cell growth. *Tetrahedron Letters*, 21, 3035–3038.

Sauder, D.N., Tamaki, K., Moshell, A.N., Fujiwara, H. and Katz, S.I. (1981) Induction of tolerance to topically applied TNCB using TNP-conjugated ultraviolet light-irradiated epidermal cells. *Journal of Immunology*, 127, 261–265.

Saijo, S., Bucana, C.D., Ramirez, K.M., Cox, P.A., Kripke, M.L. and Strickland, F.M. (1995) Deficient antigen presentation and Ts induction are separate effects of UV radiation. *Cellular Immunology*, 164, 189–202.

Saijo, S., Kodari, E., Kripke, M.L. and Strickland, F.M. (1996) UVB irradiation decreases the magnitude of the Th1 response to hapten but does not increase the Th2 response. *Photodermatology, Photoimmunology, Photomedicine*, 12, 145–153.

Schmitt, D.A., Owen-Schaub, L. and Ullrich, S.E. (1995) Effect of IL-12 on immune suppression and suppressor cell induction by ultraviolet radiation. *Journal of Immunology*, 154, 5114–5120.

Schreiber, H. (1999) Tumor immunology. In *Fundamental Immunology*, edited by W.E. Paul, pp. 1237–1270. Philadelphia: Lippincott-Raven.

Schwarz, T., Urbanska, A., Gschnait, F. and Luger, T.A. (1986) Inhibition of the induction of contact hypersensitivity by a UV-mediated epidermal cytokine. *Journal of Investigative Dermatology*, 87, 289–291.

Schwarz, T. and Luger, T.A. (1992) Pharmacology of cytokines in the skin. In *Pharmacology of the Skin*, edited by H. Mukhtar, p. 283. Boca Raton, Florida: CRC Press.

Segal, A., Taylor, J.A. and Eoff, J.C. (1968) A reinvestigation of the polysaccharide material from *Aloe vera* mucilage. *Lloydia*, 31, 423–424.

Shreedhar, V.K., Pride, M.W., Sun, Y., Kripke, M.L. and Strickland, F.M. (1998) Origin and characteristics of UV-B radiation-induced suppressor T lymphocytes. *Journal of Immunology*, 161, 1327–1335.

Simon, J.C., Tigelaar, R.E., Bergstresser, P.R., Edelbaum, D. and Cruz, P.D. (1991) Ultraviolet B radiation converts Langerhans cell from immunogenic to tolerogenic antigen-presenting cells. *Journal of Immunology*, 146, 485–491.

Simon, J.C., Mosmann, T., Edelbaum, D., Schopf, E., Bergstresser, P.R. and Cruz, P.D. (1994) *In vivo* evidence that ultraviolet B-induced suppression of allergic contact sensitivity is associated with functional inactivation of Th1 cells. *Photodermatology Photoimmunology Photomedicine*, 10, 206.

Smith, C., Lilly, S., Mann, K.P., Livingston, E., Myers, S., Kim, H., Lyerly, H.K. and Miralles, D. (1998) AIDS-related malignancies. *Annals of Medicine*, 30, 323–344.

Solar, S., Zeller, H., Rasolofonirina, N., Coulanges, P., Ralamboranto, L., Andriat-Simahavandy, A.A., Rakotovao, L.H. and Deaut, J.Y. (1979) Mis en evidence et etude des proprietes immuno-stimulantes d'un extrait isole et partiellement purifie a partir d'Aloe vahombe. *Archives Institute Pasteur Madagascar*, 47, 1–31.

Stone, J.M. and Walker, J.C. (1995). Plant protein kinase families and signal transduction. *Plant Physiology*, 108, 451.

Strickland, F.M., Pelley, R.P. and Kripke, M.L. (1994) Prevention of ultraviolet radiation-induced suppression of contact and delayed hypersensitivity by *Aloe barbadensis* gel. *Journal of Investigative Dermatology*, 102, 197–204.

Strickland, F.M. and Kripke, M.L. (1997) Immune response associated with nonmelanoma skin cancer. *Clinics in Plastic Surgery*, 24, 637–647.

Strickland, F.M., Pelley, R.P. and Kripke, M.L. (Oct. 20th 1998) Cytoprotective oligosaccharide from Aloe preventing damage to the skin immune system by UV radiation. *U.S. Patent* No. 5, 824, 659.

Strickland, F.M., Darvill, A., Albersheim, P., Eberhard, S., Pauly, M. and Pelley, R.P. (1999) Inhibition of UV-induced immune suppression and Interleukin-10 production by plant polysaccharides. *Photochemistry Photobiology*, 69, 141–147.

Strickland, F.M., Muller, H.K., Stephens, L.C., Bucana, C.D., Donawho, C.K., Sun, Y. and Pelley, R.P. (2000) Induction of primary cutaneous melanomas in C3H mice by combined treatment with ultraviolet radiation, ethanol and aloe emodin. *Photochemistry Photobiology*, 72, 407–414.

Strom, S.S. and Yamamura, Y. (1999) Epidemiology of nonmelanoma skin cancer. *Clinics in Plastic Surgery*, 24, 627–636.

Stutman, O. (1974) Tumor development after 3-methylcolanthrene in immunologically deficient athymic-nude mice. *Science*, 183, 534–536.

Stutman, O. (1979) Chemical carcinogenesis in nude mice: comparison between nude mice from homozygous matings and heterozygous matings and effect of age and carcinogen dose. *Journal of the National Cancer Institute*, 62, 353–358.

Tam, A., Pomerantz, J., O'Hagan, R. and Chin, L. (1999) Tetracyclin-regulatable expression of Ras in mouse melanocytes and melanomas. In *Cancer Biology of the Mutant Mouse: New Methods, New Models, New Insights. Proceedings of the Keystone Symposium*, Abstract B-7.

Toews, G.B., Bergstresser, P.R. and Streilein, J.W. (1980) Epidermal Langerhans cell density determines whether contact hypersensitivity or unresponsiveness follows skin painting with DNFB. *Journal of Immunology*, 124, 445–453.

Ullrich, S.E. (1994) Mechanisms involved in the systemic suppression of antigen-presenting cell function by UV irradiation: keratinocyte-derived IL-10 modulate antigen-presenting cell function of splenic adherent cells. *Journal of Immunology*, 152, 3410–3416.

Ullrich, S.E. (1986) Suppression of the immune response to allogenic histocompatibility antigens by a single exposure to UV radiation. *Transplantation*, 42, 287–291.

Ullrich, S.E., McIntyre, B.W. and Rivas, J.M. (1990) Suppression of the immune response alloantigen by factors released from ultraviolet-irradiated keratinocytes. *Journal of Immunology*, **132**, 489–498.

Ullrich, S.E., Yee, G.K. and Kripke, M.L. (1986) Suppressor lymphocytes induced by epicutaneous sensitization of UV-irradiated mice control multiple immunological pathways. *Journal of Immunology*, **158**, 185–190.

Urbach, F. (1978) Evidence and epidemiology of ultraviolet-induced cancers in man. *National Cancer Institute Monograph*, **50**, 5.

van der Leun, J.C. and Tevini, M. (November 1991) *Environmental effects of ozone depletion: 1991 update. United Nations Environment Programme, Panel Report.*

Vermeer, M. and Streilein, J.W. (1990) Ultraviolet-B-light induced alterations in epidermal Langerhans cells are mediated in part by tumor necrosis factor-alpha. *Photodermatology Photoimmunology Photomedicine*, 7, 258–265.

Vink, A.A., Strickland, F.M., Bucana, C., Cox, P.A., Roza, L., Yarosh, D.B. and Kripke, M.L. (1996) Localization of DNA damage and its role in altered antigen-presenting cell function in ultraviolet-irradiated mice. *Journal of Experimental Medicine*, **183**, 1491–1500.

Waller, T.A., Strickland, F.M. and Pelley, R.P. (1994) Quality control and biological activity of Aloe barbadensis extracts useful in the cosmetic industry. In *Cosmetics and Toiletries Manufacture Worldwide* edited by M.Caine, pp. 64–80. London: Aston Publishing Group.

Weber, R.S., Lippman, S.M. and McNeese M.D. (1990) Advanced basal and squamous cell carcinomas of the head and neck. In *Carcinomas of the Head and Neck: Evaluation and Management*, edited by C. Jacobs, pp.61–81. Norwell: Kluwar Academic.

Weber, R.S. (ed.). (1995) *Basal and Squamous Cell Carcinoma Skin Cancers of the Head and Neck.* Baltimore: Williams & Wilkins.

Welsh, E.L. and Kripke, M.L. (1990) Murine Thy-1 dendritic epidermal cells induce immunologic tolerance in vivo. *Journal of Immunology*, **144**, 883–891.

Williams, M.S., Burk, M., Loprinzi, C.L., Hill, M., Schomberg, P.J. and Nearhood, K. *et al.* (1996) Phase III double-blind evaluation of an Aloe vera gel as a prophylatic agent for radiation-induced skin toxicity. *International Journal of Radiation Oncology Biology and Physics*, **36**, 345–349.

Wolf, P. and Kripke, M.L. (1996) Sunscreens and immunosuppression. *Journal of Investigative Dermatology*, **106**, 1152–1154.

Wolf, P., Donawho, C.K. and Kripke, M.L. (1993) Analysis of the protective effects of different sunscreens on ultraviolet radiation-induced local and systemic suppression of contact hypersensitivity and inflammatory responses in mice. *Journal of Investigative Dermatology*, **100**, 254–259.

Yagi, A., Makino, K., Nishioka, I. and Kuchino, Y. (1977) Aloe mannan, polysaccharide, from *Aloe arborescens* var. *natalensis. Planta Medica*, **31**, 17–20.

Yagi, A., Hamada, K., Mihashi, K., Harada, N. and Nishioka, I.. (1984) Structure determination of polysaccharides in *Aloe saponaria* (Hill.) Haw. (Liliaceae). *Journal of Pharmaceutical Science*, **73**, 62–65.

Yagi, A., Nishimura, H., Shida, T. and Nishioka, I. (1986) Structure determination of polysaccharides in *Aloe arborescens* var. *natalensis. Planta Medica*, 213–218.

Yagi, H., Tokura, Y., Wakita, H., Furukawa, F. and Takigawa, M. (1996) TCRV-7[+] Th2 cells mediate UVB-induced suppression of murine contact photosensitivity by releasing IL-10. *Journal of Immunology*. **156**, 1824–1831.

Yarosh, D.B. and Yee, V. (1990) SKH-1 hairless mice repair UV-induced pyrimidine dimers in epidermal DNA. *Journal of Photochemistry Photobiology*, 7, 173–179.

York, W.S., Darvill, A.G. and Albersheim, P. (1984) Inhibition of 2,4-dichlorophenoxyacetic acid-stimulated elongation of pea stem segments by a xyloglucan oligosaccharide. *Plant Physiology*, **75**, 295–297.

York, W.S., Oates, J.E., van Halbeek, H., Darvill, A.G. and Albersheim, P. (1988) Location of the O-acetyl substituents on a nonasaccharide repeating unit of sycamore extracellular xyloglucan. *Carbohydrate Research*, **173**, 113–132.

York, W.S., van Halbeek, H.M., Darvill, A.G. and Albersheim, P. (1990) Structural analysis of xyloglucan oligosaccharides by ^{1}H-n.m.r. spectroscopy and fast atom-bombardment mass spectrometry. *Carbohydrate Research*, **200**, 9–31.

York, W.S., Harvey, L.K., Guillen, R., Albersheim, P. and Darvill, A.G. (1993) Structural analysis of tamarind seed xyloglucan oligosaccharides using β glactosidase digestion and spectroscopic methods. *Carbohydrate Research*, **248**, 285–301.

Young, R.C. (2000) The sounds of silence: No news about cancer prevention is good news. *Nature*, **408**, 141.

Zablackis, E., York, W.S., Pauly, M., Hantus, S., Reiter, W-D., Chapple, C.C.S., Albersheim, P. and Darvill, A. (1996) Substitution of L-fucose by L-galactose in cell walls of *Arabidopsis* mur1. *Science*, **272**, 1808–1810.

Zerp, S.F., Van Elsas, A., Peltenburg, L.T.C. and Schrier, P.I. (1999) P53 mutations in human cutaneous melanoma correlates with sun exposure but are not always involved in melanomagenesis. *British Journal of Cancer*, **79**, 921–926.

13 Aloes and the immune system

Ian R. Tizard and Lalitha Ramamoorthy

ABSTRACT

There is a moderate scientific literature on the immunological effects of extracts from plants of the genus Aloe. Unfortunately, it is difficult to assess the significance of many of these studies because of two problems. First, most studies have been undertaken using many different, poorly characterized, complex aloe extracts. Second, studies have been performed using several different *Aloe* species, making comparisons impossible. Although anecdotal reports describe a wide variety of both immunostimulating and immunosuppressive effects, controlled scientific studies have substantiated very few of these. Most studies that have been performed have focused on the clear mesophyll gel of the *Aloe vera* leaf and on its major storage carbohydrate, acetylated mannan (acemannan). Recently a unique pectin has been isolated from aloe mesophyll cell walls and appears to have unique and important properties. Some consistent properties have, however, been noted. Thus aloe gel extracts and partially purified acemannan preparations have mild anti-inflammatory activity and multiple possible pathways for this activity have been investigated. Aloe extracts also have some limited macrophage activating properties. These include the release of nitric oxide and the secretion of multiple cytokines. This macrophage activation may account for the effects of aloe extracts on wound healing, bone marrow stimulation, and their limited anti-cancer effects. Studies have also provided evidence to suggest that aloe extracts can influence apoptosis and lymphocyte function. The Madagascar species, *Aloe vahombe*(sic), has been claimed to possess a very wide array of beneficial activities, but this has not been independently confirmed.

INTRODUCTION

Two features of the *Aloe* family make them somewhat unique in the field of medicinal plants. First, there is no agreement on which ailments they might cure and second, there is no agreement on the components within the aloe plant that may exert beneficial medicinal activity. Anecdotal reports describe a bewildering array of both immuno-stimulating and immunosuppressive effects. Few controlled studies have supported these claims and to this day, much uncertainty exists as to whether aloe extracts actually have significant beneficial activity. The controlled studies that have been undertaken have tended, unfortunately, to use complex mixtures or partially purified components which makes comparative analysis difficult. Nevertheless, they have tended to support a very limited range of effects on the cells of the immune system. Indeed, the recognized effects of aloe extracts appear primarily to affect innate immune mechanisms such as inflammation, rather than acquired immunity. They also tend to be quantitatively minor in nature. Nevertheless, taken as a whole, they may well account for the 'good press' that aloes have received over the years and for the continuing interest in the use of aloe extracts as immunomodulating and anti-inflammatory agents.

Although there are more than 400 *Aloe* species recognized, one plant, *Aloe vera* (L.) Burm. f., dominates both the commercial aloe market and the research literature. (This species is sometimes referred to by the extinct name, *A. barbadensis* Miller) Other *Aloe* species that have been extensively examined, especially in Asia, include *A. arborescens* Miller, and *A. saponaria* (Ait.)Haw. Such limited comparative studies that have been undertaken have shown that the biochemical composition of extracts from different *Aloe* species varies widely (Viljoen and van Wyk, 1998). As a result, biological responses obtained with material from one species, cannot automatically be ascribed to others. Nor have any studies been undertaken to determine whether factors such as geographical location, soil quality, fertilization or plant genetics influence biological activity. It is widely believed that the treatment and storage of the harvested leaves prior to, and during, processing influences its biological activity, and although processing has been the subject of much rhetoric between competing suppliers, controlled studies have not been undertaken. The products still hover on the edge of scientific respectability (Reynolds and Dweck, 1999).

Aloes are succulent plants with characteristic thick fleshy leaves. The outer rind contains the bitter yellow sap that originates in the bundle sheath cells and is used for its purgative effects. The center of these leaves consists of a clear mesophyll gel. The mucilaginous texture of the mesophyll gel ensures that it is intrinsically soothing. As a result it has been used to treat superficial dermal inflammation resulting from a wide variety of causes. The gel has been especially useful in the treatment of radiation burns ranging from sunburn to X-ray and radium burns (Sato *et al.*, 1990; Roberts and Travis, 1995; Wright, 1936). Despite the anecdotal nature of much of this work, results are generally positive and the gel is widely employed. Indeed, it is rare to find a commercial 'soothing' preparation that does not contain 'Aloe vera' as one of its components (Grindlay and Reynolds, 1986; Reynolds and Dweck, 1999). It may also have some ill-defined antimicrobial properties (Bruce, 1967; Soeda *et al.*, 1966).

Aloe extracts are commonly made from the whole leaf, or alternatively, from the clear mesophyll gel. The composition of these two extracts differs significantly since the thick photosynthetic rind contains many components not present within the mesophyll. *Aloe vera* leaves are, however, little different from other plants in their basic composition. They consist primarily of a carbohydrate mixture with some proteins and a host of minor components such as steroids, anthraquinones, flavonoids, and chromones (Holdsworth, 1971, 1972; Hutter *et al.*, 1996; Hirata and Suga, 1977; Makino *et al.*, 1974), as well as enzymes such as carboxypeptidases (Fujita *et al.*, 1979), superoxide dismutase (Sabeh *et al.*, 1996) and glutathione peroxidase (Sabeh *et al.*, 1993).

Their cell walls contain celluloses, hemicelluloses and pectins while their major storage carbohydrate is either an acetylated mannan or a glucomannan. There may be a significant number of proteins within the gel. These have not been well characterized. From an immunological viewpoint, the most important of these proteins are the lectins

(Akev, 1999; Koike *et al.*, 1995a; 1995b; Suzuki *et al.*, 1979; Yagi *et al.*, 1985; Yoshimoto *et al.*, 1987). Their activities are discussed elsewhere in this book.

ALOE GEL

The central gelatinous portion of the aloe leaf is a translucent structure consisting entirely of mesophyll cells. These cells contain few organelles, primarily leucoplasts, and are completely filled with a very large vacuole containing a concentrated aqueous solution of a mixture of complex carbohydrates, especially acetylated mannans, and some low molecular weight solutes, especially calcium malate. The cell walls have a typical structure containing both celluloses and pectins. When aloe leaves are harvested and processed, the mesophyll cells rupture and the acetylated mannan solution is released. Soluble pectins and cell wall hemicelluloses are also present in this carbohydrate solution. Leaf extracts containing this carbohydrate mixture influence both innate immune mechanisms such as inflammation and macrophage function, as well as specific acquired immunity. These carbohydrate solutions may also contain small but significant amounts of protein.

The precise carbohydrate content of the aloe leaf gel varies according to the time of day. Aloes, as desert plants, take up carbon dioxide during the night and store it temporarily as malic acid. During the following day they convert the malic acid to carbohydrates. Thus the leaf is high in malic acid at dawn but this level drops steadily and carbohydrates rise during the day. As a result, the time at which the leaves are harvested will significantly influence the composition of the gel.

Acemannan

Acemannan is the name given to the acetylated mannan isolated from *Aloe vera* (Manna and McAnalley, 1993; Paquet and Pierard, 1996; Reynolds, 1985). The predominant storage carbohydrate of *A. vera*, it consists of long chain polydispersed β(1,4)-linked mannan polymers with random *O*-acetyl groups. Acemannan has been claimed to possess many, if not all of the important biological activities of the aloe pulp (Tizard *et al.*, 1989). However it has proven exceedingly difficult to separate acemannan from contaminating protein. Likewise there are other complex carbohydrates present in most acemannan preparations. As a result, it is by no means proven that acemannan alone posseses the biological activities ascribed to it. Another cautionary note should also be made with respect to endotoxin content. It is difficult to produce aloe carbohydrate solutions free of contaminating endotoxin. Early studies on this material contained small but significant quantities of bacterial endotoxin and it is possible that some of the biological activities ascribed to acemannan may have been endotoxin effects.

Pectins

Aloe vera cell walls contain a unique pectin (Ni *et al.*, Chapter 4, this volume). This low-methoxy pectin contains up to 90% glucuronic acid. As a result it has unusual biological properties, such as the ability to bind and stabilize certain mammalian growth factors. Its presence in crude acemannan preparations may explain, in part, why these preparations can accelerate wound healing under certain circumstances (Tizard *et al.*, 1994).

Other carbohydrates

Other complex carbohydrates may be present in aloe gel extracts in small amounts and some of these may exert significant biological activity. For example Pugh and his colleagues (Pugh *et al.*, 2000) have identified a high molecular weight polysaccharide from *A. vera* (aloeride) that is a very potent macrophage activating agent. Its molecular weight may be as large as 7 million. It contains glucose, galactose, mannose and arabinose and it is as potent as bacterial endotoxin at activating nuclear factor (NF)-κB in human macrophages. Aloeride also induces the expression of the mRNAs encoding Il(interleukin)-1β and TNF (tumor necrosis factor)-α to levels equal to those observed in cells maximally activated by bacterial endotoxin. Thus, although aloeride constitutes only 0.015% of dry weight of aloe gel juice, its potency may fully account for the macrophage stimulating activity of this juice. Limited studies have been conducted on the carbohydrates of other *Aloe species* (Vilkas and Radjabi-Nassab, 1986; Radjabi-Nassab *et al.*, 1984; Yagi *et al.*, 1977). They are discussed in detail in Chapter 4 by Ni *et al.*, in this volume.

Low Molecular weight components

While the leaf gel consists primarily of a complex carbohydrate mixture, the green plant rind contains many complex organic compounds such as chromones, flavonoids and anthraquinones. The precise composition varies greatly between aloe species (Viljoen and van Wyk, 1988). Some of these molecules, especially the chromones and flavonoids, can have significant anti-inflammatory activity (Read, 1995) or antiviral activity (Andersen *et al.*, 1991).

SPECIFIC ACTIVITIES

Anti-inflammatory effects

The ability of aloe leaf gels to reduce the severity of acute inflammation has been evaluated in many different animal models (Adler *et al.*, 1995; Davis *et al.*, 1989; Beatriz *et al.*, 1996; Davis and Maro, 1989; Davis *et al.*, 1994a; Davis *et al.*, 1994b; Saito *et al.*, 1982; Vazquez *et al.*, 1996). For example, Adler studied inflammation in the hind paw of the experimental rat induced by kaolin, carrageenan, albumin, dextran, gelatin and mustard (Adler *et al.*, 1995). Of the various irritants tested, *A. vera* was especially active against gelatin-induced and kaolin-induced edema and had, in contrast, minimal activity when tested against dextran-induced edema. Ear swelling induced by croton oil has also been used as an assay (Davis *et al.*, 1987). The swelling induced by croton oil on a mouse ear is significantly reduced by application of an aloe gel. In addition, soluble acemannan-rich extracts administered either orally or by intraperitoneal injection to mice will also reduce this swelling (Bowden, 1995). In another model, the acute pneumonia induced in mouse lungs by inhalation of a bacterial endotoxin solution is significantly reduced by systemic administration of an aloe carbohydrate solution (Bowden, 1995). In both these cases the reduction in inflammation is associated with a significant reduction in tissue infiltration by neutrophils. In general, aloe free of anthraquinones was more effective than aloe with anthraquinone. Some of this anti-inflammatory activity is due to the activities of bradykininases (Fujita *et al.*, 1976; Yagi *et al.*, 1987).

Another model that has been studied is radiation-induced acute inflammation in mouse skin (Roberts and Travis, 1995). Male mice received graded single doses of gamma radiation and aloe gel was applied daily beginning immediately after irradiation and continuing for up to five weeks. The severity of the radiation reaction was scored and dose-response curves were obtained. It was found that the average peak skin reactions of the aloe-treated mice were lower than those of the control mice at all radiation doses tested. Thus the radiation ED50 values for skin reactions of 2.0–2.75 were approximately 7 Gy higher in the gel-treated mice. The average peak skin reactions and the ED50 values for mice treated with lubricating jelly or aloe gel were similar to irradiated control values. Reduction in the percentage of mice with severe skin reactions was greatest in the groups that received aloe gel for at least two weeks beginning immediately after irradiation. There was no effect if gel was applied only before irradiation or beginning one week after irradiation. Aloe gel, but not lubricating jelly, reduced acute radiation-induced skin reactions in C3H mice if applied daily for at least two weeks beginning immediately after irradiation. This experiment can, however, be criticized on the grounds that an inappropriate control substance was used. The acemannan effect should have employed an identical gel lacking acemannan as control (excipient) since there were many other components in the gel in addition to acemannan.

Acetylated mannans from the pulp of *A. saponaria* (As mannans) (Yagi *et al.*, 1984) have also been shown to be anti-inflammatory. Thus a $\beta1 \rightarrow 4$-linked D-mannopyranose containing 18% acetyl groups inhibited carrageenin-induced hind paw edema at 50 mg/kg intraperitoneally in rats. A crude preparation of both As mannans was effective when given intraperitoneally, but not when given orally.

The effects of aqueous, chloroform, and ethanol extracts of *A. vera* gel on carrageenan-induced edema in the rat paw, and neutrophil migration into the peritoneal cavity stimulated by carrageenan has also been studied (Stuehr and Marletta, 1987), as has the ability of the aqueous aloe extract to inhibit cyclooxygenase activity. The aqueous and chloroform extracts decreased the edema induced in the hind-paw and the number of neutrophils migrating into the peritoneal cavity, whereas the ethanol extract only decreased the number of neutrophils. The aqueous extract inhibited prostaglandin E2 production from $[^{14}C]$ arachidonic acid. These results demonstrated that the extracts of *A. vera* gel have anti-inflammatory activity and suggested that some of this activity at least was due to an inhibitory action on the arachidonic acid pathway via cyclooxygenase.

A similar experiment has been conducted using an *A. vera* extract treated with 50% ethanol. The resulting supernatant and precipitate were tested for anti-inflammatory activity using the croton oil-induced ear-swelling assay (Davis *et al.*, 1991). The supernatant decreased inflammation, when applied topically, by 29.2%, while the precipitate decreased inflammation by 12.1%.

The mechanisms by which aloe extracts exert anti-inflammatory effects are multiple, and several distinct pathways have been described. For example, some evidence suggests that the activity is due to gibberellins. Thus the anti-inflammatory activities of *A. vera* and gibberellin were measured in streptozotocin-induced diabetic mice by measuring the inhibition of polymorphonuclear leukocyte infiltration into a site of gelatin-induced inflammation (Davis and Maro, 1989). Both aloe and gibberellin similarly inhibited inflammation in a dose-response manner. These data were interpreted to suggest that gibberellin or a gibberellin-like substance is an active anti-inflammatory component in *A. vera*. A second possible mechanism is due to antibradykinin activity. Thus a fraction with antibradykinin activity has been partially purified from the pulp of *A. saponaria* by gel chromatography (Yagi *et al.*, 1982). The antibradykinin-active material was probably a glycoprotein that cleaved the Gly4-Phe5 and Pro7-Phe8 bonds of the bradykinin molecule. A third possible mechanism may be due to complement depletion ('t Hart *et al.*, 1988). Thus an aqueous extract of *Aloe vera* gel was fractionated into high (h-Mr) and low (l-Mr) molecular weight fractions by dialysis. Subsequent fractionation generated two fractions with molecular weights of 320 and 200 kDa. Preincubation of human pooled serum with these fractions resulted in a depletion of classical and alternative pathway complement activity. The inhibition appeared to be due to alternative pathway activation, resulting in consumption of C3 ('t Hart *et al.*, 1989). The active fractions were mannose-rich polysaccharides.

A fourth possible mechanism may relate to the fact that mannose-rich carbohydrate solutions inhibit the activity of certain $\beta 2$ integrins and hence block neutrophil emigration into inflamed tissues (Bowden, 1995). Aloe carbohydrate solutions inhibit swelling in the mouse ear model and reduce the inflammation in a mouse lung endotoxin model. Histological staining and tissue myeloperoxidase assays show that treated tissues contain significantly fewer neutrophils than untreated control tissues. Static neutrophil adherence assays demonstrate that acemannan enriched fractions can inhibit adherence of human neutrophils to human endothelial cells. Flow adherence assays have demonstrated that this solution has no effect on leukocyte rolling (a selectin-mediated phenomenon) but does inhibit complete adherence and transmigration (mediated by integrins). By using recombinant endothelial cell lines it can be shown that the acemannan solution has no effect on selectin-mediated adherence but can inhibit adherence to the integrins MAC(macrophage)-1 (CD11b) and leucocyte function-associated antigen (LFA)-1 (CD11a). It inhibits LFA-1-mediated adherence at concentrations at least 50-fold less than required to inhibit MAC-1 mediated adherence. These reactions are not a result of neutrophil activation.

Aloe gels also contain low molecular weight components (dialysates) that can inhibit the release of reactive oxygen and hydrogen peroxide by stimulated human neutrophils ('t Hart *et al.*, 1990). The compounds inhibited the oxygen-dependent extracellular effects of neutrophils, such as lysis of red blood cells, but did not affect the ability of the neutrophils to phagocytose and kill microorganisms. The inhibitory activity of these compounds was most pronounced on the PMA(phorbol 12-myristate 17-acetate)-induced oxygen production, but this was antagonized by a Ca-ionophore, suggesting that the effect was mediated by reduced intracellular free calcium.

Macrophage activation

Aloe-based carbohydrates can activate macrophages. Consequently they stimulate antigen-processing, non-specific immunity, wound healing and resistance to infection and neoplasia. Thus macrophages from normal animals are relatively quiescent, but can be readily activated and acquire the ability to kill tumor cells or certain microorganisms (Tizard *et al.*, 1989; Tizard *et al.*, 1994). Macrophage activation can be mediated by several different pathways. For example, one major pathway is through T cells secreting the Th1 cytokines, interferon-γ (IFN-γ) and interleukin-2 (IL-2) (Bielefeldt-Ohmann and Babiuk, 1986). IFN-γ is a potent macrophage-activating agent and it is especially effective when supplemented by exposure to microbial products such as endotoxins, muramyl dipeptide or cell wall carbohydrates (glucans, mannans) (Adams and Hamilton, 1984; Lackovic *et al.*, 1970.). Thus, activation is a multi-stage process. For example, inflammatory macrophages may first be primed by interferon. In a second step bacterial products or complex carbohydrates can activate these primed macrophages. Macrophages can destroy some tumor cells only after treatment with both recombinant IFN-γ and bacterial lipopolysaccharide (LPS), suggesting that at least two stimuli are required for complete activation (Drysdale *et al.*, 1988). One of the most marked biological activities of mannans in mammals is the activation of macrophages and stimulation of T cells (Inoue *et al.*, 1983; Inoue *et al.*, 1983; Tizard *et al.*, 1989). It has been shown that each of these molecules interacts with specific high affinity receptors located on the macrophage plasma membrane (Lorsbach *et al.*, 1993).

Acemannan immunostimulant (AI) is a commercially available, partially purified carbohydrate preparation containing about 60% acetylated mannan together with other carbohydrates, especially pectins and hemicelluloses. It should not be confused with the complex carbohydrate acemannan. AI can activate macrophages (Zhang and Tizard, 1996; Merriam *et al.*, 1996; Chinnah, 1990). This macrophage activating ability is probably responsible for its activity as an adjuvant (Chinnah *et al.*, 1992; Chinnah, 1990), its pro-wound-healing activity (Tizard *et al.*, 1994), as well as its anti-tumor (Peng *et al.*, 1991) and anti-viral activity (Chinnah, 1990; Kahlon *et al.*, 1991; Sheets *et al.*, 1991; Yates *et al.*, 1992).

Calcium flux

The first step in the macrophage activation process involves endocytosis of aloe carbohydrate and a rise in intracellular calcium. Thus AI can be observed within the cytoplasm of cultured macrophages as apple green fluorescence within an hour after exposure to fluorescein-labeled carbohydrate solution. When the location of this fluorescence is compared to labels for mitochondria and lysosomes, it shows greater than 98% correlation with lysosomal distribution (Monga *et al.*, 1996; Burghardt *et al.*, 1996). A significant increase in intracellular Ca^{2+} can be detected in macrophages following exposure to $50 \mu g/ml$ AI. The Ca^{2+} flux occurs within seconds of addition of AI solution and appears as a single spike followed by a return to basal levels in less than one minute. No Ca^{2+} stimulatory activity can be detected in response to the pellet derived from centrifuged AI. Since this pellet consists of plant cell wall fragments rich in pectin (Ni, 2000) it is likely that the calcium flux is not simply due to ingestion of cell fragments by phagocytosis. When macrophage cultures are pretreated with the calcium chelating agent EGTA, the AI-induced Ca^{2+} response is completely abolished suggesting that the response to AI requires extracellular Ca^{2+}. An AI solution can increase intracellular Ca^{2+} levels, not only in macrophages but also in uterine smooth muscle but not in liver epithelial cells. Free intracellular Ca^{2+} plays a pivotal role as a second messenger involved in signal transduction, and as the initiating step in macrophage activation.

General features of macrophage activation

Normal macrophages, when cultured in medium alone, tend to be round, resting cells and usually do not spread on the glass surface until many hours have elapsed. Very few morphological changes (about 5%) occur in cells exposed to AI alone. When exposed to the macrophage activating agent IFN-γ, approximately 40% of the cells begin to spread. In contrast, when exposed to a combination of AI and IFN-γ, approximately 90% of the macrophages enlarge and develop cytoplasmic spreading. This spreading is a feature of activated macrophages, as is increased membrane activity, especially ruffling, increased formation of pseudopodia and increased pinocytosis.

Macrophage size and surface characteristics can be examined by measuring forward angle light scatter (FALS), which measures size, and orthogonal light scatter (OLS), which measures roughness, using flow cytometry (Zhang and Tizard, 1996). Normal macrophages tend to be small smooth cells. Cells exposed to AI alone were comparable to the control cells. Cells exposed to IFN-γ alone consisted of two populations. One of smallish rough cells the other of large, rough cells. In contrast, when exposed to both AI and IFN-γ, the macrophages were larger and rougher than those exposed to either AI or IFN-γ alone.

Since adhesion molecules play an important role in macrophage activation, macrophage expression of CD11a, Mac-1 and CD18 in response to AI has also been examined by flow cytometry. Surface CD11a molecules are expressed on neither resting cells nor on cells exposed to AI alone, but they are expressed when cells are treated with IFN-γ alone and greatly increased CD11a expression occurs when cells are exposed to both AI and IFN-γ. Two other integrins, CD18 and Mac-1, are constitutively expressed on resting cells. Their expression does not increase when treated with AI or IFN-γ alone but significantly increased CD18 expression-but not Mac-1 expression can be observed on cells treated with both AI and IFN-γ.

Macrophage FcR (CD64) expression has also been measured in response to AI exposure. Minimal FcR expression is observed on resting macrophages or on macrophages exposed to AI. Greater expression occurs on IFN-γ treated cells and greatly increased expression is observed on cells treated with both AI and IFN-γ.

Mannosylated bovine serum albumin (m-BSA) enhances the respiratory burst, phagocytosis, and killing of *Candida albicans* by mouse peritoneal macrophages (Stuart *et al.*, 1997). Upregulation of macrophage functions is associated with the binding of m-BSA to the macrophage mannose receptor. Experiments have been conducted to determine if a carbohydrate solution from *A. vera* can stimulate macrophages in a similar manner. Mouse peritoneal macrophages were exposed to AI for ten minutes and showed a two-fold increase in their respiratory burst, as measured by chemiluminescence, above the media controls. Macrophages were also exposed to AI, washed, and exposed to Candida. There was a marked increase in phagocytosis in the treated cells compared to controls. Macrophages exposed to AI for ten minutes resulted in a 38% killing of Candida compared to 0–5% killing in controls. Macrophages exposed to AI for 60 minutes resulted in 98% of the Candida being killed compared to 0–5% in the controls. Thus, short-term exposure of macrophages to AI upregulated the respiratory burst, phagocytosis and candidicidal activity.

The receptors on macrophages that are responsible for this activation have not been identified but they appear, in general, to be Toll-like (Toll is a *Drosophila* gene). Thus, it has been suggested that aloe carbohydrates induce macrophage activation through a specific receptor-mediated pathway. One difficulty encountered in determining the nature of the receptor involved is the fact that the precise active principle in AI solutions has not been identified. All evidence suggests that macrophage mannose receptors are not involved, but this may simply mean that acemannan, the principal component of the mixture, is not the active agent (Tizard and Ramamoorthy, unpublished observations).

Nitric oxide release

AI-activated macrophages generate nitric oxide (NO) (Zhang and Tizard, 1996). Nitric oxide mediates the bactericidal and tumoricidal activities of macrophages (Lorsbach *et al.*, 1993) and when produced in excess is responsible for the endotoxic shock and damage due to inflammation (Nathan, 1992). Two major classes of NO synthase (NOS) are known: constitutive and inducible (Sessa, 1994). In neurons and endothelial cells, NOS is expressed constitutively and is activated by calmodulin and calcium (Cho *et al.*, 1992). Macrophage NOS is, however, produced only after activation of these cells and requires transcriptional activation of the iNOS gene (Lyons *et al.*, 1992). Endogenous calmodulin is a tightly bound subunit of the macrophage iNOS, hence its activity is not regulated by Ca^{2+} levels or exogenous calmodulin (Cho *et al.*, 1992). The activity of iNOS is, however, dependent on a number of cofactors like heme, tetrahydrobiopterin and flavin nucleotides (Sakai *et al.*, 1992). The synthesis of macrophage NO synthase is induced by lipopolysaccharide (LPS) (Kolls *et al.*, 1994), tumor necrosis factor α (TNFα) (Lowenstein *et al.*, 1993) and the level of stimulation can be augmented by combining these stimuli with IFN-γ (Lorsbach *et al.*, 1993; Lowenstein *et al.*, 1993; Xie *et al.*, 1993).

In the presence of IFN-γ, AI causes an increase in the production of nitric oxide by macrophages. Macrophages were stimulated with AI, IFN-γ or various combinations of the two for 24 hours, after which accumulation of nitrite in the media was measured. Treatment of the cells with low concentrations of IFN-γ (1.0 U/ml) had no effect while higher concentrations (10 and 100 U/ml) induced NO synthesis. The effect of a combination of AI and IFN-γ on nitric oxide production was especially evident when cells were stimulated with a low concentration of IFN-γ together with AI. However, treatment of macrophages with AI alone had no effect.

Macrophages were incubated with AI, IFN-γ, or a combination of the two, and the mRNA isolated, reverse transcribed and amplified. Low concentrations of IFN-γ (0.1–1 U/ml) by itself, did not cause a significant increase in the level of iNOS mRNA. However, when cells were treated with IFN-γ in the presence of AI there was a 25-fold increase in mRNA levels.

The AI-induced increase in iNOS expression was further characterized by its time course. Macrophages were incubated with AI and IFN-γ for various times and the mRNA isolated, reverse transcribed and amplified. AI and IFN-γ together caused an increase in iNOS mRNA levels that was detectable within three hours of treatment. The mRNA levels continued to increase until 12 hours after treatment and then leveled off.

The increase in iNOS mRNA levels induced by the combination of AI and IFN-γ was reflected in an increase in iNOS protein. When macrophages were treated with AI in the presence of IFN-γ, an inducible NOS of 130 kDa appeared at around six to eight hours. The increase in iNOS protein was confirmed by immunohistochemistry where it was seen to be confined to the perinuclear cytoplasm.

Cycloheximide, a reversible inhibitor of protein synthesis, greatly suppressed the level of iNOS mRNA in cells treated with AI and IFN-γ, but had no effect on the transcription of a constitutive gene G3PD, indicating that the increased expression of iNOS caused by AI and IFN-γ was dependent on *de novo* protein synthesis (Ramamoorthy *et al.*, 1996). Pyrrolidine dithiocarbamate (PDTC), an antioxidant that is a relatively specific inhibitor of nuclear factor (NF)-κB activation, inhibited the induction of iNOS by AI in the presence of IFN-γ. Activation of NF-κB, a transcription factor, has been shown to be an essential process in the induction of NOS in macrophages. Preincubation of cells with PDTC inhibited the induction of iNOS but had no effect on the transcription of G3PD (mRNA for glyceraldehyde-3-phosphate dehydrogenase). This inhibition of iNOS mRNA was also reflected in the enzymatic assay of iNOS protein, suggesting that the induction of iNOS by AI in the presence of IFN-γ involved the activation of NF-κB.

It is known that effective induction of iNOS requires activation by both IFN-γ and LPS (Lorsbach *et al.*, 1993), however, IFN-γ also synergizes with TNF-α in the induction of NO production (Drapier *et al.*, 1988), suggesting that multiple mediators can act on the iNOS gene to achieve inducible NO synthesis. Unlike endothelial cells, neurons, and other cell types that express the NOS constitutively, NO production in macrophages is observed only after exposure of the cells to cytokines or other stimuli, and induction of iNOS represents a response associated with the activation process (Xie *et al.*, 1992; Lyons *et al.*, 1992). In this respect AI activation is very similar to the induction caused by LPS (Lorsbach *et al.*, 1993).

Various transcription factors have been implicated in the regulation of the mouse iNOS promoter (Lowenstein *et al.*, 1993). The promoter for the macrophage iNOS gene contains consensus sequence motifs for the binding of several known transcription factors involved in the induction of other cytokine-responsive genes, such as IFN-γ responsive element, γ-activated site, TNF responsive element and NF-κB site (Sessa, 1994). The induction of iNOS gene by LPS is dependent on the presence of NF-κB heterodimers, p50-c-rel and p50-rel A (Xie *et al.*, 1994), while the synergistic inductive contribution of IFN-γ requires interferon regulatory factor (IRF)-1 (Kamijo *et al.*, 1994; Tanaka *et al.*, 1994). PDTC is an antioxidant that has been shown to inhibit the activation of NF-κB by preventing the binding of NF-κB/Rel to the NF-κBd (the binding site on the iNOS promoter) (Xie *et al.*, 1994; Bedoya *et al.*, 1995). Preincubation of macrophages with PDTC completely inhibited the induction of iNOS by AI in the presence of IFN-γ. These results suggest that activation of NF-κB is involved in the induction of iNOS by AI and IFN-γ.

Maximal activation of macrophage *in vitro* requires a minimum of two signals, such as IFN-γ and LPS (Adams and Hamilton, 1987). However, NO production following four hours incubation with AI, followed by washing and continued incubation in the presence of IFN-γ, was about 80% of the full response upon AI/IFN-γ challenge. Similar results were observed in the IFN-γ pre-treated cells. Therefore, short term incubation with AI can prime macrophages but is unable to generate sufficient transmembrane signaling for NO production. The long-term incubation with AI (treated with AI/IFN-γ for 48 hours), is capable of activating macrophages for NO production. Evidence suggests that maximum NO production requires a two-signal pathway (Stuehr and Nathan, 1989). Thus, the results of NO production following transient exposure of macrophages to either AI or IFN-γ suggests that AI can increase macrophage sensitivity to the second signal for the generation of NO. Evidence from LPS and IFN-γ induced macrophage activation suggests that LPS and IFN-γ have bi-directional synergetic effects, leading to markedly higher macrophage activation at limiting concentrations of either LPS or IFN-γ (Hibbs *et al.*, 1988). This evidence, along with the absence of detectable amounts of endotoxin, suggests that the macrophage response to AI may be similar to that induced by LPS although AI's effect is not caused by endotoxin contamination.

Cultures of normal chicken spleen cells and macrophages (cell line HD11) also produce NO when exposed to AI (Karaca *et al.*, 1995). Neither cell type produce detectable amounts of NO in response to similar concentrations of yeast mannan, another complex carbohydrate, suggesting perhaps that this response is not mannan-mediated and indirectly supporting the idea that components within the mixture other than acemannan may be responsible for its biological activity. This NO production was dose dependent and inhibitable by the NOS inhibitor N^G-methyl-L-arginine. In addition, the production of NO was inhibited by preincubation of AI with the lectin concanavalin A. Concanavalin A has a high affinity for terminal mannose residues. These results suggested that AI-induced NO synthesis may be mediated through macrophage mannose receptors, and macrophage activation may be accountable for some of the immunomodulatory effects of AI in chickens. Crude acemannan preparations do have adjuvant properties. Thus Chinnah *et al.* (1992) showed that the potency of a vaccine against Newcastle disease was enhanced when AI was added to it. Interestingly, the addition of AI to another chicken vaccine against infectious bursal disease did not enhance its potency. In another study (Nordgren *et al.*, 1992) it was shown that AI enhanced the potency of an avian herpes virus vaccine (Marek's disease). In field trials associated with this vaccine it was demonstrated that chickens that received AI showed a significant improvement in production parameters. These included mortality, condemnation rate and food conversion. This result is compatible with the concept that acemannan or some component within AI preparations can activate macrophages. Non-specific macrophage activation will, in turn, improve an individual's health and resistance to infectious diseases. AI has also been investigated for its adjuvant properties against avian polyoma virus in psittacine birds and was found to induce minimal tissue reactions (Ritchie *et al.*, 1994).

Cytokine release

Activated macrophages secrete increased amounts of many proteins. Thus they release TNF-α and the interleukines IL-1α, IL-6, IL-8, and IL-12. They secrete proteases, which activate complement components. They secrete interferons, as well as thromboplastin, prostaglandins, fibronectins, plasminogen activator and the complement components C2 and B. Activated macrophages express increased quantities of major histocompatibiliy complex (MHC) class II molecules on their surface and thus have an enhanced ability to process antigen.

To determine whether AI had a direct effect on cytokine production, the amounts of IL-1β, IL-6 and TNF-α released have been measured using enzyme-linked immunosorbant assays (ELISAs) and biological assays (Merriam *et al.*, 1996). IL-6 and TNF-α were released in a dose dependent manner in response to AI stimulation. TNF was released in greater quantities than that stimulated by LPS. However, IL-1β was only produced at very low levels (≤ 20 pg) in response to $400\,\mu$g/ml of AI. This IL-1 release measured using a bioassay based on the stimulation of thymocyte blastogenesis was significant and greater than that induced by bacterial LPS. Results suggested that this cytokine production did not require the presence of IFN-γ. The presence of IFN-γ (10 U/ml) in medium did, however, increase IL-6 and TNF-α release by approximately 40%. It is intriguing to note that the activation of human macrophages by IL-2 or IFN-γ is linked to increased expression of a macrophage receptor for acetylated mannose (Zhu *et al.*, 1993). This receptor appeared to be directly involved in the ability of macrophages to kill tumor cells (myelogenous leukemia cell line, K562).

Apoptosis (programmed cell death)

Apoptosis is a form of cell death that can be induced by many normal physiological stimuli as well as by deleterious environmental conditions. Some of the characteristic features of apoptosis include cytoplasmic shrinkage associated with membrane blebbing, followed by chromatin condensation and DNA fragmentation. Although all cells undergoing apoptosis exhibit these changes sequentially, it is believed that these events occur independently and under the control of separate and distinct metabolic pathways.

It has been reported that AI in the presence of IFN-γ also induces apoptosis in macrophages and this induction appears to be by a NO-independent mechanism (Ramamoorthy and Tizard, 1998). Macrophages were treated with AI in the presence and absence of IFN-γ for 48 hours and the cells were then analyzed for the occurrence of apoptosis by a DNA fragmentation assay, gel electrophoresis and propidium iodide staining. DNA laddering characteristic of apoptosis was observed in cells treated with AI and IFN-γ. AI or IFN-γ by themselves did not cause DNA laddering. In cells treated with AI (50 μg/ml) and IFN-γ, DNA fragmentation occurred around 36 hours after treatment and by 48 hours most of the cells had undergone apoptosis. Staining of the cells with Hoechst 33342 followed by propidium iodide also indicated that apoptosis was occurring in macrophages treated with AI and IFN-γ.

Cells undergoing apoptosis lose membrane phospholipid asymmetry and expose phosphatidylserine (PS) on the outer leaflet of the plasma membrane. The detection of PS exposure by annexin V during the redistribution of the plasma membrane has been shown to be a general and early marker of apoptosis. Annexin V staining of macrophages treated with AI and IFN-γ showed that apoptosis occurred in these cells at 18 hours after treatment and by 36 hours about 65% of the cells had undergone apoptosis.

Nitric oxide is known to cause apoptosis in some cells and this can account for the apoptosis observed in these cells (Messmer *et al.*, 1995). In order to verify this, an inhibitor of NO production, L-NAME (N-nitro-L-arginine methyl ester), was used. Macrophages were treated with AI and IFN-γ in the presence and absence of L-NAME and apoptosis was studied at 48 hours using a thymidine release assay. L-NAME inhibited the production of NO by cells treated with AI and IFN-γ but did not inhibit DNA fragmentation in these cells, suggesting that NO is not involved in this induction of apoptosis caused by the combination of AI and IFN-γ.

Members of the B cell lyphoma (bcl)-2 family (bcl-2, bcl-x, bax, etc.) play a prominent role both in apoptosis and its prevention (Kroemer *et al.*, 1995; Meikrantz *et al.*, 1994; Merino *et al.*, 1995; Wang *et al.*, 1995). They are important molecular switches. Macrophages were treated with AI in the presence of IFN-γ for various times and the expression of bcl-2 was examined. The expression of bcl-2 could be detected in all cells at 12 hours, 18 hours and 24 hours post-treatment. However, bcl-2 expression could not be detected in cells treated with AI and IFN-γ for 30 hours and the effect continued at 36 hours after treatment. This was observed only when the cells were treated with AI in the presence of IFN-γ. AI and IFN-γ by themselves did not cause any decrease in the expression of bcl-2. Thus AI in the presence of IFN-γ likely caused the induction of apoptosis in macrophages by a mechanism involving the inhibition of bcl-2 expression.

Lymphocyte activation

Aloe extracts appear to influence lymphocyte function under some circumstances. For example, the ability of *A. vera* gel extract to prevent suppression of contact hypersensitivity and delayed-type hypersensitivity (DTH) responses in mice by ultraviolet (UV) irradiation has been examined in detail (Strickland *et al.*, 1994; Strickland *et al.*, 1999; Byeon *et al.*, 1998; Lee *et al.*, 1997, 1999) (See Chapter 12).

The ability of AI to enhance immune responses to antigens and to determine whether that enhancement is a macrophage driven phenomenon has been examined (Womble and Helderman, 1988, 1992). AI does not affect lymphocyte responses to syngeneic antigens in the mixed lymphocyte culture (MLC) but increases alloantigenic responses in a dose-response fashion. This effect can be detected at concentrations of *in vitro* AI achievable *in vivo*. In a similar series of experiments, Marshall *et al.* (1993) investigated the cytokines produced when AI was incubated with human peripheral blood mononuclear cells. These cells presumably consisted of monocytes and lymphocytes and the predominant lymphocytes would have been T cells. Several different cytokines were produced in response to AI. Some, such as IL-1 and TNF-α, were most likely produced by the stimulated monocytes in a manner described above. However, two Th1 cytokines were also shown to be produced in a dose dependent manner. IL-2 and IFN-γ. There is no evidence that AI has a direct effect on T cells but it is not unreasonable to suggest that these cytokines were released by T cells secondary to exposure to the monocyte-derived cytokines.

Marshall and Druck (1994) have reported that AI can stimulate natural killer (NK) cell activity. They measured NK cell activity by incubating human blood with AI overnight and subsequently measuring NK cell activity using a chromium release assay. Dose dependent stimulation of NK activity was observed. This activation could be partially inhibited by the presence of monoclonal antibodies to IL-2 and IFN-γ.

Bone Marrow Stimulation

AI directly or indirectly has significant bone marrow stimulating properties. Thus the subcutaneous administration of AI significantly increases splenic and peripheral blood cellularity, as well as hematopoietic progenitors in the spleen and bone marrow, as determined by an interleukin-3-responsive colony-forming unit culture assay and a high-proliferative-potential colony-forming-cell (HPP-CFC) assay, a measure of primitive hematopoietic precursors in myelosuppressed mice (radiated with 7 Gy). The greatest hematopoietic effect was observed following sublethal irradiation in mice receiving 1 mg AI/animal, with less activity observed at higher or lower doses. AI, administered by daily injection, had activity equal to or greater than the injection of an optimal dose of granulocyte-colony-stimulating factor (G-CSF) in these same myelosuppressed mice. Significantly greater activity was observed in splenic and peripheral blood cellularity, and in the frequency and absolute number of splenic HPP-CFC as compared to the mice receiving G-CSF at 3 mg/animal. AI, when administered to myelosuppressed animals, decreased the frequency of lymphocytes but increased the frequency of polymorphonuclear leukocytes (PMN). However, owing to the increased cellularity, a significant increase in the absolute number of PMN, lymphocytes, monocytes and platelets was observed, suggesting activity on multiple cell lineages (Egger *et al.*, 1996a).

Further analysis of this phenomenon confirmed that subcutaneous injections of 1 mg/animal of AI had equal or greater stimulatory activity for white blood cell counts and spleen cellularity, as well as on the absolute numbers of neutrophils, lymphocytes, monocytes and platelets, than did higher or lower doses of AI or an optimal dose of granulocyte-colony stimulating factor (G-CSF). Hematopoietic progenitors, colony forming unit (CFU)-C and high proliferative potential colony-forming cells (HPP-CFC) assays, were similarly increased by AI in the spleen but not in the bone marrow. The frequency of splenic HPP-CFCs and the absolute number of splenic HPP-CFCs and CFU-Cs were optimally increased by 1 mg/animal of AI. In contrast, bone marrow cellularity, frequency and absolute number of HPP-CFCs and CFU-Cs had as a optimum dosage 2 mg/animal. These parameters were similarly increased by G-CSF. In studies to determine the optimal protocol for the administration of AI we found that the hematopoietic activity of AI increased with the frequency of administration. The greatest activity in myelosuppressed mice was observed for all hematopoietic parameters, except the platelet number in mice receiving a daily administration of 1 mg/animal with activity equal to or greater than G-CSF (Egger *et al.*, 1996b).

Anti-tumor applications

Given that AI is a macrophage stimulating agent, several investigators have examined its anti-tumor effects. For example, using the implanted Norman murine sarcoma as a model, mice treated with AI showed diminished tumor growth rates and increased mouse survival (Peng *et al.*, 1991; Merriam *et al.*, 1996). Maximum survival occurred in mice receiving a single injection of 0.5 mg/kg AI i.p. at the time of tumor implantation. Tumor growth was apparently normal for 12 days after implantation. Decreased tumor growth was apparent between 12 and 15 days. Thirty-five per cent of treated animals survived by 60 days while all untreated animals were dead by 46 days. The effect of the AI was time critical. Thus it was less effective if administered 24 hours before or after implantation and had no significant effect if given 48 hours before or after implantation. Histopathology of tumors from treated animals showed extensive areas of edema, leukocytes (especially mononuclear) infiltration, necrosis and hemorrhage. Eventually the tumors were walled off by significant fibrosis. Animals which recovered from a Norman murine sarcoma transplant rejected subsequent tumor transplants. However, it is unclear how AI exerts this wide variety of effects and we believe that some of these effects are mediated through the macrophages. The combined data suggest that AI-stimulated synthesis of monokines resulted in the initiation of immune attack, necrosis, and regression of implanted sarcomas in mice. This experiment may, however, have been influenced by low levels of endotoxin in the AI preparation.

The effect of *Aloe vera* administration was studied on a pleural tumor in rat. The growth of Yoshida ascites hepatoma (AH)-130 cells injected ($2 \times 10(5)$ in 0.1 ml) into pleura of male inbred Fisher rats was evaluated at different times (7th and 14th days) (Corsi *et al.*, 1998). Winters and his colleagues (1981) have demonstrated that lectins from *A. vera* and *A. saponaria* were cytotoxic for both normal and tumor cells *in vitro*.

AI is employed clinically for the treatment of fibrosarcomas in dogs and cats (Manna *et al.*, 1993). Studies *in vitro* indicate that AI has limited anti-viral activity against herpes viruses, measles, and human immunodeficiency virus (Chinnah, 1990). It is an immunostimulant and is licensed by the United States Department of Agriculture for the treatment of fibrosarcoma in dogs and cats. In a pilot study,

AI was administered intralesionally and intraperitoneally to 43 dogs and cats suffering from a variety of spontaneous neoplasms (Harris *et al.*, 1991). Of seven animals with fibrosarcomas, five showed some sign of clinical improvement such as tumor shrinkage, tumor necrosis or both following AI treatment. In an additional study (King *et al.*, 1995) four dogs and six cats with recurrent fibrosarcoma that had failed previous treatment were treated with AI. Tumors in two of the animals shrank and disappeared. In eight animals, the tumors became edematous and enlarged rapidly. Necrosis and lymphocyte infiltration was seen in all of them. Survival times ranged from 57 to greater than 623 days with a mean tumor-free interval of 229 days. In another study, dogs and cats with histopathologically confirmed fibrosarcomas were treated with acemannan immunostimulant in combination with surgery and radiation therapy (King *et al.*, 1995). Following four to seven weekly injections of AI, tumor shrinkage occurred in four of these animals. On histology there was a notable increase in necrosis and inflammation. Following surgery and radiation and monthly AI treatment, seven of 13 animals remained alive and tumor free for 440 to 603 days. These results suggested that treatment with AI was an effective and useful adjunct to surgery in these animals.

Wound Healing

AI has been reported to promote healing of aphthous ulcers in humans (Grindlay and Reynolds, 1986; Plemons *et al.*, 1994) and to accelerated wound healing of biopsy-punch wounds in rats (Tizard *et al.*, 1994). Acemannan immunostimulant enhances wound healing in rats (Tizard *et al.*, 1994; Tizard *et al.*, 1994). Macrophages play a crucial role in wound healing, as it is involved in both the inflammatory and debridement phases (Hunt *et al.*, 1983). Macrophage-derived nitric oxide modulates angiogenesis (Koprowski *et al.*, 1993). Macrophages are a source of growth factors and play the most important role in removing damaged tissues. Animals that lack macrophages fail to heal.

In general, it is difficult to accelerate the healing of uninfected wounds in healthy young animals, given that their healing process is fully functional and probably proceeding at maximal speed. However, that is not the case in old animals. Wound healing slows steadily as an individual animal ages and takes approximately 50% longer, three as opposed to two weeks, in rats over two years of age. Using this model and administering AI by local injection it is possible to accelerate this time course so that treated wounds will heal in two weeks. It is possible therefore that this effect is due to macrophage activation. Activated macrophages influence fibroblast function and promote the deposition of collagen in wounds. On the other hand, it is equally likely that the effect is due to growth factor stabilization by the pectins in the mixture.

The effects of *Aloe vahombe* (sic)

In an extensive series of papers published in the local literature, Ralamboranto and his colleagues have cataloged a remarkable series of effects of *A. vahombe* (=*A. vaombe* Decorse et Poiss.), a plant endemic in the south of Madagascar (Brossat *et al.*, 1981, Ralamboranto *et al.*, 1982; Radjabi-Nassab *et al.*, 1984; Vilkas and Radjabi-Nassab, 1986; Ralamboranto *et al.*, 1987). Because it is not generally available to other investigators, it is difficult to determine the significance of this plant. For example, a partially purified extract of leaves of *A. vahombe*, administered intravenously to mice, protects them against infection of the bacteria *Listeria monocytogenes, Yersinia pestis,*

Plasmodium berghei and the yeast *Candida albicans* (Brossat *et al.*, 1981). The protective fraction must be administered two days before inoculation of the pathogenic agent. In addition, when the mice were injected with an unrefined extract from *A. vahombe*, they were protected against *Klebsiella septicaemia* (Solar *et al.*, 1980). Neither bactericidal nor bacteriostatic activity has been detected in this aloe extract. Nevertheless, the anti-infectious activity was proportional to the dose of extract injected, and the protective activity was the greatest when the mice were treated with aloe two or three days prior to infection.

A fraction extracted from *A. vahombe*, was studied for its effect on experimental fibro-sarcomas and melanomas in mice (Ralamboranto *et al.*, 1982). 'Cures' were observed only in the case of the McC3-1 tumor but the rate of growth of tumors in animals which were treated was slower than in untreated animals. The active fraction was identified as a water soluble, thermostable, polysaccharide with a molecular weight of more than 30 kDa.

CONCLUSION

There is no doubt that extracts of aloe leaf gels can exert significant anti-inflammatory activity. At the same time similar gels can activate macrophages and perhaps other cells involved in the defense of the body. Unfortunately the aloe preparations used for these studies have tended to be complex, ill-defined mixtures and it is therefore difficult to elucidate the biochemical pathways involved.

REFERENCES

Adams, D.O. and Hamilton, T.A. (1984) The cell biology of macrophage activation. *Annual Review of Immunology*, 2, 283–318.

Adams, D.O. and Hamilton, T.A. (1987) Molecular transductional mechanism by which IFN gamma and other signals regulate macrophage development. *Immunological Reviews*, 97, 5–27.

Adler, H., Frech, B., Thony, M., Pfister, H., Peterhans, E. and Jungi, T.W. (1995) Inducible nitric oxide synthase in cattle. Differential cytokine regulation of nitric oxide synthase in bovine and murine macrophages. *Journal of Immunology*, 154, 4710–4718.

Akev, N. and Can, A. (1999) Separation and some properties of *Aloe vera* L. leaf pulp lectins. *Phytotherapy Research*, 13, 489–493.

Andersen, D.O., Weber, N.D., Wood, S.G., Hughes, B.G., Murray, B.K. and North, J.A. (1991) In vitro virucidal activity of selected anthraquinones and anthraquinone derivatives. *Antiviral Research*, 16, 185–196.

Beatriz, V., Avila, G., Segura, D. and Escalante, B. (1996) Anti-inflammatory activity of extracts from Aloe vera gel. *J. Ethnopharm*, 55, 69–75.

Bedoya, F.J., Flodstrom, M. and Eizirik. D.L. (1995) Pyrrolidine dithiocarbamate prevents IL-1 induced nitric oxide synthase mRNA, but not superoxide dismutase mRNA, in insulin producingcells. *Biochemical and Biophysical Research Communications*, 210, 816–822.

Bielefeldt-Ohmann, H. and Babiuk, L.A. (1986) Alteration of some leukocyte functions following *in vivo* and *in vitro* exposure to recombinant bovine alpha- and gamma-interferon. *Journal of Interferon Research*, 6, 123–136.

Bowden, R.A. (1995) The effect of a mannose-rich extract on integrin expression on vascular endothelial cells. *PhD Thesis*, Texas A&M University, Houston, USA.

Brossat, J.Y., Ledeaut, J.Y., Ralamboranto, L., Rakotovao, L.H., Solar, S., Gueguen, A. and Coulanges, P. (1981) Immunostimulating properties of an extract isolated from Aloe vahombe.

2. Protection in mice by fraction F 1 against infections by *Listeria monocytogenes, Yersinia pestis, Candida albicans* and *Plasmodium berghei. Archives de l'Institut Pasteur de Madagascar*, 48, 11–34.

Bruce, W.G. (1967) Investigations of antibacterial activity in the aloe. *South African Medical Journal*, 7, 984.

Burghardt, R.C., Barhoumi, R., Stickney, M., Ku, C.-Y. and Sanborn, B.M. (1996) Correlation between Connexin 43 Expression, Cell-Cell Communication, and Oxytocin-Induced Ca^{2+} Responses in an Immortalized Human Myometrial Cell Line. *Biology of Reproduction*, 55, 433–438.

Byeon, S.W., Pelley, R.P., Ullrich, S.E., Waller, T.A., Bucana, C.D. and Strickland, F.M. (1998) *Aloe barbadensis* extracts reduce the production of interleukin-10 after exposure to ultraviolet radiation. *Journal of Investigative Dermatology*, 11, 811–817.

Chinnah, A.D. (1990) Evaluation of antiviral, adjuvant and immunomodulatory effects of a β-(1,4)-linked polymannose (acemannan). *Doctoral Dissertation*; Texas A&M University, Houston, USA.

Chinnah, A.D., Baigh, M., Tizard, I.R. and Kemp, M.C. (1992) Antigen dependent adjuvant activity of a polydispersed (1,4)-linked acetylated mannan (acemannan). *Vaccine*, 10, 551–558.

Cho, H.J., Xie, Q.-W., Calaycay, J., Mumford, R.A., Swiderek, K.M., Lee, T.D. and Nathan, C. (1991) Calmodulin is a subunit of nitric oxide synthase from macrophages. *Journal of Experimental Medicine*, 176, 599–604.

Corsi, M.M., Bertelli, A.A., Gaja, G., Fulgenzi, A. and Ferrero, M.E. (1998) The therapeutic potential of *Aloe vera* in tumor-bearing rats. *International Journal of Tissue Reactions*, 20, 115–118.

Davis, R.H., DiDonato, J.J., Hartman, G.M. and Haas, R.C. (1994a) Anti-inflammatory and wound healing activity of a growth substance in *Aloe vera. Journal of the American Podiatric Medical Association*, 84, 77–81.

Davis, R.H., DiDonato, J.J., Johnson, R.W. and Stewart, C.B. (1994b) *Aloe vera*, hydrocortisone, and sterol influence on wound tensile strength and anti-inflammation. *Journal of the American Podiatric Medical Association*, 84, 614–621.

Davis, R.H., Leitner, M.G. and Russo, J.M. (1987) Topical anti-inflammatory activity of Aloe vera as measured by ear swelling. *Journal of the American Podiatric Medical Association*, 77, 610–612.

Davis, R.H., Leitner, M.G., Russo, J.M. and Byrne, M.E. (1989a) Anti-inflammatory activity of *Aloe vera* against a spectrum of irritants. *Journal of the American Podiatric Medical Association* 79, 263–276.

Davis, R.H. and Maro, N.P. (1989) Aloe vera and gibberellin. Anti-inflammatory activity in diabetes. *Journal of the American Podiatric Medical Association*, 79, 24–26.

Davis, R.H., Parker, W.L., Samson, R.T. and Murdoch, D.P. (1991) The isolation of an active inhibitory system from an extract of *Aloe vera. Journal of the American Podiatric Medical Association*, 81, 258–261.

Davis, R.H., Rosenthal, K.Y., Cesario, L.R. and Rouw, G.A. (1989b) Processed Aloe vera administered topically inhibits inflammation. *Journal of the American Podiatric Medical Association*, 79, 395–397.

Drapier, J.C., Wietzerbin, J. and Hibbs, J.B. Jr (1988) Interferon-gamma and tumor necrosis factor induce the L-arginine-dependent cytotoxic effector mechanism in murine macrophages. *European Journal of Immunology*, 18, 1587–1592.

Drysdale, B.E., Agarwal, S. and Shin, H.S. (1988) Macrophage-mediated tumoricidal activity: mechanisms of activation and cytotoxicity. *Progress in Allergy*, 40, 111–161.

Egger, S.F., Brown, G.S., Kelsey, L.S., Yates, K.M., Rosenberg, L.J. and Talmadge, J.E. (1996a) Hematopoietic augmentation by a beta-(1,4)-linked mannan. *Cancer Immunology and, Immunotherapy*, 43, 195–205.

Egger, S.F., Brown, G.S., Kelsey, L.S., Yates, K.M., Rosenberg, L.J. and Talmadge, J.E. (1996b) Studies on optimal dose and administration schedule of a hematopoietic stimulatory beta-(1,4)-linked mannan. *International Journal of Immunopharmacology*, 18, 113–126.

Fujita, K., Ito, S., Teradaira, R. and Beppu, H. (1979) Properties of a carboxypeptidase from aloe. *Biochemal Pharmacology*, 28, 1261–1262.

Fujita, K., Teradaira, R. and Nagatsu, T. (1976) Bradykinase activity of aloe extract. *Biochemical Pharmacology*, 15, 205.

Grindlay, D. and Reynolds, T. (1986) The *Aloe vera* phenomenon: a review of the properties and modern uses of the leaf parenchyma gel. *Journal of Ethnopharmacology*, 16, 117–151.

Harris, C., Pierce, K., King, G., Yates, K.M., Hall, J. and Tizard, I.R. (1991) Efficacy of acemannan treatment of canine and feline spontaneous neoplasms. *Molecular Biotherapy*, 3, 207–213.

Hibbs, J.B. Jr, Tainer, R.R., Vavrin, Z. and Paachlin, E.M. (1988) Nitric oxide: a cytotoxic activated macrophage effector molecule. *Biochemical and Biophysical Research Communications*, 157, 87–94.

Hirata, T. and Suga, T. (1977) Biologically active constituents of leaves and roots of *Aloe arborescens* var. *natalensis*. *Zeitschrift für Naturforschung*, C, 32, 731–734.

Holdsworth, D.K. (1971) Chromones in aloe species. I. Aloesin—a C-glucosyl-7-hydroxy-chromone. *Planta Medica*, 19, 322–325.

Holdsworth, D.K. (1972) Chromones in aloe species. II. Aloesone. *Planta Medica*, 22, 54–58.

Hunt, T.K., Knighton, D.R., Thakral, K.D., Goodson, W.H. and Andrews W.S. (1983) Studies on inflammation and wound healing: angiogenesis and collagen synthesis stimulated *in vivo* by resident and activated macrophages. *Surgery*, 96, 48–54.

Hutter, J.A., Salman, M., Stavinoha, W.B., Satsangi, N., Williams, R.F., Streeper, R.T. and Weintraub, S.T. (1996). Antiinflammatory C-glucosyl chromone from *Aloe barbadensis*. *Journal of Natural Products*, 59, 541–543.

Inoue, K., Nakajima, H. and Kohno, M. (1983) An antitumor polysaccharide produced by *Microellobosporia grisea*: preparation, general characterization, and antitumor activity. *Carbohydrate Research*, 114, 164–168.

Inoue, K., Kawamoto, K.K. and Kadoya, S. (1983) Structural studies on an antitumor polysaccharide DMG, a degrade D-mannose glucan from *Microellobosporia grisea*. *Carbohydrate Research*, 114, 245–256.

Kahlon, J.B., Kemp, M.C., Carpenter, R.H., McAnalley, B.H., McDaniel, H.R. and Shannon, W.M. (1991) Inhibition of AIDS virus replication by acemannan in vitro. *Molecular Biotherapy*, 3, 127–135.

Kamijo, R., Harada, H., Matsuyama, T., Bosland, M., Gerecitano, J., Shapiro, D., Le, J., Im, K.S., Kimura, T., Green, S., Mak, T.W., Taniguchi, T. and Vilcek. J. (1994) Requirement for transcription factor IRF-1 in NO synthase induction in macrophages. *Science*, 263, 1612–1615.

Karaca, K., Sharma, J.M. and Nordgren, R. (1995) Nitric oxide production by chicken macrophages activated by acemannan, a complex carbohydrate extracted from *Aloe vera*. *International Journal of Immunopharmacology*, 17, 183–188.

King, G.K., Yates, K.M., Greenlee, P.G., Pierce, K.R., Ford, C.M., McAnalley, B.H. and Tizard, I.R. (1995) The effect of acemannan immunostimulant in combination with surgery and radiation therapy on spontaneous canine and feline fibrosarcomas. *Journal of the American Animal Hospital Association*, 31, 439–447.

Koike, T., Beppu, H., Kuzuya, H., Maruta, K., Shimpo, K., Suzuki, M., Titani, K. and Fujita, K. (1995a) A 35 kDa mannose-binding lectin with hemagglutinating and mitogenic activities from 'Kidachi Aloe' (*Aloe arborescens* Miller var. *natalensis* Berger). *Journal of Biochemistry*, 118, 1205–1210.

Koike, T., Titani, K., Suzuki, M., Beppu, H., Kuzuya, H., Maruta, K., Shimpo, K. and Fujita, K. (1995b) The complete amino acid sequence of a mannose-binding lectin from 'Kidachi Aloe' (*Aloe arborescens* Miller var. *natalensis* Berger). *Biochemical and Biophysical Research Communications*, 214, 163–170.

Kolls, J., Xie, J., LeBlanc, R., Malinski, T., Nelson, S., Summer, W. and Greenberg, S.S. (1994) Rapid induction of messenger RNA for nitric oxide synthase 11 in rat neutrophils *in vivo* by endotoxin and its suppression by prednisolone. *Proceedings of the Society for Experimental Biology and Medicine*, 205, 220–225.

Koprowski, H., Zheng, Y.M., Heber-Katz, E., Fraser, N., Rorke, L., Fu, Z.F., Hanlon, C. and Dietzschold, B. (1993) In vivo expression of inducible nitric oxide synthase in experimentally induced neurological diseases. *Proceedings of the National Academy of Sciences USA*, **90**, 3024–3027.

Kroemer, G., Petit, P., Zamzami, N., Vayssiere, J.L. and Mignotte, B. (1995). The biochemistry of programmed cell death. *FASEB Journal*, **9**, 1277–1287.

Lackovic, V., Borecky, L., Sikl, D., Masler, L. and Bauer, S. (1970) Stimulation of interferon production by mannans. *Proceedings of the Society for Experimental Biology and Medicine*, **134**, 874–879.

Lee, C.K., Han, S.S., Shin, Y.K., Chung, M.H., Park, Y.I., Lee, S.K. and Kim, Y.S. (1999) Prevention of ultraviolet radiation-induced suppression of contact hypersensitivity by *Aloe vera* gel components. *International Journal of Immunopharmacology*, **21**, 303–310.

Lee, C.K., Han, S.S., Mo, Y.K., Kim, R.S., Chung, M.H., Park, Y.I., Lee, S.K. and Kim, Y.S. (1997) Prevention of ultraviolet radiation-induced suppression of accessory cell function of Langerhans cells by *Aloe vera* gel components. *Immunopharmacology*, **37**, 153–162.

Lorsbach, R.B., Murphy, W.J., Lowenstein, C.J., Snyder, S.H. and Russell, S.W. (1993) Expression of the nitric oxide synthase gene in mouse macrophages activated for tumor cell killing. *Journal of Biological Chemistry*, **268**, 1908–1913.

Lowenstein, C.J., Alley, E.W., Raval, P., Snowman, A.M., Snyder, S.H., Russel, S.W. and Murphy, W.J. (1993) Macrophage nitric oxide synthase gene: two upstream regions mediate induction by interferon γ and lipopolysaccharide. *Biochemistry*, **90**, 9730–9734.

Lyons, C.R., Orloff, G.J. and Cunningham, J.M. (1992) Molecular cloning and functional expression of an inducible nitric oxide synthase from a murine macrophage cell line. *Journal of Biological Chemistry*, **267**, 6370–6374.

Makino, K., Yagi, A. and Nishioka, I. (1974) Studies on the constituents of *Aloe arborescens* Mill. var. *natalensis* Berger. II. The structures of two new aloesin esters. *Chemical and Pharmaceutical Bulletin*, **22**, 1565–1570.

Manna, S. and McAnalley, B.H. (1993) Determination of the position of the O-acetyl group in a β-(1–4)-mannan ('acemannan') from *Aloe barbardensis* Miller. *Carbohydrate Research*, **24**, 317–319.

Marshall, G.D. and Druck, J.P. (1994) *In vitro* stimulation of NK activity by acemannan. *Journal of Immunology*, **150**, 241A.

Marshall, G.D., Gibbons, A.S. and Parnell, L.S. (1993) Human cytokines induced by acemannan. Abstract 619. *Journal of Allergy and Clinical Immunology*, **91**, 295.

Meikrantz, W., Gisselbrecht, S., Tam, S.W. and Schlegel, R. (1994). Activation of cyclin A-dependent protein kinases during apoptosis. *Proceedings of the National Academy of Sciences USA*, **91**, 3754–3758.

Merino, R., Grillot, D.A., Simonian, P.L., Muthukumar, S., Fanslow, W.C., Bondada, S. and Nunez, G. (1995). Modulation of anti-IgM-induced B cell apoptosis by Bcl-xL and CD40 in WEHI-231 cells. Dissociation from cell cycle arrest and dependence on the avidity of the antibody-IgM receptor interaction. *Journal of Immunology*, **155**, 3830–3838.

Merriam, E.A., Campbell, B.S., Flood, L.P; Welsh, C.J.R., McDaniel, H.R. and Busbee, D.L. (1996) Enhancement of immune function in rodents using a complex plant carbohydrate, which stimulates macrophage to secretion of immunoreactive cytokines. In *Advances in Anti-Aging Medicine*, **1**, 181–203. New York: Mary Ann Liebert, Inc.

Messmer, U.K., Lapetina, E.G. and Brune, B. (1995). Nitric oxide-induced apoptosis in RAW 264.7 macrophages is antagonized by protein kinase C- and protein kinase A-activating compounds. *Molecular Pharmacology*, **47**, 757–765.

Monga, M., Ku, C.-Y., Dodge, K. and Sanborn, B.M. (1996) Oxytocin-stimulated responses in a pregnant human immortalized myometrial cell line. *Biology of Reproduction*, **55**, 427–432.

Nathan, C. (1992) Nitric oxide as a secretory product of mammalian cells. *FASEB. Journal*, **6**, 3051–3064.

Nordgren, R.M., Steward-Brown, B. and Rodenberg, J.H. (1992) The role of acemannan as an adjuvant for Marek's disease vaccine. *Proceedings of the. XIX. World's Poultry Congress*, pp. 165–169. Amsterdam, The Netherlands.

Paquet, P. and Pierard, G.E. (1996) Interleukin-6 and the skin. *International Archives of Allergy and Applied Immunology*, 109, 308–317.

Peng, S.Y., Norman, J., Curtin, G., Corrier, D., McDaniel, H.R. and Busbee, D. (1991) Decreased mortality of Norman murine sarcoma in mice treated with the immunomodulator, Acemannan. *Molecular Biotherapy*, 3, 79–87.

Plemons, J.M., Rees, T.D., Binnie, W.H., Wright, J.M., Guo, I. and Hall, J.E. (1994) Evaluation of acemannan in the treatment of recurrent aphthous stomatitis. *Wounds*, 6, 40–45.

Pugh, N., Ross, S.A., ElSohly, M.A. and Pasco, D.S. (2001) Characterization of Aloeride, a new high molecular weight polysaccharide from *Aloe vera* with potent immunostimulatory activity. *Journal of Agricultural and Food Chemistry*, 49, 1030–1034.

Radjabi-Nassab, F., Ramiliarison, C., Monneret, C. and Vilkas, E. (1984) Further studies of the glucomannan from *Aloe vahombe* (liliaceae). II. Partial hydrolyses and NMR 13C studies. *Biochimie*, 66, 563–567.

Ralamboranto, L., Rakotovao, L.H., Coulanges, P., Corby, G., Janot, C. and Le Deaut, J.Y. (1987) Induction of lymphoblastic transformation by a polysaccharide extract of a native Madagascar plant *Aloe vahombe*: ALVA. *Archives de l'Institut Pasteur de Madagascar*, 53, 227–231.

Ralamboranto, L., Rakotovao, L.H., Le Deaut, J.Y., Chaussoux, D., Salomon, J.C., Fournet, B., Montreuil, J., Rakotonirina-Randriambeloma, P.J., Dulat, C. and Coulanges, P. (1982) Immunomodutating properties of an extract isolated and partially purified from *Aloe vahombe*. 3. Study of antitumoral properties and contribution to the chemical nature end active principle. *Archives de l'Institut Pasteur de Madagascar*, 50, 227–256.

Ramamoorthy, L. and Tizard, I.R. (1998) Induction of apoptosis in macrophage cell line RAW264.7 by acemannan, a β-(1,4)-acetylated mannan. *Molecular Pharmacology*, 53, 415–421.

Ramamoorthy, L., Kemp, M.C. and Tizard, I.R. (1996) Acemannan a β-(1,4)-acetylated mannan induces nitric oxide production in a macrophage cell line RAW264.7. *Molecular Pharmacology*, 50, 878–884.

Read, M.A. (1995) Flavonoids, Naturally occurring anti-inflammatory agents. *American Journal of Pathology*, 147, 235–237.

Reynolds, T. (1985) The compounds in *Aloe vera* leaf exudates: A review. *Botanical Journal of the Linnean Society*, 90, 157–159.

Reynolds, T. and Dweck, A.C. (1999) Aloe vera leaf gel: a review update. *Journal of Ethnopharmacology*, 68, 3–37.

Ritchie, B.W., Niagro, F.D., Latimer, K.S., Pritchard, N., Greenacre, C., Campagnoli, R.P. and Lukert, P.D. (1994) Antibody response and local reactions to adjuvanted avian polyomavirus vaccines in psittacine birds. *Journal of the Association of Avian Vetinaries*, 8, 21–27.

Roberts, D.B. and Travis, E.L. (1995) Acemannan-containing wound dressing gel reduces radiation-induced skin reactions in C3H mice. *International Journal of Radiation Oncology, Biology, Physics*, 32, 1047–1052.

Sabeh, F., Wright, T. and Norton, S.J. (1993) Purification and characterization of a glutathione peroxidase from the *Aloe vera* plant. *Enzyme Protein*, 47, 92–98.

Sabeh, F., Wright, T. and Norton, S.J. (1996) Isozymes of superoxide dismutase from Aloe vera *Enzyme Protein*, 49, 212–221.

Sakai, N., Kaufman, S. and Milstien, S. (1992) Tetrahydrobiopterin is required for cytokine-induced nitric oxide production in a murine macrophage cell line (RAW 264). *Molecular Pharmacology*, 43, 6–10.

Sato, Y., Ohta, S. and Shinoda, M. (1990) Studies on chemical protectors against radiation. XXXI. Protection effects of *Aloe arborescens* on skin injury induced by X-irradiation. *Yakugaku Zasshi*, 110, 876–884.

Saito, H., Ishiguro, T., Imanishi, K. and Suzuki, I. (1982) Pharmacological studies on a plant lectin aloctin A. II. Inhibitory effect of aloctin A on experimental models of inflammation in rats. *Japanese Journal of Pharmacology*, 32, 139–142.

Sen, S. and D'Incalci, M. (1992) Apoptosis: biochemical events and relevance to cancer chemotherapy. *FEBS Letters*, 307, 122–127.

Sessa, W.C. (1994) The nitric oxide synthase family of proteins. *Journal of Vascular Research*, 31, 131–143 .

Sheets, M.A., Unger, B.A., Giggleman, G.F.J. and Tizard, I.R. (1991) Studies of the effect of acemannan on retrovirus. *Molecular Biotherapy*, 3, 41–45.

Soeda, M., Otomo, M., Ome, M. and Kawashima, K. (1966) Studies on anti-bacterial and anti-fungal activities of Cape aloe. *Japanese Journal of Bacteriology*, 21, 609–614.

Solar, S., Zeller, H., Rasolofonirina, N., Coulanges, P., Ralamboranto, L., Andriatsimahavandy, A.A., Rakotovao, L.H. and Le Deaut, J.Y. (1980) Mise en evidence et etude des proprietes immunostimulantes d'un extrait isole et partiellement purifie apartir d'*Aloe vahombe*. *Archives d e l'Institut Pasteur de Madagascar*, 47, 9–39.

Strickland, F.M., Darvill, A., Albersheim, P., Eberhard, S., Pauly, M. and Pelley, R.P. (1999) Inhibition of UV-induced immune suppression and interleukin-10 production by plant oligosaccharides and polysaccharides. *Photochemistry and Photobiology*, 69, 141–147.

Strickland, F.M., Pelley, R.P. and Kripke, M.L. (1994) Prevention of ultraviolet radiation-induced suppression of contact and delayed hypersensitivity by *Aloe barbadensis* gel extract. *Journal of Investigative Dermatology*, 102, 197–204.

Stuart, R.W., Lefkowitz, D.L., Lincoln, J.A., Howard, K., Gelderman, M.P. and Lefkowitz, S.S. (1997) Upregulation of phagocytosis and candidicidal activity of macrophages exposed to the immunostimulant acemannan. *International Journal of Immunopharmacology*, 19, 75–82.

Stuehr, D.J. and Marletta, M.A. (1987) Synthesis of nitrite and nitrate in murine macrophage cell lines. *Cancer Research*, 47, 5590–5594.

Stuehr, D.J. and Nathan, C.F. (1989) Nitric oxide, a macrophage product responsible for cytostasis and respiratory inhibition in tumor target cells. *Journal of Experimental Medicine*, 169, 1543–1555.

Suzuki, I., Saito, H., Inoue, S., Migita, T. and Takahashi, T. (1979) Purification and characterization of two lectins from *Aloe arborescens* Mill. *Journal of Biochemistry*, 85, 163–171.

Tanaka, N., Ishihara, M., Kitagawa, M., Harada, H., Kimura, T., Matsuyama, T., Lampher, M.S., Aizawa, S., Mak, T.W. and Taniguchi, T. (1994) Cellular commitment to oncogene-induced transformation or apoptosis is dependent on the transcription factor IRF-1. *Cell*, 77, 829–839.

't Hart, L.A., Nibbering, P.H., van den Barselaar, M.T., van Dijk, H., van den Berg, A.J. and Labadie, R.P. (1990) Effects of low molecular constituents from *Aloe vera* gel on oxidative metabolism and cytotoxic and bactericidal activities of human neutrophils. *International Journal of Immunopharmacology*, 12, 427–434.

't Hart, L.A., van den Berg, A.J., Kuis, L., van Dijk, H. and Labadie, R.P. (1989) An anti-complementary polysaccharide with immunological adjuvant activity from the leaf parenchyma gel of *Aloe vera*. *Planta Medica*, 55, 509–512.

't Hart, L.A., van Enckevort, P.H., van Dijk Zaat, R., de Silva, K.T. and Labadie, R.P. (1988) Two functionally and chemically distinct immunomodulatory compounds in the gel of Aloe vera. *Journal of Ethnopharmacology*, 23, 61–71.

Tizard, I.R., Carpenter, R.H., McAnalley, B.H. and Kemp, M.C. (1989) The biological activities of mannans and related complex carbohydrates. *Molecular Biotherapy*, 1, 290–296.

Tizard, I.R., Busbee, D., Maxwell, B. and Kemp, M.C. (1994) Effects of acemannan, a complex carbohydrate, on wound healing in young and old rats. *Wounds*, 6, 201–209.

Tizard, I.R., Maxwell, B. and Kemp, M.C. (1994) Use of macrophage stimulating agents to stimulate wound healing in aged rats. *Ostomy/Wound Management*, 40, 3.

Vazquez, B., Avila, G., Segura, D. and Escalante, B. (1996) Antiinflammatory activity of extracts from Aloe vera gel. *Journal of Ethnopharmacology*, 55, 69–75.

Vilkas, E. and Radjabi-Nassab, F. (1986) The glucomannan system from *Aloe vahombe* (*liliaceae*). III. Comparative studies on the glucomannan components isolated from the leaves. *Biochimie*, 68, 1123–1127.

Viljoen, A. and van Wyk, B.E. (1988) Are chemical compounds reliable taxonomic signposts in the genus Aloe? Two case studies from series Mitriformes and section Kumara. *Alo*, 35, 62–66.

Wang, Z., Karras, J.G., Howard, R.G. and Rothstein, T.L. (1995) Induction of bcl-x by CD40 engagement rescues slg-induced apoptosis in murine B cells. *Journal of Immunology*, 155, 3722–3725.

Winters, W.D., Benavides, R. and Clouse, W.J. (1981) Effects of aloe extracts on human normal and tumor cells *in vitro*. *Economic Botany*, 35, 89–95.

Womble, D. and Helderman, J.H. (1988) Enhancement of allo-responsiveness of human lymphocytes by acemannan (Carrisyn). *International Journal of Immunopharmacology*, 10, 967–974.

Womble, D. and Helderman, J.H. (1992) The impact of acemannan on the generation and function of cytotoxic T-lymphocytes. *Immunopharmacology and Immunotoxicology*, 14, 63–77.

Wright, C.S. (1936) Aloe vera in the treatment of roentgen ulcers and telangiectasis. *Journal of the American Medical Association*, 106, 1356–1364.

Xie, Q.W., Cho, H.J., Calaycay, J., Mumford, R.A., Swiderek, K.M., Lee, T.D., Ding, A., Troso, T. and Nathan, C. (1992) Cloning and characterization of inducible nitric oxide synthase from mouse macrophages. *Science*, 256, 25–228.

Xie, Q.-W., Kashiwabara, Y. and Nathan, C. (1994) Role of transcription factor NF-kB/Rel in induction of nitric oxide synthases. *Journal of Biological Chemistry*, 269, 4705–4708.

Xie, Q.-W., Whisnant, R. and Nathan, C. (1993) Promoter of the mouse gene encoding calcium-independent nitric oxide synthase confers inducibility by interferon γ and bacterial lipopolysaccharide. *Journal of Experimental Medicine*, 177, 1779–1784.

Yagi, A., Harada, N., Shimomura, K. and Nishioka, I. (1987) Bradykinin-degrading glycoprotein in *Aloe arborescens* var. *natalensis*. *Planta Medica*, 53, 19–21.

Yagi, A., Harada, N., Yamada, H., Iwadare, S. and Nishioka, I. (1982) Antibradykinin active material in *Aloe saponaria*. *Journal of Pharmaceutical Science*, 71, 1172–1174.

Yagi, A., Hamada, K., Mihashi, K., Harada, N. and Nishioka, I . (1984) Structure determination of polysaccharides in *Aloe saponaria* (Hill.) Haw. (Liliaceae). *Journal of Pharmaceutical Science*, 73, 62–65.

Yagi, A., Machii, K., Nishimura, H., Shida, T. and Nishioka, I. (1985) Effect of aloe lectin on deoxyribonucleic acid synthesis in baby hamster kidney cells. *Experientia*, 41, 669–671.

Yagi, A., Makino, K., Nishioka, I. and Kuchino, Y. (1977) Aloe mannan, polysaccharide, from *Aloe arborescens* var. *natalensis*. *Planta Medica*, 31, 17–20.

Yates, K.M., Rosenberg, L.J., Harris, C.K., Bronstad, D.C., King, G.K., Biehle, G.A.,Walker, B., Ford, C.R., Hall, J.E. and Tizard, I.R. (1992) Pilot study of thhe effect of acemannan in cat infected with feline immunodeficiency virus. *Veterinary Immunology and Immunopathology*, 35, 177–189.

Yoshimoto, R., Kondoh, N., Isawa, M. and Hamuro, J. (1987) Plant lectin, ATF1011, on the tumor cell surface augments tumor-specific immunity through activation of T cells specific for the lectin. *Cancer Immunology and Immunotherapy*, 25, 25–30.

Zhang, L. and Tizard, I.R. (1996) Activation of a mouse macrophage cell line by acemannan: the major carbohydrate fraction from *Aloe vera* gel. *Immunopharmacology*, 35, 119–128.

Zhu, H.-G., Zollner, T.M., Klein-Franke, A. and Anderer, F.A. (1993) Activation of human monocyte/macrophage cytotoxicity by IL-2/IFNg is linked to increased expression of an anti-tumor receptor with specificity for acetylated mannose. *Immunology Letters*, 38, 111–119.

14 Bioactivity of *Aloe arborescens* preparations

Akira Yagi

ABSTRACT

Bioactive properties of *Aloe arborescens* (a well-known Japanese aloe, Kidachi aloe) are discussed based on the classification of low molecular weight, high molecular weight and whole leaf extract. Both clinical studies and side-effects of the whole leaf extract are also mentioned. The application of *Aloe arborescens* preparations is quite different from that of *Aloe barbadensis*, the former of which is available as both an over-the counter drug and health food in Japan.

INTRODUCTION

Bioactivity of *Aloe arborescens* preparations

Arborescens refers to the tree-like habit and *A. arborescens* var. *natalensis* (Kidachi aloe in Japanese, which is also called 'doc-buster'), is naturalized on the west coast (West Sonogi penisula in Nagasaki) and the south coast (Izu penisula and Shikoku island) of Japan. *A. arborescens* hybridises readily with other species of *Aloe* with which it co-occurs. Morphologically, it is hard to discriminate *A. arborescens* from A. *arborescens* var. *natalensis*, and we are studying a phylogenetic analysis of *Aloe* species in an internal transcribed spacer 1 in *A. arborescens* and *A. arborescens* var. *natalensis*. The results up indicate a high sequence similarity matrix between them (unpublished data). The preparations from Kidachi aloe gel are not as popular as those of *A. vera* gel on the market in the U.S.A., because the gel part of Kidachi aloe leaf (leaf length, width and thickness are $200 \times 50 \times 20$ mm on average) is comparatively smaller than the rind part and it is difficult to separate only the gel part from the leaf without any contamination of the yellow bitter barbaloin in the rind. Kidachi aloe preparations have traditionally been applied to cuts and wounds as a folklore medicine. They have also been used as a health food to treat constipation and were registered as a dietary supplement, together with *A. vera* gel preparations by the Japanese health food and nutrition food association (JHNFA) in March 1997. Some preparations of Kidachi aloe are also available on the market as over-the counter (OTC) drugs for the acceleration of gastric secretion, cathartic and dermatological cosmetic uses. Thus, quality control of the preparations was standardized for foods and OTC drugs in Japan. Kidachi aloe preparations are classified into the materials which are composed of low molecular weight, high molecular weight and the freeze-dried powders of whole leaf extract by physicochemical procedures. Following this classification, the bioactivity of *A. arborescens* preparations, together with the side-effects, is presented in this chapter.

LOW MOLECULAR WEIGHT MATERIALS

As low molecular weight materials, phenolic bitter components, such as barbaloin and homonataloin, have been widely examined from the chemical and pharmacological points of view. Chemically, the structures of barbaloin and homonataloin were established as anthrone-*C*-glycosides (Hay and Haynes, 1956; Haynes and Henderson, 1960), which was a rare component in those days. The structure of aloeresin, a *C*-glycosyl chromone, was also determined by Haynes *et al.* in 1970. Pharmacologically, as a cathartic component in aloe extract, barbaloin was established by Mapp and McCarthy (1970) and a *p*-coumaric acid-containing compound (aloeresin A) was found as an antimicrobial component by Dopp (1953), though the structural skeleton of aloesin was not established. Since 1970, several modified compounds related to barbaloin and aloesin have been reported from the *Aloe* species (Figure 14.1).

Isolation of barbaloin

McCarthy reported that the cathartic effect of aloe leaf was caused by yellow oxidizable aloin, and Haynes demonstrated the structure of barbaloin which was equated with aloin, as an anthrone-*C*-glycoside. We first isolated and identified barbaloin from fresh Kidachi aloe leaves in a yield of 0.007% (Makino *et al.*, 1973).

Cathartic effect

Barbaloin administered orally was decomposed to aloe-emodin-9-anthrone and aloe-emodin in the rat large intestine. The metabolites were found to cause an obvious increase of water content in the large intestine. Aloe-emodin-9-anthrone inhibited rat colonic Na^+, K^+ adenosine triphosphatase *in vitro* and increased the paracellular permeability across the rat colonic mucosa *in vivo*. Therefore, it seemed that the increase in water content of the rat large intestine produced by aloe-emodin-9-anthrone was due to both inhibition of absorption and stimulation of secretion without stimulation of peristalsis. Charcoal transport was significantly accelerated at both 3.5 and 6.5 hours after the administration of barbaloin. At 6.5 hours, diarrhoea instead of normal feces was observed. Moreover, at one hour before the acceleration of charcoal transport, a marked increase in the relative water content of the large intestine was observed. It appears that the increase in water content of the large intestine induced by barbaloin precedes the stimulation of peristalsis attended by diarrhea. Therefore, it is suggested that the increase in water content is a more important factor than the stimulation of peristalsis in the diarrhea induced by barbaloin (Ishii *et al.*, 1994).

Antihistamine and antiinflammatory activities of barbaloin

Inhibitory effect (ID 50) of barbaloin on antigen (dinitrophenyl-bovine serum albumin) and compound 48/80-induced histamine from mast cells degranulation was shown to

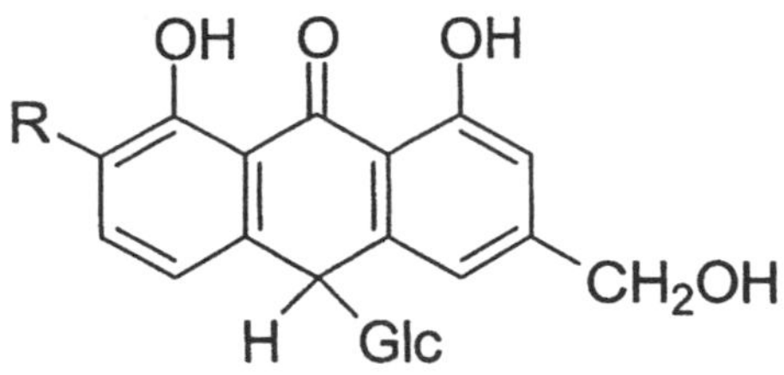

R=H, Barbaloin; R=OH, homonataloin

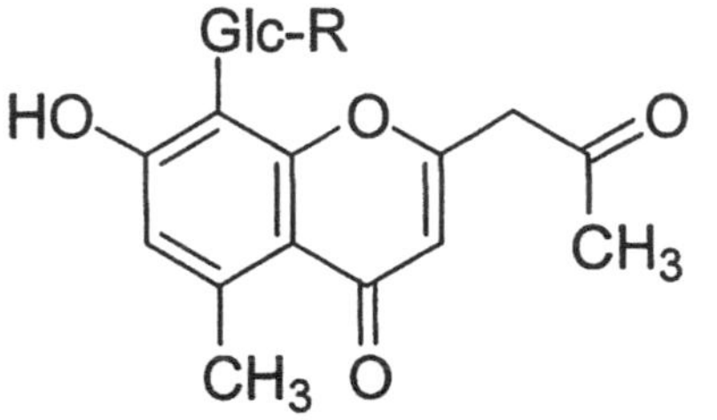

R=H, Aloesin; R=*p*-coumaroyl, Aloeresin A

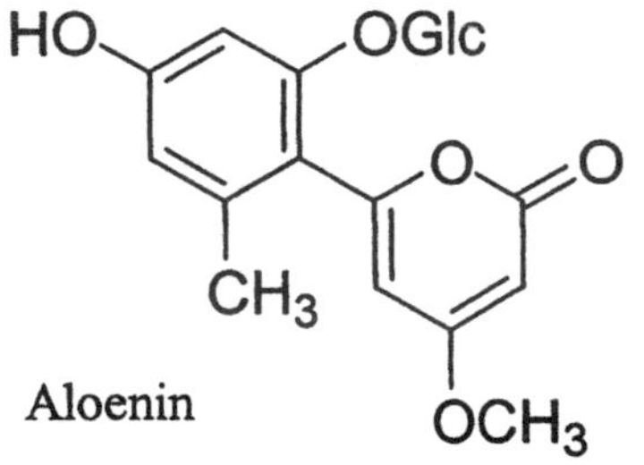

Figure 14.1 Structure of barbaloin, aloesin and aloenin.

be 0.02 and 0.06 mg/ml, respectively, and barbaloin was focused as an important constituent in Kidachi aloe. Anti-inflammatory activity of barbaloin was exhibited by the monitoring of carrageenin-induced edema in rats and 28.8% inhibition was reported, while that of indomethacin was 48.2% (Yamamoto *et al.*, 1991).

Structure of aloesin esters

In a course of study on cathartic components, aloeresin together with barbaloin was shown to be antibiotic. Aloeresin was hydrolyzed into p-coumaric acid, ferulic acid and aloesin. Aloeresin A indicated a p-coumaroyl group located at C-2″ position in aloesin (Makino *et al.*, 1974). Since then, a series of aloesin and aloesol (which is hydrogenated at a ketone group in aloesin), was isolated from *A. vera* leaf extract, and isoaloeresin D showed anti-inflammatory and antioxidative activities (Hutter *et al.*, 1996; Lee *et al.*, 2000).

Skin-whitening aloesin esters

It is well known that Kidachi aloe extract shows skin-whitening activity and the active component is barbaloin. Recently, aloesin esters as well as barbaloin, were found as the active inhibitory components against mushroom tyrosinase, which converts tyrosine into melanin. The active components were determined to feruloyl- and *p*-coumaroyl aloesin (Yagi *et al.*, 1987a). The skin-whitening activity of semi-synthetic aloesin derivatives was demonstrated by monitoring on rat melanoma cells (KB cells) and in clinical tests (Jones, 1999).

Structure of aloenin and its dermatological studies

As one of the phenolic constituents in Kidachi aloe extract, aloenin (0-glycoside of a substituted phenylpyran-2-one) was isolated (Hirata and Suga, 1977). Aloenin was found in 12% of all *Aloe* species, particularly in Section Eualoe, subsection Magnae and Grandes, and utilized as a marker substance of Kidachi aloe preparations in JHNFA. Aloenin promoted hair growth in depilated mice and had recuperative effects on tape-stripped human skin as determined from parameters such as shape factor of corneocytes, thick abrasion, nuclear ghosts and cellular arrangement of corneocytes. Administration of aloenin before the carragenin-induced rat inflammatory edema by interperitoneal route inhibited swelling by 41.7% (Yamamoto *et al.*, 1993).

Acceleration of gastric secretion by barbaloin

Effect of Kidachi aloe extract on the gastric secretion in rats after pyloric ligation was examined by oral administration of the powder, mainly contained barbaloin. The results showed that powdered aloe leaf significantly increased total acid amount and acid output in gastric secretion, suggesting gastrostimulating activity of the extract (Tanizawa *et al.*, 1980).

Antihistamine activity of aloeulcin

The chemical properties and inhibition on histamine synthetic enzyme of a new substance, aloeulcin, were demonstrated. The action of aloeulcin, which was fractionated by ethanol-precipitation followed by chromatographical separation, was observed in both *in vitro* and *in vivo* experiments (Yamamoto, 1970).

Gastrointestinal function of aloeulcin

Aloeulcin, a new principle of Cape aloe (corrected to Kidachi aloe by author), and its gastrointestinal function, was demonstrated and aloeulcin inhibited gastric secretion, ulcer index and pH, on the L-histamine-induced aggravation of Shay rat's ulcer (Yamamoto, 1973).

Anti-inflammatory active components

The effect of aloenin, barbaloin and aloe extract was demonstrated on mast cell degranulation and porcine platelet aggregation. Barbaloin showed inhibitory activity on histamine released from mast cells induced by compound 48/80 at a concentration of 1 μg/ml, IC50 of barbaloin was about 40 μg/ml (Nakagome *et al.*, 1986).

HIGH MOLECULAR WEIGHT MATERIALS

In a previous section bioactive low molecular weight materials from the fresh leaf extract were demonstrated. High molecular weight materials such as polysaccharides and glycoproteins, however, are included in the gelatinous pulp. Thus, the gel part has to be mechanically separated from the green rind and water soluble portion or buffer solution of the pulp is purified by dialysis, repeated precipitation with ethanol, or acetone, to provide non-dialyzable freeze-dried materials.

Anti-tumor aloemannan

A main polysaccharide (aloemannan) isolated from Kidachi aloe extract in a pure state showing a single symmetrical peak in the sedimentation diagram (S20,w = 1.55 S) was proved to be a partially acetylated β-D-mannan. The molecular weight was calculated to be approximately 15,000 daltons by equilibrium ultracentrifugation and aloemannan inhibited the anti-tumor activity (inhibition ratio: 48.1%) against the implanted sarcoma-180 in ICR (Institute of Cancer Research, Philadelphia) mouse by intraperitoneal (i.p.) route at a concentration of 100 mg/kg for a 10 days administration. No body change was observed (Yagi *et al.*, 1977).

Structural determination of polysaccharides and their effect on phagocytosis and nitroblue tetrazolium (NBT) reduction

Neutral polysaccharides and a glycoprotein were isolated by gel filtration from the nondialysable fraction. These were shown to contain the following: a linear polymer of a (1 → 6)-O-linked α-D-glucopyranose (molecular weight, 15 Kda); a branched polymer of an arabinogalactan (MW, 30 Kda); a molar ratio of galactose to arabinose = 1:1.5 with two principle linear chains of (1 → 2)-O-L-arabinopyranose and (1 → 2)-O-D-galacto-pyranose at O-2 and O-6 of the D-galactopyranose residue; a linear polymer of a (1 → 4)-O-linked β-D-mannopyranose (MW 40 Kda) with 10% acetyl group; a glyco-protein (MW, 40 Kda; protein, 57%; hexosamine, 4% and carbohydrate, 34%) consisting of Glc, Man, Gal, GlcN, GalN, GlcNAc (2:2:1:1:4:1) and amino acids. The linear chain of (1 → 4)-O-linked β-D-mannopyranose with 10% acetyl group and the glycoprotein enhanced phagocytosis in adult bronchial asthma patients (Yagi *et al.*, 1986).

Arborans A and B with hypoglycemic activity

Arboran A and B, fractionated by gel permeation, showed the following characteristics: arboran A, MW 12 Kda, containing peptide moiety 2.5%; neutral sugar components Rha, Fuc, Ara, Xyl, Man, Gal, Glc; 0.3, 0.2, 0.1, 1.0, 0.2, 1.0, 0.3 in molar ratio; 16.7% of acetyl group; arboran B, MW, 57 Kda, containing peptide moiety 10.4%; neutral components Man, Glc, 10.3:1.0 in molar ratio. When administered IP (inter-peritoneal) to normal mice, both arboran A and B diminished plasma glucose levels in a dose-dependent manner, and showed the marked hypoglycemic effects 24 hours after their administration IP at a concentration of 100 mg/kg. In alloxan-produced hyper-glycemic mice, both compounds also exerted significant blood sugar-lowering activity (Hikino *et al.*, 1986).

Aloctins A and B

Two lectins were isolated by salt precipitation, pH-dependent fractionation and gel filtration. Aloctin A has a molecular weight of c.18 Kda, consists of two subunits and contains more than 18% by weight of neutral carbohydrate. The smaller subunit has a molecular weight of c.7.5 Kda and the larger subunit a molecular weight of c.10.5 Kda. Aloctin B has a molecular weight of c.24 Kda, consisting of two subunits with a molecular weight of c.12 Kda and contains more than 50% by weight of carbohy-drates. The amino acid composition of aloctin A and B has a high proportion of acidic amino acids. Aloctin A showed mitogenic activity on lymphocytes precipitate-forming reactivity with serum proteins, and C3 activating activity via the alternative pathway. Aloctin B showed a strong hemagglutinating activity (Suzuki *et al.*, 1979).

A plant lectin, ATF 1011

A plant lectin, ATF 1011, which was separable from aloctin A and B, lectins purified from the same origin on Sephadex G-100 column and characterized by the molecular weight and the lack of hemagglutination activity towards human erythrocytes, was investigated. ATF 1011 augmented tumor-specific immunity on the tumor cells surface through activation of T cells specific for the lectins (Yoshimoto *et al.*, 1987).

Pharmacological studies of aloctin A

Aloctin A markedly inhibited the growth of a syngenic transplantable fibrosarcoma of mice in ascites form. There is evidence that the inhibition mechanism is host-mediated and is not a direct effect on the tumor cell (Imanishi *et al.*, 1981).

Aloelectin on DNA synthesis in baby hamster kidney cells

A homogeneous glycoprotein (MW 40 Kda) containing 34% carbohydrate was isolated from Kidachi aloe leaf. The glycoprotein was shown to stimulate deoxyribonucleic acid synthesis in baby hamster kidney cells at a concentration of 5 µg/ml and to have the properties of a lectin which reacts with sheep blood cells. The chemical and physical properties of the glycoprotein, aloelectin, are also demonstrated (Yagi *et al.*, 1985).

Biological and pharmacological activity of aloctin A

Aloctin A prepared by Suzuki's method showed the following biological and pharma-cological activities: hemagglutinating activity; cytoagglutinating activity; mitogenic activity on lymphocytes; precipitate-forming reactivity with α2-macroglobulin; com-plement C3-activating activity; inhibition of heat-induced haemolysis of rat erythro-cytes; antitumor effect; antiinflammatory effect; and inhibition of gastric secretion and gastric lesions (Saito, 1993).

Aloctin A as an immunomodulator

Aloctin A, an active substance isolated from Kidachi aloe leaf, showed a range of biological and pharmacological activities. Aloctin A also affected *in vitro* and *in vivo* treatments on the immune response of murine and human lymphoid cells. Immuno-modulatory activity of aloctin A is of importance for the cure of cancer, as well as preventing life-threatening secondary infectious diseases resulting from immuno-suppression (Imanishi, 1993).

A 35 Kda mannose-binding lectin

A novel lectin was isolated from the leaf skin, not pulp, of Kidachi aloe. The native lectin had a molecular weight of 35 Kda and was composed of two subunits of 5.5 Kda and 2.3 Kda, in addition to a major subunit of 9.2 Kda. The native lectin showed hemagglutinating activity toward rabbit but not human and sheep erythrocytes, and especially bound to mannose, indicating that the aloe and snowdrop lectins-the latter of which also bound with mannose-are structurally and functionally similar proteins. The 35 Kda mannose-binding lectin showed strong mitogenic activity toward mouse lymphocytes (Koike *et al.*, 1995a).

The amino acid sequence of a 35 Kda mannose-binding lectin

The complete amino acid sequence of a 35 Kda mannose-binding lectin from the leaf of Kidachi aloe was determined, and the location of an intrachain disulfide bond and *in vivo* proteolysis site of the subunit with molecular mass of 12.2 Kda of the 35 Kda mannose-binding lectin were also reported (Koike *et al.*, 1995b).

Bradykinin-degrading glycoprotein

A homogeneous glycoprotein (MW 40 Kda), containing 50.7% protein, was isolated, from Kidachi aloe extract by precipitation with 60% ammonium sulfate. Aloe glyco-protein had bradykinin-degrading activity on an isolated guinea pig ileum *in vitro*. Peptide analysis using a reversed-phase HPLC coupled with amino acid analysis showed that aloe glycoprotein cleaves the Pro7-Phe8 and Phe8-Arg9 bonds of the bradykinin molecule. The proteolytic action suggests that aloe glycoprotein has carboxy-peptidase N- and P-like activity. These results may provide a pharmaceutical basis for the anti-inflammatory action of Kidachi aloe (Yagi *et al.*, 1987b).

USES OF WHITE LEAF EXTRACT

Whole leaf extract of Kidachi aloe is used not only for over-the-counter drugs and health foods but also for cosmetics and non-official drugs. Thus, the extract and freeze-dried powders of whole leaf was widely studied in Japan.

Bactericidal activity of aloe powder

Bactericidal activity of Kidachi aloe powder and the fraction obtained by ethanol-precipitation was demonstrated against Gram positive and negative microbials. Aloe powder showed the inhibition against *Pseudomonas aeruginosa*, *Trichophyton interdigitale* and *T. asteroides* at a concentration of 125, 5, and 0.5 μg/ml, respectively (Soeda *et al.*, 1960).

Antitumor activity of alomicin

Antitumor activity of alomicin fraction obtained from frozen and ethanol precipitated fraction, followed by Sephadex G 25 gel permeation chromatography, was examined *in vivo* for sarcoma 180 and Ehrlich ascites (EAC) cancers. Alomicin inhibited 60% of EAC at a concentration of 2.5 mg/kg twice, by the IP route, and 100% of sarcoma 180 at a concentration of 100 mg/kg by the IP route in DDS (dorsal dark stripe) mice (Soeda, 1969).

Effect of kidachi aloe extract for irradiation leucopenia

The intravenous and subcutaneous administrations of aloetin for five to seven days at a dose of 1 mg/kg in rabbit, subsequent to the whole body irradiation (182 gamma) from cobalt, resulted in prompt improvement of the white blood cell counts (Soeda *et al.*, 1964).

Bradykininase activity

Bradykininase activity of the fraction consisting of the components of molecular weight higher than 10 Kda was estimated by biological assay on the guinea pig ileum. The results indicated that aloe bradykininase may hydrolyze bradykinin mainly between Gly4 and Phe5 (Fujita *et al.*, 1976).

Immunoreaction of aloe extract with serum proteins of various animals

Aloe extract contains a lectin-like substance which reacts with serum proteins of various animals. In human serum two proteins, α_2-macroglobulin and α_1-antitrypsin, were shown to be reactive with aloe extracts (Fujita *et al.*, 1978a).

Antifungal activity against *Trichophyton mentagrophytes*

The antifungal activity of a lyophilized powder containing aloe leaf homogenate against *Trichophyton mentagrophytes* was investigated to show the minimal inhibition concentration (MIC) of 25 mg/ml by the agar dilution method. The powder containing components with a molecular weight higher than 10 Kda, prepared by filtration followed by dialyzer concentrator, showed MIC against three strains of *T. mentagrophytes* were all 10 mg/ml. The inhibitory activity was fungicidal and lost by heating at 100° C for 30 minutes. Both the whole-leaf powder and the high molecular weight component powder induced various morphological abnormalities in spore and hyphae by the inhibition of spore germination and development of hyphae (Fujita *et al.*, 1978b).

Inhibitory activity on induction of preneoplastic focal lesions in the rat liver

Inhibitory effects of whole leaf powder on induction of preneoplastic glutathione S-transferase (GST-P) positive hepatocyte foci GST-P+ were studied in male rats. The results indicated that 30% Kidachi aloe extract inhibits the promotion and possibly even retards the initiation stage of hepatocarcinogenesis (Tsuda *et al.*, 1993).

Decreased levels of IQ-DNA adducts

To assess mechanisms of chemoprevention of hepatocarcinogenesis by trans-(3)-carotene, DL-α-tochopherol and freeze-dried whole leaves of Kidachi aloe, a formation of 2-amino-3-methylimidazo [4,5-t]quinoline (IQ)-DNA adducts was measured by ^{32}P-post-labeling analysis. CYP1Al(cytochrome P450 1A1) and CYP1A2 protein levels were analyzed by enzyme-linked immunosorbent assay (ELISA). In conclusion, β-carotene and possibly also α-tocopherol and Kidachi aloe were shown to have the potential to reduce IQ-DNA adduct formation, presumably as a result of decreased formation of active metabolites. The results may explain, at least in part, the previously observed inhibitory effects of these compounds on induction of preneoplastic hepatocellular lesions (Uehara *et al.*, 1996).

In vivo effects on experimental *tinea pedis* in guinea-pig feet

Tricophytosis was induced in guinea-pigs and the antifungal effects of Kidachi aloe were evaluated in comparison with lanoconazol, a commercially available antifungal agent. Trichophytosis was induced by inoculation of arthrospores of *Trichophyton metagrophytes* cephalic strain SM-110 on to the plantar part of guinea pig feet. Culture studies after application of 30% freeze-dried Kidachi aloe for ten days showed a 70% growth inhibition compared with the untreated animals. In an *in vitro* experiment, the fraction with a molecular weight of less than 10 Kda and barbaloin showed growth inhibition of *Trichophyton* at a minimum concentration of 75 mg/ml and 200 μg/ml, respectively (Kawai *et al.*, 1998).

Protection on skin injury induced by X-ray irradiation

Protective effects of Kidachi aloe on mouse skin injury induced by soft x-ray-irradiation were examined. The mechanisms on radiation protection were further investigated by measuring scavenger activity of activated oxygen, protective effects of nucleic acid and induction of antioxidative protein. The active fraction, S6–3-b, protecting the skin injury significantly, showed scavenger activity of hydroxyl radicals generated by the Haber-Weiss reaction, suppression of the changes of activity in superoxide dismutase and glutathione peroxidase at seven days after soft x-ray irradiation. Metallothionein, which is a protective low molecular weight protein containing a high cysteine against x-ray irradiation, was induced in the skin and liver against normal mice at 24 hours after administration of fraction S6–3-b (Sato *et al.*, 1990).

Inhibition of hepatoma (HepG 2) cells growth by alomicin

The fraction corresponding to alomicin isolated by Soeda (1969) was investigated against HepG 2 *in vitro*. Inhibition of HepG 2 growth and a decrease in the cells at S-stage (synthesis stage of DNA replication) were observed (Okada *et al.*, 1994a).

Inhibition of HepG 2 cells growth by Kidachi aloe extract

The fraction corresponding to alomicin was further examined against HepG 2 growth by comparison with hepatocyte growth factor and direct inhibition of the fraction was reported (Okada *et al.*, 1994b).

Biochemical properties of carboxypeptidase

A carboxypeptidase was partially purified from Kidachi aloe. Aloe carboxypeptidase was inhibited by metal ions and sulphydryl reagents, indicating a serine carboxypeptidase (Ito *et al.*, 1993).

Mechanism of anti-inflammatory and anti-thermal burn action of carboxypeptidase in rats and mice

Aloe carboxypeptidase was administered intravenously to female ICR mice with inflammation. The enzyme preparation exhibited significant analgesic effects and inhibited vascular permeability in the abdominal region. It also revealed an anti-thermal burn action on the hind paw when was administered to female Wistar rats intravenously (Obata *et al.*, 1993).

Antigastric ulcer effects in rats

The antigastric ulcer effects of Kidachi aloe extract with high molecular weight component(s) A over 5000 d. and B over 50,000 d. were examined by changes of ulcer index with three different experimental models in rats. The suppressive effects were evaluated according to stress, ligation of pylorus and acetic acid test. Both fractions showed a slight suppressive effect on the stressed animals. However, in the groups with pyrolus ligation, aloe extract A significantly disclosed suppressive effects. The same fraction heated at 100 °C for 15 minutes also demonstrated a significant effect on animals with pylorus ligation. Rats induced with acetic acid exhibited a significant healing effect. However, fraction B produced only a slight effect. These results suggest the existence of both suppressive and healing substances on gastric ulcers in the aloe extract within the range of molecular weight 5000–50,000 d. These effects were confirmed by micro- and macroscopic findings (Teradaira *et al.*, 1993).

Hypoglycaemic and antidiabetic effects in mice

Two different components were separated from Kidachi aloe which exhibited hypoglycaemic activity in spontaneously diabetic mice and normal mice. One component was separated from the succulent layer of the leaf (leaf pulp). This component decreased the blood glucose level by the IP and *per os* (by mouth) routes, and maintained the lower level for about 24 hours. On the other hand, administration of the superficial layer of the leaf skin to streptozotocin (pancreatic islets β-cells toxin)-induced diabetic mice significantly depressed hyperglycaemia and examination of the tissue section under the light microscope revealed less denaturation and necrosis of islets β-cells. These results indicate that Kidachi aloe relieves the diabetic condition by direct hypoglyceration and activates β-cells (Beppu *et al.*, 1993).

CLINICAL USES

Clinically, Kidachi aloe extract has been applied for cathartic and topical uses as an OTC-drug in the Japanese market, and one of the active components is barbaloin. Apart from the medical uses of barbaloin, the following clinical study was reported.

Clinical report of hepato-cirrhosis on the freeze-dried whole leaf extract

α-Fetal protein content continuously and significantly decreased after one to three years administration (5–10 ml/day) of Kidachi aloe extract. Although liver enzyme level (glutamic-oxaloacetic transaminase (GOT) and glutamic-pyruvic transaminase (GPT)) were not decreased, the levels of albumin, cholinesterase and hepaplastin were recovered after administration of the extract (Yukawa *et al.*, 1990).

Effect of high molecular weight substances on peripheral phagocytosis in adult bronchial asthma patients

Oral administration of Kidachi aloe extract for six months showed efficacy for chronic bronchial asthmatic patients of various ages as well as intrinsic types. The extract was not effective for the patients who had previously been administered a steroid drug. An active fraction of the extract was separated by gel permeation chromatography and glycoprotein, and polysaccharide fractions clearly enhanced activity of both phagocytosis and nitroblue tetrazolium reduction in a dose-dependent fashion. These findings suggest that polysaccharide and glycoprotein fractions act as an immunoprotector (Shida *et al.*, 1985).

Effect of amino acids on phagocytosis by peripheral neutrophil in adult bronchial asthma

The dialysable material from Kidachi aloe fresh leaf extract was examined in phagocytosis and a phagocytic killing test using *Candida albicans*. The results from the assay and comparative study on amino acid compositions in the active fraction separated, showed a positive participation by cysteine and proline in phagocytosis. A mixture of these two amino acids (1:1) significantly enhanced the depressed phagocytosis of neutrophils in adult bronchial asthmatic patients (Yagi *et al.*, 1987c).

Clinical effects of aloe extract (ECW) cream on dry skin

Clinical effects of the preparation (cream) composed of 1% Kidachi aloe (ECW), which was obtained by precipitation with ethanol followed by charcoal treatment, (0.2% dl-tocopherol and 0.1% gamma-olizanol) were studied in comparison with active placebo (AP) cream. The results of the application of ECW cream and AP cream in a two-week period showed 71% and 58% in usefulness, respectively. ECW cream was significantly superior to AP cream on the improvement of scale formation and dryness of the skin. No side-effect was seen at all (Nagashima *et al.*, 1987).

Enhancement of acetaldehyde metabolism by Kidachi aloe extract in human liver

Seventeen healthy volunteers (aged between 20–35 years, 5 of which were male, 12 female) were divided into non-flusher (n=10) and flusher (n=7) groups. The tablets made from Kidachi aloe freeze-dried powders (Yurikaron R) and colured placebo tablets were administered, and after 30 minutes alcohol was administered at a concentration of 0.5 g/kg within five to ten minutes. Blood alcohol and acetaldehyde concentrations were measured by using blood alcohol UV-test and the acetaldehyde UV-method, respectively, together with several blood tests. The results suggested that the aloe tablets significantly promote acetaldehyde metabolism of a flusher group in liver (Kawai *et al.*, 1996).

DELETERIOUS EFFECTS

As plant sources for allergic contact dermatitis, exudates of ginkgo and lacquer tree are well known in Japan and Kidachi aloe gel portion is topically applied to cure an itchy dermatitis. However, several case studies of Kidachi aloe jelly portion were reported on allergic contact dermatitis as follows.

A case study of contact dermatitis to Kidachi aloe

A 66-year-old Japanese male with allergic dermatitis to Kidachi aloe leaf gel had generalized nummular and papular dermatitis after 20 days use of topical aloe jelly. Patch testing with aloe jelly gave a strong positive reaction in the patient but was negative in eight controls. Thus, the positive patch test for aloe jelly is proof of allergic hypersensitivity to Kidachi aloe (Shoji, 1982a, 1982b).

Contact dermatitis from Kidachi aloe

A 7-year-old boy had no personal or family history of contact dermatitis, eczema, asthma or hay fever. When he used the leaf gel of Kidachi aloe, because of a scaly eruption, he had an itchy, erythematous, buroring, papular and edematous eruption around the mouth. A patch test with the fresh leaf gel was positive after 48 hours. Six control subjects were all negative. He made a rapid recovery following the use of topical cortico-steroid. A mechanical irritation caused by crystals (calcium oxalate) was speculated to be the cause (Nakamura and Kotajima, 1984).

Kidachi aloe extract was topically applied to four patients with pruritis cutanea who showed negative patch test results for the leaf extract and red papular dermatitis was caused after 10–20 days use. It was again suggested that the contact dermatitis was caused by crystals, such as calcium oxalate, in the leaf extract (Kono *et al.*, 1981).

Irritant contact dermatitis from Kidachi aloe

A case of irritant contact dermatitis from Kidachi aloe was reported. Calcium oxalate crystals in the leaf gel were proved to be a potent irritant substance and it caused an irritant contact dermatitis resembling an allergic dermatitis (Kubo *et al.*, 1987).

The Japanese health food and nutrition food association (JHNFA)

JHNFA is a public service corporation approved by the Minister of Health and Welfare and was established in 1980 with the aim of encouraging people to develop and maintain healthy dietary habits. JHNFA recognized 47 business items including 1081 commodities as health foods until March, 2000. Both *Aloe arborescens* and *Aloe vera* processed preparations were registered with the serial number of 44 as a health food by JHNFA on March, 1997. What follows is a public announcement about the standard of *Aloe arborescens* processed food and the standard of *Aloe vera* processed food translated by Prof. Dr Yagi.

The standard of *Aloe arborescens* processed food:-

A. Applying range

This standard is applied to the powder, granule, and pill type of *Aloe arborescens* Miller var. *natalensis* Berger (abbreviated to *A. arborescens*) food and *A. arborescens* processed food and the liquid type of *A. arborescens* juice and *A. arborescens* drink.

B. Definition

A. arborescens in this standard is the leaf part of *A. arborescens* (Asphodelaceae):

1 *A. arborescens* powder is the powder of dried *A. arborescens*.
2 *A. arborescens* liquid is the liquid prepared from extraction followed by filtration of *A. arborescens*.
3 *A. arborescens* food is the food containing not less than 80% of *A. arborescens* powder.
4 *A. arborescens* processed food is the food containing not less than 50% of *A. arborescens* powder and not more than 80% of *A. arborescens* powder.
5 *A. arborescens* extract is the drink containing not less than 95% of *A. arborescens* liquid.
6 *A. arborescens* drink is the drink containing not less than 50% of *A. arborescens* liquid and not more than 95% of *A. arborescens* liquid.

C. The standard of product

1 Appearance and description:
There are no strange tastes, smells or contamination by foreign substances, except for the original taste and smell from *A. arborescens*.
2 Identifying substance(s) and content:
 a The content of *A. arborescens* powder and *A. arborescens* liquid must be not less than the indicated quantity.
 b On assay of the products, aloenin, a marker substance, corresponding to the quantity indicated must be contained. Quantification of aloenin is demonstrated at the amount of 0.5 g of the test liquid sample after lyophilization and 3.0 g of the test solid sample by HPLC.
3 Arsenic: Not more than 2 ppm, as the liquid type products do not detect arsenic (Limitation: 0.4 ppm).
4 Heavy metal: Not more than 20 ppm as lead. The liquid type products should not detect lead and cadmium (Limitation: Pb 0.4 ppm; Cd 0.1 ppm).
5 Pheophorbide (degradation of toxic products from chlorophyll):
 a Existing pheophorbide: Not more than 30 mg% in solid samples and 10 mg% in liquid.
 b Total pheophorbide: Not more than 50 mg% in solid samples and 15 mg% in liquid.
6 Residual agricultural chemicals:
 a Must not detect Endrin and Dieldrin including Aldrin.
 b BHC: Not more than 0.2 ppm.
 c DDT: Not more than 0.2 ppm.
7 General bacteria: Not more than 5×10^4 cells/g.
The liquid type products do not detect not more than 3×10^2 cells/g.
8 Coliform bacteria: Negative ~

D. The standard for manufacturing and processing, etc is omitted.

E. The standard for claim and advertisement, etc is omitted.

The standard of *Aloe vera* processed food

A. Applying range

This standard is applied to the powder, granule and pill type of *Aloe vera* food and *Aloe vera* processed food, and the jelly or liquid type of *Aloe vera* juice and *Aloe vera* drink.

B. Definition

Aloe vera in this standard is the leaf part of *Aloe barbadensis* Miller(synonym, *A. vera* (L.) Burm.f. Asphodelaceae)

1 *Aloe vera* gel:
Aloe vera gel is the jelly of the pulp removed the skin (rind) of the leaf.
2 *Aloe vera* powder:
Aloe vera powder is the powder of dry *Aloe vera* gel.
3 *Aloe vera* food is the food containing not less than 80% of *Aloe vera* powder.

4   *Aloe vera* processed food:
    Aloe vera processed food is the food containing not less than 50% of *Aloe vera* powder
    and not more than 80% of *Aloe vera* powder.
5   *Aloe vera* juice:
    Aloe vera juice is the drink containing not less than 95% of *Aloe vera* gel.
6   *Aloe vera* drink:

Aloe vera drink is the drink containing not less than 50% of *Aloe vera* gel and not more
than 95% of *Aloe vera* gel.

C. *The standard of product*

1   Appearance and description:
    There are no strange tastes, smells or foreign substances except for the original
    taste and smell from *Aloe vera*.
2   Identifying substance(s) and content:
    a   The content of *Aloe vera* powder and *Aloe vera* gel must be not less than the indi-
        cated quantity.
    b   On assay the product mannose oriented from polysaccharide corresponding to
        the quantity indicated must be contained.
    The content of mannose in *Aloe vera* gel material must not be less than 0.04%, the
    content of mannose in *Aloe vera* powder must not be less than 1.3%, and the content
    of glucose must not be more than 1/10 of the content of mannose.
3   A marker substance, β-sitosterol-3-O-β-D-glucopyranoside (abbreviated to SG),
    which is an original material in *Aloe* species, should be detected.
4   No detection of aloenin.
5   No detection of barbaloin.
6   Negative response for iodine-starch reaction.
7   Arsenic: Not more than 2 ppm as arsenic. The liquid type products do not detect
    arsenic (Limitation, 0.4 ppm.).
8   Heavy metal: Not more than 2 ppm as lead. The liquid type products should not
    detect lead and cadmium (Limitation: Pb, 0.4 ppm; Cd, 0.1 ppm)
9   Residual agricultural chemicals.
    a   Must not detect Endrin and Dieldrin including Aldrin.
    b   BHC: Not more than 0.2 ppm.
    c   DDT: Not more than 0.2 ppm.
10 General bacteria: Not more than 5×10^4 cells/g. The liquid type products do not
    detect not more than 3×10^2 cells/g.
11 Coliform bacteria: Negative.

D. *The standard for manufacturing and processing, etc is omitted.*

E. *The standard for claim and advertisement, etc is omitted.*

F. *Assay methods for mannose and glucose, SG and palmitoyl-glucoside of β-sitosterol glucoside (SPG), aloenin, barbaloin and iodine-starch reaction.*

1   Assay of mannose and glucose:
    Enzyme assay with F-Kit glucose/fructose (Boehringer Ingerheim) is performed.

2   Identification of SPG and SG:
    Thin-layer chromatography (TLC) is carried out with standard samples of SPG and SG, and the ratio in the distance of SG and developed solvent is about 0.150 ± 0.05, and the ratio in the distance of SPG and developed solvent is about 0.4 ± 0.1.

3   Identification of aloenin:
    No detection of aloenin at the amount of 0.5 g of test liquid sample after lyophilization and 3.0 g of test solid sample on TLC.

4   Identification of barbaloin:
    No detection of barbaloin at the amount of 29 g of test sample on TLC.

5   Identification by iodine-starch reaction:
    No colour reaction for the solution of test sample 109 g or test solution 10 ml with iodine reagent.

SUMMARY

Davis showed in his book 'Aloe vera- a scientific approach' that the total biological activity in *Aloe vera* comes from a synergistic modulation of the compounds involved rather than from a single component (Davis, 1997). This is commensurate with Kampo (Oriental) medicine which contains both positive and negative activating materials and acts to normalize or produce a balance in a biological, specially immunoreaction system as an immunomodulator (Yagi *et al.*, 1986; Sato *et al.*, 1990). Kidachi aloe has been widely utilized for several diseases as a folklore medicine in Japan. Some ingredients in water media of the leaf antagonize or synergize with each other to produce a harmonizing efficacy between wound-healing and anti-inflammatory activity through an immunoreaction. This balance in Kidachi aloe gives us a good understanding as to why it is known as of 'doc-buster'. Kidachi aloe extract is not a only a foodstuff but it is also registered as both a OTC-drug (aloin content: about 3 mg/tablet, a daily up-take for children, 1–2 tablets) and a health and nutrition food in Japan. However, even a small amount of aloin contained in 'Kidachi aloe candy', may produce a pharmacological side-effect in children who consume several pieces. In fact, aloin was included in the list of twelve restricted components for which the European Economic Community (EEC) Council Directive 88/388 fixed the maximum allowable concentrations (MAC) in food products and beverages owing to their pharmacological activities. When aloin concentration in Japanese candy was compared with that in the EEC MAC, an exceedingly high level of aloin (0.04–0.2 mg/g; 0.1–0.8 mg/a piece of candy) was found in Japanese candy (Yamamoto *et al.*, 1985). Since the ingestion amount of Kidachi aloe extract as food and candy is not restricted in Japan, the minimum concentration of aloin in Kidachi aloe food preparations and candy does, at least, need to be labelled.

ACKNOWLEDGEMENT

The author appreciates very much the advice of Prof H. Kuzuya and Dr H. Beppu, Fujita Memorial Division of Pharmacognosy, Institute for Comprehensive Medical Science, School of Medicine, Fujita Health University.

REFERENCES

Beppu, H., Nagamura, Y. and Fujita, K. (1993) Hypoglycaemic and antidiabetic effects in mice of Kidachi aloe. *Phytotherapy Research*, 7, S37–S42.

Davis, R.H. (1997) *Aloe vera-a scientific approach*. New York: Vantage Press Inc.

Dopp, W. (1953) Die tuberkulostatische wirkung der *Aloe* und ihrere wichtigsten inhaltsstoffe in vitro. *Arzneimittel-Forschung*, 3, 627–630.

Fujita, K., Teradaira, R. and Nagatsu, T. (1976) Bradykininase activity of aloe extract. *Biochemical Pharmacology*, 25, 205.

Fujita, K., Suzuki, I., Ochiai, J., Shimpo, K., Inoue, B. and Saito, H. (1978a) Specific reaction of aloe extract with serum proteins of various animals. *Experientia*, 34, 523–524.

Fujita, K., Yamada, Y., Azuma, K. and Hirozawa, S. (1978b) Effect of leaf extracts of *Aloe arborescens* subsp.*natalensis* on growth of *Trichophyton metagrophytes*. *Antimicrobial agents and chemotherapy*, 14, 132–136.

Hay, J.E. and Haynes, L.J. (1956) The aloins. Part I. The structure of barbaloin. *Journal of the Chemical Society*, 3141–3147.

Haynes, L.J. and Henderson, J.I. (1960) Structure of homonataloin. *Chemistry and Industry*, 50.

Haynes, L.J., Henderson, J.I. and Russell, R. (1970) *C*-Glycosyl compounds. Part VI. Aloesin, a *C*-glucosylchromone from *Aloe* sp. *Journal of the Chemical Society*, 2581–2586.

Hikino, H., Takahashi, M., Murakami, H., Konno, C., Mirin, Karikura, M. and Hayashi, T. (1986) Isolation and hypoglycemic activity of Arboran A and B, glycans of *Aloe arborescens* var.*natalensis*. *International Journal of Crude drug Research*, 24, 183–186.

Hirata, T. and Suga, T. (1977) Biologically active constituents of leaves and roots of *A. arborescens* var. *natalensis*. *Zeitschrift fur Naturforschung*, 32 c, 731–734.

Hutter, J.H., Salman, M., Stavinoha, W.B., Satsangi, N., Williams, R.F., Streeper, R.T. and Weintraub, S.T. (1996) Antiinflammatory *C*-glucosylchromone from *A.barbadensis*. *Journal of Natural Products*, 59, 541–543.

Imanishi, K. (1993) Aloctin A, an active substance of *Aloe arborescens* as an immunomodulator. *Phytotherapy Research*, 7, S20-S22.

Imanishi, K., Ishiguro, T., Saito, H. and Suzuki, I. (1981) Pharmacological studies on a plant lectin, Aloctin A, I. Growth inhibition of mouse methylcholanthrene-induced fibrosarcoma (Meth A) in ascites form by Aloctin A. *Experientia*, 37, 1186–1187.

Ishii, Y., Tanizawa, H. and Takino, Y. (1994) Studies of Aloe V. Mechanism of cathartic effect. *Biological & Pharmaceutical Bulletin*, 17, 651–653.

Ito, S., Teradaira, R., Beppu, H., Obata, M. and Fujita, K. (1993) Biochemical properties of carboxypeptidase from *Aloe arborescens* var.*natalensis*. *Phytotherapy Research*, 7, S26–S29.

Jones, K. (1999) *Skin whitening activity of aloesin and derivatives*. 18th Annual Scientific Seminar International Aloe Science Council.

Kawai, K., Chihara, T., Beppu, H., Shimeta, Y., Ito, M. and Nagatsu, T. (1996) Effects of *Aloe arborescens* var.*natalensis* on acetaldehyde metabolism in human liver. *Medicine and Biology* (Japanese), 132, 43–48.

Kawai, K., Beppu, H., Shimpo, K., Chihara, T., Yamamoto, Nagatsu, T., Ueda, H. and Yamada, Y. (1998) *In vivo* effects of *Aloe arborescens* var. *natalensis* Berger (*Kidachi aloe*) on experimental tinea pedis in guinea-pig feet. *Phytotherapy Research*, 12, 178–182.

Koike, T., Beppu, H., Kuzuya, H., Nurata, K., Shimpo, K., Suzuki, M., Titani, K. and Fujita, K. (1995a) A 35KD mannose-binding lectin with hemagglutinating and mitogenic activities from Kidachi aloe. *Journal of Biochemistry*, 118, 1205–1210.

Koike, T., Titani, K., Suzuki, M., Beppu, H., Kuzuya, H., Maruta, K., Shimpo, K. and Fujita, K. (1995b) The complete amino acid sequence of a mannose-binding lectin from Kidachi aloe. *Biochemical Biophysical Research Communications*, 214, 163–170.

Kono, A., Saruta, T., Okuma, S. Nakamizo, Y. and Ichikawa, T. (1981) Contact dermatitis from Aloe arborescens. *Rinshyo derma (Tokyo)*, 23, 173–178.

Kubo, Y., Nonaka, S. and Yoshida, H. (1987) Irritant contact dermatitis from Aloe arborescens. *Skin Research*, **29**, 209–212.

Lee, K.Y., Weintraub, S.T. and Yu, B.P. (2000) Isolation and identification of a phenolic anti-oxidant from *A. barbadensis*. *Free Radical Biology & Medicine*, 28, 261–265.

Makino, K., Yagi, A. and Nishioka, I. (1973) The structure of aloearbonaside, a glucoside of a new naturally occurring chromone. *Chemical & Pharmaceutical Bulletin*, 21, 149–156.

Makino, K., Yagi, A. and Nishioka, I. (1974) The structures of two new aloesin esters. *Chemical & Pharmaceutical Bulletin*, 22, 1565–1570.

Mapp, R.K. and McCarthy, T.J. (1970) The assessment of purgative principles in aloes. *Planta Medica*, 18, 361–365.

Nagashima, M., Shiohara, T., Nagashima, T., Kameyama, S. and Hayashi, T. (1987) Clinical effects of aloe extract (ECW) cream on dry skin. *Skin Research*, 29, 989–994.

Nakagome, K., Oka, S., Tomizuka, N., Yamamoto, M., Masu Nakazawa, H. (1986) A novel biological activity in Aloe components: Effect on mast cell degranulation and platelet aggregation. *Report of Fermentation Research Institute*, 63, 3–29.

Nakamura, T. and Kotajima, S. (1984) Contact dermatitis from aloe arborescens. *Contact Dermatitis*, 11, 51.

Obata, M., Ito, S., Beppu, H. and Fujita, K. (1993) Mechanism of antiinflammatory and anti thermal burn action of Cpase from Kidachi aloe in rats and mice. *Phytotherapy Research*, 7, 530–533.

Okada, K., Doi, R., Mimura, K., Nishigan, S., Ueda, H. *et al*. (1994a) Inhibition hepatoma cells (HepG 2) growth by Aloe component (alomicin). *The Journal of Japanese Society for Cancer Therapy*, 29, 1191.

Okada, K., Doi, R., Mimura, K., Ideya, I., Ueda, H., Fujiwara, S. *et al*. (1994b) Inhibition of HepG 2 cells growth by Aloe extract. *Acta Hepatologica Japonica*, 35, S270.

Saito, H. (1993) Purification of active substances of *Aloe arborescens* Miller and their biological and pharmacological activity. *Phytotherapy Research*, 7, S14–S19.

Sato, Y., Ohta, S. and Shinoda, M. (1990) Studies on chemical protectors against radiation. XXXI. Protection effects of *Aloe arborescens* on skin injury induced by x-irradiation. *Yakugaku Zasshi*, 110, 876–884.

Shida, T., Yagi, A. Nishimura, H. and Nishioka, I. (1985) Effect of *Aloe* extract on the peripheral phagocytosis in adult bronchial asthma. *Planta Medica*, 51, 273–275.

Shoji, A. (1982a) A case study of contact dermatitis from *Aloe arborescens*. *Japanese Journal of Clinical Dermatology*, 29, 761–763.

Shoji, A. (1982b) Contact dermatitis to *Aloe arborescens* (jelly). *Contact Dermatitis*, 8, 164–167.

Soeda, M., Otomo, M., Aoume, E. and Kawashima, K. (1960) Bactericidal activity of Aloe powder. *Japanese Journal of Bacteriology*, 21, 1609–1613.

Soeda, M., Fujiwara, M. and Otomo, M. (1964) Studies on the effect of Cape Aloe (correct to *A. arborescens*) for irradiation leucopenia. *Nippon Acta Radiology*, 24, 1109–1112.

Soeda, M. (1969) Studies on the anti-tumor activity of Cape aloe (correct to *A. arborescens*). Isolation of alomicin with an anti-tumor activity. *Journal of Medical Society of Toho University, Japan*, 16, 365–369.

Suzuki, I., Saito, H., Inoue, S., Migita, S. and Takahashi, T. (1979) Purification and character-ization of two lectins from *Aloe arborescens*. *Journal of Biochemistry*, 85, 163 –171.

Tanizawa, H., Konagai, M. and Takino, Y. (1980) Effect of Aloe on the gastric secretion in rats. *Kiso to Rinshyo*, 14, 35–38.

Teradaira, R., Shinzato, M., Beppu, H. and Fujita, K. (1993) Antigastric ulcer effects in rats of Kidachi aloe extract. *Phytotherapy Research*, 7, S34–S36.

Tsuda, H., Matsumoto, K., Ito, M., Hirono, I., Kawai, K., Beppu, H., Fujita, K. and Nagao, M. (1993) Inhibitory effect of *Aloe arborescens* var. *natalensis* on induction of preneoplastic focal lesions in the rat liver. *Phytotherapy Research*, 7, S43–S47.

Uehara, N., Iwahori, Y., Asano, M., Baba-Toriyama-Baba, H., Ligo, M., Ochiai, M. *et al.* (1996) Decreased levels of 2-amino-3-methylimidazo[4,5-f]quinoline-DNA adducts in rat treated with beta-carotene, alpha-tocopherol and freeze-dried aloe. *Japanese Journal of Cancer Research*, **87**, 342–348.

Yamamoto, M., Sugiyama, K., Yokota, M., Maeda, Y. and Inaoka, Y. (1993) Study of possible pharmacological actions of *A. arborescens* on mouse, hamster and human skin. *Japanese Journal of Toxicology and Environmental Health*, **39**, 409–414.

Yamamoto, M., Ishikawa, M., Masui, T., Nakazawa, H. and Kabasawa, T. (1985) Liquid chromatographic determination of barbaloin (aloin) in foods. *Journal of the Association of Official Analytical Chemist*, **68**, 493–494.

Yamamoto, M., Masui, T., Sugiyama, K., Yokota, M., Nakagomi, K. Nakazawa, N. (1991) Anti-inflammatory active constituents of *A. arborescens Agricultural & Biological Chemistry*, **55**, 1627–1629.

Yamamoto, I. (1970) A new substance, aloeulcin, its chemical properties and inhibition on histamine synthetic enzyme. *Journal of Medical Society of Toho University, Japan*, **17**, 361.

Yamamoto, I. (1973) Aloeulcin, a new principle of Cape Aloe and gastrointestinal function, especially experimental ulcer in rats. *Journal of Medical Society of Toho University, Japan*, **20**, 342–347.

Yagi, A., Makino, K., Nishioka, l. and Kuchino, Y. (1977) Aloemannan, polysaccharide from *Aloe arborescens* var. *natalensis*. *Planta Medica*, **31**, 17–20.

Yagi, A., Nishimura, H., Shida, T. and Nishioka, I. (1986) Structure determination of polysaccharide in *Aloe arborescens* var. *natalensis, Planta Medica*, **52**, 213–218.

Yagi, A., Kanbara, T. and Morinobuyu, N. (1987a) Inhibition of mushroom-tyrosinase by *Aloe* extract. *Planta Medica*, **53**, 515–517.

Yagi, A., Harada, N., Shimomura, K. and Nishioka, I. (1987b) Bradykinin-degrading glycoprotein in *Aloe arborescens* var. *natalensis, Planta Medica*, **53**, 19–21.

Yagi, A., Machii, K., Nishimura, H., Shida, T. and Nishioka, I. (1985) Effect of Aloe lectin on deoxyribonucleic acid synthesis in baby hamster kidney cells. *Experientia*, **41**, 669–671.

Yagi, A., Shida, T. and Nishimura, H. (1987c) Effect of amino acids in *Aloe* extract on phagocytosis by peripheral neutrophile in adult bronchial asthma. *Allergy*, **36**, 1094–1101.

Yoshimoto, R., Kondoh, N., lsawa, M. and Hamuro, J. (1987) Plant lectin, ATF 1011, on the tumor cell surface augments tumor-specific immunity through activation of T cells specific for the lectin. *Cancer Immunology and Immunotherapy*, **25**, 25–30.

Yukawa, S., Hasegawa, S., Kawano, K. and Nomoto, T. (1990) Clinical report of Kidachi aloe extract for hepato-cirrohosis. *Journal of Medical and Pharmaceutical Society for Wakan Yaku*, **7**, 504.

Part 4

Aloe biology

15 The chromosomes of *Aloe* – variation on a theme

Peter Brandham

ABSTRACT

The chromosome complement of the genus *Aloe* has long been considered to be exceptionally stable. All species have a single basic chromosome number ($x=7$) and a large, strongly bimodal karyotype always comprising three short chromosomes and four much longer ones in the haploid set. Nevertheless, major and minor changes in chromosome morphology occur frequently, together with a two-fold range of interspecific variation in overall nuclear DNA amount, with the low values in primitive species. Polyploidy is now known to be more frequent than was believed earlier, with a record of triploidy ($2n=21$) and several of tetraploidy ($2n=28$) added recently to the single well-known example of hexaploidy in the genus ($2n=42$). The evolutionary significance of these numerical and structural changes in the chromosomes of *Aloe* will be discussed in this chapter.

INTRODUCTION

The genus *Aloe*, together with its relatives *Gasteria, Haworthia, Lomatophyllum, Astroloba* and a few mostly monotypic minor genera, belongs to a natural assemblage of species formerly known as the Tribe Aloineae of the Family Liliaceae sens. lat. The Tribe was elevated to family status, the Aloaceae, but is now included in the Family Asphodelaceae (Brummitt, 1992; Adams *et al.*, 2000a).

In *Aloe*, together with many other petaloid monocotyledons, the chromosomes are large and relatively easy to prepare from root tips or from meiotic anthers in immature flower buds. Consequently, the genus was a popular subject for early chromosome research, e.g. by Resende (1937) but apart from a few instances of polyploidy and of variation in the positions of the secondary or nucleolar–organising constrictions, *Aloe* chromosomes appeared to be rather uniform from one species to another. Interest in them therefore waned in favour of more 'interesting' genera. Nevertheless, because the living collections of aloes in the Royal Botanic Gardens Kew have been very large for many years and comprise plants mostly of documented wild origin, they were clearly promising material for research into the mechanisms of chromosomal evolution in a group that appears at first sight to be completely stable. Together with my extensive field-collections they have formed the basis of an extensive cytological survey of chromosome variability in the genus that has been made in the Jodrell Laboratory at Kew from the 1960s onwards. The work has been ably supported by my colleague Margaret Johnson and by students Mary-Jo Doherty and Jon West. Recently additional colleagues, notably Stuart Adams, have extended the study in other directions with the use of fluorescence microscopy to reveal fine details of chromosome structure. They have found a level of intra- and inter-specific chromosomal variability much higher than was previously known to occur in the genus. This variability will be surveyed below and its significance indicated.

CHROMOSOME NUMBER AND SIZE

Cytologically, *Aloe* and its close relatives are remarkable for the uniformity of the basic number and the gross morphology of their chromosomes. Every species that has been investigated cytologically (the great majority of the described taxa) has the basic number $x=7$ and a karyotype that is always bimodal (Brandham, 1971). The haploid karyotype comprises three short acrocentric chromosomes and four long ones, of which three are acrocentric and one, the L_1, submetacentric with a more substantial short arm. All of the chromosomes are large compared with those of most angiosperms, with the short ones averaging 1.5–3 µm in length, and the long ones almost exactly 4.6 times larger in a sample of *Aloe* species (Brandham and Doherty, 1998).

Brandham and Doherty (1998) showed that there is considerable interspecific variation in overall chromosome length in *Aloe*, as reflected in the nuclear DNA C-value, which ranges in diploid species from $4C=41.78$ picograms (pg) in *A. tenuior* Haw. to 95.4 pg in *A. peckii* Bally et Verdoorn. Although the number of species examined was not large, there was a strong indication that the primitive species, e.g. *A. tenuior*, had the lowest nuclear DNA amount (and thus the smallest chromosome set), and that the amount of DNA and overall chromosome size increased with evolutionary advancement. It was noted by Brandham and Doherty (1998) that despite the addition of so much nuclear DNA to the chromosomes of advanced species the ratio of the sizes of their long and short chromosomes was the same as that of the primitive species.

NUCLEOLAR CONSTRICTION AND rDNA SITE VARIATION

Since the work of Resende (1937), the nucleolar organising or secondary constrictions of *Aloe* chromosomes have been known to vary in position from one chromosome to another in different species. They usually occur distally on the long arms of one, sometimes two pairs of the long chromosomes, but not always on the same member(s) of the set of four. Less commonly, they are found distally on the short arms of the short chromosomes (Brandham, 1971). These correspond to the 18S-5.8S-26S rDNA sites that have been demonstrated in the same positions by Adams *et al.* (2000a) using fluorescence *in situ* hybridization stain technology.

Adams *et al.* (2000a) also demonstrated numerous other smaller 18S-5.8S-26S rDNA sites that could not be resolved as non-staining gaps with the Feulgen staining method used in earlier studies. These sites were found in several different places in the karyotype, revealing a degree of inter-specific variation in the detailed structure of the chromosomes that had not been suspected earlier.

Adams *et al*. (2000a) further examined the distribution of 5S rDNA sites and found that, in contrast to the above, all studied species had one in each haploid set placed interstitially on the long arm of the L2. The chromosomes of *Aloe* are thus very conserved with respect to this feature.

TELOMERE IRREGULARITY

In the majority of plant families the telomeres of the chromosomes are of the *Arabidopsis* type, containing many repeat copies of the 5'-TTTAGGG-3' DNA sequence (Cox *et al*., 1993). This feature has been demonstrated to be absent from the chromosomes of *Aloe*, a peculiar characteristic known to be shared only with some members of the Alliaceae (Adams *et al*., 2000b). The occurrence of this phenomenon is currently under investigation in genera and families related to *Aloe* to a greater or lesser extent, and is expected to be of great significance in the familial classification of this group of monocotyledons.

MAJOR STRUCTURAL CHANGES

Major structural changes are remarkably frequent in the chromosomes of *Aloe* (Brandham, 1976; Brandham and Johnson, 1977). These are sometimes paracentric or pericentric inversions, but more frequently take the form of inter-chromosomal movements of segments of material of different lengths. They are either transposed from one chromosome to another or mutually exchanged between chromosomes, and are often large enough to alter the gross morphology of the chromosomes involved. They can also appear as Robertsonian interchanges, in which two of the acrocentric or submetacentric chromosomes break at or near the centromere and re-join to produce a short metacentric from the two short arms and a very long metacentric from the two long arms (Brandham, 1976).

Although these structural changes have been shown to be very common in wild populations of some species, as found in mapped and analysed populations of *A. pubescens* Reynolds and *A. rabaiensis* Rendle, they are never found in the homozygous condition (Brandham and Johnson, 1977). They are always in the heterozygous condition, with a single long fusion chromosome and a single short one, the homologues retaining their normal appearance. There seems to be selection against the formation of homozygotes in the wild, even though they have been produced in cultivation (Brandham, 1983). A homozygote for a Robertsonian change involving two pairs of long chromosomes of a diploid aloe will comprise three pairs of short acrocentrics as normal, two pairs of long acrocentrics/submetacentrics, one pair of short metacentrics derived from the fused short arms and one pair of very long metacentrics derived from the fused long arms. Such an individual would thus have a karyotype that is new for *Aloe*, even though the basic number of chromosomes remains unaltered. It would be bivalent-forming at meiosis, fully fertile and genetically isolated from non-interchange plants. If it were to appear in the wild, it would form the origin of a new group with the potential to develop into new species with a karyotype not typical of the genus. Being always heterozygous in the wild, these major structural changes in *Aloe* chromosomes do not stabilise widely; in fact, they do not contribute to the genetic differentiation of species and should therefore be regarded as 'evolutionary noise' (Brandham, 1983; Brandham and Doherty, 1998).

POLYPLOIDY

Polyploidy is uncommon in *Aloe* and is not uniform in its geographical distribution in the genus. In southern Africa, where large numbers of species occur (Reynolds, 1950), only *A. ciliaris* Haw. is known to be polyploid, with *A. ciliaris* var. *ciliaris* being hexaploid (2n=42). This is the only known case of hexaploidy in the entire genus, and has been known for many years, since it was first discovered by Resende (1937). More recently, other varieties of *A. ciliaris* were found not to be hexaploid. *A. ciliaris* var. *redacta* S. Carter is tetraploid (2n=28). This and the hexaploid have been shown by Brandham and Carter (1990) to be naturally-occurring intraspecific autopolyploids derived from *A. ciliaris* var. *tidmarshii* Schonl. (also known as *A. tidmarshii* (Schonl.) Muller ex Dyer), which is diploid. As expected in this situation, the morphological differences between the varieties are quantitative rather than qualitative.

Polyploidy is more common among the East African aloes, with Somali *A. inermis* Forsk. and *A. cremnophila* Reynolds, Kenyan *A. juvenna* P. Brandham & S. Carter and Ethiopian *A. jacksonii* Reynolds being tetraploid with 2n=28 (Brandham and Carter, 1982; Brandham *et al.*, 1994) and can be regarded mostly as the products of sporadic doubling of the chromosome numbers of unrelated diploid species that have stabilised through natural selection to produce tetraploid species, albeit rare and local ones. *A. cremnophila*, for instance occurs only on a single cliff in northern Somalia and is quite probably a single clone spreading vegetatively, as suggested on chromosomal and chemical evidence by Brandham *et al.* (1994). Since *A. cremnophila* and *A. jacksonii* show considerable morphological similarity and grow quite close together, there is a possibility that they might be derivatives of the doubling of the chromosome number of a single diploid ancestor and therefore could be very closely related. Within a species, doubling of the diploid chromosome number produces an autotetraploid, but doubling of the chromosome number of a diploid interspecific hybrid produces an allotetraploid. It is not clear which of these alternatives applies to the origin of the above four species.

A more substantial instance of evolutionarily–significant doubling of chromosome number followed by diversification to produce a group of polyploid species has been found by Cutler *et al.* (1980). It was shown that of the shrubby aloes of Kenya, the small and local diploid species *A. morijensis* S. Carter & P. Brandham, or a closely related one, gave rise to a tetraploid that has become vigorous and widespread. This tetraploid has differentiated into a group of related tetraploid species growing in habitats that differ in humidity and altitude but are geographically quite close to each other. The tetraploid group comprises *A. cheranganiensis* S. Carter & P. Brandham, *A. dawei* Berger, *A. elgonica* Bullock, *A. kedongensis* Reynolds, *A. ngobitensis* Reynolds and *A. nyeriensis* Christian, which occupy the Rift valley of central and northern Kenya and the higher ground and mountains on either side of it. This local burst of evolution, involving doubling of chromosome number followed by extensive species differentiation, is the only such case recorded to date in *Aloe*, and is thus not a significant pathway as far as the genus as a whole is concerned.

Interestingly, in an analysis of a population of *A. elgonica*, Brandham and Johnson (1977) found some specimens having 2n=29, with either an extra long or an extra short chromosome. This type of aneuploidy is occasional and is to be expected in tetraploid angiosperms as a result of the irregular 3–1 segregation of quadrivalents during meiosis (rather than 2–2), with subsequent survival of the aneuploid gametes to fuse with normal ones, but it has no evolutionary significance.

There are a few reports of triploidy in *Aloe* (2n=3x = 21). A triploid is a sexually sterile but vegetatively vigorous individual usually originating sporadically from diploid parents following the fusion of a normal gamete with a non-reduced one. Less commonly, triploids arise from diploid-tetraploid hybridisation, but there is no evidence for this occurrence in *Aloe*. Triploids do not persist for long in the wild, except in apomicts and in some long-lived perennials. In *Aloe*, one example of triploidy is in *A. jucunda* Reynolds (Brandham, 1971). This plant was collected in the 1950s by Peter Bally from a single site in Somalia and grown in the Royal Botanic Gardens, Kew, where it continues to flourish today. It is much larger than its small diploid progenitor and closely resembles the more robust related species, *A. somaliensis* W. Watson. It was found again at the same site, growing together with smaller diploids and was confirmed to be triploid by Carter *et al.* (1984). Another earlier record of triploidy in *Aloe* is in *A. humilis* (L.) Mill. (Mitra, 1964). In all probability, this represents a case of the horticulturist's involuntary selection of a vigorous seedling (caused by its triploidy) from a group of normal diploid ones in cultivation.

The above examples of polyploids in *Aloe* comprise all of the cases known to date. With the exception of the case of *A. ciliaris*, polyploidy is not known among the aloes of sub-equatorial Africa, nor of West Africa, Madagascar or the Arabian Peninsula, although many species occurring in these areas have been examined cytologically.

CONCLUSION

Chromosome evolution in *Aloe* is a mixture of stability and variation. It has been shown that polyploidy is uncommon in the genus and is a minor contribution to the evolution of its species. The highly-conserved, orthoselected karyotype of *Aloe* displays a uniform absence of the telomere structure that is otherwise widespread, but shows variability of distribution of nucleolar constrictions and rDNA sites (Adams *et al.*, 2000a). It does not stabilise other major structural chromosome changes as homozygotes, even though they are common as heterozygotes.

Despite the uniformity of the gross karyotype, interspecific variation in its size, as expressed by total nuclear DNA amount, has been demonstrated. Nuclear 4C DNA values increase from lower levels in primitive species to up to twice those levels in advanced species. It has been suggested by Brandham and Doherty (1998) that to maintain the uniformity of relative chromosome size within the karyotype the main sequence of meiotically-pairable chromosome regions in different-sized homeologues (i.e. the same chromosome in different species) is the same in all aloes, but in between these regions many sequences of non-coding DNA are amplified to differing degrees. In advanced species with higher DNA amounts, the larger amplifications further separate the regions capable of crossing over. The enlarged sequences are distributed evenly within all of the chromosomes in proportion to the length of each chromosome. Thus they perpetuate the characteristic gross morphology of each of the seven members of the chromosome set, irrespective of total size, and maintain the constant proportion of total length and DNA volume between the seven chromosomes in the basic *Aloe* set that was found by Brandham and Doherty (1998). Karyotypic orthoselection and evolutionary increase in DNA amount are therefore reconciled in this genus.

REFERENCES

Adams, S.P., Leitch, I.J., Bennett, M.D., Chase, M.W. and Leitch, A.R. (2000a) Ribosomal DNA evolution and phylogeny in *Aloe* (Asphodelaceae). *American Journal of Botany*, 87, 1578–1583.

Adams, S.P., Leitch, I.J., Bennett, M.D. and Leitch, A.R. (2000b) *Aloe* L. – a second family without (TTTAGGG)$_n$ telomeres. *Chromosoma*, 109, 201–205.

Brandham, P.E. (1971) The chromosomes of the Liliaceae. II. Polyploidy and karyotype variation in the Aloineae. *Kew Bulletin*, 25, 381–399.

Brandham, P.E. (1976) The frequency of structural change. In *Current Chromosome Research*, edited by K. Jones and P.E. Brandham, pp.77–87. Amsterdam: Elsevier.

Brandham, P.E. (1983) Evolution in a stable chromosome system. In *Kew Chromosome Conference II*, edited by P.E. Brandham & M.D. Bennett, pp. 251–260. London: George Allen & Unwin.

Brandham, P.E. and Carter, S. (1982) *Aloe juvenna* rediscovered. *Cactus and Succulent Journal of Great Britain*, 44, 70.

Brandham, P.E. and Carter, S. (1990) A revision of the *Aloe tidmarshii/A. ciliaris* complex in South Africa. *Kew Bulletin*, 45, 637–645.

Brandham, P.E., Carter, S. and Reynolds, T. (1994) A multidisciplinary study of relationships among the cremnophilous aloes of northeastern Africa. *Kew Bulletin*, 49, 415–428.

Brandham, P.E. and Doherty, M.-J. (1998) Genome size variation in the Aloaceae, an angiosperm family displaying karyotypic orthoselection. *Annals of Botany*, 82, (Suppl. A), 67–73.

Brandham, P.E. and Johnson, M.A.T. (1977) Population cytology of structural and numerical chromosome variants in the Aloineae (Liliaceae). *Plant Systematics and Evolution*, 128, 105–122.

Brummitt, R.K. (1992) *Vascular Plant Families and Genera*. Royal Botanic Gardens Kew, UK.

Carter, S., Cutler, D.F., Reynolds, T. and Brandham, P.E. (1984) A multidisciplinary approach to a revision of the *Aloe somaliensis* complex (Liliaceae). *Kew Bulletin*, 39, 611–613.

Cox, A.V., Bennett, S.T., Parokonny, A.S., Kenton, A., Callimassia, M.A. and Bennett, M.D. (1993) Comparison of plant telomere locations using a PCR-generated synthetic probe. *Annals of Botany*, 72, 239–347.

Cutler, D.F., Brandham, P.E., Carter, S. and Harris, S.J. (1980) Morphological, anatomical, cytological and biochemical aspects of evolution in East African species of *Aloe* L. (Liliaceae). *Botanical Journal of the Linnean Society*, 80, 293–317.

Mitra, K. (1964) Chromosome number in some plants. *Science and Culture*, 30, 344–345.

Resende, F. (1937) Karyologische Studien bei den Aloinae. II. Das Auftreten von spontanen Mutationen und die Entstehung der SAT-Typen. *Boletim da Sociedade Broteriana*, Ser. 2, 12, 119–137.

Reynolds, G.W. (1950) *The Aloes of South Africa*. Cape Town: The Aloes of South Africa Book Fund.

16 Aloe leaf anatomy

David F. Cutler

ABSTRACT

This paper describes how, in *Aloe* leaves, the appearance of surface sculpturing of the epidermis, as seen by scanning electron microscopy, can be used to help, identify sterile plants or leaf fragments and to indicate relationships between species. Sometimes surface sculpturing may provide some information on habitat preference of the particular species.

INTRODUCTION

The study of aloe leaf anatomy can provide information that is valuable to taxonomists, plant physiologists and biochemists, and ecologists among others. Sometimes flowers and fruits are not available to help in the identification of an aloe. However, the examination of the general form of the plant, together with a detailed look at the appearance of the epidermal (skin) cells of the leaf at a magnification of between 300 and 100× could uncover sufficient characteristics to make an identification. Sometimes identification can be made at species level.

Internal anatomy

The leaf anatomy of aloes was first studied for taxonomic purposes (Cutler, 1969). Later, when it became clear that aloe exudates could be of medicinal interest, anatomy was used to try and locate the cells or tissues in which particular substances arose or were stored. (Beaumont *et al.*, 1985, 1986).

The internal anatomy of aloe leaves is fairly constant, regardless of the species. This means that apart from defining about three groups of miscellaneous species, transverse sections provide little information of wider systematic significance. Thicker leaves have more parenchymatous ground tissue, but the outer chlorenchyma layers, containing the green chloroplasts, and the flattened ring of vascular bundles just to the inner side of the chlorenchyma, are common to all. Initially, interest was focussed on the features that indicated adaptation to particular habitats. For example, thicker-leaved aloe species tend to be found in the harsher environments, where water supply is limited (e.g. *Aloe dichotoma* Masson, *A. pegleri* Schonl.). Thinner leaves occur in species from more moist habitats, for example *A. haemanthifolia* A.Berger et Marloth. Figure 16.1 shows a cross section of a typical leaf of *A. kedongensis* Reynolds.

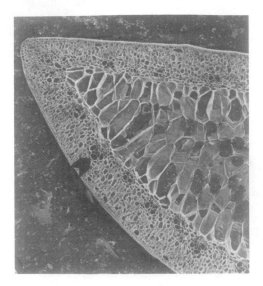

Figure 16.1 Aloe kedongensis: transverse section of part of leaf. Arrow head shows thin walled cells of bundle cap. Magnification × 20.

Of particular interest is the cap of thin-walled, wide cells to the outer side of each vascular bundle. These strands contain exudates that may be coloured, and which are discussed in Chapter 3. The mucilage, which is of medicinal and cosmetic interest, is mainly found in the central cells of the parenchyma. Its function in the plant may be to store water. Beaumont *et al.* (1985, 1986) looked at the origins of the substances present in the bundle cap cells at the phloem poles. In the plant during life, it seems they are colourless, but on exposure to air, they often become coloured. The particular colour, purple, yellow or brown, for example, is typical of the species in which it is found. Granular particles had been observed in the bundle cap cells, and a suggestion is that these may represent precursors to the exudates. However, it still remains to be determined whether the exudates are synthesised in the cap cells themselves, or in other, surrounding cells. The exudates may dry, sealing a cut or damaged leaf. Aloe leaves are generally not damaged by insects or herbivores, so the exudates may also have properties that could deter browsing or insect attack.

Anatomy of leaf surfaces

Initial studies of aloe leaf surfaces with the light microscope indicated a possible rich source of diagnostic characteristics. There appeared to be a range of types of very fine surface sculpturing. This was difficult to see, because the very thick cuticle covering the epidermis had a rough interface with the outer wall of the epidermal cells. The interface appeared granular, and interfered with the interpretation of the fine surface features. However, the scanning electron microscope (SEM) proved ideal for such surface studies, and the beautiful but complex nature of these surfaces became readily understood. Since the electron beam does not penetrate deeply, a clear image of the surface itself could be produced. In addition to the sculpturing of the cuticle itself, the varied nature of epicuticular wax could also be determined (Cutler, 1979). The surface sculpturing was smooth, or composed of fine striations arranged in various ways, or of micropapillae. The size and frequency of the micropapillae are often diagnostic. These sculpturings are superimposed on the general form of the outer wall of the epidermal cells. These walls can be level, domed to varying degrees, or slightly sunken. When they are sunken, the outer edges of the side (anticlinal) walls of the epidermal cells may appear raised. A new terminology had to be developed.

The discovery of this rich source of information of potential taxonomic use was followed by a set of rigorous experiments involving anatomical and cytological studies, and crossing of species. Observations from field collected specimens of *A. ferox* Miller and *A. africana* Miller, which have a very wide distribution, showed little variation in their leaf surface sculpturing (Figure 16.3). Experiments have shown that this epidermal sculpturing is under close genetic control (e.g. Brandham and Cutler, 1978; Cutler *et al.*, 1980; Carter *et al.*, 1984). In crossing experiments between diploid and tetraploid species, it was possible to demonstrate the arms of particular chromosomes having coding for the normal sculpturing characters. Further experiments crossing *Aloe* with *Gasteria* and *Haworthia* species (Cutler, 1972; Cutler and Brandham, 1977) confirmed the strong genetic control of the characteristic sculpturing features. Cuticular sculpturing can therefore be used as a diagnostic tool, helping in the identification of species, or groups of similar species (Cutler, 1985). Two factors come into play. The first of these is relatedness. Closely related species tend to share similar epidermal features. Second, there is some genetic adaptation to environment. Normally this is expressed by surface roughness increasing in line with the ability of the species to withstand dryer and dryer habitats. Both factors influence the final appearance, and sometimes it may be that environmental adaptation overrides characters of relatedness (Cutler DF, 1982). Even so, looking at leaf surfaces under a high power epi-illuminating light microscope, or preferably an SEM, it is often possible to say which *Aloe* species are closely related to one another.

Stomata are usually deeply sunken in species that normally grow in exposed, water-stressed conditions. The stomata are overarched by four well developed lobes, one from each of the four surrounding epidermal cells. Below these is an elliptical, dome like structure, with an axial slit in it. This is formed of cuticle. Only when we look deeper into the stoma do we see the guard cells themselves, and the aperture that can be opened or closed, according to prevailing conditions.

It is easy to detect those species that grow in areas that are generally more moist, such as *A. haemanthifolia*, a species that grows among grasses and other herbaceous vegetation, and *A. ciliaris* Haw., a species that scrambles through shrubs and trees on river banks, since their stomata are not sunken and have no overarching lobes (Cutler, 1982).

The results of these studies confirmed that in aloe leaves the surface sculpturing could be used safely to:

- help identify sterile plants, or leaf fragments;
- indicate relationship status of given species;
- provide some information on habitat 'preference'.

The plates illustrate selections from the range of surface features. Some show the similarity between closely related species, and others striking adaptation to habitat preference. Note that the cuticular sculpturing is sometimes obscured by surface wax. This can be removed chemically to expose the sculpturing beneath, but it too is of diagnostic significance and can provide further clues to the identity of an unknown plant lacking flowers.

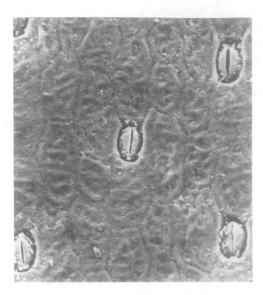

Figure 16.2 Aloe gracilis: leaf surface showing superficial stomata without pronounced surrounding lobes. Anticlinal walls are indicated by shallow grooves; sculpturing consists of striations variously jointed into a loose reticulum. Magnification × 300.

Figure 16.3 Aloe africana: leaf surface with raised ridges above anticlinal walls and small micropapillae. The stoma is deeply sunken and the four surrounding lobes almost occlude the cavity above the guard cells. Magnification × 300.

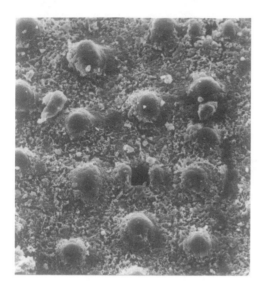

Figure 16.4 Aloe pegleri: leaf surface with copious wax particles partly obscuring the surface sculpturing. Each epidermal cell has a prominent central papilla and the lobes surrounding a sunken stoma are well developed. Magnification × 300.

Figure 16.5 Aloe mutabilis: leaf surface, each epidermal cell has a domed outer wall with smaller micropapillae surrounding larger central ones. Lobes surrounding the sunken stomata are well developed. Fine wax particles cover the surface. Magnification × 300.

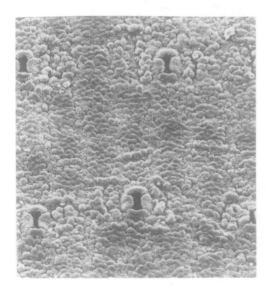

Figure 16.6 *Aloe arborescens*: leaf surface with fine micropapillae covering the flat outer walls of the epidermal cells. As in *A. mutabilis*, to which it is thought to be closely related, the lobes surrounding the sunken stoma are well developed. Magnification × 300.

REFERENCES

Beaumont, J., Cutler, D.F., Reynolds, T. and Vaughan, J.G. (1985) The secretory tissue of aloes and their allies. *Israel Journal of Botany*, **34**, 265–82.

Beaumont, J., Cutler, D.F., Reynolds, T. and Vaughan, J.G. (1986) Secretory tissues in the East African shrubby aloes. *Botanical Journal of the Linnean Society*, **92**, 399–403.

Brandham, P.E. and Cutler, D.F. (1978) Influence of chromosome variation on the organisation of the leaf epidermis in a hybrid Aloe (Liliaceae). *Botanical Journal of the Linnean Society*, **77**, 1–16.

Carter, S., Cutler, D.F., Reynolds, T. and Brandham, P.E. (1984) A multidisciplinary approach to a revision of the *Aloe somaliensis* complex (*Liliaceae*). *Kew Bulletin*, **39**, 611–33.

Cutler, D.F. (1969) Cuticular markings and other epidermal features in Aloe leaves. *Notes from the Jodrell Laboratory*, **VI**, 21–7.

Cutler, D.F. (1972) Leaf anatomy of certain Aloe and Gasteria species and their hybrids. pp.103–22. In *Research trends in plant anatomy* edited by A.K.M. Ghouse and M. Yunus, pp.103–122. Bombay & New Delhi: Tata McGraw-Hill Publishing. Co.

Cutler, D.F. (1979) Leaf surface studies in Aloe and Haworthia species (Liliaceae): taxonomic implications. *Tropisch und Subtropisch Pflanzenwelt*, No. **28**, 1–29.

Cutler, D.F. (1982) Cuticular sculpturing and habitat in certain Aloe species (Liliaceae) from southern Africa. In *The plant cuticle* edited D.F. Cutler, K.L. Alvin and C.E. Price, pp. 425–444. London, etc.: Academic Press.

Cutler, D.F. (1985) Taxonomic and ecological implications of leaf surface features in Aloe and Haworthia (Liliaceae) as seen with the SEM. *Proceedings of the Royal Microscopical. Society*, **20**, 23.

Cutler, D.F. and Brandham, P.E. (1977) Experimental evidence for the genetic control of leaf surface characters in hybrid Aloineae (Liliaceae). *Kew Bulletin*, **32**, 23–32.

Cutler, D.F., Brandham, P.E., Carter, S. and Harris, S.J. (1980) Morphological, anatomical cytological and biochemical aspects of evolution in East African shrubby species of Aloe L. (Liliaceae). *Botanical Journal of the Linnean Society*, **80**, 293–317.

17 Pests of aloes

Monique S.J. Simmonds

ABSTRACT

Cultivated aloes are more susceptible to arthropod pests than those growing in their natural habitats. The main arthropod pests include mealy bugs, scales, beetles, mites and aphids. When grown under glass the dominant pests are mealy bugs, scales, aphids and thrips. A survey of aloes growing at the Royal Botanic Gardens, Kew, showed that the susceptibility of different species of *Aloe* to these pests varied. *Aloe ferox* and *A. marlothii* were the most susceptible species.

The role of aloe-derived compounds in aloe-insect interactions is not known. Although many compounds, especially phenolics, have been isolated from aloes their potential ecological role has not been studied. Extracts from aloes can stimulate *Locusta migratoria* and *Spodoptera littoralis* to feed. This suggests that the extracts contain compounds that could act as phagostimulants. One of the compounds, a dihydroisocoumarin glucoside isolated from *A. hildebrandtii*, did not significantly influence the feeding behaviour of *Spodoptera littoralis*, although it was a potent anti-feedant against *Pieris brassicae*. and *Plutella xylostella*. Some aloe extracts also deterred *S. littoralis* from feeding. A comparison of the compounds in the aloes that vary in their susceptibility to insect herbivory might provide an insight into the role these compounds have in the ecology of aloes.

INTRODUCTION

In their natural environment aloes suffer very little damage from insect pests. Their main herbivore predators are mammals and include baboons (Figure 17.1), elephants and cattle.

However, when grown under high densities in glasshouses or in fields they can be susceptible to damage from scale insects, beetles and mealy bugs. Although species of *Aloe* in cultivation or grown as part of a botanical collection can be damaged by pests it is uncommon for them to be totally destroyed.

Very little has been reported on any specific aloe-insect interactions. It is known that insects, such as solitary bees, can be involved in their pollination but it is not know whether there is any specificity in these interactions. Nectar of some aloes are rich in phenolic compounds and these could deter insects. For example, a phenolic-rich extract from the nectar of *A. littoralis* Baker deterred the honeybee, *Apis mellifera* L., from feeding (Hagler and Buchmann, 1993).

The majority of the estimated 420 species of *Aloe* are not grown commercially and thus have not been grown in conditions that might expose them to horticultural pests.

Figure 17.1 Damaged plants of *Aloe hildebrandtii* Baker with the stems chewed by baboons for their moisture but with the leaves untouched because of their bitterness. (Photo P. Brandham).

In fact, many horticulturists looking after botanical collections that can harbour a range of pests say that they rarely have to use insecticides to control pests on the aloes. This suggests that the plants must be resistant to many forms of pests. Whether this resistance is associated with the diversity of compounds in their leaves or to their morphology has not been established.

Due to the observed lack of herbivory on aloes, researchers have investigated whether aloe-derived extracts could be used to control insect pests. For example, extracts from *Aloe barbadensis* Miller, *A. ferox* Miller and *A. succotrina* Lam. have general anti-insect activity, whereas extracts from *A. striata* Haw. were reported to attract fleas (Grainge and Ahmed, 1988). Extracts from *A. barbadensis* deterred feeding of *Athalia proxima* Klug and repelled *Popillia japonica* Newman (Jacobson, 1990). Ash from *A. marlothii* A.Berger was able to control the maize weevil, *Sitophilus zeamais* Motschulsky (Achiano *et al.*, 1999). Mosquito coils containing *A. vera* (L.) Burm.f. had moderate knock-down activity against mosquitoes after a 24-hour period but caused very little mortality. When D-*trans*-allethrin was added to the coil, mosquito mortality increased (Jantan *et al.*, 1999). In these studies, the compounds in the aloes associated with the respective activities were not identified.

Some aloes, particularly the spotted group (Saponariae), are prone to rust (*Uromyces aloes*) (Figure 17.2), especially in cultivation where it is damp and shady (Jeppe, 1969). Recovery is slow even after treatment with fungicides (Van Wyk and Smith, 1996).

Figure 17.2 Leaf showing discoloration and erupting pustules of aloe rust (*Uromyces aloes*), a severe disease of aloes, particularly Series Saponariae. (Photo P. Brandham) (*see Colour Plate 10*).

Black leaf spots can be caused by a number of fungi and can be controlled by fungicides (Jeppe, 1969). Bacterial infections show up as rots, wilts and blights and sometimes crown galls (Jeppe, 1969).

ARTHROPOD PESTS

The most frequently encountered arthropod pests that damage aloes grown commercially include mealy bugs, scales, beetles and mites (Jeppe, 1969; Beyleveld, 1973). Other groups of insects that can damage aloes include locusts and aphids (Beyleveld, 1973). At the Royal Botanic Gardens, Kew, thrips have also been observed feeding on the flowers. The fact that aphids and thrips are feeding on aloes is of concern as they are vectors for a range of pathogenic viruses. If these viruses became established in aloes than the loss of plants could start to increase.

Mealy bugs

The citrus mealy bug, *Planococcus citri* Risso, is commonly found as a pest of aloes (Jeppe, 1969). Adults and nymphs of this mealy bug are often found on leaves, stems, rosettes and roots. *Aloe dichotoma* Masson is reported to be very susceptible to mealy bugs (Chips Dreilinger, email communication).

Scale insects

Leaf surfaces can be covered with bodies of white scales that make the plants unsightly (Jeppe, 1969). Different species of scales are known to be able to breed on aloes including the Brown Soft Scale, *Coccus hesperidum* L. The adult lives underneath a protective scale and is very difficult to kill with contact insecticides. However, the immature stages can be killed with insecticides or solutions of soft soap.

Mites

Jeppe (1969) reported that the red spider mite *Tetranychus cinnabarinus* Meyer can cause damage to aloes. When aloes are under stress then the susceptibility to mites often increases. Mite damage is observed most frequently on the underside of the lower outer leaves. White-yellow small spots occur on the leaves. These leaves are covered with a thin layer of webbing. The presence of the webbing is a key to the fact that the damage was or is associated with mites. An examination of the leaves will usually reveal the cast skins of developing mites. Some produce fasciation (Figure 17.3).

Figure 17.3 Detached branch of *Aloe nyeriensis* held by P. Brandham showing mite-induced fasciation, producing a disorganised cluster of thin elongated leaves.

Beetles

Weevil-like beetles belonging to the genus *Brachycerus* can damage aloes (Beyleveld, 1973). The yellow-brown 10–15 mm long adult lays her eggs on leaves near the centre of the plant. On emergence the larvae burrow into the leaves and start to consume the centre of the plant. Another smaller beetle, *Mecistocersus aloes*, also causes damage to aloes (Beyleveld, 1973). The brown-black 6–8 mm long adult beetle feeds on the leaves and females lay eggs on the flowers. On emergence the larvae burrow into the stem. Beetle larvae are often not observed until the aloe starts to rot, and the leaves loose their firmness and collapse. The rotting tissue attracts flies that also start to breed on the plant.

In America, the *Agave*-snout weevil can damage aloes. This pest normally infests *Agave americana* L. and can lead to the total destruction of the plant.

ARTHROPODS ON ALOES IN A BOTANICAL COLLECTION UNDER GLASS

In the mid 1980s the Royal Botanic Gardens, Kew (RBG Kew), changed their pest control strategy from one that was based on the use of broad-spectrum pesticides to a more Integrated Pest Management strategy that included beneficial organisms such as parasitoids, predators and nematodes. As part of this strategy a survey was undertaken of the different types of pests in the high diversity glasshouses at RBG Kew. This survey included monitoring the pests feeding on the aloe collections in the dry zones of one of the main glasshouses at RBG Kew, the Princess of Wales Conservatory, and in some of the research glasshouses.

A summary of the susceptibility of 58 species of *Aloe* to insects is presented in Table 17.1. The susceptibility of the different species of *Aloe* to insect pests varied greatly and sometimes within a species there was a variation among the different accessions (Table 17.1). The dominant pest attacking the aloes was the mealybug, *Planococcus citri*. Leaves of a few species of *Aloe* had scales insects and the flowers of some species supported aphids and thrips. Mites were not observed on any of the aloes monitored during the survey.

The initial survey was undertaken over a year. However, the distribution of pests on some of the aloes has been monitored for over seven years. The results show that, overall, the diversity of pests has not changed but the occurrence of thrips has increased. The control strategies being used in the glasshouses are working as the level of pest infestation on many species of *Aloe* has decreased. In fact, during these studies the pests on aloes have never reached levels that required a targeted control strategy, although some plants were 'spot' treated with insecticides to control mealy bugs and others were removed because they harboured high numbers of mealy bugs.

Aphids

In the glasshouses at RBG Kew, aphids can be found on aloe flowers in the autumn and through the winter until spring (Table 17.1). It was rare to find populations of aphids establishing themselves on the leaves of aloes and they were not observed breeding on

Table 17.1 Insect pests found on some of the aloes growing at the Royal Botanic Gardens, Kew.

Species	RBG, Kew Accession Number	Scales[1]	Mealy bugs[2]	Aphids[3]	Thrips[4]
A. adigratana Reynolds	1973–2897	0	0	0	0
A. arborescens Miller	1973–2483	0	0	0	0
A. babatiensis Christian & Verd.	1974–4463	0	0	0	1 (Fo)
A. ballyi Reynolds	1974–4467	0	0	0	0
A. broomii Schönland	1986–1132	0	1 (all leaves)	0	0
A. bussei A. Berger	1990–1816	0	0	0	0
A. calidophila Reynolds	1974–4199	0	1 (all leaves)	0	0
A. cameroni Hemsl.	1973–2843	0	0	0	0
A. canarina S. Carter	1977–3888	0	0	0	0
A. chabaudii Schönl.	1957–35705	0	0	0	0
A. classenii Reynolds	1977–3983	1	1 (older leaves)	0	1 (Fo + Hh)
A. dawei Berger	1951–35710	0	0	0	0
A. descoingsii Reynolds	1981–6431	0	0	0	0
A. dichotoma Masson	1980–2081	0	0	0	0
A. divaricata Berger	1980–2084	0	1 (edge of leaves)	0	0
A. dolomitica Groenew.	1973–4091	0	2 (rosette)	0	1 (Fo)
A. elgonica Bullock	1973–3979	0	0	0	0
A. excelsa Berger	1986–1125	0	4 (all parts)	1	1 (Fo)
A. ferox Miller	1973–3407	2 (all parts)	3 (all parts)	0	1 (Fo)
A. ferox Miller	1973–3487	2 (all parts)	5 (all parts)	0	1 (Fo + Hh)
A. ferox Miller	1973–3549	1 (leaves)	1 (older leaves)	0	0
A. glauca Miller	1973–4058	0	2 (all parts)	0	0
A. hemmingii Reynolds & P.R.O. Bally	0848–01059	0	1 (edge of leaves)	0	0
A. jucunda Reynolds	1987–63006	0	0	0	0
A. lateritia Engl.	1974–4462	0	0	0	0
A. lateritia Engl.	1977–806	0	0	0	0
A. leachii Reynolds	1990–1820	0	1 (edge of leaves)	0	0
A. leptosiphon Berger	1990–1812	0	1 (edge of leaves)	0	0
A. marlothii Berger	1973–2467	1 (leaves)	2 (rosette, older leaves and stem)	1	1 (Tt + Hh)
A. marlothii Berger	1973–2497	1 (leaves)	5 (all parts)	0	0
A. megalacantha Baker	1981–790	1 (leaves)	1 (rosette)	0	1 (Fo + Hh)
A. megalacantha Baker	1981–722	0	0	0	0
A. microdonta Chiov.	1966–12803	0	3 (all parts)	0	1 (Fo)
A. microdonta Chiov.	1981–660	0	3 (all parts)	0	0
A. mutabilis Pillans	1973–4078	1 (leaves)	3 (all parts)	1	3 (Tt + Hh)
A. ngongensis Christian	1975–903	0	1 (leaves)	0	1 (Fo + Tt)
A. ngongensis Christian	1975–980	0	1 (leaves)	0	0
A. nyeriensis Christian	1973–2040	0	0	0	0
A. nyeriensis Christian	1974–4111	0	0	0	0
A. otallensis Baker	1953–10601	0	0	0	0
A. parvidens M.G. Gilbert & Sebsebe	1985–4219	0	1 (edge of leaves)	0	0
A. peckii P.R.O. Bally & Verd.	1981–900	0	0	0	0

		Scales	Mealy bugs	Aphids	Thrips
A. *peckii* P.R.O. Bally & Verd.	1981–870	0	0	0	0
A. pienaarii Pole-Evans	1977–788	0	0	1	2 (Fo)
A. *plicatalis* (L.) Miller	1978–3337	0	0	0	0
A. *pluridens* Haw.	1973–3751	0	4 (rosette)	1	1 (Fo)
A. *pubescens* Reynolds	1958–29526	0	0	0	0
A. *pubescens* Reynolds	1974–4171	0	0	0	0
A. *rabaiensis* Rendle	1975–903	0	0	0	0
A. *rabaiensis* Rendle	1979–3431	0	1 (lower leaves)	0	0
A. *rigens* Reynolds & P.R.O.Bally	1981–792	0	1 (rosette)	0	0
A. *rigens* Reynolds & P.R.O. Bally	1981–800	0	2 (older leaves)	0	0
A. *rigens* Reynolds & P.R.O. Bally	1981–802	0	1 (all leaves)	0	1 (Fo)
A. *rivae* Baker	1974–4204	0	1 (base of leaves)	0	0
A. *scabrifolia* L.E. Newton & Lavranos	1977–3749	0	1 (older leaves)	0	0
A. *schelpei* Reynolds	427–64–42705	0	1 (edge of leaves)	0	0
A. *scobonifolia* Reynolds & P.R.O. Bally	084–81–01110	0	1 (edge of leaves)	0	0
A. *scobonifolia* Reynolds & P.R.O. Bally	1981–865	0	0	0	0
A. *sinkatana* Reynolds	1987–4087	0	0	0	1 (Fo)
A. *sinkatana* Reynolds	1987–4089	0	0	0	0
A. *somaliensis* W. Watson	1949–38701	0	0	0	0
A. *somaliensis* W. Watson	1957–63009	0	0	0	0
A. *somaliensis* W. Watson	1981–3676	0	0	0	0
A. *speciosa* Baker	1973–3122	0	1 (edge of leaves)	0	0
A. *speciosa* Baker	1973–3208	0	0	0	0
A. *speciosa* Baker	1957–14506	0	0	0	0
A. *steudneri* Schweinf.	1987–4090	0	1 (edge of leaves)	0	0
A. *tenuior* Haw.	1951–35702	0	0	0	0
A. *tomentosa* Deflers	1976–453	1 (leaves)	2 (edge of leaves)	1	1 (Fo + Hh)
A. *tweediae* Christian	1970–1752	0	1 (edge of leaves)	0	0
A. *vacillans* Forssk.	1977–3128	0	0	0	0
A. *vanbalenii* Pillans	1973–2847	0	0	0	0
A. *vaombe* Decorse & Poisson	1986–1117	1 (leaves)	5 (all parts)	1	1 (Fo)
A. *vera* (L.) Burm.f.	1975–3937	0	0	0	0
A. *volkensii* Engl.	1977–3980	0	0	0	0
A. *wollastonii* Rendle	1973–1961	1 (leaves)	4 (all leaves, especially older leaves)	1	1 (Fo, Tt)
A. *wredfordii* Reynolds	1977–4192	0	0	0	0
A. *yavellana* Reynolds	1974–4193	0	1 (leaves)	0	0

Notes

The levels (1–5) represent the populations of insects observed on the plants.

1   Scales present on the plants: 1 = 1–25 scales.

2   Mealy bugs on the plants: 1 = 1–25, 2 = 26–50, 3 = 51–75, 4 = 76–100, 5 = >100 insects.

3   Aphids on the flowers: 1 = 1–10.

4   Thrips on the flowers: 1 = 1–10, 2 = 11–20, 3 = 21–30. Fo = *Frankliniella occidentalis*, Tt = *Thrips tabaci*, Hh = *Heliothrips haemorrhoidalis*.

the stems. The species of aphids varied but were usually *Aphis gossypii* Glover, *Aphis fabae* Scopoli or *Myzus persicae* Sulzer. On some aloes a proportion of the aphids were parasitised by species of *Aphidius*. These parasitoids were released to control aphids on other plants growing in different zones of the glasshouses and breeding populations have become established. No parasitoids were released specifically to control the aphids on aloes. Ants were seen farming the aphids and mealy-bugs on a few plants. It was observed that the number of mummified aphids (i.e. parasitised aphids) was lower on the ant-infested plants than on the plants with no ants. This could be because the ants were deterring the beneficial insects from attacking the pests on the aloes (Beyleveld, 1973).

The following species of *Aloe* supported aphids: *A. excelsa, A. marlothii, A. mutabilis, A. pienaarii, A. pluridens , A. tomentosa*s, *A. vaombe* and *A. wollastonii* (Table 17.1).

Thrips

The survey showed that the Western Flower Thrips, *Frankliniella occidentalis* (Pergande), can occur on the flowers of many of the species of *Aloe* that flower during the winter months. *Thrips tabaci* Lindemann (Onion Thrips) and *Heliothrips haemorrhoidalis* (Bouché) (Greenhouse Thrips) were also found on flowers. The number of thrips on flowers during the summer months was very low and during the year only a few thrips were observed on aloe leaves. The occurrence of *F. occidentalis* on aloes is a concern as this species of thrips can be an effective vector for different plant viruses, including the tomato spotted wilt virus (TSWV) that can attack a range of different species of plants. *Thrips tabaci* acts as a vector for tospoviruses. Flowers of two species of *Aloe, A. mutabilis* and *A. pienaari*, supported relatively high numbers of thrips, whereas the flowers on another 16 species contained a few thrips adults and nymphs (Table 17.1).

Mealy bugs

Adults and nymphs of the mealy bug *Planococcus citri* were found on 31 of the 58 species of *Aloe* surveyed in Table 17.1. High numbers of adults were found on the underside of the leaves and in the rosettes. Young nymphs were often found aggregating around spines at the edge of the leaves. Three species, *A. ferox, A. marlothii* and *A. vaombe*, growing in the research glasshouse were very heavily invested with all stages of the mealy bug and would not have been suitable for growing in the public glasshouses. The leaves were covered with honeydew and fungi. Most of the different accessions of *A. ferox* at RBG Kew are susceptible to mealy bugs, although the population levels vary.

Scales

Eight species of *Aloe* were contaminated with scales (Table 17.1) (Figure 17.4). However, on examination, the majority of scales did not contain live insects. However, *A. ferox* was badly attacked by the soft scale, *Coccus hesperidum* L., and the hemispherical scale, *Saissetia coffeae* Deplanche. It was noted that the accessions of *A. ferox* infested with scales also had high numbers of mealy bugs.

Figure 17.4 Severe infestation of aloe plant with scale showing bluish-white overall appearance. Close-up
shows more scale insects than leaf surface. (Photo P. Brandham) (*see Colour Plate 11*).

EFFECTS OF ALOE-DERIVED EXTRACTS AND COMPOUNDS ON INSECTS

Many plants produce compounds that inhibit some herbivores, especially insects, from
feeding. A few of these compounds will deter and be toxic, whereas others deter feeding
but are not toxic. Insects vary in their responses to these compounds and a compound
that deters one species of insect could stimulate another. In fact, some compounds
produced by plants are used as 'sign' compounds by insects to assist them identify a
host plant. Although some research has been done on the chemistry of aloes, especially
phenolics (Chapter 3), the role of these compounds in the ecology of aloe-insect interactions has not been studied. There is some literature on the potential role of the
phenolics in herbivore-aloe interactions but this is theoretical not experimental
(Gutterman and Chauser-Volfson, 2000a,b).

Thus there is very little experimental information about the role aloe-derived phenolics play in aloe-insect interactions and whether they influence the susceptibility of
plants to insect pests. If the compounds are to play a role in host selection then insects

need to be able to perceive the compounds. For example, compounds on the leaf surface could influence host selection but as yet there is no literature on which compounds occur on the surface of aloe leaves. Most chemical studies have concentrated on the types of compounds that occur in the leaf exudates. The same compounds might exude on to the leaf surface but this has not been confirmed. Previous studies showed that the vascular bundles in aloes leaves contain thin walled secretory cells known as aloin cells. These cells surround the vascular bundles near the phloem (Cutler *et al.*, 1980; Beaumont *et al.*, 1986). Thus, insects that either probe into the vascular bundle or bite into the leaf could damage these cells and encounter some of the phenolic compounds stored in the cells.

Gutterman and Chauser-Volfson (2000a, b) showed that with *A. arborescens* Miller the content of barbaloin, aloenin and aloeresin vary within a leaf. The contents are higher along the leaf margins and the adaxial surface contains higher levels of the compounds than the abaxial surface. They found that the levels of these phenolics increased after removing parts of the leaves. For example, when cuttings are taken from the same leaf of *A. arborescens* at monthly intervals for three months the levels of barbaloin increase three-fold from 11.8% to 20% to 25.8 to 37%. They suggested that this could prevent the remaining leaf being consumed. Previous researchers have shown that the concentration of secondary compounds, including phenolics, can increase in leaves after herbivory (Chadwick and Goode, 1999). Increased phenolic content has also been correlated with a decrease in herbivory (Dudt and Shure, 1994).

The accessions of *A. arborescens* at RBG Kew were free of insects pests (Table 17.1). Whether this is related to the concentrations of compounds like barbaloin, aloenin and aloeresin is not known. However, it would be interesting to compare the concentration of these compounds in *A. arborescens* with the concentration in *A. ferox* and *A. marlothii* where the pests were frequently found feeding on the sides and adaxial surface of the leaves.

In order to look at the potential role of aloe-derived compounds on insect behaviour we have tested extracts from aloes on the caterpillar *Spodoptera littoralis* Boisduval and the locust *Locusta migratoria* L. At RBG Kew, *S. littoralis* is used as a model insect to test whether plant-derived extracts or compounds influence the feeding behaviour of insects (Simmonds, 2000). It is a polyphagous species that can detect a range of compounds that either stimulate or deter feeding. *L. migratoria* is an oligophagous species feeding mostly on grasses but it was included in the bioassays as locusts have been recorded to feed on aloes.

Extracts from 47 species of *Aloe* were tested in the binary choice feeding bioassay (Table 17.2). Extracts from 26 species stimulated the locusts to feed and no extracts deterred feeding. In contrast, extracts from only 14 species stimulated *S. littoralis* to feed and extracts from eight species deterred feeding. Of the extracts that stimulated feeding, eight stimulated both species, including extracts from *A. ferox*, one of the plants shown to be susceptible to many of the pest insects.

The extract of *A. hildebrandtii* stimulated *S. littoralis* to feed (Table 17.2). A phytochemical study of this species identified a dihydroisocoumarin glucoside that when tested against *S. littoralis* caused the larvae to feed slightly more on the compound-treated disc than the control disc but the response was not significant (Veitch *et al.*, 1994). However, the dihydroisocoumarin was a potent deterrent against two species of Cruciferae-feeding caterpillars, *Pieris brassicae* L. and *Plutella xylostella* L. (Veitch *et al.*, 1994). This dihydroisocoumarin is closely related to that of the corresponding aglycone, feralolide, that occurs in *A. arborescens*. Jacobson (1990) reviewed the deterrent activity of a range of plants and has only one report for aloes: aloin isolated from *A. barbadensis* was not an effective mothproofing agent.

Table 17.2 Effect of *Aloe* extracts on the feeding behaviour of *Spodoptera littoralis* and *Locusta migratoria*. (see foot note for details about bioassay)

Species	R.B.G.Kew Accession Number	Feeding Index (Binary choice feeding bioassay)	
		Spodoptera littoralis	Locusta migratoria
A. *adigratana* Reynolds	1973–2897	#	#
A. *arborescens* Miller	1973–2483	0	#
A. *ballyi* Reynolds	1974–4467	0	#
A. *broomii* Schönland	1986–1132	0	#
A. *calidophila* Reynolds	1974–4199	*	0
A. *cameroni* Hemsl.	1973–2843	*	0
A. *chabaudii* Schöenl.	1957–35705	*	0
A. *classenii* Reynolds	1977–3983	0	0
A. *descoingsii* Reynolds	1981–6431	0	0
A. *dichotoma* Masson	1980–2081	0	0
A. *divaricata* Berger	1980–2084	0	0
A. *dolomitica* Groenew.	1973–4091	0	0
A. *elgonica* Bullock	1973–3979	*	0
A. *excelsa* Berger	1986–1125	#	#
A. *ferox* Miller	1973–3407	#	#
A. *ferox* Miller	1973–3487	#	#
A. *ferox* Miller	1973–3549	#	#
A. *glauca* Miller	1973–4058	#	#
A. *hildebrandtii* Baker	1981–885 + 1981–887	#	NT
A. *jucunda* Reynolds	1987–63006	0	#
A. *lateritia* Engl.	1974–4462	0	0
A. *lateritia* Engl.	1977–806	0	0
A. *marlothii* Berger	1973–2467	0	0
A. *marlothii* Berger	1973–2497	0	0
A. *megalacantha* Baker	1981–790	0	0
A. *megalacantha* Baker	1981–722	0	0
A. *microdonta* Chiov.	1981–660	#	0
A. *mutabilis* Pillans	1973–4078	#	0
A. *ngongensis* Christian	1975–903	#	0
A. *ngongensis* Christian	1975–980	#	#
A. *nyeriensis* Christian	1973–2040	#	#
A. *nyeriensis* Christian	1974–4111	0	#
A. *otallensis* Baker	1953–10601	#	#
A. *peckii* P.R.O. Bally & Verd.	1981–900	#	#
A. *peckii* P.R.O. Bally & Verd.	1981–870	#	#
A. *pienaarii* Pole-Evans	1977–788	0	#
A. *plicatalis* (L.) Miller	1978–3337	0	#
A. *pluridens* Haw.	1973–3751	0	0
A. *pubescens* Reynolds	1958–29526	0	#
A. *pubescens* Reynolds	1974–4171	#	#
A. *rabaiensis* Rendle	1979–3431	#	0
A. *rigens* Reynolds & P.R.O. Bally	1981–792	#	#
A. *rigens* Reynolds & P.R.O. Bally	1981–800	#	#

Table 17.2 (Continued).

Species	R.B.G.Kew Accession Number	Feeding Index (Binary choice feeding bioassay)	
		Spodoptera littoralis	Locusta migratoria
A. rigens Reynolds & P.R.O. Bally	1981–802	#	#
A. rivae Baker	1974–4204	*	0
A. scabrifolia L.E. Newton & Lavranos	1977–3749	*	0
A. scobonifolia Reynolds & P.R.O. Bally	1981–865	*	0
A. sinkatana Reynolds	1987–4087	*	0
A. sinkatana Reynolds	1987–4089	*	0
A. somaliensis W. Watson	1949–38701	0	0
A. somaliensis W. Watson	1957–63009	0	#
A. somaliensis W. Watson	1981–3676	0	#
A. speciosa Baker	1973–3122	0	#
A. speciosa Baker	1973–3208	0	#
A. speciosa Baker	1957–14506	0	#
A. tenuior Haw.	1951–35702	0	#
A. tomentosa Deflers	1976–453	0	#
A. vacillans Forssk.	1977–3128	0	#
A. vanbalenii Pillans	1973–2847	0	#
A. vaombe Decorse & Poisson	1986–1117	0	#
A. vera (L.) Burm.f.	1975–3937	0	#
A. wollastonii Rendle	1973–1961	0	#
A. yavellana Reynolds	1974–4193	0	#

Notes

Feeding Index: * = significant antifeedant activity $P < 0.05$, # significant phagostimulant activity $P < 0.05$, 0 = no significant effect, NT = not tested (Wilcoxon matched pairs test). Methanol extracts from the aloes leaves were tested at 1000 ppm in a binary choice bioassay. Larvae (*S. littoralis*, 5th instar) or nymphs (*L. migratoria*, 5th instar) were placed singly in sealed containers with two weighed glass-fibre discs. Both discs had been treated with sucrose (0.05 M) and the treatment disc was also treated with 100 µl of one of the test extracts. The insects were removed after 12 hours or earlier if they had eaten 50% or more of one of the discs (Simmonds *et al.*, 1990). The discs were reweighed and the amount eaten of control and treatment discs calculated. Each extract was tested against five to ten insects.

Further research into the effects of the many compounds currently being isolated from aloes (Viljoen and van Wyk, 2000; Viljoen *et al.*, 1999, 2001, 2002) on insect feeding might assist in clarifying the ecological importance of these compounds in the ecology of aloe-insect interactions.

ACKNOWLEDGEMENTS

I am grateful to Tom Reynolds who introduced me to the chemistry of aloes and to Dr Renee Grayer who translated the relevant parts of the book by G.P. Beyleveld. I also acknowledge the dedication of the horticultural staff at RBG Kew for maintaining the aloe collection.

REFERENCES

Achiano, K.A., Giliomee, J.H. and Pringle, K.L. (1999) The use of ash from *Aloe marlothii* Berger for the control of maize weevil, *Sitophilus zeamais* Motschulshy (Coleoptera: Curculionidae) in stored maize. *African Entomology*, 7, 169–172.

Beaumont, J., Cutler, D.F., Reynolds, T. and Vaughan, J.G. (1986) Secretory tissues in the East African shrubby aloes. *Botanical Journal of the Linnean Society*, 92, 399–403.

Beyleveld, G.P. (1973) *Aalwyne in die Tuin*. Pretoria, South Africa: Muller & Retief.

Chadwick, D.J. and Goode, J.A. (1999). *Insect-plant interactions and induced plant defence* p. 281. New York, Chichester, etc: John Wiley & Sons Ltd.

Cutler, D.F., Brandham, P.E., Carter, S. and Harris, S.J. (1980) Morphological, anatomical, cytological and biochemical aspects of evolution in East African shrubby species of *Aloe*. L. (Liliaceae). *Botanical Journal of the Linnean Society*, 80, 293–317.

Dudt, J.F. and Shure, D.J. (1994). The influence of light and nutrients on foliar phenolics and insect herbivory. *Ecology*, 75, 86–98.

Grainge, M. and Ahmed, S. (1988). *Handbook of plants with Pest-Control properties*, p. 470. New York :John Wiley & Sons.

Gutterman, Y. and Chauser-Volfson, E. (2000a) The distribution of the phenolic metabolites barbaloin, aloeresin and aloenin as a peripheral defense strategy in the succulent leaf parts of *Aloe arborsecens*. *Biochemical Systematics and Ecology*, 28, 825–838.

Gutterman, Y. and Chauser-Volfson, E. (2000b) A peripheral defense strategy by varying bar-baloin content in the succulent leaf parts of *Aloe arborescens* Miller (Liliaceae). *Botanical Journal of the Linnean Society*, 132, 385–395.

Hagler, J.R. and Buchmann, S.L. (1993) Honey-bee (Hymenoptera, Apidae) foraging responses to phenolic-rich nectars. *Journal of the Kansas Entomology Society*, 66, 223–230.

Jacobson, M. (1990) *Glossary of Plant-derived insect deterrents*, p. 213. Florida: CRC Press.

Jantan, I., Zaki, Z.M., Ahmad, A.R. and Ahmad, R. (1999) Evaluation of smoke from mosquito coils containing Malaysian plants against *Aedes aegypti*. *Fitoterapia*, 70, 237–243.

Jeppe, B. (1969) *South African aloes*, pp. xix-xxii. Cape Town, Johannesburg, London: Purnell.

Simmonds, M.S.J. (2000) Molecular- and chemo-systematics: Do they have a role in agrochemical discovery? *Crop Protection*, 19, 591–596.

Simmonds, M.S.J., Blaney, W.M. and Fellows, L.E. (1990) Behavioural and electrophysiological study of antifeedant mechanisms associated with polyhydroxy alkaloids. *Journal of Chemical Ecology*, 16, 3167–3196.

Van Wyk, B.-E. and Smith, G. (1996) *Guide to the aloes of South Africa*, p. 20. Pretoria: Briza Publications.

Veitch, N.C., Simmonds, M.S.J., Blaney, W.M. and Reynolds, T. (1994) A dihydroisocoumarin glucoside from *Aloe hildebrandtii*. *Phytochemistry*, 35, 1163–1166.

Viljoen, A.V. and van Wyk, B-E. (2000) The chemotaxonomic significance of the phenyl pyrone aloenin in the genus *Aloe*. *Biochemical Systematics and Ecology*, 28, 1009–1017.

Viljoen, A.V., van Wyk, B-E. and Newton, L.E. (1999) Plicataloside in *Aloe* – a chemotaxonomic appraisal. *Biochemical Systematics and Ecology*, 27, 507–517.

Viljoen, A.V., van Wyk, B-E. and Newton, L.E. (2001) The occurrence and taxonomic distribu-tion of the anthorones aloin, aloinoside and microdontin in *Aloe*. *Biochemical Systematics and Ecology*, 29, 53–67.

Viljoen, A.V., van Wyk, B-E. and van Heerden, F.R. (2002) The chemotaxonomic value of the diglucoside anthone homonataloside B in the genus *Aloe*. *Biochemical Systematics and Ecology*, 30, 35–43.

Cut leaves of *Aloe ferox* placed in a pile with cut surfaces inwards for the collection of exudates. (Photo, P. Brandham) (*see Colour Plate 12*).

Index

acemannan® 60, 62, 81, 174, 213, 214,
218, 219, 221, 223, 243, 245, 273,
285–90, 293, 311, 313–18, 321, 325
acetylated mannan 60–2, 75, 83, 111, 119,
121, 122, 145, 200, 242, 281, 288, 289,
298, 311–13, 315, 317
activity-guided fractionation 120
adhesion molecules 318
adjuvant properties 61, 93, 218, 221, 266,
317, 321
adulteration 142, 184, 188, 194, 198, 199,
201, 218, 270, 286
agglutinin 89, 90, 93
agronomy 139, 142, 143, 147, 149, 150,
151, 156
alditol acetates 117
Alexander the Great 128, 211
aloctins 62, 74, 91–5, 98, 101, 102, 104,
107, 108, 213, 214, 222, 223, 260, 338,
339, 376
Aloe adigratana 372, 377
A. africa 42, 45, 104, 107, 209, 365
A. albiflora 8
A. alooides 7
A. ankoberensis 7
A. arborescens 10, 12, 21, 30, 42, 46, 50, 52,
54, 61, 62, 63, 64, 65, 78, 79, 82, 84, 89, 91,
97, 98, 99, 103, 104, 105, 106, 107, 108,
112, 128, 130, 133, 140, 153, 157, 209, 210,
213, 222, 224, 260, 283, 284, 288, 312, 333,
334, 335, 345, 346, 347, 366, 372, 376, 377
A. aristata 10
A. babatiensis 372
A. bakeri 46
A. ballyi 9, 10, 46, 372, 377
A. barbadensis see *A. vera*
A. barberae 10
A. berhana 43, 47
A. boscawenii 7
A. boylei 10
A. brevifolia 21
A. breviscapa 7

A. broomii 42, 51, 52, 372, 377
A. buettneri 4, 7
A. bulbifera 9
A. calcariophila 7
A. calidophila 372, 377
A. cameroni 372, 377
A. canarina 372
A. candelabrum 209
A. castanea 42
A. chabaudii 372, 377
A. cheranganiensis 8, 358
A. chinensis 104, 209
A. chrysostachys 7, 10, 11
A. ciliaris 358, 359, 366
A. classenii 7, 372, 377
A. claviflora 52
A. confusa 10, 39
A. conmixta 21
A. cooperi 10
A. cremnophila 40, 42, 50, 358
A. dawei 10, 46, 358, 372
A. descoingsii 372, 377
A. dichotoma 7, 101, 372, 377
A. divericata 372, 377
A. dolomitica 372, 377
A. ecklonis 12
aloe emodin 43, 47, 129, 160, 171, 175,
198, 203, 228
A. elgonica 45, 52, 358, 359, 372, 377
A. eminens 4, 7
A. excelsa 42, 372, 374, 377
A. falcata 7
A. ferox 10, 12, 21, 40, 42, 43, 45, 46, 51, 61,
62, 63, 78, 83, 112, 131, 133, 148, 209, 210,
213, 365, 367, 368, 372, 374, 376, 377
A. fibrosa 3
A. forbesii 209
A. glauca 21, 372, 377
A. globuligemma 4
A. graminicola 43, 47
A. haemanthifolia 7, 361, 366
A. hemmingii 372

A. hereroensis   50
A. hildebrandtii   54, 367, 376, 377
A. humilis   21, 359
A. inermis   358
A. jacksonii   42, 358
A. jucunda   359, 372, 377
A. juvenna   358
A. kedongensis   10, 11, 46, 358
A. khamiesensis   46
A. kilifiensis   7
A. kraussii   10, 12
A. kulalensis   4, 5
A. lateritia   7, 9, 46, 50, 372, 377
A. leachii   372
A. leotosiphon   372
A. littoralis   51, 367
A. lutescens   42, 48
A. macrocarpa   10
A. maculata   9, 12, 21
A. marlothii   10, 42, 45, 52, 367, 368, 372,
   374, 376, 377
A. mawii   4
A. megalacantha   10, 43, 372, 377
A. microdonta   52, 372, 377
A. microstigma   52
A. minima   10
A. morijensis   7, 358
A. murina   5
A. mutabilis   48, 50, 366, 372, 374, 377
A. myriacantha   4, 5
A. ngobitensis   358
A. ngongensis   372, 377
A. nyeriensis   48, 50, 54, 358, 372, 377
A. officinalis   6
A. otallensis   372, 377
A. parvidens   372
A. patersonii   9
A. peckii   356, 372, 377
A. peglerae   42, 361
A. penduliflora   4
A. perfoliata   21
A. perryi   21, 42, 209, 210
A. pienaarii   372, 374, 377
A. pillansii   7
A. platylepis   209
A. plicatilis   21, 55, 61, 78, 79, 283, 284,
   288, 372, 377
A. pluridens   372, 374, 377
A. polyphylla   7
A. pretoriensis   45
A. pubescens   49, 357, 372, 377
A. pulcherrima   43, 50
A. rabaiensis   42, 48, 357, 372, 377
Aloe Research Foundation (ARF)   139, 143–5,
   147, 151, 154, 157, 164, 167, 168, 174,
   179, 180, 184, 188, 193, 199, 200, 203,
   265, 267, 270–81, 289–91

A. rigens   372, 377
A. rivae   4, 12, 43, 372, 377
A. rubroviolacea   42
A. rupestris   41, 42
A. ruspoliana   9
A. saponaria   9, 43, 46, 48, 52, 55, 61, 62,
   78–80, 84, 96, 104, 107, 112, 224, 283,
   284, 288, 312, 315, 316, 324
A. saunderisae   4
A. scabrifolia   372, 377
A. schelpei   372
A. scobinifolia   372, 377
A. secundiflora   7
A. sinkatana   372, 377
A. somaliensis   359, 372, 377
A. speciosa   42, 210, 372, 377
A. spicata   209
A. steudneri   7, 372
A. striata   368
A. succotrina   21, 51, 368
A. suzannae   8, 13
A. tenuior   18, 356, 372, 377
A. tidmarshii   358
A. tomentosa   4, 372, 374, 377
A. torrei   7
A. tricosantha   4
A. tweediae   372
A. vacillans   372, 377
A. vahombe   *see vaombe*
A. vanbalenii   61, 78, 79, 283, 284, 288, 372, 377
A. vaombe   46, 61, 78, 79, 80, 118, 283, 284,
   288, 325, 326, 372, 374, 377
A. vaotsanda   10
A. variegata   7, 21
A. vera   6, 9, 12, 13, 15, 21, 40, 41, 42, 43, 44,
   45, 46, 48, 50, 51, 55, 56, 58, 60, 62, 63, 64,
   65, 77, 78, 79, 80, 81, 82, 83, 84, 95, 96, 101,
   102, 104, 105, 107, 108, 111, 112, 114, 115,
   117, 121, 128, 131, 139, 171, 184, 185, 189,
   191, 192, 201, 209, 210, 211, 212, 218, 221,
   223, 228, 229, 230, 239, 240, 241, 242, 243,
   245, 259, 260, 266, 270, 271, 283, 286, 289,
   311, 312, 313, 314, 315, 316, 318, 323, 324,
   333, 335, 345, 346, 347, 348, 368, 373, 378
A. veseyi   4
A. volkensii   7, 372
A. wollastonii   372, 374, 377
A. wrefordii   372
A. yavellana   372, 377
A. zebrina   10
aloebarbendol   47, 48
aloechrysone   47
aloenin   52, 54, 335, 336, 346–9, 376
aloeresin A   43, 133, 334, 335
aloeresin B   42, 43
aloeresin C   42
aloeresin D   42, 335

aloeresin E   42, 43, 133
aloeresin F   42, 335
aloeride   314
aloesaponarin I   46, 47
aloesaponarin II   43, 46, 47
aloesaponol I   47, 48
aloesaponol II   47, 48
aloesaponol III   47, 48
aloesaponol IV   47, 48
aloesin   40–3, 120, 133, 171, 172, 175,
    203, 213, 334–6
aloesol   42
aloeulcin   336
aloin   *see* barbaloin
aloin cells   375
alomicin   340, 342
aneuploidy   359
anti-feedant activity   154, 367, 368, 375, 376, 385
anti-inflammatory activity   91, 93, 108, 109,
    120, 121, 145, 182, 218, 219, 223, 228,
    239, 240, 242–6, 256, 259, 260, 271, 277,
    281, 311–15, 326, 334–6, 338
α1-antitrypsin   89, 91, 93, 105, 340
anti-tumor activity   94, 95, 98, 99, 104, 223,
    224, 245, 285, 317, 324, 337, 338, 340
aphids   367, 369, 371, 374
aphthous ulcers   224, 322, 325
apigenin   54, 56
apoptosis   223, 267, 311, 320, 322
arabinan   60, 75, 77, 83, 111
arborans   61, 79, 84, 337
ARF   *see* Aloe Research Foundation
Aristotle   211
asthma   224, 337, 343, 344
Astroloba   16, 17, 21–5, 31
ATF 1011   62, 98, 99, 105, 107, 338

baboons   7, 96, 367, 386
Barbados   131, 209, 211, 212
barbaloin (aloin)   39, 48–53, 127–33, 171,
    175, 195, 213, 214, 216, 246, 303, 333–7,
    343, 347–9, 376
*iso*barbaloin   50–1, 214, 335
beetles   367, 369, 371
bitter aloes   39, 127, 128, 134
black leaf spot   369
bone marrow   311, 323, 324
bradykininase   213, 218, 248
bronchial asthma   337, 343, 344
bulbil formation   9
bundle cap cells   363, 364, 366
bundle sheath cells   212, 312

calcium flux   317
Cape aloe   40, 43, 45, 46, 54, 56, 131, 209,
    212, 214, 222, 336
Cape of Good Hope   26, 44

carboxypeptidase   63, 119, 260, 312, 339, 342
carcinoma   96, 219, 223, 227, 265–8
Carrington Laboratories   75, 111, 285
carrisyn™   *see* acemannan®
cathartic effect   195, 212, 224, 228, 333–5, 343
cellulase   81, 121, 122, 145, 158, 169, 174,
    178, 180–3, 199, 200, 265, 271, 280, 281,
    283, 288, 291–5, 297
Chortolirion   16, 17, 19, 22–5, 29, 30
chromones   40, 42, 43, 54, 120, 147,
    153, 154, 156, 172, 175, 195, 199,
    261, 312, 314
chrysophanol   43, 45–7, 52
CITES   3, 13
Clifford, H.T.   17, 18, 19, 33
conductivity   188, 193, 194, 203
γ-coniceine   9, 56
conservation   3, 12, 13
constipation   212, 214, 216, 217, 225, 236, 333
contact dermatitis   219, 220, 228, 229, 239,
    270, 344, 345
cosmetic uses   9, 12, 111, 121, 139, 151, 156,
    164, 168, 171, 195, 258, 270, 333, 362
counter current chromatography   130
crassulacean acid metabolism (CAM)   3, 113,
    114, 120, 121
cremnochromone   40, 41, 43
Cronquist, A.   17, 18, 19
Curaçao Aloe   209, 214
cutaneous contact hypersensitivity (CHS)   182,
    272–81, 291, 292, 294, 295, 297, 299
cuticle   3, 154, 160, 365
cyclo-oxygenase   213, 315
cytokine cascade   281
cytokines   198, 219, 241, 243, 265, 266,
    277, 281, 299, 311, 316, 320, 321, 323
cytoprotective oligosaccharide   145, 183, 265,
    281, 289, 295, 298

Dahlgren, R.   18, 19, 33
Davis, R.H.   120, 181, 188, 218, 219,
    224, 242, 243, 246, 251, 271, 277, 281,
    314, 348
decolorization   165, 171, 173, 175, 176
decorative use   10
delayed type cutaneous hypersensitivity
    (DTH)   269, 272, 275, 276, 278–81,
    295, 297–9, 323
deoxyerythrolaccin   46
diabetes   221, 222
dihydro*iso*rhamnetin   54, 56
6, 8-dihydroxy*iso*coumarin   *see* feralolide
dioscorides   9, 128, 211, 229
dispersal   7
distribution   4–6, 16, 39, 127, 162, 184,
    194, 287, 293, 358, 365, 371
DNA C-values   356

E peak   *see* malic acid
Ebers papyrus   211, 229
edema   220, 253, 254, 259, 277, 281, 314, 324, 325, 344
eicosandoids   243
endemic species   5, 6
endemism   5, 6
erythrocytes   88–92, 95, 96, 100–2, 105, 107, 338, 339

fast blue B salt   132
feralolide   54, 58, 376
feroxidin   56
feroxin A   56, 59
feroxin B   56, 59
fibrosarcoma   81, 94, 95, 98, 213, 223, 324, 325, 326, 338
filleting   139, 143, 153, 155–8, 175, 181, 182
free radicals   31, 34, 35, 38, 41, 44, 47, 50, 56, 59, 62, 64, 67, 70, 73, 75, 78, 81
freeze-drying   143, 157, 179
frosts   25, 147, 149, 150, 240, 251, 253, 254
FTIR spectroscopy   198, 285, 287, 288
furaloesone   40, 41

Galen   211, 229
gas chromatography   115, 130
Gasteria   16, 17, 20–4, 29, 30, 48, 355, 365
gastric lesions   93, 94, 338
gastric ulcer   209, 222, 229, 342
gel filtration chromatography   60, 61, 78, 99, 100, 115, 198, 284, 290, 337
glucomannan   60, 61, 75–81, 83, 85, 111, 139, 145, 169, 178, 198–200, 218, 241, 245, 246, 281–6, 288, 289, 312
glycoprotein   39, 58, 62, 63, 75, 77, 84, 88–90, 97, 98, 101, 103, 104, 111, 120, 213, 221, 222, 224, 260, 316, 337–9, 343
gram +ve cocci   147, 159, 161–3
gram +ve rods   164
gram −ve rods   147, 161, 163, 164, 166, 193
guacamole   157, 158, 171, 178

habitat   3, 7, 9, 12, 13, 23, 25, 26, 211, 358, 361, 365, 366, 367
Hakomori method   115, 118
Haworthia   16, 17, 20, 22–8, 30, 48, 355, 365
helminthosporin   45, 46, 48
hemagglutenin   88, 89, 91
hemagglutinating activity   88, 91, 92, 94–102, 104, 108, 260, 338, 339
herbivory   367, 368, 376
herpes simplex   209, 221, 229, 321, 324
hind paw edema   93, 218, 284, 315, 335, 336
HIV   221, 245, 265
homonataloin   46, 48, 50, 128, 133, 334

HPLC separation   49, 51, 115–17, 120–2, 132, 133, 161, 167, 171, 193–5, 197, 218, 339, 346
5-hydroxyaloin (5-hydroxybarbaloin)   49, 50, 52, 173, 214, 371
7-hydroxyaloin (7-hydroxybarbaloin)   49, 52, 131, 173, 371
10-hydroxyaloin (10-hydroxybarbaloin)   51, 52
hypersensitivity   219, 265, 269, 272, 274, 275, 323, 344

IASC   *see* International Aloe Science Council
immunostimulation   81, 111, 120–2, 214, 224, 242, 243, 245, 279, 285, 317, 324, 325
in situ hybridization   356
insulin   221, 242
Integrated Pest Management   371
interleukin-1   81, 223, 243, 253
interleukin-2   81, 224, 253, 268, 316
interleukin-10   289
International Aloe Science Council (IASC)   127, 139–44, 147, 153, 154, 167, 174, 175, 178, 184, 193–5, 198, 201
iodophors   170

Japanese health food and nutrition food association (JHNFA)   333, 345

karyotype   16, 18, 355, 356, 357, 359
Kew Gardens   26
Kidachi aloe   333–45, 348

laccaic acid D methyl ester   46
Langerhans cells   271, 272, 298
laxative abuse   225, 226, 227, 228, 230, 231
laxative activity   149, 209, 211–15, 217, 225–9, 231, 270
leaf surface   160, 365, 366, 370, 375, 386
leaf washing   151, 171, 202
light scattering   115, 117
lignaloe   211
lipid components   63, 64, 76, 112, 119, 221, 222, 229, 256
littoraloin   51, 52
littoraloside   52, 54
Lomatophyllum   4, 6–8, 20, 22, 23, 25, 27, 28, 30, 43, 46, 355
lymphocytes   89, 93, 100, 107, 223, 243, 269, 299, 323, 324, 338, 339

α2-macroglobulin   91–3, 105, 223, 338, 340
macrophages   93, 145, 219, 223, 225, 231, 242–5, 253, 291, 299, 314, 316–22, 324–6
magnesium lactate   64, 65, 213, 218, 243, 260
malic acid (E peak)   3, 64, 113, 114, 120, 121, 123, 124, 142, 147, 169, 170, 193–5, 201, 313, 376

maltodextrin   179, 180, 184, 188, 193–5, 197–202, 218, 286
mass spectrometry   115, 130
mast cells   260, 334, 337
mealy bugs   367, 369–71, 374
melanoma   62, 228, 261, 265, 267–9, 303, 326, 336
melanosis coli   225, 227
membrane filtration   167
mesophyll cells   114, 154–6, 167, 202, 311–13
mesophyll gel   311, 312
methylation analysis   118
micrococcus   159, 161, 164, 194, 203
microdontin   52, 53
microstigmin   52, 53
mites   367, 369–71
mitogenic activity   92, 93, 98, 100, 104, 105, 107, 223, 338, 339
mucilage   3, 58, 60, 77, 83, 114, 212, 213, 217, 244, 362

naringenin   54, 56
Nataloin   49, 128
natural killer cells (NK)   94, 243, 277, 323
nitric oxide (NO)   215, 219, 311, 319, 322, 325
nitric oxide synthase (NOS)   215, 319–21
NO   *see* nitric oxide
NOS   *see* nitric oxide synthase

oncogenes   266, 267
oral administration   195, 209, 214, 219, 222, 225, 228, 229, 314, 315, 334, 336, 343
ornamentals   62, 128, 155
overgrazing   12, 44
overprocessing   168–70

PAF   *see* platelet-activating factor
pasteurization   121, 164–8, 170, 171, 173–5, 180, 183, 271
pectic substances   60, 76, 82, 83, 112
pectins   76, 77, 82, 145, 169, 178, 182, 281, 283, 311–13, 317, 325
periodate test   50, 129, 132, 172
phagocytosis   93, 219, 245, 285, 317, 318, 337, 343
phenol-sulfuric acid assay   116
phytohemagglutenin   89, 93, 97
platelet-activating factor (PAF)   215
platelet aggregation   224, 336
platelets   245, 253, 323, 324
plicatiloside   55
Pliny the Elder   211
Poellnitzia   16, 17, 19, 21, 22, 24, 25, 28, 30
poisonous aloes   9, 56

pollination   7, 8, 367
polymorphonuclear leucocytes   218, 246, 315, 322, 323
prechrysophanol   47
prostaglandins   45, 64, 218, 243, 314, 321, 340
pseudoplasticity   153, 158, 167, 174, 293
psoriasis   209, 220, 221, 229
purgative activity   129, 134

quality control   82, 127, 128, 133, 139, 140, 141, 150, 151, 164, 165, 170, 177, 183, 193–6, 201, 281, 333
quaternary ammonium compounds   170

Rabaichromone   44
radiation burns   212, 312
'red compound'   175, 176, 177
red spider mite   370
Reynolds, G.W.   17, 18, 21–3, 26, 27, 43, 46, 47, 50, 56, 58, 357, 358, 377
Rio Grande Valley (RGV)   139, 142, 149, 150, 151, 153, 161, 202
Rule of Nines   257, 263
rust fungus   368

scale insects   367, 369–71, 374
SDS-PAGE   97, 101, 103, 119
serum complement   94
sitosterol   63, 64, 112, 120, 214, 243
size exclusion chromatography   81, 115, 117, 122, 130, 197, 284, 292
skin immune system   181, 182, 265, 266, 269, 271, 272, 276–8, 281, 289, 292, 293, 296, 299
Smith, Gideon F.   13, 15–19, 21–6, 34, 51, 368
sodium hypochlorite   152, 160, 170, 202
soil pH   149, 150
specific rotation   118
spoilage   156–9, 161, 164, 171, 179, 194
spray drying   167, 178–80, 184, 188, 193, 194, 199, 218
stomata   3, 365, 366
succulence   3
sunscreens   266, 270, 274, 277, 281, 298
sycamore oligosaccharide   296
synonymy   16, 19, 24
synovial pouch model   218, 224

T cells   69, 99, 226, 242, 269, 271–3, 298, 299, 316, 317, 323, 338
Tamarind polysaccharide   296, 299
telomeres   357, 359
tetrahydroanthracenes   55
tetrahydroanthracenones   46, 48
tetraploids   355, 358, 359

thermal injury   62, 251, 253, 270
thin film vacuum evaporation   177, 271
thrips   367, 369, 371, 374
thromboxane   224, 245, 246, 253, 259,
    260, 322
TLC separation   116, 117, 119, 171, 172,
    195, 203, 218, 348
toxicity   56, 223, 225, 228, 245, 344
trichophytosis   341
trimethylsilyl ethers   117, 130
triploids   355, 359
tumor suppression   266, 267, 268
tumorigenesis   267

ulcers   94, 209, 211, 219, 222, 225, 229,
    239, 270, 325, 336, 342
UV absorption   129, 132
UVB radiation   62, 219, 266, 269, 271–81,
    293, 294, 296–9

Van Wyk, B.-E.   16, 17, 19, 21, 22, 25, 26,
    34, 43, 46–8, 51, 52, 54, 55, 63, 65, 132,
    134, 312, 314, 368, 376
vascular bundles   3, 16, 39, 114, 154, 212, 361, 375
venous leg ulcers   225
verectin   103, 104, 107
*iso*vitexin   54, 56

wax   3, 64, 154, 160, 365, 366
weevils   9, 368, 371
whole leaf extract (WLE)   119, 153, 157,
    166, 322, 333, 339, 343
WLE   *see* whole leaf extract
wound healing   96, 105, 108, 111, 121, 209,
    217–20, 228, 229, 239, 240–5, 259, 270,
    289, 311, 313, 316, 317, 325, 348

*iso*xanthorin   45, 46, 48
x-ray burns   312, 341